D0812082

GeoDestinies

The Inevitable Control of Earth Resources Over Nations and Individuals

Walter Youngquist

National Book Company
Portland, Oregon

Second Printing

0-89420-299-5

371440

Library of Congress Cataloging-in-Publication Data

Youngquist, Walter Lewellyn, 1921-
GeoDestinies : the inevitable control of earth resources over nations and individuals / Walter Youngquist.
p. cm.
Includes bibliographical references and index.
ISBN 0-89420-299-5 (cloth)
1. Power resources—Social aspects. 2. Mines and mineral resources—Social aspects. 3. Natural resources—Social aspects. I. Title.
HD9502.A2Y684 1997
304.2'8—dc21 97-4825
CIP

To the important few

who make available to the many

the vital materials from the Earth

on which all civilization exists

Acknowledgments

The able, very time-consuming but cheerfully given critical review work by Mr. William E. Eaton, Dr. Laurence R. Kittleman, and my former colleague at the University of Oregon, Professor Ewart M. Baldwin, each of whom read the more than 1,000 manuscript pages, is deeply appreciated.

I am particularly indebted to Professor David Pimentel of Cornell University for his always prompt and most informative replies to my many inquiries on agricultural economics related to energy source and use, soils, and the general question of carrying capacity of the Earth. His studies have been widely drawn upon in this volume. Professor Peter J. Reilly of Iowa State University provided information on energy economics of ethanol.

For global political aspects of mineral resources I am indebted to my long-time friend the late Dr. Arthur A. Meyerhoff, who, through both his publications and conversation, in large measure impelled me to pursue this present study. Worldwide basic petroleum statistics were generously provided through publications and patient replies to my numerous letters by Joseph P. Riva, Jr., Earth Science Specialist, Science Policy Research Division of the Congressional Research Service, The Library of Congress. His studies have been used extensively. Similarly, by conversation and correspondence, L. F. Ivanhoe, student of world petroleum reserves and trends, greatly enhanced my view of the future of petroleum.

From his more than 30 years' experience in the region, the late Liston Hills, retired Chairman of the Board of ARAMCO, provided perceptive insights on the history and the petroleum development of the Persian Gulf region, particularly Saudi Arabia. Dr. Fred Stross, by personal memo, shared his intimate first-hand knowledge and scientific perspective on the Nile River and the influence which the dams have had on the ecology of the region. Professor William Purdom of Southern Oregon State College provided technical information on radon gas and its geological associations. Professor Thomas Stitzel, School of Business, Boise State University, reviewed the manuscript and offered his perspectives as an economist.

The extensive North Slope Alaska experience of Richard A. Bray, retired geologist of Sohio Oil Company, was generously contributed through conversation and manuscript notes. Dr. H. H. Hall's many years as managing resident geologist for units of Exxon Corporation in Europe, the Middle East, and Southeast Asia which I drew upon by correspondence and conversation provided valuable information on petroleum prospects in these regions. He also reviewed the entire manuscript.

The expertise and observations of Dr. Albert Bartlett, retired physicist, University of Colorado, and Dr. Garrett Hardin, retired human ecologist, University of California, Santa Barbara on the relation of population to natural resources, through conversation, correspondence, and publications have been both factually helpful and very stimulating.

Lester R. Brown and Hal Kane of the Worldwatch Institute were most obliging to my several inquiries. The numerous timely publications of that organization have been extensively cited.

I also express my gratitude to my former employers who gave me the opportunity to study both domestic and international mineral resource economics and supplies, and also view the national and international social, political, and environmental problems of mineral development. These include the U.S. Geological Survey, my consulting clients, the Sun Oil Company, the Shell Oil Company, Amoco, the Minerals Department of Exxon Corporation, and my full-time employer overseas, International Petroleum Company, Ltd. of Peru. A 19-year consulting relationship on geothermal energy for the Eugene Water & Electric board gave me reason to study that interesting energy source both here and abroad. Carlos E. Milner Jr., by computer, redrafted and improved several of the text figures, and kindly pursued points of information for me on the Internet.

Finally, I am most appreciative for initial acceptance of the manuscript proposal and subsequent encouragement and advice through all phases of book production by Carl Salser, Editor, and Mark Salser, Associate Editor of the National Book Company. I could not have worked with more capable, considerate, and helpful people.

Contents

Tables

Page

Figures

Page

Preface

Earth's past sets the stage for the present and future

The slow sinking of a sag in the Earth's crust which now includes the Persian Gulf, and the accumulation over millions of years of organic-rich sediments which, by geological processes, ultimately produced oil, has profoundly influenced the destinies of many nations and millions of people. Most people are unaware of oil's geological history and the great length of time it took to make the petroleum products they use every day. But the geologic framework of the Earth developed over eons of time, with its varied resources distributed unevenly over the globe, established the environment which we are inevitably destined to accept. Human history, the present, and the human future have been, are, and will be subject to geology.

Myriad past geological events affect us now in many ways

Glaciers deposited rich unleached mineral soil across the northern United States making it a breadbasket for the world, affecting millions of people. Gold and other mineral deposits have caused great human migrations and the settlement of new lands. Some nations, such as those in the Persian Gulf region, are almost entirely dependent on mineral resources to support their economies. The energy minerals, coal, oil, and uranium, have been decisive factors in warfare, and fuel the industrial world.

The destinies of all nations and all people are in many ways bound up with the mineral and energy mineral resources of the Earth. Events of the geologic past have richly endowed some nations with valuable Earth resources, whereas others have very few. The result is markedly different destinies for different nations. How Earth resources have affected the peoples of the past, how they influence our lives now, and how they will determine our futures is the study of *GeoDestinies*.

Geodestinies — an ongoing flow

Set in motion by the first human to make use of a rock, and now greatly accelerated by our modern high mineral-consuming industrial society, Geodestinies are an on-going flow of events which continually relate nations and individuals to the Earth resources on which they depend for survival.

This book is a current snapshot of the human condition, the pursuit and achievement of various degrees of affluence by the nations of the world, and what the future may hold for humanity, all of which are based on the mineral and energy resources of the Earth.

Minerals and civilizations

Stone tools show that from early times of human existence, minerals have been the materials on which civilizations have depended, and by which they have advanced. Possession of the bronze sword and later iron weapons of war carried the Roman Empire to its peak of strength. Building on a remarkable variety and abundance of mineral and energy resources, the United States rose from a wilderness to the most affluent nation in the world in less than 200 years. In less than 50 years, Saudi Arabia, solely by oil, made the leap from an obscure, undeveloped country to a thoroughly modern nation with global economic influence.

The rate at which various civilizations have progressed has largely been a function of how many mineral resources were available and knowledge of how to use them. The Stone Age has persisted in parts of New Guinea until fairly recent times, and tribes in remote parts of South America likewise have existed in a primitive state because they did not have use of the energy and mineral resources which other nations did.

Mineral possession changes nations

The Industrial Revolution began chiefly in England based on coal and iron. But, when oil became the chief fuel of industrial nations, and the diverse array of high quality mineral and energy resources in the United States began to be developed, the center of industrial power moved west across the Atlantic. More recently the entire Persian Gulf region has been markedly transformed in only a few decades by oil.

Energy minerals do much of the work

The work of the world has been lightened tremendously by the use of various energy minerals. The back-breaking labor of plowing, planting, and harvesting once done by humans, sometimes with the aid of animals, has been replaced in many countries by machines fueled by gasoline or diesel oil. Humans, relieved of these tasks, have been able to pursue science and technology which in turn have moved civilization forward even further.

Special minerals for special purposes

Highly specialized needs of new technologies demand many minerals previously not used. Computers use beryllium, gallium, germanium, lithium, platinum, palladium, quartz crystals, rhenium, selenium, silicon, strontium, tantalum, and yttrium. The modern jet airplane demands high quality aluminum for the body and wings. It uses chromium, titanium, manganese, nickel, cobalt, and tantalum in its jet engines, and in this use, these minerals have no known substitutes. The hull of the spaceship that carried men to the moon was made of the heat resistant, light metal, titanium, which appears to be the only material which could have performed that mission.

Some special purpose minerals such as cobalt, platinum, palladium, manganese, and chromium are relatively rare and exist in commercial deposits in only a few places in the world. This limited availability and various political situations may create significant international trade and national security problems, as they did in South Africa and adjacent regions a decade ago.

In energy industries, particularly petroleum and in nuclear power, more than 30 non-fuel minerals are used. Telephone systems use 42 different minerals, and the electrical industry overall uses 85 different minerals. In medicine, surgical instruments, chemotherapy, radiation treatments, and many different diagnostic tests require a great variety of minerals. These materials must be obtained from somewhere in the Earth and made into useful products for the benefit and welfare of all people.

Minerals enter into the lives of everyone every day. Homes, factories, offices, highways, and myriad kinds of equipment are made in large part from minerals. Today, coal and oil as energy for transportation literally move the human race and all the supplies needed to support it.

Minerals number one in world trade

The importance of minerals to the modern world is evident in many ways, but one simple statistic sums it up. The value of world crude mineral production (not refined end-products which would add much more value) currently amounts to more than 1.7 trillion U.S. dollars a year. This is by far the largest element in total world trade. Oil accounts for most of it, but other minerals, particularly coal, copper, iron, sulfur, potash, and phosphate are also important.

Rising expectations demand more minerals

The increasing demand for energy and mineral supplies results not only from the basic annual increase in population, but from the rising expectations of peoples who are endeavoring to move from a relatively undeveloped status to a more advanced civilization. As the lesser developed nations industrialize, they require more oil and other resources which are in finite supply. This is now a significant trend in India, Southeast Asia, and China. This region is home to half the world's people. If China alone used oil on a per capita basis as does the United States, China would use 10 million barrels more oil each day than the present total daily oil production of the entire world. The rising economic giant which is China and Southeast Asia will increasingly demand and be able to compete for more Earth resources. China now has a program to greatly expand automobile production and enlarge its highway system. China may soon be the world's second largest economy. It will need much more oil. So will the rest of the world.

Nations' concern: access to minerals

Control of, or access to mineral and energy mineral resources is critical to maintaining the standard of living which the more developed nations now enjoy. In the past, all means including war have been used to obtain resources, or to protect them. If the chief export of Kuwait had been broccoli, there would have been little world-wide concern about that tiny country when it was invaded by Iraq in 1990. But because Kuwait, in its 6,880 square miles, holds nearly four times the total oil reserves of the United States in its three and one-half million square miles, Kuwait was and remains a very rich prize, to which the industrialized world speedily rose to defend.

The desire of growing populations of industrial nations to maintain and perhaps even increase their standard of living, and the hopes of many more people in lesser developed nations to appreciably raise theirs, intensifies the demand on the world's mineral resources. This desire combined with world population growth at a rate (nearly a million every four days) which each year adds to the Earth's human inhabitants a number equal to the combined size of France and Canada further increases competition for resources. Sharing Earth resources with lesser developed countries' rising expectations and with growing populations will be an on-going challenge to world peace and prosperity. And, as regions of the world exceed their carrying capacity, some people starve, and others move. The huge flow of immigrants, both legal and illegal, into the United States is an example of how people go to the nations which still have the natural resources to provide an adequate standard of living.

Resources, environment, and population

As the population increases, there is now also a globally heightened concern for the environment. It must clearly be recognized that in obtaining the Earth resources which each of us needs to live, there is an inevitable environmental impact that must be accommodated. The more humans, the greater the impact. The environmental impact may be mitigated by various means to some extent, but each new arriving human makes an impact for life.

Continuing to demand the use of minerals but prohibiting their production in the region where they are used for environmental concerns, simply means the environmental impact is shipped off to another area. Out of sight and out of mind does not eliminate the impact. Somewhere the Earth has to be disturbed to yield the resources by which we all live.

***GeoDestinies:* a multidisciplinary study**

GeoDestinies considers the global impact of the present huge demand for minerals in its many aspects. It is a multidisciplinary view which includes national and international economics, politics, technology, demography, history, sociology, engineering, and geology. Issues of ethics must also be a part of this study. Humans live with one another in an increasingly crowded world demanding more and more resources. How are these resources to be divided? Who has the right to use them, and what claims upon them do we recognize for generations yet to come? Do we provide for such claims? If so, how?

Mineral income and social structures

In many countries newly rich from mineral development as are the oil-endowed nations, a variety of social and economic programs have been put in place. Can they be sustained when the Earth resources on which they are based is diminishing and then are gone? What social and/or political upheavals may occur, or military ventures result from the stress of increasing populations and decreasing resources on which to live? Already in places such as Venezuela, civil disturbances have arisen because increasing population and the costs of maintaining their social programs have now outstripped the ability of the national income which is largely oil-derived to pay for them. Iran is also in this situation, and other countries probably including Saudi Arabia are on this path. In the Middle East, sand dunes cannot ultimately be a substitute for oil.

Political and economic power

GeoDestinies also concerns what changes in the balance of political and economic power among countries occur when resources in one area are depleted, and production moves elsewhere. After 1970, the United States was not able to supply its own oil needs, and could no longer control the world price of oil. As a result it lost control of an important part of its economic destiny. This vulnerability was quickly recognized by those who had abundant oil resources, as the 1973 oil crisis demonstrated. Can technology somehow make the United States again energy self-sufficient as it once was? If so, what are the choices, or will the United States remain hostage to those who have the present chief source of U.S. energy — oil? What will supply all the world's energy demands when the very brief time of petroleum is past? What will be the situation of Japan, a highly industrialized society with almost no domestic mineral resources of any sort, when their energy pipeline of oil tankers from the Persian Gulf no longer exists?

The problem of "non-renewable"

The story of the discovery, development, and exhaustion of mineral deposits has already been played out many times in the western United States and other mining regions. Ghost towns remain as monuments to once affluent times, now gone. Similarly, as petroleum production declines in parts of the United States, formerly flourishing oil towns are now having economic downturns because of gradual loss of their natural resource base. Similarly, where underground water supplies are being produced faster than they can be recharged by nature, a vital resource is lost and agriculture suffers.

Will this cycle be played out on a national scale in nations now largely dependent on diminishing mineral resources for their income? These resources eventually will be exhausted. Minerals can be harvested from the Earth only once. What then?

Nature provides many examples where organisms have exceeded the long term carrying capacity of the land. A population increases for a time based on an abundant but depletable resource. When the resource is exhausted, the species in some instances has been reduced by up to 90 percent. What happens to the human population in a nation in which the same circumstance develops?

Informed, realistic decisions

The diminishing natural resource base called upon to supply a growing world population demands that public and private resource decisions henceforth be made on a rational, factually informed basis. We no longer have the luxury of playing politics with the future. Individuals and policymakers now need to clearly recognize the vital influence of mineral and energy resources on nearly every aspect of life and make realistic plans and decisions accordingly.

Reality: the facts, the possible, and the improbable

In considering the historic, current, and future importance of mineral resources, there is a crucial need to clearly know the possibilities and also the limitations which Earth resources offer for the future. Until now, more knowledge combined with abundant energy and mineral resources have allowed industrialization to expand and produce rising standards of living for increasing segments of world population. Because of this past success, most people do not understand the concept of resource depletion. When asked, "What will we use for a convenient energy source once oil is gone" the common response is "the scientists will think of something." We all hope they will. *GeoDestinies* examines such prospects realistically to see if the "something" may possibly be achieved. If so, what might it be, and are there Earth materials available to supply the needs demanded by the use of new technologies which might do good things?

Technology cannot make something out of nothing. Some Earth resources must be used to provide the necessities of life, and many more resources are needed to achieve a good standard of living. In this context the daunting realities of providing Earth resources, minerals, energy, soil, and water, on a sustainable basis to meet the needs of the future are critically examined.

"History is subject to geology."

Will and Ariel Durant
The Lessons of History

Editorial notes

The term "petroleum" in that industry and as used here, includes both oil and gas. The term "gas" means natural gas, not gasoline. Also, the fossil fuels, petroleum, coal, shale oil, and oilsands are all minerals like bauxite or iron ore, but they are sometimes split off from other minerals by terming them "energy minerals." This is frequently done in this book, but for the sake of brevity at some places the term "mineral" has been used for all energy and other Earth-derived resources. Elements, such as copper, lead, and silver are regarded as minerals as are the ores in which these materials are found. The terms "resources" and "reserves" have different meanings in the strict sense. Resources are all the existing raw materials in the Earth. Reserves are those known portions of the resources which can be produced economically. Because technology can change mineral production economics, the distinction between reserves and resources has not been made at all times in this text.

The bibliography following each chapter lists the works which have been consulted for that topic. References pertinent to more than one chapter are included in each bibliography. References are listed in the bibliographies which express divergent views. Each has been consulted but may not be cited. These are included so that varied opinions may be examined if the reader wishes to pursue the subject.

GeoDestinies is a thoroughly revised and much enlarged successor to *Mineral Resources and the Destinies of Nations,* published in 1990, now out of print.

Walter Youngquist
Eugene, Oregon
April 1997

CHAPTER 1

Minerals Move Civilization

History has quite properly recorded the progress of civilization in terms of various ages which have commonly been designated as the Stone Age, the Copper Age, the Bronze Age, and the Iron Age. The use of minerals has provided the material basis for the development of civilization, and has been a major factor in the rise and fall of communities, empires, and nations from the Stone Age to the present.(2,10,20,23,24,25,26)

The Stone Age

A rock thrown at a game animal or bounced off the skull of an enemy was probably the first use of minerals by humans. Eventually the notion occurred to someone to chip a rock into something which could be used as a knife or scraper, and flint, because it can be easily flaked into such tools, became important. Flint occurs geologically in a variety of associations but it is most readily found and extracted as nodules in limestone. In certain valleys of France, especially in the La Claise River Valley, flint nodules are present in great abundance for several miles in the bordering limestone cliffs. Here, ancient peoples developed extensive flint workings, chiefly near the present village of Grand Pressigny. These rocks probably made this area the most important economic district in Europe at that time.

During the Stone Age, flint was among the more valuable possessions a human could have. In various cultures the Stone Age persisted for different lengths of time. In the Great Plains of the United States among the native people the Stone Age lasted until just a few hundred years ago as seen in the great quantities of arrowheads and other stone implements which were made and used until relatively recent times. And, in a remote area in New Guinea, while one man was walking on the moon, another man was cutting down a tree with a stone axe.

The Copper Age

In parts of both the Old and New World, deposits of native copper were discovered. Before that, gold was used as an ornament, because gold is commonly found native, that is, in pure metallic "native" form uncombined with other elements, and is therefore conspicuous and easily recognized. Bright and attractive, gold can readily be beaten into desired shapes, and this was done at an early date in human history. But the first "working metal" used by humans was copper. Copper was extensively mined by early people in the Sinai Desert, and later on Cyprus.(24) The deposits on Cyprus were so highly valued that war followed war in bloody contests for the metal. The island passed under successive control of many groups from the Egyptians through the Romans. The Romans gave us the word from which "copper" is derived, by shortening *aes cyprium* (Cyprium copper) to *cuprum.*

In Upper Michigan (the Keweenaw Peninsula portion) numerous prehistoric pits in the extensive native copper deposits testify to the widespread use of this metal by the native populations.(24) The mining and working of copper was a large enterprise. More than 10,000 individual mines have been discovered in this area. It is estimated that it took at least 1,000 miners a minimum of 1,000 years to produce all the copper workings now visible. The presence of abundant native copper, which could easily be shaped into knives and other tools, was of great importance to these first miners who would otherwise have to use stones which were much more difficult to work. The copper could be hammered into all kinds of implements, including arrowheads, chisels, hooks, and axes.

This copper was also apparently a major basis for trade. Tools of native copper derived from this region have been found in early habitations throughout the Upper Mid-West, and down the length of the Mississippi River Valley.

Elsewhere, native copper tools helped shape the stones for the pyramids of Egypt, and copper was the first metal employed as a shaped weapon in Old World warfare. Copper ores are relatively easy to smelt, so copper metallurgy developed early and copper became the first metal to be used extensively by several civilizations. Its use marked a transition from the Stone Age into the age of metals.(26)

In modern times, without copper the electrical age may never have been possible, or at least considerably delayed, for copper has been the first and primary workhorse of the electric industry. Copper has the highest electrical conductivity of any metal, except silver. It has only recently being partially displaced by aluminum, and glass fibers which transmit information by means of light. However, the production of aluminum has depended on vast amounts of electricity produced by copper coil-wound generators, and initially transported by copper wire to the aluminum smelters. Without copper, we might still be reading by candlelight or oil lamps.

The Bronze Age

The Bronze Age was a logical successor to the Copper Age as bronze is simply copper with some tin added. The metallurgist who discovered that tin, combined with copper, would make a much harder metal is not recorded. The first true tin bronze wherein the amount of tin is high enough to indicate that the tin was an intentional addition to the copper appears about 3000 B.C. in Mesopotamia.(24) During succeeding centuries, bronze objects were widely made in that region. Later, the Romans made extensive use of the copper/tin mixture to produce numerous bronze weapons. In order to obtain adequate supplies of tin the Romans, by force, took possession of the great tin deposits of Cornwall in England. Roman weapons, first made of bronze, and later of iron, conquered much of the then known western world.

The Iron Age

Although iron was known much before Roman times, it had only very limited use, as the metallurgy of iron is difficult due to the high temperature required to melt it. Because of this, for a long time iron was not well known nor widely used. However, there is a record of iron being employed as far back as 1450 B.C., and about 1385 B.C. the Hittites manufactured a substantial number of weapons from iron. With this superior weapon they subdued the

Assyrians, and then drove into northern Syria and Palestine. There they fought the Egyptians, and ultimately established the Hittite Empire as a major political and military power in western Asia.

But the iron which equipped the Hittites for military success was eventually used against them. The Hittites had jealously guarded their secret of iron metallurgy, because of its great military importance. But the secret got out and ultimately the enemies of the Hittites were also equipped with the same metal, and the Hittite Empire was besieged and finally disintegrated.

The Iron Age and the Industrial Revolution

The Iron Age, as we now know it, came much later as iron came into widespread use in industry and construction fairly recently. Iron has been of great and continual permanent importance since beginning of the Industrial Revolution led by Great Britain starting in the 18th century. The Industrial Revolution made iron and coal the most valuable mineral resources a country could have. Great Britain had the geological good fortune of having supplies of both coal and iron in close geographic proximity so they could easily be transported (at first by horse and wagon) to produce the iron and steel (an alloy of iron) to build the factories and machines necessary to Britain's Industrial Revolution. This fortunate geological circumstance can hardly be over emphasized for without it, it is quite unlikely that Britain could have led the Industrial Revolution. Britain also had significant deposits of zinc, lead, tin, and copper, which greatly strengthened Britain's economic situation. In the nineteenth century Britain was successively the world's largest source of coal, iron, lead, tin, and copper. During that time she was the wealthiest nation in the world and supplied more than half the world's demand for some of these metals. From 1700 to 1850 Britain mined more than 50 percent of the world's lead, and from 1820 to 1840 she produced 45 percent of the world's copper. From 1850 to 1890 Britain increased iron production from one-third to one-half of the entire world supply.(16)

But this rapid exploitation of the minerals was followed by the inevitable decline of the industry due to increasing costs with deeper mining, and the gradual exhaustion of the deposits. Peak production of lead was in 1856, copper 1863, tin 1871, iron ore, 1882, and coal in 1913. Britain's coal industry continues to contract. In 1992, British Coal, the government coal company after the industry was nationalized in 1947, announced that 31 of its remaining 50 mines would be closed, eliminating more than half the industry that once fueled Britain's Industrial Revolution. In 1947 there were 718,000 coal miners. Now only about 48,000 people are employed in coal production.

During the same time in the late nineteenth century, as Britain's mining declined, the United States was discovering and developing even richer and much larger mineral deposits. By the end of World War I, which put a heavy demand on Britain's mineral resources, the United States had replaced Great Britain as the largest industrial power. The large new demand for oil which the United States was discovering widely distributed and in large quantities, also caused the world industrial center to shift to the United States. Britain had no oil production at the time.

Industrial Revolution and metal demand.

Although metals since the Copper Age have been important, for a long time they were used in relatively small quantities. It was not until the Industrial Revolution that there was large demand for minerals. Earlier, the economies were mostly agriculturally based. It was rich, fertile land that was the resource prize. Productive land is still prized, but to work it economically metals and energy resources have become important, and nations are much more interested in the supplies of these materials.

Coal

Coal also moved civilization rapidly forward during this time, because iron needed large quantities of coal in the smelting process. The ability to produce iron in quantity and relatively cheaply compared with smelting iron in the past without the use of coal, allowed the British to develop new uses for this hard material with a high melting point, which could contain fire.

The beginning of mass rapid transport

One significant result of cheap iron was the invention of the steam engine and the railroad. The ability to move vast quantities of materials cheaply over long distances very literally and technologically moved Britain forward, and indeed has done so ever since for much of the world. The advent of the railroad has been one of the great steps forward for civilization. Iron in quantity made possible by abundant coal did it. And coal, in turn, for many years was the energy which moved the railroads.

The Iron/Aluminum Age

The second half of the Twentieth Century is sometimes called the Atomic Age (made possible by the metal uranium). But we do not construct buildings nor make cars, or other machines of uranium but rather of iron, and its alloy, steel, and aluminum. So in the real and practical sense we are in the Iron/Aluminum Age now and likely will remain so for many years to come.

To some degree we also are living in a plastic age, for plastics have replaced metals in many uses. Plastics are chiefly petroleum-based, so the advent of the widespread use of petroleum has also modified the composition of many structures we use, from toys to electronics to car parts.

Cultures Move at Different Rates

All of civilization does not move at the same rate. Much of Africa went from the Stone Age to the Iron Age without the intervening Copper or Bronze ages. With the arrival of the white man in North America, the Native Americans who were not in the copper country of northern Michigan or had access to it by trade, were still in the Stone Age (flint or obsidian arrowheads, knives, and scrapers). They suddenly found themselves in the Iron Age, facing rifles.

How fast a culture has moved forward has largely been determined by what mineral and energy mineral resources it had available to it and to what degree these were used.

Saudi Arabia

There is hardly a more striking example of how mineral resources have moved a segment of civilization so far and so fast as Saudi Arabia and adjacent oil-rich areas of the Middle East — Iraq, Iran, Qatar, the United Arab Emirates, Kuwait, Bahrain, and Oman.

The market for sand dunes is not great. A hundred years ago a loosely organized group of nomadic tribes, a few small farm areas, and some little fishing villages occupied the Arabian Peninsula. The economic impact of this relatively small number of isolated people was negligible in the world economy. There was no indication then that things would ever be any different. But less than a hundred years later, Saudi Arabia has the largest and most modern airport in the world, a telecommunications system second to none, and first class hospitals and universities. The standard of living compared to a century ago is vastly different and incomparably better. The oil beneath the sand dunes changed the destiny of the region to what has become a world-influencing economic power.

These people, still a very minor percentage of the world population, economically shook the world's greatest industrial nation, the United States, with the oil embargo of 1973. This was possible only because Saudi Arabia had oil. Without the oil, Saudi Arabia and the other Arab states would have had no such influence.

Now there is, and will continue to be for many years, a very considerable concern by industrialized nations about the happenings on the Arabian Peninsula and in all other countries around the Persian Gulf. Oil brought this area great prosperity in a phenomenally short time. Until the oil wells run dry, these relatively small nations will have a very large say in world economics. And this influence will increase in the next several decades as other oil-producing areas pass their peak of production, as has the United States, and the Mid-East holds an increasing percentage of the remaining oil reserves. It has the makings of an interesting economic situation.

Energy Mineral Resources the Key

The importance of mineral and energy mineral resources cannot be overestimated. Most critical among the resources is energy. Energy is the key which unlocks all other natural resources.(2) Without it the wheels of industry do not turn, no metals are mined and smelted. No cars, trucks, trains, ships or airplanes could be built, and if built, they could not move without energy. Without energy, houses would remain cold and unlighted, food would be uncooked. Fields could not be plowed nor planted with the ease and on the vast scale they are today by means of relatively little human labor. Military defense as we know it today would not exist. Without energy resources we would literally be back in the Stone Age. And without the use of energy and metals as we use them today, it is probable that the world's population would be reduced at least one-half, some estimates say 90 percent.

In the industrialized world we take our high standard of living very much for granted, without realizing what a great debt we owe every moment to mineral and energy resources. We convert minerals to metals and the metals into tools and machines. Many machines are run with mineral fuels. By these means we produce easily and in great quantities the goods which previously either did not even exist or took large amounts of hand labor to produce. Minerals and energy (in the form of the steel plow and the tractor and fuel to move them) together with petroleum-based fertilizer and pesticides, have enabled two percent of the working population in the United States to feed all its citizens and some of the rest of the world.

In contrast, 60 percent of the people of China still till the fields. When 60 percent of the population tills the soil, it indicates an economic existence close to simply subsistence for most of the people. Only when the burden and necessity of making a living from the soil by dawn to dusk labor is at least partially lifted can human societies develop sufficient numbers of scientists, engineers, doctors, and technicians to move society forward in a significant way.

Tilling the fields precludes going to school which provides the foundation for nearly all progress. Unnecessary manual labor in the fields and factories limits human potential. Minerals and energy have allowed those parts of the world which have been able to obtain and use these resources to ease the physical burdens of existence and provide the time and the opportunity for the human mind to become educated. That is where the forward movement of civilization begins, and continues.

Energy supplies and employment

In the industrial nations' factories and research laboratories, as well as in the service industries, energy is all important. Employment and energy consumption when graphed are essentially parallel lines. This relationship between energy supply and employment was strikingly demonstrated during the 1973-74 Arab oil embargo against the United States. As a result of that partial cutoff of oil to the United States, the Federal Energy Administration estimated that the nation's gross national product declined $20 billion, and a half million American workers lost their jobs.

Energy and goods distribution

Oil also allows products of our farms and factories to be widely and inexpensively distributed by truck, train, and plane. It would do no good to produce quantities of appliances, clothing, machinery, farm products, and all the other products which form the basis of our material standard of living, if these could not be easily and inexpensively transported to the population at large.

Energy for each person

For many years the United States has surpassed all other nations in terms of per capita consumption of energy and mineral resources. The United States with about five percent of the world's population uses about a third of the globe's annual energy supplies. In ancient days, the capture and use of slaves was one of the chief sources of wealth. Slaves were kept for their labor, and their labor simply represented valuable usable energy.

Now we capture slaves in the form of barrels of oil from an oil well, cubic feet of gas from a gas well, tons of coal, or pounds of uranium ore from a mine. These, in a very real sense, are our modern slaves. Each year every U.S. citizen, on the average, uses 8,000 pounds of oil, 4,700 pounds of natural gas, 5,150 pounds of coal, and 1/10th of a pound of uranium. In terms of energy, visualize, how many slaves it would take push your automobile up to the mountains or to the coast for a roundtrip of perhaps 200 miles. And human slaves would surely not be able to do it at 60 miles per hour. But this is done for you now by less than 10 gallons of "gasoline slaves" in your car's tank. The richest ancient Egyptian or medieval European king never had it so good.

Some interesting calculations have been made with regard to these energy "slaves" we draw on each day. The figures are based on the estimate that one "person-power" (PP) equals .25 horsepower = 186 Watts = 635 Btu/Hr.

For example, if the energy requirements of the United States were met by person-power, it would take more people than presently exist in the world — almost three times as many. An interesting set of calculations by Fodor estimates that in the United States "...each person consumes an average of 58 person-power continuously 24 hours a day! To visualize this, think of 58 Olympic athletes pedalling exercise bikes (connected to electric generators) at maximum speed without tiring or resting. If one prefers to visualize their energy servants working only an eight-hour day, then each individual would require the equivalent of 174 person-power to meet his/her energy needs. Surely a Roman nobleman would be extremely rich if he had 174 tireless personal servants working at maximum output on his behalf every day of the year. Yet, through our harnessing of energy, each one of us has an equivalent amount of power working on our behalf."(11)

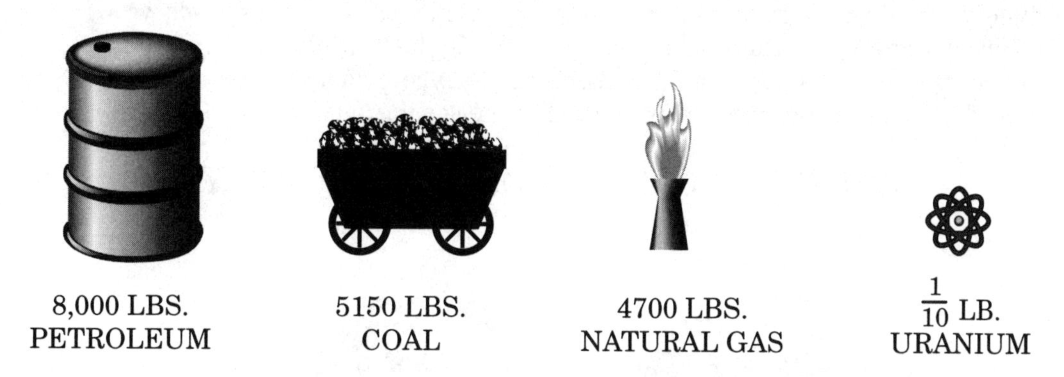

Energy equivalent to 300 persons working around the clock for each U.S. citizen

Figure 1. U.S. per capita annual energy mineral consumption.

(Source: U.S. Bureau of Mines, 1990)

In more detail, a light bulb takes ½ PP, the TV takes 1 PP, hair dryer uses 8 PP, gasoline lawn mower uses 14 PP, a water heater takes 79 PP. The study goes on, "Driving is the most energy-intensive activity commonly engaged in. It takes about 500 person-power to keep the average auto at 55 miles per hour. Our person-power athlete would have to pedal continuously for an entire year (8760 hours) to produce the amount of energy contained in a single barrel of crude oil. If we purchased the energy in a barrel of oil at the same price we pay for human labor ($5/hour), it would cost us over $45,000! Fortunately, a barrel of oil is an incredible value at just $25 or less. Get yours while they last!"(11)

Non-energy Materials We Use

The U.S. Bureau of Mines calculates that each U.S. citizen annually accounts in consumption for about 1300 pounds of steel and iron, 65 pounds of aluminum, 25 pounds of copper, 15 pounds of manganese, 15 pounds of lead, 15 pounds of zinc, and 35 pounds of other metals such as cobalt, a critical material, incidentally, without which a jet airplane could not fly.

Mineral supplies needed each year

Including sand, gravel, cement, dimension stone, clay, and the energy and metal supplies already listed, more than five billion tons of new minerals are now needed each year in the

U.S. economy. This amount grows each year with the increase in population. Next year another more than five billion tons of minerals will be required. These demands add up to more than 20 tons of raw energy mineral and mineral supplies which have to be produced each year for every man, woman, and child in the United States, if our material standard of living is to be maintained. And, importantly, these demands increase each year as our population grows by natural native increase, and by both legal and illegal immigration. Producers of these resources face a tremendous task whose vital role in the daily life of the individual is little understood or appreciated by the general public.

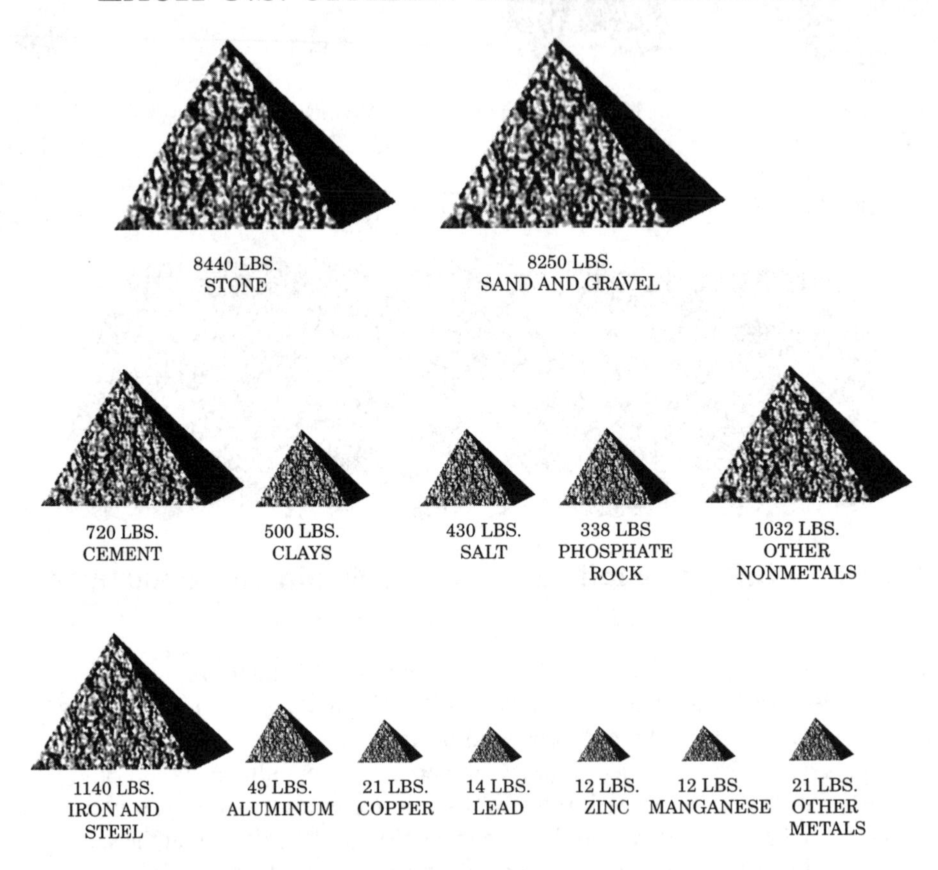

Figure 2. U.S. per capita annual mineral consumption

(Source: U.S. Bureau of Mines, 1990)

These consumption figures shown in Figure 2, do not mean that each person directly uses this quantity of each material. But the total amount of these supplies used in the United States annually divided by the population gives the average figures. Steel, for example, may go into building construction for homes, apartment houses, supermarkets, shopping malls, roads, and bridges. Railroads, cars, and trucks which in many ways serve citizens every day are made from steel and other metals.

This is Our Material Standard of Living

This total, then, represents our current material standard of living. As we add population, we have two choices in providing these resources. Either we dig up and produce more of these minerals domestically or we import more each year, or by a combination of the two we obtain that total of 20 tons of these materials for each resident, or we reduce our material standard of living.

Problem: Increasing population against finite resources

Compounding the problem is the fact that in the United States and worldwide, the most easily recovered, higher grade mineral and energy mineral deposits are used first. Thus there is increased demand each year against resources which are declining in quality and cost more to obtain.

Again, it should be emphasized that these are annual rates of consumption. Each 12 months another 20 tons per person must be obtained to maintain the status quo. The task is huge and grows larger each year as population increases.

The desire for an increased standard of living coupled with the continuing rise in population create an exponential demand on mineral resources. In the first fifty years of the twentieth century, the total production of minerals and mineral fuels in the world was far greater than the total of these materials produced during all previous history. Then, in the following twenty years, this production was exceeded again by approximately 50 percent. These statistics when graphed show an exponential curve which is now beginning to steepen into a vertical line. Such a rate of consumption, of course, cannot be sustained. This sharply rising line representing energy mineral and mineral use also represents the marked rise in the standard of living for the world in general, and especially for the industrialized nations. This steep curve of resource consumption means the domestic sources of these materials in the industrialized nations have been drawn upon very heavily. As a result, many of the remaining and generally higher quality reserves of some materials are now held by the so-called under-developed nations which are still in, or have just recently come from an agrarian or nomadic economy.

Phenomenal Rise of the United States

North America, and particularly the United States, initially had a resource variety and abundance virtually unmatched by any other area of the world. This marvelous spectrum of mineral and energy resources allowed the United States to rise from a wilderness to become the richest and most powerful nation in the world in less than 300 years — an event unequaled in all history, and probably never to be repeated anywhere again. Just prior to this, and in part overlapping the start of the rise of the United States, Great Britain, with moderate quantities of coal, iron, tin, and lead in relatively close convenient geographic association, led the way into the Industrial Revolution. But Great Britain rapidly depleted its modest mineral supplies.

The much greater abundance of these resources in the United States, and also the generally higher quality deposits (especially of coal, and iron), as well as the ultimate discovery of vast oil resources, allowed the United States to quickly overtake Britain and then surpass it in industrial development. By the late 1920s the United States, with six percent of world population, was producing 70 percent of the world's oil, almost 50 percent of its copper, 46 percent of its iron, and 42 percent of its coal.(13) The ability to produce and use these basic industrial materials and energy resources was the key to the phenomenal rise of the United States.

Mineral Measure of Standard of Living

The physical standard of living can be quite well measured by noting the consumption of two basic resources — iron and oil. Oil is the largest single energy source of the industrial nations. Iron is the chief metal used. The consumption of oil is taken up in the Chapter 12, *The Petroleum Interval.* In the case of iron, the 18 developed nations with a total of about 700 million people now use iron (and its alloy, steel) at rates ranging from 680 pounds to 1,400 pounds per person per year. This contrasts with some 1.8 billion people in a portion of the undeveloped world who use less than 55 pounds of iron and steel per person per year.

Minerals in Everyday Life

Mineral resources have moved civilization from caves into houses, office buildings, and factories. We build with materials from the Earth.(8) The highways on which civilization moves in a very literal sense, are made either of concrete (limestone and sand and gravel with some gypsum and clay) or asphalt (from an oil well) with rocks mixed in for durability. An average asphalt road is about ten percent tar. Without the tar it would just be a gravel road, but even the gravel road is made from minerals.

Energy and mineral resources are vital to almost all sectors of an economy from traditional farming, construction, and transportation to the newer high-technology areas of communications, where satellites and electronic networks encompass the world. All this is made of minerals of various sorts and uses vast amounts of energy. The initial process of xerography which revolutionized the world of printing, was made possible by the metal, selenium. Tungsten is critical in high speed hard tool steel used for cutting other steel. Cylinder blocks of automobiles are bored with tungsten carbide steel, and tungsten is the usual filament in light bulbs.

Whether it is air conditioning which makes life livable and productive in hot, humid climates, or heat which keeps us comfortable in cold weather and in cold climates, energy is vital. Many parts of the world would not be habitable to large populations if it were not for energy supplies in quantity.

Minerals and Your House

The basic house, since it first became a reasonably comfortable place with indoor plumbing, comes largely out of mines. The foundation is probably of concrete which is made from limestone, clay, sand, and gravel. The exterior walls may be made of stone or brick (clay). The insulation may be glass wool (quartz sand, feldspar, and trona — a sodium carbonate which is mined). The lumber is put together with screws and nails of steel and zinc. The wallboard which forms the interior walls of many homes is made chiefly of gypsum. The roof is probably covered with asphalt shingles. The asphalt came out of an oil well, and the filler in the asphalt shingles is a variety of colored silicate minerals. The fireplace is brick or stone with a steel fire box. The sewer pipe is made of clay or iron pipe or may be plastic from material out of an oil well. The electrical wiring is copper or aluminum. The sanitary facilities are made of porcelain which is clay. Plumbing fixtures are made of brass (copper and zinc) or stainless steel (nickel and chrome with iron). Roof gutters are of galvanized steel (iron and zinc) or plastic from an oil well. The paint has mineral fillers and pigments from minerals. Windows are glass which are made from trona, quartz sand, and feldspar —all mined. Door knobs, locks, and hinges are of brass (copper and zinc) or steel (alloy of iron), and finally the mortgage, if it is not simply written on newsprint, is written on quality paper made from wood or cloth fibers and filled with clay.(8)

National Defense

In national defense, everything from tanks to submarines to supersonic planes to rockets and intercontinental ballistic missiles require minerals for their production, and energy to power them. In modern warfare it has virtually become a matter of which side can deliver the greatest amount of energy in some destructive fashion against the other side that determines who wins. A demonstration of this was made in Japan twice in August of 1945, by use of the energy mineral, uranium.

"Service" Economies Still Based on Minerals

It is sometimes said that the United States is becoming a "service economy," but it is doubtful that we can really survive as an important nation by serving hamburgers to one another, and exchanging computer printouts. The computers are made at least partly out of metal and the energy to run them and fry hamburgers must come from somewhere. Hamburgers and computer printouts do not a great nation make. And service economies are not an effective national defense. Nations cannot protect themselves with word processors or ham and cheese sandwiches. Minerals and energy minerals win wars, build factories, and supply our daily basic needs.

Energy resources and minerals are the foundation of industrialized society. Nations which have these resources (or have access to them) will be the important nations of the future, even as they are now. A nation with a variety of minerals and energy supplies within its own borders is much better off than those which have to import supplies. Free access to these materials is fine as long as they can be imported. But being largely dependent on imported raw materials makes for a vulnerable situation. Japan is very much in that position.

Minerals Needed to Produce Minerals

It is important to remember that the production of energy and mineral resources requires that other energy and mineral resources be used. Imagine trying to drill a 10,000 foot deep oil well by hand or breaking up iron ore with a pick and shovel (although even in this instance you still are using steel in the pick or shovel). To drill wells, blast out ore deposits, and load and transport them, to manufacture drilling pipe and drill bits, to make oil well casing, to build turbines and generators for power plants, to make solar collectors, and to make and run the chain saw used simply for cutting firewood, energy and metals have to be used.

Change of Sources — Shift of Power

Of special significance is that as energy and mineral resources are depleted and costs rise in one area, sources for these materials change; witness the shift of major world oil production from the United States to the Middle East. With these geographic shifts of resource centers also go balances of economic power among regions and nations. Hibbing, Minnesota, was once the iron ore capital of the world, but the high grade ore was mined out. Much of the iron used by the United States to make weapons to defeat Japan came from the Hibbing mining district. Today, Japan gets much of its iron ore from excellent deposits in relatively nearby Australia, and residents of Hibbing buy Japanese cars made from Australian iron. How things change!

Shifts in mineral source areas have occurred in the past and will continue to occur in the future. The United States is gradually experiencing this truth, as it has to import more and more raw materials in which it was once self-sufficient. The long term effects of this trend will have significant implications for the future of the United States. The process will have a continual and substantial effect upon the lives of all U.S. citizens, but many persons do not realize that in the United States the era of high grade energy mineral and mineral resource abundance is gone. Every nation throughout history has sought the basic necessities of life, and to try to upgrade its standard of living. Earth resources have been the chief factor in raising living standards, and will continue to be so. Civilizations in the past moved at irregular speeds in their progress, in large part due to the availability or lack thereof of mineral resources. As resources are depleted in a given country, will that impede its progress as more and more raw materials have to be imported? Will that country be able to maintain its standard of living or will it decline in a material sense? The United States is beginning to enter the time when the answer to those questions will be determined.

Nations With Few Mineral Resources

However, some nations which have no substantial mineral or energy mineral base have been able to do very well. Japan, with very few such resources, has become the economic miracle of the latter half of the Twentieth Century.

But Japan's ability to survive and prosper is due to its being able to import huge quantities of mineral and energy supplies. To pay for this, Japan must be a strong exporting nation, which is why Japan is so defensive about its export position. Can Japan's miracle be sustained or will Japan be the first major industrial nation to demonstrate the fatal flaw in having no domestic sources of these vital materials? It was the lack of these energy and mineral resources and the cutting off of foreign supplies of oil in particular which caused Japan to go to war in 1941. Will Japan ultimately lose the industrial war because of lack of raw materials? If the presently rising competition from other southeast Asia nations hurt Japan's exports, Japan will be in serious difficulty.

Minerals, including energy minerals, are the basis for our modern civilization. Nations not possessing these are either doomed to stay at a relatively low standard of living, or they have to get these resources in raw or finished form by trade in some fashion. If denied, they will have to resign themselves to a second class position or go to war. This was precisely the decision which Japan faced in 1941. Can free trade and access to raw materials be maintained from now on so that the "Japan decision" will not ever be repeated? To the present, however, minerals and energy minerals have markedly altered the course of civilization through warfare.

In World War I, it was said that the Allies "floated to victory on a sea of oil." In World War II, a previously obscure metal, uranium, in the form of the atomic bomb, dramatically ended that struggle. But can a future be achieved where war can be replaced by diplomacy and free trade of vital raw materials, even as these materials become increasingly scarce and costly?

Education and Technology Can Compensate — Partly

Countries can succeed economically with few mineral resources, if they have the capability through exports or other means such as providing financial services, to import their mineral needs. Japan and Switzerland demonstrate that an educated work force and a technologically advanced country can compensate for lack of material resources. But, they still must have access to materials from somewhere. Japanese cars and Swiss hydro-electric turbines, or watches cannot be made from nothing. They are made of metals dug from somewhere on the Earth.

Can a Global Economy Move Everyone Along?

It remains to be seen whether today's emerging global economy can smooth out the differences in economic development of various countries as now measured by their standards of living. The differences are currently large, and the present over-population of some areas, and the ever-increasing world population as a whole will make this objective difficult to achieve.

Mineral resources have been, are now, and will be in the foreseeable future a major factor in the progress of individual nations, and in the broad sweep of civilization. Four classic works which clearly describe this are Lovering, Park, Poss, and Raymond.(16,23,24,25) The economic geologist Lovering many years ago observed: "The countries within whose boundaries such mineral treasures lie possess temporarily an asset that gives them a greater commercial advantage over their less fortunate trade rivals...Productive mineral deposits are among the most valuable resources that a nation may have; but such an asset will be widely coveted, and serious international problems may be engendered by the mineral policies of the government that controls the deposits."(16)

U.S. Senator Larry Craig summed up the importance of minerals to each of us:

> "Without the thousands of materials derived from minerals, society would be unable to provide food, fuel, shelter, clothing, potable water, treatment and disposal of sewage and industrial wastes, garbage removal and processing, medicines and medical services, communications, facilities to produce and deliver electricity, construction and manufacturing facilities and equipment to build and maintain our places of business and our jobs, and the transportation systems that deliver these products to our homes and businesses and then remove our wastes. Moreover, we would be unable to maintain law and order and a national defense. Most of us would die."(6)

How mineral resources can continually be supplied in increasing quantities to a growing population, and to a world where more and more nations seek a rising standard of living is the task ahead. It will also be a challenge to do this in an orderly manner without conflict.

BIBLIOGRAPHY

1 AGNEW, A.F., 1983, International Minerals. A National Perspective: American Association for the Advancement of Science, Washington, D.C., 164 p.

2 AYRES, EUGENE, and SCARLOTT, C. A., 1952, Energy Sources — the Wealth of the World: McGraw-Hill Book Company, Inc. New York, 344 p.

3 BARNETT, H. J., and MORSE, CHANDLER, 1963, Security and Growth: The Economics of Natural Resource Availability: Johns Hopkins Univ. Press, Baltimore, 238 p.

4 BATES, R. L., and JACKSON, J. A., 1982, Our Modern Stone Age: William Kaufman, Los Altos, California, 132 p.

5 CHILDE, GORDON, 1954, What Happened in History: Penguin Books, Baltimore, Maryland, 288 p.

6 CRAIG, L. E., 1996, Government Support for Geologic Research Faces Uncertain Future: Geotimes, March, p. 29-32.

7 DAVIES, OLIVER, 1935, Roman Mines in Europe: Oxford Univ. Press, London, 291 p.

8 DORR, ANNE, 1987, Minerals — Foundations of Society (Second Edition): American Geological Institute, Alexandria, Virginia, 96 p.

9 EHRENREICH, R. M., (ed.), 1991, Metals in Society: Theory Beyond Analysis: The Museum Applied Science Center for Archaeology (MASCA). The Univ. Museum, Univ. of Pennsylvania, MASCA Research Papers in Science and Archaeology, Philadelphia, v. 8, Part II, 92 p.

10 FLEMING, S. J., and SCHENCK, H. R. (eds.), 1989, History of Technology: The Role of Metals: The Museum Applied Science Center for Archaeology (MASCA), The Univ. Museum, Univ. of Pennsylvania, MASCA Research Papers in Science and Archaeology, Philadelphia, 87 p.

11 FODOR, EBEN, 1991, Looking at Energy From the Human Perspective: Southern Willamette Alliance, Eugene, Oregon, May, p. 5.

12 FRANK, TENNY, 1927, An Economic History of Rome: Johns Hopkins Univ. Press, Baltimore, 519 p.

13 GRONER, ALEX, (ed.), 1972, The American Heritage History of American Business & Industry: American Heritage Publishing Company, New York, 384 p.

14 KELLY, W. J., et al., 1986, Energy Research and Development in the USSR: Duke Univ. Press, Durham, North Carolina, 417 p.

15 KENNEDY, PAUL, 1987, The Rise and Fall of the Great Powers: Random House, New York, 677 p.

16 LOVERING, T. S., 1943, Minerals in World Affairs: Prentice-Hall, Inc., New York, 394 p.

17 MACZAK, ANTONI, and PARKER, W. N.,(eds), 1978, Natural Resources in European History: Resources for the Future, Washington, D.C., 226 p.

18 McDIVITT, J. F., and MANNERS, GERALD, 1974, Minerals and Men: Resources for the Future, Washington, D. C., 175 p.

19 McLAREN, D. J., and SKINNER, BRIAN, (eds.), 1987, Resources and World Development: John Wiley and Sons, New York, 940 p.

20 MIRSKY, ARTHUR, and BLAND, E. L., 1996, Influence of Geologic Factors on Ancient Civilizations in the Aegean Area: Journal of Geoscience Education, v. 44, n. 1, p. 25-35.

21 NATIONAL ACADEMY OF SCIENCES, 1972, The Earth and Human Affairs: Committee on Geological Sciences, Division of Earth Sciences, National Research Council, National Academy of Sciences, Washington, D. C., published by Canfield Press, San Francisco, 138 p.

22 PARK, C. F., Jr., 1968, Affluence in Jeopardy. Minerals and the Political Economy: Freeman, Cooper and Company, San Francisco, 368 p.

23 PARK, C. F., Jr., 1975, Earthbound. Minerals, Energy, and Man's Future: Freeman, Cooper and company, San Francisco, 279 p.

24 POSS, J. R., 1975, Stones of Destiny: Keystones of Civilization: Michigan Technological Univ., Houghton, Michigan, 253 p.

25 RAYMOND, ROBERT, 1986, Out of the Fiery Furnace. The Impact of Metals on the History of Mankind: The Pennsylvania State Univ. Press, University Park, Pennsylvania, 274 p.

26 ROCKARD, T. A., 1932, Man and Metals. A History of Mining in Relation to the Development of Civilization: (2 vol.) McGraw-Hill, New York, 1068 p.

27 ROSTOW, W.W., 1978, The World Economy. History & Prospects: Univ. Texas Press, Austin, 833 p.

28 SHIMKIN, D. B., 1953, Minerals — A Key to Soviet Power: Harvard Univ. Press, Cambridge, Massachusetts, 452 p.

29 WOLFE, J. A., 1984, Mineral Resources. A World Review: Chapman and Hall, New York, 293 p.

CHAPTER 2

Minerals Move People

Along with the influence of mineral supply on the rise and progress of civilization, is a parallel story of how the search for, and discovery of minerals from salt to gold and silver has caused mass migrations of people. In many cases the minerals were the basic cause of the opening of new lands. It has been said that "the flag follows the miner's pick." The quest for gold and silver lured Spaniards to the New World resulting in the conquests of Mexico, Colombia, Peru, and some of the adjacent lands.

These movements continue to the present. Migrations of people today to resource areas, or to nations which can obtain resources are perhaps greater than anytime in the past. Seeking jobs and a higher standard of living largely based on mineral resources, people are today moving, both legally and illegally, by the millions.

Salt and Trade

Except for the occasional travels by native peoples to certain localities to obtain chert or obsidian with which to make tools and weapons, probably the first mineral to cause people to travel substantial distances was common salt. No doubt even before humans arrived on the Earth, animals traveled considerable distances to salt licks, even as they do today. Trails made by animals to salt licks in the eastern United States were some of the first trails the early settlers used.

History records the caravans and traders who moved salt in ancient times over great distances. Some of these salt routes are still used. In the 6th century salt was the chief item of trade for Venice, which developed a salt monopoly that extended over parts of the Mediterranean, and Venetian salt traders traveled widely in their commerce.(13)

Gold Seekers

Gold was the first metal used by humans as it is bright and attractive in the native (pure) form, in which it commonly occurs. It can easily be worked into many shapes and does not tarnish. Gold nuggets in stream beds attracted attention very early. This attraction for gold probably precedes humans, as pack rats and some birds will pick up small gold nuggets and put them in their nests. Silver is rarely found in native form and it tarnishes easily. However, many silver ores can be smelted readily so silver, too, has a long history of being sought by humans. It was put into use among earlier peoples chiefly as ornaments, and later as coinage.

Egypt

Gold was one of the earliest reasons for conquest and exploration. What is perhaps the first map ever made is a papyrus map of the Rammessides which shows a route to the Coptos gold mines along the eastern border of Egypt fringing the Red Sea. Egyptians took great quantities of gold from this area.

The gold occurred as native gold in white quartz veins, and in placers (sand and gravel deposits) downstream in the valleys below. But even richer gold deposits were in the upstream areas of the Nile, outside of Egypt in the Nubian Desert. The name "Nubian" comes from the Egyptian word "nub" which means gold. The Nubian gold lured the Egyptians to exploration and conquest of that area. The first gold was simply taken by plunder from the natives who already had it. Eventually the Pharaohs sent miners and established regular gold mining camps in Nubia. This pattern of exploration for gold plunder and later the establishment of permanent gold camps was repeated many times in history.

Spanish conquests

It was said that the Spaniards had the "gold disease," but the sad fact is they had other diseases also, such as smallpox. Many of the native cultures the Spaniards encountered were virtually destroyed by the contagious diseases which the gold and silver seekers brought with them. In that way, too, gold and silver altered the history of the native nations of the western hemisphere.

Gold Rushes

California

The '49 gold rush to California is one of the great epochs of migration caused by a mineral discovery. Ships went around Cape Horn, and others went to Panama where the passengers hiked across the Isthmus to pick up another ship on the other side. Still others took the overland route to California. Before 1840, migration across the Mississippi River to the west was "hardly more than a trickle." But, with the gold discovery of 1848 about 40 miles from present day Sacramento, by 1850 California's population was 96,000, and larger than the State of Delaware.(4) California was opened up by the discovery of gold, and it went on to become one of the great states of the United States, developing an economy of global significance. Gold was the initial catalyst destined to open up California.

Australia

The discovery of gold in Australia opened up large areas of that continent which had been previously ignored. The early gold rushes to Victoria and New South Wales caused many changes in Australia. In 1850, there were only about 400,000 people in all of Australia, but with gold discovery by 1861 there were more than a million. Melbourne got its start during the gold rush, and at one time was the richest colony in the British Empire. Many of the fine gardens and some of the buildings (such as Government House) are legacies from what the discovery of gold the middle of the nineteenth century did for Australia.

Raymond emphasized the importance of gold migration of people in world trade:

> "The finding of gold in Australia, as in California, had a profound effect on the nation's economy, and would do so in other parts of the world where gold was soon to be discovered: New Zealand, South Africa, and Alaska. The gold rushes, wherever they occurred, brought new settlers, new ideas, new vigor, and created new wealth. Without the enormous amounts of gold that were produced in the

latter half of the nineteenth century the commerce of the modern world could never have reached the proportions that it has today. Only after the gold rushes was it possible to speak of something called world trade."(16)

South Africa

The discovery of diamonds in South Africa in 1867 brought thousands of immigrants to that region. The subsequent discovery of many other important minerals, especially gold, led to the transformation from what was largely an agricultural economy in South Africa to the present day urban-industrial economy. The continuing international interest in South Africa over the years has stemmed largely from its possession of several strategic minerals. It remains a potentially valuable prize.

Siberia

Siberia had no gold rush in the classic sense, but with a need for acceptable foreign exchange, the Russians over the years have established gold camps and gradually moved people into the Kolyma gold province of Siberia.(12) These camps have been expanded into permanent settlements, and now there are twelve cities in the region with a total population of more than half a million. Russia does not plan to allow these towns to be ghost towns when the gold is gone. There is lots of gold so it will last for some time, but planners have gradually established manufacturing facilities there. In this way, they expect to keep the gold country settled permanently. Gold was simply the basis for moving people there initially and getting the settlements started.

More recently the Siberian city of Norilsk has been built 200 miles north of the Arctic Circle. Temperatures there reach 40 below zero and for two months it is dark. Minerals are the only reason for the city which is based on what is probably the richest ore body in the world. It contains an estimated 35 percent of the world's nickel, 10 percent of the copper, 14 percent of the cobalt, 55 percent of the palladium, and 20 percent of the platinum. The mine, even with no additional discoveries, can continue to produce at the present rate for at least 40 years. The city will be home to the mine's 155,000 employees and their families far into the 21st Century.

Smaller gold rushes

To a modest degree, South Dakota, Colorado, and Georgia experienced gold rushes which brought numbers of people into new territories. In South Dakota it was a military expedition which first found gold in the Black Hills. With this discovery the gold miners moved into what was Native American territory, causing numerous bloody conflicts.

In Colorado, an uninhabited broad upland valley in a few short months became Cripple Creek, which grew from a population of 15 people in 1891 to 50,000 by 1900.(20) Similar growth occurred in several other areas of Colorado where gold was discovered such as Central City.

The earliest discovery of gold in the United States appears to have been in North Carolina in 1799 when a boy found a shiny rock in a creek. But it was not recognized as gold until in 1802 when a traveling jeweler saw it. By 1820 people from many parts of the world had come to the area, and at one mine 13 languages were spoken. Most of these people stayed and contributed to the growing population of the state.

In 1829, gold was discovered in what became the town of Dahlonega in northern Georgia and a new gold rush was on. Some of the land involved was Cherokee Indian territory, but with the influx of gold miners the demand for the land grew and ultimately the Cherokees lost out. In 1835 the Cherokees were forced to give up all their lands east of the Mississippi River and were ordered to move westward. However, about 14,000 refused to leave, and in 1838 were forced out militarily during which time some 4,000 died. The cause of this displacement was the discovery of gold, and ultimately resulted in the uprooting of the Cherokee Nation.

Native Americans

Usually, it was the trappers and miners who came first to the more remote areas of the west, rather than the settlers who were largely farmers. It was often the miners who first settled on the land occupied by the Native Americans. The Sioux knew of gold in the Black Hills and had shown specimens of it to Father De Smet before Custer's soldiers found it in French Creek.(23) Ultimately, word got out about the gold in the Black Hills. Although the area had been set aside by the government for the Native Americans, this was ignored and miners flocked in. The initial discovery of gold on French Creek was in Native American land which by the terms of the treaty of 1868 was off limits to white settlement. But miners persisted, and when restrictions were lifted during the year 1875-1876, 11,000 miners entered the Black Hills. This white invasion led the Sioux to resist and resulted in the famous Battle of the Little Bighorn when General Custer and his men were massacred on June 25, 1876. By September of that year, however, the Sioux were forced to sign a treaty giving up the Black Hills. Gold had moved out the Sioux. All across the west, Native Americans came into conflict with the miners and had to give up territory.

The general result of the invasion of gold miners was loss of lands by the native populations, and a great weakening of their economic and political positions.

Alaska and the Yukon

The Yukon and Alaska gold rush of 1897-1898 was the last gold rush of the Nineteenth Century, but it had all the excitement and problems of previous gold rushes, and it too opened up virgin territory. It had its origin when two prospectors, Robert Henderson and George Carmack, were salmon fishing in a tributary of the Yukon River. The tributary was later called the Klondike. These men saw the glint of gold in the stream bed late in the summer of 1896, but news of the discovery did not get out until 1897. Then the rush was on.

Dawson City grew from almost nothing to a population of 25,000 within a single year. By February, 1898, 41 ships made regular runs between San Francisco and Skagway, Alaska, the port nearest the gold fields. From Skagway, the prospectors had to go over Chilkoot Pass or White Pass to the Yukon. During the winter of 1897-1998, 22,000 people were checked through the border between Canada and the United States on these trail routes. The interior of Alaska was largely opened up on the basis of gold.

The town of Valdez at the head of Prince William Sound was a little fishing village until the Alaska gold rush started. Although it was not the shortest route to the goldfields, it was a route which did not cross into Canada and therefore avoided border inspection. Twenty thousand people flooded into Valdez. But in a few years the gold was gone, and by the 1930s the population was down to about five hundred. The population remained small until it was determined that the Alaska pipeline would terminate at Valdez and once again Valdez boomed. Now, with the steady work which the pipeline terminal affords, the population of Valdez has settled to about 4,000. Thus Valdez has seen two major movements of population, one caused by gold and one by oil. And after oil?

Fairbanks has had a history similar to that of Valdez. It got its start from the gold rush, then dwindled in population. But the oil discovery at Prudhoe Bay 390 miles north of Fairbanks brought a second boomtime to Fairbanks. It was the logical place to build Prudhoe Bay support facilities such as warehouses, and it was the halfway point on the Alaska pipeline. Fairbanks' population doubled in five years.(21)

The northernmost road in North America ends at the northernmost and largest oil field in North America, Prudhoe Bay, for which reason the road was built. Although the road now allows only partial access to the region by the general public, that portion of it which is open to the public has caused some minor migration. The oil pipeline maintenance and the beginning of the freight road to Prudhoe Bay, and all that goes with that, keeps Fairbanks busy.

More recently the discovery of microscopic gold in black shales principally in Nevada (the so-called "Carlin-type" gold deposit) has caused a number of sizable communities to be established. So named the "Silver State" from the Comstock Lode and numerous other silver mines, Nevada is now the major gold producer of the United States and the silver mining is largely gone.

Silver

This metal, being much less valuable per ounce than gold, has generally not attracted nearly so much interest as gold. But, discovery of the Comstock Lode(3,16) and the many smaller subsequent silver strikes in western Nevada did bring many people into the area, including quite a few from the Welsh mining areas of Great Britain. They became know as "cousin jacks," and a walk through the cemetery at Virginia City will show many tombstones noting the British origins of those who lie there. The influx of population and the great wealth from the silver also was a major factor in moving Nevada into early Statehood (1864) compared with the adjacent states of Utah (1896) and Arizona (1912), and only 14 years later than California (1850).

Chinese

In many mining areas of the west, significant numbers of Chinese arrived largely to do the menial tasks. They also had the patience to rework mine dumps left by the original miners. The Chinese were not generally accepted as part of the community, and so formed communities of their own, some literally underground. Remains of such communities are still in evidence beneath the streets of Pendleton, Oregon and Idaho City, Idaho. As the mines gave out, some Chinese returned to China. Others stayed and established themselves in farming and various small businesses now scattered across the west.

Modern gold rushes

Twentieth Century gold rushes include the black shale Carlin-type Nevada developments, and a much bigger gold rush in Brazil where tens of thousands of people in recent years have moved into the Amazon River Basin. In remote areas, roads and airfields have been built, and towns established where a short time ago there was only jungle. Kane has described these events: "...in the late seventies, gold prices rose sharply, and gold mining in Brazil took off. Tens of thousands of landless workers left low-paying jobs in the coastal areas to move inland and prospect for the metal. They cleared virgin lands, opened large pits, and often forced out indigenous peoples, some of whom then had no choice but to migrate themselves."(7)

In addition to affecting the local area environmentally and impacting the social structure of the native populations, miners, who were once simply scattered people throughout Brazil, have now been brought together by gold to form a cohesive group, and have become at least a modest force in Brazil's political scene.

The current Brazilian gold rush, however, was preceded several centuries earlier by the movement of people to Brazil caused by the discovery of gold by the Portuguese, who first held colonial control of that country. They left the legacy of Portuguese as the national language. In 1682, gold was discovered in northeastern Brazil in a province which, because of its rich mineral resources, was named Minas Gerias — General Mines. For many years the taxes imposed on Brazilian gold production kept the Portuguese treasury alive. But when the gold supplies declined, so also went the economy of Portugal. Gold briefly made that nation rich, and when the gold was gone, Portugal never again has enjoyed such affluence.

Somewhat smaller gold rushes moving people into new areas have recently occurred on the southern Philippine Island of Mindanao, and also in New Guinea, where perhaps the largest single gold deposit now known in the world is located. In the relatively primitive wilderness of New Guinea this event has had profound effects in the movement of people, with the construction of a large mining camp and supporting transport facilities into previously remote areas.(17)

Canada and its northland

With more than half the people of Canada living within a hundred miles of the U.S. border, there are vast northern areas of this second largest nation in area in the world which are even today very sparsely inhabited. These regions are largely lakes, swamps, flat low-lying tundra plains, and some slightly more upland areas with forests of small trees. Small ranges of hills occur in some parts, and some mountainous areas exist in northeastern and northwestern Canada. The northernmost portion of mainland Canada is chiefly a featureless plain.

Aside from limited hunting, trapping, and fishing, the only major economic resources are minerals and energy minerals. Hunting, fishing, and trapping did not provide an economic base to justify road-building, but minerals did. In Alberta, the northernmost road leads to the great Athabasca oilsands deposits near Fort McMurray, a city now almost entirely sustained by the two large oilsands processing facilities.

In Saskatchewan, the two roads which go farthest north are those which go to mines. The eastern one leads to the Rabbit Lake uranium mine. The western road goes to the Key Lake and other uranium mines in the vicinity. What is now the world's most important uranium mining region is what has opened up northern Saskatchewan, and moved people there.

In Manitoba, the north country was opened up by the great nickel discoveries in the vicinity of the now thriving town of Thompson, population more than twenty thousand. The road going the farthest north from there leads to the Lynn Lake mining district of northwestern Manitoba.

In Ontario the great gold mines at Timmins, which once were the world's largest producers, brought people and large towns to that area, forming a complex of communities today with more than 50,000 people.

In Quebec, the northernmost road makes a big loop, and at the top of the loop are the gold mines of the Chibougama area. Gold was the reason the road was built and is now maintained.

Along the Quebec-Labrador border tremendous iron ore deposits are the only reason for Labrador City, and the road and railroad to that area.

Ghost Towns and Some that Survive

Just as minerals move people into areas, exhaustion of these deposits may cause an outward migration. The presence of many ghost towns in the western United States as well as in other parts of the world are monuments to the fact that minerals are a one-crop resource.

The economic cycle is the discovery, development, and then decline and exhaustion of the one-time crop which is minerals. People accordingly will move into developing mineral resource areas. Then, as the mineral base gradually declines, people begin to move out. There are examples of this now in partially abandoned mining towns, and the decline of once rich oil-producing areas. This can be seen even now in parts of the once oil giant, Texas. However, in some instances, once people have come to an area because of the mineral wealth, they find other ways to survive after the minerals are gone. Some areas become farming or ranching communities. Others become tourist attractions like Virginia City, site of the once fabulously rich silver Comstock Lode, in Nevada. Some have become gambling communities like Cripple Creek and Central City in Colorado.

Oil

The opening up and settlement of new lands by the oil industry is also a cause of the movement of people. In the northern Sechura Desert of Peru, the town of Talara with about 4,000 inhabitants exists solely because of the oilfield there. Rains come only once every several years or less. Without the oilfield, there would be little basis for a settlement.The now thriving city of Maracaibo, Venezuela located near the southeast end of the semi-desert Guajira Peninsula is there only because of the rich oil deposits in the Maracaibo Basin (a similar geologic oil-forming setting as the Persian Gulf, but not so large).

The oil discovery in the United States in 1859 by “Col.” Drake (assumed title, he was actually a railroad conductor) near Titusville, Pennsylvania, brought a “Fifty-niner” rush. Some 30,000 people almost immediately moved into the area, and then spread out to other areas to search for more oil.

Oil in the Persian Gulf, opened work opportunities that brought tens of thousands of people to the region. Indians, Pakistanis, Filipinos, Palestinians, and Egyptians were among the major groups represented. When the 1991 Gulf War broke out, more than half the people living in Kuwait were foreigners. Today, in Dubai, of the United Arab Emirates, the population is only 15 percent native. Neighboring Qatar sits atop the world’s largest single natural gas deposit (12 percent of world proved reserves) and has oil exports which give it a per capita income of $13,600 a year. This robust economy has attracted foreign workers to the extent that they make up more than two-thirds of Qatar’s current population of about 520,000. Qatar is a desert peninsula with an annual rainfall of only about 2½ inches. For centuries fishing was a principal occupation of the relatively few inhabitants of the region who led a subsistence existence. Petroleum made profound changes, and caused a large migration of people to the country.

Fertile Topsoil

Gold rushes are a strikingly visible demonstration of how minerals move people and make for romantic history, but far more people have moved because of the availability of new lands with fertile topsoil to cultivate. In North America, Canada and the United States were widely settled first by farmers, after which came migration to the cities and the large development of cities today. The rich lands of South Africa, the pampas of Argentina, and the great wheat growing plains of Australia also encouraged mass migrations of people.

In the case of the United States, the Homestead Act, which granted tracts of public land to settlers who would first claim and then farm the land moved people westward. On at least one occasion this was a rather dramatic movement, when lands in Oklahoma were about to be released for settlement. There was a fixed opening date and a firm boundary behind which people were to assemble, and at a signal go in and claim land parcels. But some people did not wait and sneaked into the territories to be offered before the opening date. They arrived sooner and thus the nickname for Oklahoma as the "Sooner" State.

Railroads were granted lands, usually alternate sections (a square mile), in a strip ten miles wide along each side of the tracks to encourage them to build the railroad. Once built, the railroads advertized free lands and were eager to move people to them, so the railroads in turn could generate business by hauling settlers and their goods west, and then haul agricultural products east. Availability of rich, fertile land was the prime mover.

A somewhat similar situation existed in Canada and resulted in the settlement of the last great prairie agricultural area of the world, the southern portions of Alberta, Saskatchewan, and Manitoba.

Move to the Cities

Even as the rich, fertile land resource initially brought immigrants to North America (and also to other places such as the pampas of Argentina), energy resources have since moved much of the population off the farms and into the cities. Today, only about two percent of America's population farm the land, where a hundred years ago the majority of people did. But energy slaves (chiefly oil) now do the farm work, causing another great migration of people to the industrial areas, which, in turn, exist because of abundant energy to run the factories. This has also been true in Europe, and is now taking place in China and southeast Asia. Mineral and energy resources have caused and continue to cause great human migration, both within countries and across international boundaries.

The Continued Invasion of North America

In a broad sense, the migration to North America was caused by the availability of fertile lands and other mineral resources. As they were discovered and developed, they produced an ever-higher standard of living, luring more people to come. This attraction continues. Currently, the United States annually has more legal and illegal immigrants than all the rest of the world combined. The natural resources which are still available in the United States together with those which can now be imported provide the basis for the standard of living envied and sought by much of the rest of the world.

The persistent attempts by Haitians to enter the United States is resource based. Haiti, which occupies the western third of the Island of Hispaniola, has an area of about 10,500 square miles. It has no oil, gas, coal, water power, nor any other appreciable mineral resources. When one flies into the capital, Port-au-Prince, there is a noticeable brown ring in the ocean shore area where rivers are depositing silt and clay from the already highly eroded and continually eroding hilly and mountainous regions which make up most of Haiti. What fertile soil remains continues to be lost at a disastrous rate.

Nearly seven million people crowd Haiti's limited area. Once nearly entirely wooded, Haiti is now almost treeless, and people are digging up roots for fuel. The contrast of this situation to the abundance of resources available to citizens of the United States is enormous. Energy supplies and mineral resources available for an educated technologically developed society to work with make the difference. Haiti has virtually no mineral or energy resources, and does not now have the capability of importing them to sustain any sort of advanced manufacturing economy.

As a result, there is an overwhelming desire to migrate from a land with virtually no resources, to a land which has resource wealth to provide even its lowest economic segment of society a better standard of living than the average Haitian has. With Haiti's population growth rate of about 2.8% annually — one of the highest in the western hemisphere — the problem will only increase. At that rate, the population will double in about 25 years which can become an absolute disaster. Supplying more and more "outside" food to such a situation with no heed to population control, simply treats the symptoms and not the cause, ensuring even greater problems in the future. Some reasonable relationship between population and the resource base a country has or can import must be established. Otherwise, the people there will either starve or be on permanent international welfare. To continue to export population problems cannot be the ultimate solution either, as fewer and fewer countries are now willing or are ultimately able to continue to be the safety valve for migrating population pressures. France has completely closed its borders to all immigration. Japan accepts virtually no immigrants, and Sweden for the first time has been turning them away. Germany has been expelling foreign nationals. Resources to house and feed the immigrant flood is the critical factor.

Increasing urge to migrate

With rapidly growing populations, and mineral resources becoming more costly as the higher quality and easily accessible sources are exhausted, problems of resource distribution will grow. The pressure for populations to migrate to areas which still have resources available will increase. By both legal and illegal means, mass migrations of people continue, and is already a problem which has the potential of great social unrest. Such is already happening. In 1995, the decision by President Clinton to return all Cuban refugees to Cuba, led to riots in Miami.

The refugee problem is not decreasing. What is decreasing is the willingness of many countries to accept more refugees. In contrast, the United States now has a very liberal immigration policy, allowing two million newcomers in each year. It also has a relatively porous border which lets in another estimated half million or more illegal immigrants. The result is that some states are beginning to resist this burden. Because of the impact of illegal immigrants upon their resources, the states of California, Texas, and Florida in 1994 filed lawsuits against the U.S. Government. The suits assert that lack of enforcement of federal immigration laws had resulted in an intolerable drain of resources from the states. In California, the recent growth of that state has been largely due to foreign immigration. As a result, to accommodate the increase just in children, in 1995 one new schoolroom had to be built each hour, and one new school each day.(2) In 1994, California passed Proposition 187 providing that illegal immigrants would be denied a variety of services, including schooling. This has caused numerous protests and demonstrations.

It is important to remember that the physical standard of living which is high in the United States and which attracts immigrants, legal and illegal, is based on availability of mineral resources. To maintain that standard of living, each year each person in the United States must be provided with some 20 tons of mineral resources. As the United States becomes more and more dependent on importing these resources, the balance of international payments problem grows. Being able to supply resources to immigrants is what draws them, but at some point this in turn will adversely affect the future of the United States if the balance of payments problem will no longer allow increased resources to be imported. The much valued U.S. standard of living will fall, and its destiny will be altered. A nation which does not control its borders loses its sovereignty, and control of its economic future.

In a study of why people migrate, Kane has observed, "Water tables will continue to be drawn down far faster than they can replenish themselves in many countries; soils continue to erode, and new people will react to these pressures in the future by leaving their homes." Describing Ethiopia, with a present population of 57 million, Kane states:

> "...the nation faces a colossal increase of 106 million during the next forty years, based on current growth rates. It is almost impossible to imagine how Ethiopia could possibly feed so many more people. It has some of the world's most severely eroded soils, much of its cropland is on steep slopes, and its tree cover stands at a mere 3 percent. Many in Ethiopia's next generation will probably have to choose between emigration and starvation."(7)

Kane calls the tendency of people to migrate "The push of poverty, the pull of wealth." He concludes that a major and increasing cause of human migration is the exhaustion of natural resources:

> "Many countries, particularly in Africa, but increasingly elsewhere as well have been living off their capital — consuming their foreign reserves, their forests, soils, and freshwater aquifers, and the patience of their citizens —in order to survive. As these reserves are diminished, pressures and conflicts mount and more and more people are forced to flee. The number of people on the move today has reached its highest point in history. But if nations do not shift spending priorities from military security to investments in the long-term environmental and social health of their citizens, these numbers may be dwarfed by the tide yet to come."(9)

Kane estimates that if present trends continue the world's refugee population will have doubled to 46 million within five years. Brown has drawn the same conclusions as Kane.(1)

The End of the Physical Frontier

Today there are no large unoccupied rich Earth resource areas to absorb migration. During the last 500 years the major waves of human population have generally migrated westward. When the final big wave hit the Pacific shores of North and South America, beyond lay Asia and India with their teeming billions. The circle was complete. The globe was filling up. New lands with untouched resources were no more.

With no new geographic frontiers in which to expand, today's nations jostle for position within the well populated and fully explored world. The jostling through migration and perhaps military conflicts will increasingly be over access to Earth's remaining resources, of energy, water, fertile soil, and other minerals, for these are the bases for simply surviving. Making rational and successful adjustments between population and resources will determine the destiny of the human race, and populations must recognize that destiny imposed upon them by geology.

Globalization of Carrying Capacity Problems

Because resources and population are unevenly distributed, the current trend is for people to move from distressed areas to areas which have more resources, or for wealthier nations to send basic resources to the impoverished regions.

If emigration from distressed areas to areas which still have resources continues, without eventually stabilizing population, will there ultimately be world-wide prosperity or a world-wide slum? Hardin considers this matter: "...the production of human beings is the result of very localized human actions; corrective action must also be local. Globalizing the 'population problem' would only insure that it would never be solved." Hardin adds:

> "Some social experiments have had very bad outcomes indeed. For this reason Kenneth Boulding wisely said: 'There are catastrophes from which there is recovery, especially small catastrophes. What worries me is the irrevocable catastrophe. That is why I am worried about the globalization of the world. If you have only one system, then if anything goes wrong, everything goes wrong.' The wisdom is very old: Don't put all your eggs in one basket. Given many sovereign nations it is possible for humanity to carry out many experiments in population control. Each nation can observe the successes and failures of the others. Experiments that have a good outcome can be copied and perhaps improved upon; unsuccessful experiments can be noted and not repeated. Such learning by trial and error is perilous if the borderless world created by unrestricted migration converts the entire globe into a single huge experiment."(6)

The "lifeboat ethics" concept

Hardin's observations are a facet of his "lifeboat ethics."(5) A ship is sinking, and there is one lifeboat. It is launched and filled to its stated capacity of 50 people, but there are still 100 people in the water. Do you take on the additional 100 from the water and have everyone drown, or do you preserve the one lifeboat and its passengers so they can get to the far shore and survive? Do you convert the entire world to a giant slum by unrestricted immigration and no population control? Or do you restrict immigration and insist that individual nations do something about population, so that at least some of them who are successful survive? At present, a number of nations are trying to export their population problems which ultimately will, if not checked, become a global disaster. However, it will have the merit of equality — poverty will be universal.

Continued population migration will surely make this concept of "lifeboat ethics" a serious consideration. Responsible and firm action may be required to prevent "lifeboat nations" from being swamped and sunk. Lucas and Ogletree relate this to world hunger.(10) Pimentel and Giampietro have an implied "lifeboat" role for the United States in their statement "Self-sufficiency in food production and other basic resources should be viewed as a strategy to guarantee a continued high standard of living and national security to U.S. citizens in the face of turbulence that can be expected around the world in the next decades. There is no time for delay, choosing not to change the current pattern of high immigration and rapid population growth means moving into the Malthusian trap in the United States."(15)

BIBLIOGRAPHY

1 BROWN, L. R., 1995, Vital Signs 1995. The Trends That Are Shaping Our Future: Worldwatch Institute, Washington, D.C., 176 p.

2 CARRYING CAPACITY NETWORK, 1995, Clearinghouse Bulletin, Washington, D.C., v.5, n. 2, p. 1.

3 De QUILLE, DAN, 1947, The Big Bonanza: Alfred A. Knopf, New York, 436 p.

4 GRONER, ALEX, (ed.), 1972, The American Heritage History of American Business & Industry: American Heritage Publishing Company, New York, 384 p.

5 HARDIN, GARRETT, 1974, Living on a Lifeboat: BioScience, v. 24, p. 561-568.

6 HARDIN, GARRETT, 1993, Living Within Limits. Ecology, Economics, and Population Taboos: Oxford Univ. Press, New York, 339 p.

7 KANE, HAL, 1995, Leaving home: Society, May/June, p. 16-25.

8 KANE, HAL, 1995, The Hour of Departure: Forces That Create Refugees and Migrants: Worldwatch Paper 125, Worldwatch Institute, Washington, D. D., 56 p.

9 KANE, HAL, 1995, Refugees: World Watch, November/December, p. 35.

10 LUCAS, G. R., Jr., and OGLETREE, T. W., 1976, Lifeboat Ethics. The Moral Dilemmas of World Hunger: Harper & Row, Publishers, New York 159 p.

11 MARTINEZ, LIONEL, 1990, Gold Rushes of North America: The Wellfleet Press, Secaucus, New Jersey, 192 p.

12 MOWAT, FARLEY, 1973, Sibir. My discovery of Siberia: McClelland and Stewart Limited, Toronto, 313 p.

13 MULTHAUF, R. P., 1978, Neptune's Gift. A History of Common Salt: The Johns Hopkins Univ. Press, Baltimore, 325 p.

14 PAUL, R. W., 1963, Mining Frontiers of the Far West: Holt, Reinhard and Winston, New York, 236 p.

15 PIMENTEL, DAVID and GIAMPIETRO, MARIO, 1994, Food, Land, Population and the U.S. Economy: Carrying Capacity Network, Washington, D.C., 81 p.

16 RAYMOND, ROBERT, 1986, Out of the Fiery Furnace. The Impact of Metals on the History of Mankind: The Pennsylvania State Univ. Press, University Park, Pennsylvania, 274 p.

17 SHARI, MICHAEL, et al., 1995, Gold Rush in New Guinea: Business Week, November 20, p. 66, 68.

18 SHINN, C. H., 1992, The Story of the Mine: As Illustrated By the Great Comstock Lode of Nevada: Univ. Nevada Press, Reno, 277 p.

19 SMITH, D. A., 1967, Rocky Mountain Mining Camps: Univ. Nebraska Press, Lincoln, 304 p.

20 SPRAGUE, MARSHALL, 1953, Money Mountain. The Story of Cripple Creek Gold: Ballentine Books, New York, 241 p.

21 STROHMEYER, JOHN, 1993, Extreme Conditions. Big Oil and the Transformation of Alaska: Simon & Schuster, New York, 287 p.

22 WATKINS, T. H., 1971, Gold and Silver in the West: American West Publishing Company, Palo Alto, California, 287 p.

23 WOLLE, M. S., 1953, The Bonanza Trail: The Swallow Press, Inc., Chicago, 510 p.

CHAPTER 3

Minerals and War, and Economic and Political Warfare

Minerals have been involved both directly and indirectly in warfare. More importantly, they often have been the *cause* of war, as nations that wanted them took them or tried to take them by conquest. Minerals are used by nations in economic warfare, by withholding access to minerals, or by cutting off the supplies of minerals by an embargo to accomplish some political end.(3) Economic warfare can be waged by forming a cartel and fixing the price of oil, as the Organization of Petroleum Exporting Countries (OPEC) has successfully done on occasion and may do with probable greater success in the future.

Minerals Used in Warfare

For early humans, hand to hand combat with perhaps a few minerals in the form of rocks were the weapons of warfare. Then the stone axe and knife were invented, followed by the longer knives called swords and spears.

The first knife was stone, but when native copper was discovered and hammered into knives as was done in several places in the world, the metal age of warfare began, and metals in war have been important ever since. Arrows had a similar history. Arrows were first tipped with stone materials such as flint and obsidian, and later they were metal tipped.

Early use of oil in warfare

Oil in its various forms has long played a key role in warfare. In modern times its role has been to fuel warplanes, ships, tanks, and the great variety of motorized transport vehicles of warfare. But oil had a very early history in war. About the year 670, the Byzantine navy was besieged by the Muslims, but a native of Syria named Kallinikos defected to the Byzantines, and brought with him the formula for making a petroleum mixture which would burn even on water.(3) Armed with this information the Emperor Constantine IV built a siphon-like structure on the bow of some of his ships so that the flaming mixture could be squirted onto enemy vessels and on the water near them. During the seventh year of the siege this weapon was used with remarkable success against the Muslim navy, in what was called the Battle of Kyzikos. The entire fleet, manned chiefly by Syrians and Egyptians, was destroyed. One historian estimated the losses at 30,000 men. A modern modification of this use of oil in warfare is in the form of a flame thrower extensively used in World War II, particularly on some of the rocky and cavernous islands of the South Pacific.

A somewhat similar use of oil in a form slightly refined approximately to the stage of naphtha, was employed by the Muslims in an instrument called a mangonel. This was a heavy-duty catapult designed to bombard enemy fortifications with projectiles of flaming naphtha. The Muslims use this weapon in a number of military campaigns. One of the more famous was in the siege of Makkah (Mecca) in the year 683, by one Muslim faction against another. Naphtha fires raged around and in the city for more than half a year until the besieging forces finally prevailed. From various accounts of those times it is clear that Syrian Muslim forces had oil sources, and that they were able to transport it anywhere they wished for military purposes.

Between 809 and 813, two brothers fought against one another in a war between Syria and Persia. Ultimately one of the brothers trapped the other in Baghdad. With hundreds of catapults (mangonels) he bombarded the city with barrels of burning naphtha, which eventually destroyed the city so completely that it was not until six years later that its reconstruction was begun.

Oil, in the form of naphtha, continued to be important in Middle East conflicts. In 1167, the Christian crusader king in Jerusalem, Amalric I, decided to annex Egypt. He moved against Cairo with his army. Unable to successfully resist, the city commander resolved to reduce Cairo to rubble. This he did with the aid of thousands of naphtha pots, some of which were apparently modified to become primitive explosive grenades containing a kerosene jelly, nitrates, and sulfur. An Egyptian, historian, al-Maqrizi, reported that the city commander, "Shawar sent 20,000 naphtha pots and 10,000 lighting bombs and distributed them throughout the city. Flames and smoke engulfed the city and rose to the sky in a terrifying scene. The blaze raged for 54 days..."(3)

Alexander the Great, in his military ventures in the Middle East, was given a warmer reception than he anticipated when the natives of one area poured naphtha on a downslope toward Alexander's tent and then set it on fire.

Silver and war

Silver was discovered in many areas of the ancient world, but in one particular area, it played an important role affecting the course of western civilization. In the limestone hills near the town of Laurium, and also near the village of Plaka, about 30 miles northeast of Athens, large deposits of silver were discovered. For a thousand years, Athens and Greek culture in general flourished in part because of the wealth taken from these mines. Each citizen of Athens was given an annual share of this treasure, recovered at great effort and loss of life, by thousands of slaves working in the mines. But threatening the wealth and culture enjoyed by the Greeks, were the Persians. They first moved against Greece in the year 492 B.C., under Mardonius, who had a substantial fleet and many men. This Persian fleet was largely wrecked by a storm as it approached southern Greece, and the survivors returned home. But the threat did not disappear. It only grew larger as the Persians reassembled and attacked Attica in 490 B.C. with a fleet of 600 ships.

After this invasion of southern Greece, the Persians landed on the beach at the plain of Marathon, only a few miles from Athens. Here, 20,000 Persians were met by only 9,000 Athenians and 1,000 Plateans. But the Greeks, with great courage, staged a mass running attack against the Persians which demoralized them and they fled in confusion. Reportedly, some 6,400 Persians were killed, whereas the Greeks lost only 192 soldiers.

But even with this defeat, the Persian menace did not vanish. The Persians planned yet another invasion and this time silver played a crucial role. Themistocles, a perceptive and foresighted Greek, had persistently argued for a substantial navy. To accomplish this he suggested that the Athenians forego their annual dividend from the great silver mines near Athens, and that the money be used to build ships. At the time, the Greeks had only about 70 ships, but the Athenians heeded Themistocles. Ultimately 130 more ships were built, paid for by the silver.(15)

The Persians did come again with a great fleet and several hundred thousand men. After a series of skirmishes, the Persians overran and sacked Athens. It now remained for the Persian fleet to destroy the Greek Navy, which consisted of about 300 ships, 200 of them Athenian, 130 of which were built with the silver revenues from the Laurium/Plaka area. These were the newest and best ships, designed for speed and maneuverability.

In late September of 480 B.C., the Persian commander, Xerxes, took a seat on a prominence overlooking the Bay of Salamis, just west of Piraeus, the port to Athens. Here he confidently prepared to witness the destruction of the Greek navy.

The ships of the Greeks were smaller and outnumbered about three to one by the Persian vessels. But the Greek ships were fast and easy to maneuver, and they were fitted with battering rams. They quickly moved through the Persian fleet, shearing off their oars, ramming them, and leaving them dead in the water, waiting to be further rammed and sunk. The Persian fleet was destroyed.

Figuratively and literally the Battle of Salamis marked the high water mark for the Persians in Greece, and eventually the Greeks freed themselves from the Persians. Greece continued to flourish and provided the foundation of western civilization in the form of its democratic ideals, and the corresponding liberties in individual freedom and in economics. Without the silver mines to provide the ships with which the Greeks destroyed the Persian fleet, the course of western civilization would have been markedly different.

About 150 years later, Philip of Macedonia (a province in northern Greece) captured the Athenian town of Amhibolis, the key to an area rich in gold and silver. Great quantities of silver and gold were taken from this region by Philip. With these precious metals he hired and trained a large professional army. He used this army to move south and ultimately united much of Greece, and then began his plan to carry the fight against the Persians into Persia. In 336 B.C., however, Philip was murdered, and his son, later to be known as Alexander the Great, took over the leadership. Alexander continued to draw on the mines in Macedonia and Thrace, and with silver and gold from these sources paying his army, he moved eastward against the Persians.

The Persians, in their conquests, had accumulated immense wealth, much of it in precious metals. Alexander wanted this prize, and ultimately he did capture the rich Persian treasury. Plutarch reported that it took 20,000 mules and 5,000 camels to transport it. These precious metals were carried from the eastern world to the western Greek world and made into coins which financed a great expansion of trade and commerce, and, in turn, the arts.

Immense quarries were opened in the marble hills of Greece. The resulting —and now world famous — sculptures made from these rocks record the Golden Age of Greece, made possible, first, by the silver from Laurium and vicinity which enabled the Greeks to defeat the Persian invasion.

Salt, Rome, and Carthage

Salt has been used as a final act of warfare. After the long series of the Punic wars with Carthage from 264 B. C. to 146 B. C., Rome finally prevailed. It utterly destroyed Carthage, plowed the site of the city and its fields, and sowed salt on the fields to destroy their fertility.

Minerals and Sweden's military ventures

The exploitation of the silver, copper, and iron deposits in Sweden about 1600 coincided with an increase in the wealth and energy of that country. Money derived from the Swedish minerals helped finance the Thirty Years' War, and the invasion of Germany in 1630 by the Swedish King Gustavus Adolphus. Ultimately, in Alliance with Cardinal Richelieu of France, the Hapsburg empire was destroyed and Germany was cut into more than three hundred independent states, principalities, and free towns.(15)

Equipment of war

From ancient times to the present, various metals, but especially iron, copper, tin (to make bronze), and aluminum, have been used to make the equipment of war — chariots, swords, spears, guns, bullets and shells, trucks, tanks, ships, planes, bombs, and missiles. The ability of the Romans to ultimately develop an iron-based weaponry (after their bronze weapons), finally led to the defeat of Greece, and the transfer of what was left of the Greek treasury to Rome.

The first metal used to make both weapons and utensils was copper as it was found in several places in native (that is, pure) form, and it ores (mineral compounds) are for the most part easily smelted. Very early it was discovered that copper could be alloyed with tin to make a much harder substance — bronze. The helmet of the Greek Warrior, Achilles, was made of bronze.

Aluminum in modern times (it was not isolated as a metal until 1827) has become exceedingly important as the critical metal for aircraft. Flying a plane made of iron would be a bit difficult. But aluminum made aerial warfare and long range bombing possible. An aluminum plane delivered the atomic bomb.

Many specialty metals are very important in war. For example, magnesium is used in flares to illuminate enemy positions. Without cobalt and vanadium, the jet engine would be impossible.

A particularly useful metal employed in equipment of war as well as in civilian uses such as in automobile sheet steel, is molybdenum. It makes steel tough, rather than brittle. Prior to World War I, the Germans, who were good metallurgists, had obtained and developed the world's largest molybdenum deposit at Climax, Colorado. "Mount Moly," it has been called. These properties were, of course, confiscated during World War I. But the Germans had already taken quantities of molybdenum from "Mount Moly" before the outbreak of hostilities, for they knew that molybdenum made especially tough steel. They used this property of the metal in building the longest range cannon ever constructed. It was nick-named "Big Bertha" by the Allies and was used by the Germans to hurl shells into Paris from a distance of 75 miles.

The Germans knew that the physical damage from this sort of bombardment from just one cannon would be minor, but they hoped that the shelling would have a demoralizing effect on the French. It did not, and was largely regarded as just a nuisance. Molybdenum remains a critical metal in many steel products, civilian and military. Without it neither the ships and guns of the Navy, nor the tanks and guns of the Army could be built.

Uranium — the ultimate weapon

World War II in the Pacific was triggered over oil supplies. World War II in Europe was over land and the resources the land contained. Hitler especially wanted the huge metal and petroleum resources of the Soviet Union. But it was a metal which at the beginning of World War II was not much more than a laboratory curiosity that ultimately ended the war, and forever changed the world. This mètal is uranium which before World War II did not have much of a market, and, in fact, was regarded as an undesirable impurity in some ores. Silver ores in the Coeur d'Alene mining district of northern Idaho, especially those from the Sunshine Mine, contained some uranium and this ore was penalized at the smelter. Some of the rock from the mine which contained uranium was simply used to make and maintain part of the road to the mine.

But the possession of uranium by the United States allowed the U.S. to dramatically end World War II by means of two atomic bombs dropped on Japan. Japan's demand for oil started the Pacific war and uranium ended it. This single mineral moved warfare a quantum leap forward in just six years. The war started with the use of horse drawn cannon and mounted cavalry (in Poland), and ended in a nuclear flash over Japan.

Oil and the European war

Oil may have been the decisive factor, in ending World War II in Europe. Germany, with virtually no oil of its own, had stockpiled some before the outbreak of hostilities. Subsequently small amounts came from the conquest of Poland, Austria, and Hungary. Romania was taken in 1940 which became Germany's major oil source, but the German main drive was to the huge Baku oil fields in southern Russia.(26) When this was thwarted, and the Romanian oil fields came under intense bombing from the Allies, Hitler's oil supplies began to fail.

German motorized divisions toward the end of the war suffered markedly from lack of oil. When General George Patton was finally on the move across France with the Germans in full retreat, pipeline specialists from Texas (where else!) followed Patton's tanks and laid pipeline at the rate of up to 50 miles a day. And there was oil to fill those pipelines because during World War II the Allies controlled 86 percent of the world's oil supply.

Oil and the Iraq/Iran war

Oil was a major factor in the Persian Gulf hostilities in the 1980's between Iran and Iraq. In this instance, it was the ability to market the oil and thus obtain foreign exchange by means of which to buy the weapons of war which made oil so important. At first, Iraq, greatly outnumbered in terms of fighting men and equipment, had a difficult time against larger and more powerful Iran. The materials of war had to be purchased by both sides with their oil resources. Iran had the Kharg oil terminal on the Persian Gulf, and largely controlled access to the Gulf through which almost all of Iraq's export oil had traveled. This put Iraq at a great initial disadvantage in selling oil abroad to pay for the war. Iraq survived this early crisis because of massive aid from friendly Arab nations who used their own oil to buy war materials for Iraq. (Iraq is both Muslim and Arab; Iran is Muslim but not Arab).

With help chiefly from Saudi Arabia, Iraq eventually built a pipeline to the Mediterranean followed shortly by a second line. Then a pipeline was built from Iraq to the Saudi Arabian port of Yanbu on the Red Sea which Iran could not control. By these means Iraq was able to increase oil exports from virtually nothing to more than a million and a half barrels a day to buy war materials. This turned the tide in the war. Mineral resources often affect the destinies of nations through the procurement of much needed foreign exchange, and the materials it buys. In the case of Iraq, it literally meant its survival.

Later, when Iraq became stronger, it attacked Iran's Kharg oil terminal rendering it at least partly useless as an oil outlet. Iran had no pipeline alternative to markets as did Iraq now. As a result, Iran's ability to sell oil abroad was substantially reduced. The initial advantage which Iran had over Iraq was lost and the war became an eventual stalemate. Furthermore, some popular support for the war was lost within Iran because without substantial oil revenues Iran could no longer subsidize its domestic economy as much as it had in the past. Large imports of foodstuffs (chiefly wheat and rice) had earlier been distributed at reduced prices to the general population. The lack of Iran's ability to market oil was the swing factor in this conflict.

Because Kuwait and Saudi Arabia both supported Iraq, Iran attacked oil shipping from those countries. The importance of Persian Gulf oil to the industrialized world was emphasized by the ultimate presence of naval forces of the United States, Great Britain, France, Italy, and the Soviet Union in the Gulf. A number of Kuwait tankers were even re-flagged with the U.S. flag, and given U.S. naval escort through the Gulf. It was variously estimated that, with the cost of maintaining the naval presence in the Gulf, the real cost of each barrel for the United States was about $135. In fact, however, most of this expensive oil went to western Europe and Japan who were more dependent at that time on Arab oil than was the U.S.

Minor military actions and oil

In some less developed countries, where oil production is an important part of the economy, military factions will frequently try to take control of the oil facilities. In 1990, in the west African nation of Gabon, anti-government forces battled to gain control of the oil-producing center of Port Gentil. Gabon earns 80 percent of its income from oil, and who controls the oil production controls Gabon. The government forces kept control in this skirmish.

Water and war

The resource which all sides in warfare must have is water, and control of water supplies has been used in warfare in various ways. In the 6th Century B.C., the King of Syria seized water wells as part of his campaign against Arabia. The Inca conquered the desert coastal cities of Peru such as Chan Chan by cutting off their water supply. In feudal times in Europe, castles or other fortresses under siege were vulnerable if they did not have water supplies within their walls. The enemy quickly determined if such was the case. Moats filled with water were part of the standard military protection of the day.

During World War II and in the Korean War, hydroelectric facilities were bombed, both to destroy electric power generating capability, and to destroy the water reservoirs behind the dams. During the Kuwait war, the water supplies of Iraq were repeatedly attacked. In 1986 North Korea announced their intent to build the Kumgansan hydroelectric dam on a tributary of the Han River upstream from Seoul, South Korea. South Korea estimated that if the North Koreans opened the flood gates or destroyed the dam, the level of the Han River through Seoul would rise by more than 50 meters and would destroy most of the city. South Korea built a series of levees to try to prevent this possibility.(8)

During the recent conflict with Iraq over Kuwait, there were discussions among the Allied Gulf Forces concerning the possibility of having Turkey use its dams on the Euphrates River to cut off water to Iraq. This action was not implemented, but water was clearly considered as a weapon, and in time of war, a military objective.(8) In peacetime, water can be used as a form of economic warfare. More that 30 nations receive one-third or more of their water from outside their borders. In the case of Egypt it is 97 percent, Hungary 95 percent, Syria 79 percent, and Iraq 66 percent. With their surface waters chiefly in the control of other countries, these and 16 other nations which have more than 50 percent of their surface waters coming from outside their borders, are vulnerable to economic water warfare.

A British observer, Sir Crispin Tickell, in considering the future stated:

> "Fresh water is a particular problem. The global demand for water doubled between 1940 and 1980, and is expected to double again by the year 2000. More people need more water. Competition for water was a prime source of conflict in the past, and will be in the future: for example over the Nile, which flows through nine states, each with its own interests and demands; over the Euphrates and the Jordan which nourish Turkey, Syria, Iraq, Jordan and Israel, and over the Colorado, now no more than a sickly salty stream when it finally reaches the sea."(34)

Oil was recently a source of conflict in the Middle East, and may be again, but in the longer term it is likely that the last battle will be fought over water.

Wars to Obtain Materials for War

Native Americans fought for possession of Obsidian Cliff in what is now Yellowstone National Park — probably one of the earliest battles fought for the possession of mineral resources in the Western Hemisphere. Obsidian, a volcanic glass, makes razor-sharp arrowheads and is easy to shape. Those who had obsidian had a great advantage in both hunting and in warfare.

Phoenicians and later the Romans fought for the possession of the tin mines at Cornwall in England after it was discovered that tin added to copper made a very hard alloy for weapons. Earlier, numerous bloody conflicts had been fought over possession of the large copper deposits on the Island of Cyprus in the eastern Mediterranean.

German expansion

In the middle of the Nineteenth Century, Germany, once a loose organization of independent principalities, became one nation. It early became clear to the German military that their operations were definitely dependent on having raw materials for the conduct of warfare, and Germany was not self-sufficient in this regard. To obtain markets for finished industrial goods, and to obtain raw materials, chiefly minerals, for building armaments, and supply the industrial complex, Germany began to look hungrily at adjacent territories.

After the Franco-Prussian War of 1871, Germany annexed the iron deposits of Alsace-Lorraine. Unfortunately they later found that the boundaries they set did not include the bulk of the iron deposits because of the geologic structure there. They apparently had a good military department but a poor geology department. Germany fought the First World War in part "to correct the error of 1871." Germany, however, in losing World War I, had to give back this territory.

The experiences of World War I, made both the Allies and the Central Powers (Germany and its allies) keenly aware of the need for minerals with which to conduct military operations. The slogan for the Allies became "never again," referring to the mineral supply problems they had experienced during the conflict. Immediate and extensive post-war mineral exploration and development programs were begun by Britain and France, for both foreign and domestic resources. Because these countries, especially Great Britain, still had extensive colonial holdings there was a lot of territory to explore. Germany, in contrast, had lost all her foreign lands, which was one of the circumstances that precipitated World War II. As Germany prepared for World War II in the 1930s under Hitler, it had been unable to gain back its colonies so it began to annex its immediate neighbors who possessed some useful resources. Austria was next door and had some small iron deposits. It was the first to be taken. Czechoslovakia contained some rather good and famous mining districts with fairly large iron deposits, and a wide variety of other metals. Czechoslovakia was taken. With the early defeat of France in World War II — the Maginot Line proved very leaky — Germany again gained control of the iron of Alsace-Lorraine.

Self-sufficiency, or as close to it as possible in mineral and energy supplies, became one of the cornerstones of both British and French military and foreign policy. By the late 1930s, Germany and Italy were demanding return of their colonies, and some military actions had begun. Italy had invaded Ethiopia (then called Abyssinia).

Japan

This emerging Pacific power was increasingly in need of raw materials, particularly metals and coal of which Japan had virtually none, with which to equip the military. As the military gradually obtained resources by which to build its strength, in turn, it seized raw materials for Japan's expanding civilian economy. Japan invaded parts of the Asiatic mainland, notably Manchuria, which they renamed Manchuko. Here, deposits of two vital industrial minerals, iron and coal, were present in substantial quantities, and Japan used them to build more armaments.

Global reach for minerals for war and peace

In the 1930s, the world was seeing for the first time a global series of political and military moves to obtain raw materials, chiefly minerals and energy minerals. Growing industrialization, with its huge demand for raw materials, growing populations, and a desire for a higher standard of living were the immediate causes for the intensified search for resources. And in the background was the thought that should another war come, the materials must be available to survive and win it.

Hitler covets Russia's resources.

Once Germany had conquered France and driven the British back across the English Channel, Hitler turned east. Hitler had long held the view that only the Soviet Union had adequate land and minerals to take care of Germany's needs.(26) He said about taking this region that "We shall become the most self-supporting state in every respect in the world. Timber we will have in abundance, iron in unlimited quantities, and the greatest manganese-ore mines in the world, and oil — we shall swim in it." Hitler turned his armies toward the Urals and the Ukraine. This region, along with the Donets Basin, collectively contain extensive deposits of hematite (high grade iron ore), excellent coking coal, and limestone — the fundamental three ingredients for steel-making.

Germany, Hitler thought, was now on its way to becoming a permanent world power, and his troops headed into the Soviet Union. But they were stopped at Stalingrad, and also at Leningrad (now, once again, called St. Petersburg) by heroic Russian defenders. The Russian winter also did its part.

Oil

It was not only metals which became so clearly evident as important tools of warfare in World War I. Energy, chiefly in the form of that relative newcomer, oil, was obviously going to be very significant. Some of the Allied ships bringing troops and supplies across the Atlantic from the United States to Europe were coal-fired and some were oil-fired. The oil-powered ships were the better and faster vessels, greatly speeding delivery of both troops and equipment. Oil grew more important in warfare. Gasoline-powered tanks made an appearance on the military scene in World War I. Airplanes, primitive as they were, also came into the war in a limited fashion fueled by gasoline. For the first time in a major conflict, trucks replaced horse-drawn vehicles on a large scale. Recognizing this, world military establishments in the time between World War I and World War II began to give serious consideration to oil supplies. This is why Hitler wanted so much to move into the oil fields of the southern USSR.

During World War I, the British Empire was still intact, and because the Allies controlled the sea lanes, raw materials for warfare were funneled in from all parts of the globe. As World War II began to loom as a possibility, Britain began strategic planning to insure the proper flow of war materials. The United States, although at a rather late date, made a similar move in 1939 when the Congress authorized funds to start an emergency stockpile of certain critical materials. Fortunately, Britain was still in greater or lesser contact with its possessions and former possessions around the world. Also, to a large extent, the British still had control of the sea lanes so materials could be brought in to both the United States for processing into finished war material products, and also into Britain.

By the time of full military engagement in World War II — both Japan and the German-Italian Axis becoming involved — the Japanese and German submarines made shipping lanes much more hazardous. The situation then and during the rest of the war was summed up by Simon Strauss, an officer of the Metals Reserve Company set up by the United States government just before World War II for this emergency. Strauss stated:

> "By the time Pearl Harbor broke out, all of Europe was under the domination of Hitler. Africa, Latin America, Australia, India, and even parts of China remained accessible to the U.S. The only competitive customer for the mineral exports of these vast areas was the United Kingdom. An agency called the Combined Raw Materials Board was created by the U.S. and the U.K. and a coordinated buying program was launched. It was possible to fight World War II without an actual shortage of these critical materials. But the key was the fact that we retained access to Latin America, to all of Africa, to most of Asia and to Australia. Had we been denied access to them, we would have been in trouble."

The United States and Britain had access to large supplies of raw materials at the outset of World War II. This was not true for Japan.

After Japan was opened to the rest of the world in the late 1800's, they resolved to become an empire, but they had very few natural resources.(1) The nearest significant coal and iron deposits were in northern China (Manchuria), and became the initial objective. The Japanese invaded Manchuria. Later they went into southeastern Asia to obtain raw materials. This continuing Japanese expansion to obtain control of new territory and therefore resources ultimately brought the United States and Japan into conflict in World War II.

This over-riding Japanese concern for their vulnerability in the matter of energy and mineral resources is evident in information which came out after the war. Minutes of the Japanese General Staff meetings just prior to Pearl Harbor show the principal reason for their expansion was to correct, as far as possible, their deficiencies in raw materials. This included the invasion of China and other parts of mainland Asia. Between 1931 and the time of Pearl Harbor (1941) the U.S. supported the concept of "Open Door" trade. It also supported China's independence from foreign invasion. But against the Japanese invasion of China, the United States could only be indignant. However, this indignation rose to a peak in 1940. Until then the Japanese had little fear because the United States had a relatively small navy which was split between two oceans. Japan, for centuries a sea-faring nation, had a relatively large navy, and it was all in one ocean.

Japan continued to expand its sphere of influence, largely through military action. The United States finally decided to invoke economic sanctions, and in July of 1940 put an embargo on aviation fuel. But this was only a nuisance to Japan because other grades of gasoline could be cracked in Japanese refineries to produce aviation fuel. In September of 1940, the United States also embargoed scrap iron and steel shipments to Japan. Until then the United States had been chiefly concerned with matters in Europe. Asia was a secondary consideration, but when the Tripartite Pact was signed in September 1940 among Germany, Italy, and Japan the shape of things to come began to emerge. With the signing of this pact which said that each of the signatories would go to the aid of any other in time of war, the war which was already in progress in Europe since September 1, 1939, became fused with activities in the Far East. This made the United States more anti-Japanese than before.

But all was not well for the Japanese. They had hoped that the Germans would soon finish off Britain which would open up the areas of British influence in Southeast Asia to Japan. Instead, Germany declared war on the Soviet Union, a country with whom Japan wanted to have improved relations. With this turn of events, Japan began to realize it would have to go it alone. Japan increased preparations for war with the United States, the only potential enemy of consequence in the Pacific region.

Oil: Chief concern of the Japanese

Records of meetings of the Japanese General Staff clearly show the great concern about Japan's ability to survive a war with the resources they could have at hand at the start of hostilities, and those which they thought they could obtain immediately thereafter.(12) The major concern was oil. The United States and what was at that time the Dutch East Indies (now Indonesia) were the main suppliers of petroleum products to Japan. In July 1941, Japanese troops invaded southern Indo-China. Britain, the Dutch East Indies, and the United States immediately embargoed all exports to Japan including oil, which reduced Japanese oil imports to about 10 percent of their previous volume. Immediately, Japan saw its storage tanks of oil as an hour-glass gradually running down. Something had to be done and done very quickly.

The Japanese General Staff now had to make some critical decisions.(12) If Japan were to retain its position as a substantial power, and maintain or improve its standard of living, it had to have access to raw materials. Without oil to fuel its merchant ships and its navy this would be impossible. Japan would have to retreat to its own islands and remain an island-bound limited economy, as the Japanese government at that time had no vision of what Japan might be eventually able to do industrially. The hour-glass of oil would last two years or less. Under conditions of intensive warfare the oil might last only six months. "An outline plan for carrying out the national policy of the empire" was drawn up and presented to a meeting of the Liaison Council on September 3, 1941. Faced with the alternatives of either beginning to withdraw or trying to move ahead, Japan decided to go to war with the United States if an agreement on oil could not be reached by October 15. The date was later extended to November 25 and then November 30. Japan's consumption of 12,000 tons of oil a day made oil the critical resource.(35) If Japan had possessed adequate oil deposits within its own borders the Japanese probably would not have gone to war with the United States. Japan's military was determined to seize the Dutch and British oil fields in southeast Asia. To do this they had to neutralize the U.S. fleet at Pearl Harbor.

The record shows, however, that there was a split opinion among the Japanese leaders. Some thought that diplomacy could accomplish Japan's goals, and therefore there was an increased effort in Washington on the part of some Japanese to change the position of the United States. This faction of the Japanese government tried very hard up to the last minute to achieve a change in U.S. policy. These efforts explain why Japanese diplomats were in Washington at the very hour of the attack on Pearl Harbor. It was not because the Japanese were trying to be deceptive, as has been commonly thought in the United States. It simply was a genuine effort on the behalf of some of the Japanese to avoid war. But the United States chiefly under the direction of Secretary of State Cordell Hull, insisted that Japan essentially abandoned all its positions in Southeast Asia and Manchuria, and also other parts of China. The Japanese offered to simply withdraw from Southeast Asia, but Cordell Hull would have none of that, although President Roosevelt was said to have been at least mildly interested in the idea.

Pearl Harbor: Key to Japan's access to oil

Japanese Premier Tojo was the deciding influence on the decision to go to war, arguing that the fate of Japan hung in the balance because of its lack of oil.(35) Faced with having to abandon all thoughts of an Imperial Japanese empire and being reduced to their island position alone, the Japanese attacked Pearl Harbor on December 7, 1941. The military alternative had prevailed over diplomacy. Some say Japan had no choice. The key to the decision was oil.

When the decision for war was made, and December 7th came, Japan had two inter-related immediate objectives. The first was to destroy or at least neutralize the United States fleet in the Pacific. The second was to immediately move south and seize the oil fields of the Dutch East Indies. To do this the U.S. Fleet had to be sufficiently disabled so that it could not protect the Dutch East Indies. This is what happened, and Japan did take the oilfields from the Dutch. Had the United States Fleet been able to intercept the Japanese as they headed for the Dutch oil fields, the Pacific war probably would have been over in less than a year. Oil was the key and the Japanese got their needed supplies, at least for awhile.

But as the war continued and the United States was gradually able to cut off oil shipments to Japan, the matter of fuel became increasingly desperate for the Japanese. In the waning days of the war Japan resorted to "kamikaze" action —suicide pilots who would crash their planes into the ever-closer ships of the U.S. Navy. This was in large part a matter of fuel efficiency, for the planes would be given just enough fuel to fly one way. They were not expected to return and the pilots were told so. More than four thousand young Japanese men were sacrificed this way. It was a desperate way of stretching fuel supplies, and, in theory at least, was very efficient. If each plane did hit a U.S. Navy ship, this would be a highly effective use of a small amount of fuel. Some planes did get through the anti-aircraft barrages and did considerable damage, but many more kamikaze planes were shot down. Finally there was not enough oil to keep either the Japanese navy or air force operational. After the war, some hidden supplies of fuel for these kamikaze planes were found. So desperate were the Japanese for fuel that some of the fuel was made from roots of pine trees. Testing showed this was exceedingly bad fuel which fouled the engines beyond use. Also, when General Douglas MacArthur drove from Tokyo down to Yokohama where the Battleship Missouri was anchored, to be the site of the Japanese surrender, the trip of 20 miles took two hours. The reason was that "...the battered vehicles, the best that the Japanese could provide, were powered not by gasoline, because there was no gasoline, but by charcoal. They repeatedly broke down."(35) Oil was Japan's Achilles Heel in World War II.

Wars Simply to Gain Minerals

Huns and salt

In the vicinity of Salzburg, Austria there are ancient salt mines. Although these did not provide minerals for war, they provided a necessity of everyday life. As such they were richly prized and various clans of the Huns fought for their possession.

Polish zinc, lead, and coal

Battles were fought by principalities over the possession of the lead and zinc mines of Poland. As part of the settlement after World War II, Poland received some of the coal mines which were previously in German territory. These coal mines are very important to the economy of Poland.

Chile and nitrates

For a long time there was a boundary question between Chile and Bolivia, but no one really cared about the exceedingly desolate northern Atacama desert territory. That is, not until nitrates were discovered there in great quantity. These were the cause of the Nitrate War. Chile declared war on both Peru and Bolivia on April 5, 1879. The Chileans were victorious and obtained all of the Atacama Desert area by the treaty of Acon in 1883.

The victory was of great economic value to Chile. From 1879 to 1889 the duty on nitrate exports alone reached more than $557 million dollars, a very considerable sum in those days. The total value of nitrate exports in that period exceeded $1.4 billion. Nitrates have continued to be an important Chilean export, although the synthetic production of nitrates has reduced their value.

Recent and Current Conflicts

Morocco, Algeria, and phosphate

Currently there is a very real shooting war of which few people are aware. Sometimes called the "forgotten war," it has been going on for more than 20 years between Morocco and Algerian-backed forces called the People's Front for the Liberation of Saguiat el Hamra and Rio de Oro, otherwise known as the Polisario Front.(11) The actual battlefront is more than 1,000 miles long, lined with rocks, minefields, and radar and artillery emplacements. The argument is over 103,000 square miles of the western Sahara that contains very large uranium-bearing phosphate deposits. A supporter of the Algerian-backed forces says that the region could be the "Kuwait of Africa" with an annual per capita income of $16,000 from the phosphates, plus the coastal fishing potential. This is clearly a contemporary war over mineral resources, with no agreement presently in sight.

Falkland Islands

The Falkland Islands are British possessions which geologically lie on a shallow shelf extending east from Argentina. On April 2, 1982, Argentina invaded the islands. On May 21st, the British landed troops there, and on June 14th Argentina surrendered. This British action was ostensibly to protect the sovereignty of Britain in the Falklands. There were, however, other interests as the Falklands planned to offer offshore oil concessions for bid. Argentina, trying to do what they could not do militarily, offered to pay the Falklanders for citizenship by suggesting that they might pay up to $100,000 for each islander and up to $800,000 for each family if they voted to transfer sovereignty to Argentina. The Argentine foreign minister, Guido di Tella, stated that Argentina wants the islands back for their "intrinsic value," not for their potential oil and gas resources.(19) As there are only about 2,100 islanders, $100,000 for each person would cost $210 million for Argentina to buy the islands. It would only take a modest oil discovery to make that a very good investment. The Falkland Islanders, however, declined to take the offer, which was probably a good economic decision, regardless of their British loyalty.

In 1996, The Falkland government offered 400,000 square kilometers of adjacent offshore acreage for oil company bid. This area is larger than Britain's area of the North Sea. The Falkland government stated in summarizing the tax provisions that they would impose a tax of between 40 and 50 percent. They presented an illustration of what a 500 million barrel oilfield would produce in income to the producer and tax revenue to the government after exploration and development costs and assuming a figure of $18 barrel oil, and a 46 percent tax on net project proceeds. On this basis, the Falkland government would receive close to $2 billion, a substantially larger sum than the $210 million offered for the country by the Argentine government.

Some Peaceful Settlements

Can we ever finally leave these conflicts behind, or will the only increase in the future as mineral resources become more scarce? In a hopeful trend, more recently some disputes over minerals have been settled peaceably, or at least continue in negotiation form.

Australia and Indonesia

In 1989, Australia and Indonesia began negotiations over disputed ownership of certain waters off the Island of Timor. No one had been interested in this area until oil was discovered. Then the important question of ownership was raised. It will apparently be settled across a table.

Venezuela/Guyana

Like the Australian/Indonesian question, prospects for oil have renewed a dispute between Venezuela and Guyana over the Essequibo territory claimed by both countries. The Guyana Natural Resources Agency issued exploration permits to several oil companies for this region, but Venezuela disputes Guyana's right to the territory. This matter is still under peaceful discussion.

Texas/Oklahoma

Although it was very unlikely to ever become a shooting war, a dispute between Texas and Oklahoma occurred when oil was struck in the valley of the Red River between these two states. Then, what had been the more or less accepted boundary came into dispute. The precise legal boundary was the river at the time Texas became a state in 1845, but the river had changed channels since then. The boundary was finally determined by students of tree rings who located where the channel had been in 1845.

Future Conflicts?

South China Sea

This is a region where oil has recently been discovered, and the potential for more discoveries is good. The Spratly Islands dot this area and consist of about a hundred coral reefs, tips of rocks, and 21 slightly submerged landforms. The emergent land is less than two square miles in total. In 1996, China's foreign ministry said that China held, "indisputable sovereignty" over all of the Spratly Islands. Ownership in the South China Sea is asserted in whole or in part by nine nations, chiefly China which claims at least 80 percent, and Vietnam. All nine nations have set up little outposts on various of these rocks. In spite of lack of clear ownership, both Vietnam and China have issued lease blocks to oil companies. In the meantime, oil-short China is building up its navy and expanding its presence in the South China Sea. China needs much more oil as it makes plans to greatly enlarge its road system and increase motor vehicle production.

Some shots have already been fired in the South China Sea. The first Vietnam-China conflict was in 1974, and since 1982 these two nations have had additional clashes. In 1995, the Chinese briefly clashed with Philippine ships. As more oil is discovered, the ownerships in this region will have to be resolved. Indications are that may be accomplished by military presence or actual force. In that situation, China would almost certainly prevail.

In 1996, Vietnam awarded Conoco Incorporated the right to explore for oil and gas in an area of the South China Sea about 190 miles from the Vietnamese coast. Vietnam calls this area blocks 133 and 134 and has stated, "No one has the right to interfere or obstruct this." China awarded oil exploration rights in the same area which it calls Wan'an Bei 21 to Denver-based Crestone Energy Corporation in 1992.

Persian Gulf region border disputes

Asher Susser, a Middle East expert at Tel Aviv University states "The average Middle East border is only 70 years old, which is nothing in the context of the history of the region. In the eyes of many Arab countries, the region's borders remain fundamentally illegitimate, the artificial creations of foreign powers."(17)

The main reason for the border disputes which exist in the region today is the geography of its broad relatively featureless desert areas that were occupied by nomadic tribes. There was little concern about boundaries and national borders were rarely surveyed and marked. "The discovery of oil and other natural resources changed all this. Suddenly a mile here or there could determine who got an outlet to the sea, possession of scarce water resources, or control of oil wells. Remote islands became important because they offered access to untapped oil reserves in the surrounding seabed. But when two countries finally did try to demarcate borders, they often discovered that the official maps drawn up by the former colonial powers that once controlled the region were unfinished or inadequate when it came to resolving the many complicated claims."(17)

Iraq's stated reason for the invasion of Kuwait was based on a disputed border crossing a very large oil field. More recently, United Arab Emirates officials say Iran is threatening military action over three islands in the Persian Gulf with probable oil potential. With the large number of border disputes in the Middle East, some believe it may take as much as 30 years to settle them all. But some countries are using the arrangement of joint development of natural resources to solve the problem. Whether this approach can resolve all the border disputes remains to be seen. Iran is very actively building up its military strength and all the Persian Gulf nations are increasing their spending on armaments.

Other disputes

Several other areas are now also under dispute which, before there was oil involved, were of no concern. Five nations, Iran, Russia, Azerbaijan, Turkmenistan, and Kazakhstan share the coastline of the oil-rich Caspian Sea region. In 1996, a conference among these nations was held. Iran and Russia insisted that the oil resources beneath the water of the Caspian Sea be shared equally among the five nations. The other nations want ownership to be related to the sectors which lie off their respective shores. There is at present no resolution of this matter

Also in 1996, the United States and Mexico were in disagreement about two segments of the Gulf of Mexico. These have been called "doughnut holes" by government officials because they fall outside previously agreed upon boundary areas. To complicate matters, the eastern "doughnut hole" ownership also involves Cuba. These areas are in waters with depths of from 6,000 to 13,000 feet. So far, the deepest water in which drilling has been conducted is 7,625 feet, but technology is expected to be developed which will enable drilling to be done in deeper waters. Gary Lore of the U.S. Minerals Management Service office, which is in charge of federal offshore leases, has stated, "Just about every field in the shallower waters of the continental shelf has been drilled. So companies are pushing to explore new areas, and the deepwater gulf is the last undrilled frontier in North America." This dispute at present also remains unresolved.

Minerals in Economic and Political Warfare

With the invention of the atomic bomb, it may be that intensive global military action by major powers is not so likely in the future as it is probable that atomic warfare would result in everyone's losing. Therefore it seems more likely that actual warfare will be a limited nature, to control specific mineral resources such as the case of the Persian Gulf War over Kuwait. Other "wars" will be in the form of economic warfare. The side which prevails will be that which can produce and maintain the most robust economy. But here again, command of mineral resources is vital to winning such conflicts. Considerable economic and political control of other nations can be achieved by means of control of resources vital to them. With the importance of oil and natural gas in the world economy today, these become particularly important elements in economic warfare.

The first use of oil in economic warfare occurred when Britain tried to maintain control of the Suez Canal in the 1950's. In support of Egypt, Saudi Arabia cut off oil supplies to Britain, but at that time, the United States still had surplus oil producing capacity and simply opened the oil well valves wider and supplied Britain's needs.

The Arabs lost that economic skirmish (although the Suez Canal eventually did go to Egypt). But in the 1970's the Arabs and Iran realized that the United States was no longer self-sufficient in oil. As a result, the Persian Gulf Muslim nations were able to do two things.

They could exert economic/political influence on the United States (with regard to the Israeli situation) and they also realized they could begin to take control of their own economic destinies as industrial competitors with the western world.(3)

Between 1940 and 1970, the price of Middle East crude oil ranged between $1.45 and $1.80 per barrel. This price was set in dollars not adjusted for inflation which meant that the price of oil actually had gone down during that period. But with the realization that the western world, especially the United States, was no longer self-sufficient in oil, the Gulf countries, and all OPEC (Organization of Petroleum Exporting Countries) members, raised the price to more than $35 per barrel in less than eight years. Colonel Qadaffi of Libya was the one who broke the price barrier. When he seized control of Libya in 1969, he pressured one Libyan oil producer to give a 30 cent a barrel price hike. Other Libyan producers soon did the same. Then other OPEC members followed, and the price rise began. In 1970, the United States could no longer supply its own oil needs and began to depend on imports, and a weapon of oil was handed to the Persian Gulf nations. Against the military might of the West they would be powerless, but their oil weapon verified the saying that "power now does not come out of the barrel of a gun but out of a barrel of oil." When the Arab-Israeli fighting began in 1973, the oil producers of the Middle East declared a selective boycott against consuming countries, in particular the United States, because of its long-standing support of Israel. OPEC, using its new found economic power, raised the price of oil from three dollars a barrel in September 1973 to $11.65 in December, a near quadrupling of price. Henceforth neither the oil companies nor their western governments could control the price of oil. The best that could be done was to convince the oil producing nations that their interests were intertwined with those of the West, and that reasonable oil prices were vital to preserving their oil markets. For the most part, this logic has since prevailed.

But influencing the price of oil by the OPEC nations is a form of economic warfare which continues to the present, and will persist and could be increasingly effective as OPEC gradually will hold control of more and more the world's exportable oil supplies. There is, however, another kind of economic warfare developing, which might, in less strong terms, simply be called "economic competition." This is in the form of the industrial developments in some of the nations of the Persian Gulf. They include the large petrochemical complexes of Saudi Arabia and the United Arab Emirates (UAE), and the aluminum refining capacity of the UAE, competing directly with similar facilities in the West. This economic aggression will continue and probably intensify as the petroleum producing capabilities of the world become more and more concentrated in the Middle East, and the nations which hold most of the petroleum continue to develop more petrochemical and refining industries to upgrade the oil and gas to higher value end products. The oil-rich countries have been endowed by geology with the resources to be able to do this, and this is guiding their economic development. At the same time it transfers jobs from the up-to-now dominant petrochemical industry of the industrialized countries, to the Middle East.

Saudi Arabia and OPEC

Another way Saudi Arabia has used its oil as an economic weapon has, at times, been against its fellow OPEC members. OPEC periodically agrees to production quotas for its members in order to achieve a particular world oil price. But some members will occasionally cheat on their quotas. Saudi Arabia, with its huge oil reserves, has threatened to, and in some instances actually has used its oil capacity to pump excess oil into the market thereby driving down the price of oil. In this way, Saudi Arabia has tried to discipline the other members into adhering to their production quotas.

The United States and Rhodesia

There has been an interesting variation tried at times in using minerals in economic warfare. Rather than a producing country withholding a mineral resource, an importing country puts a ban on importing it to try to influence in some way the producing country. The United States at one time embargoed the importation of chromium from Rhodesia (now Zimbabwe) to influence the then white Rhodesian government with respect to its apartheid racial policies.

But the United States still needed chromium, and the only other substantial supplier was the USSR. The Soviets promptly tripled the price of the chrome they sold to the United States. What they could not supply themselves they bought from Rhodesia, also tripling the price of that chrome when it was re-sold to the United States. The United States finally had to give up on the embargo.

As a footnote to that sad but educational experience, it is interesting to observe that the 1988 U.S. Congress passed a law banning all purchases from, and investment in South Africa. This again, as in the case of Rhodesia, was to try to influence the government's racial policies. But the Congress was careful to include an exemption for purchase of strategic minerals. They had learned.

USSR and Lithuania

In the final days of the former Soviet Union, the Baltic Republics, and particularly Lithuania, showed intentions of wanting to break away. Fearing the precedent this might create, President Gorbachev of the Soviet Union in April 1990, threatened to cut off all energy supplies to Lithuania. Almost all of Lithuania's coal, oil, and gas comes from what is now Russia. This was blatant economic warfare. It imposed a temporary hardship which the Lithuanians endured without shaking their resolve to leave the Soviet Union. When the USSR came apart, supplies were restored. Russia needed the income.

Embargo on Iraq oil

An oil embargo has been used against the United States by Middle East nations. Following the Gulf War, the tables were turned, and the United Nations, led by the United States and the other Gulf coalition countries, imposed an embargo on Iraq's oil sales to try to force Iraq to fully comply with the provisions of the surrender agreement. This action met with only moderate success.

Embargo on oil to Haiti

Following the military coup which deposed the duly elected president of Haiti in 1987, the United States imposed a blockade of oil going into Haiti in an effort to change the political control of that nation. The embargo was only partially successful because gasoline was smuggled into Haiti from the adjacent Dominican Republic. Ultimately it was the threat of military action and eventually an occupation of Haiti by U.S. forces which accomplished the change in administration and the return of the elected president, Aristide.

United States and Libyan Oil

Libya, under Col. Qaddafi, has been less than an international model citizen. Once, in protest against repeated Libyian terrorism, the United States made an air strike on Libya. But there is a large unresolved matter. In 1988, a U.S. Pan American jet, while flying over Lockerbie, Scotland, exploded, killing all 259 people aboard, and 11 people on the ground. Evidence linked this action to Libya, and the United States has been trying ever since to have the Libyan suspects extradited to the United States, but to no avail. To try to accomplish this, the United States has repeatedly tried to place a global embargo on Libyan oil through the United Nations. So far this effort has been unsuccessful. It probably will not be successful because a number of UN countries do some business with Libya based on Libya's foreign exchange from its sale of oil. However, the United States' attempts to use oil as the lever on Libya illustrates how important oil resources are to Libya. The U.S. chose the right weapon but probably will not be allowed to use it. Libya's oil money buys it protection. If the oil embargo weapon could be used, it would very quickly cripple Libya which has almost no other product by which to support the present oil-export based economy. It has no significant manufacturing capabilities.

The Future

Economic warfare clearly exists in various forms today. It will continue, and may even intensify as resources become scarcer against rising industrial demand around the world, especially as less developed nations like China become more industrialized and add to the competition for energy and minerals.

Successful economic warfare using energy and mineral resources as weapons can only be employed by those who possess these resources in some significant quantity. Today, there are only two places where this situation exists. One is the Persian Gulf region, and the other is Russia, which still includes about 80 percent of the former Soviet Union with its vast mineral wealth.

Persian Gulf

Oil export embargoes have already been used by Persian Gulf nations against western nations. As oil in other regions is depleted, and more and more of the remaining world oil resources become concentrated in the Persian Gulf area, the opportunity for effective economic warfare using these resources increases. However, Persian Gulf nations are not a solid block. Iraq has been a renegade, and it has huge reserves. And Iran and Saudi Arabia have generally been at odds. Whether all the Persian Gulf nations could present a solid front in the future for economic warfare is a question. Their greatest unifying element is that they are all Muslim. The fundamentalist Muslim movement in the Gulf region is the unpredictable factor in that region's relationship with the western oil-consuming nations.

Russia

This largest country in area in the world has seen two political upheavals in the Twentieth Century. First was the rise of communism and the Soviet State following the fall of the Czar after World War I. Second, was the disintegration of the Soviet Union in the early 1990's with Russia as the remaining dominant unit in a rather uncertain political and economic condition. It is difficult to predict what kind of leadership will emerge in this situation, or what the intentions of those leaders will be. But one thing is certain, Russia has huge natural resources within its borders which can be used in economic warfare in the future.

Russia has the world's largest deposits of coal and natural gas; extensive and still mostly undeveloped copper resources; and large deposits of virtually all other important minerals. But it has been difficult to develop this wealth because of limited access. To correct this, the Soviets decided years ago to build another railroad through Siberia, to augment the heavily overloaded original Trans-Siberian line. This new line, termed the Baykal-Amur Magistral (the BAM), was recently completed, and has opened up a mineral-rich region that will allow

Russia to export a variety of important resources.(28) As a result, the Russians will be in a position to influence world mineral markets well into the next century.

Slow to develop, but much left

Russia has been slow in developing its great natural resources, with a considerable disadvantage to the standard of living of its citizens. But this delayed development also means that the present Russia, which includes about 80 percent of the former Soviet Union, still has these resources. In contrast, the United States developed its resources early and rapidly, which propelled the United States to the top of the world's living standards in record time. But in the process, the richest U.S. energy mineral and mineral resources have been used. Russia still has much of its resources undeveloped, and this will be a factor in economic warfare of the future.

Economic warfare was foreseen by Lenin, the Father of Communism, when he wrote in 1916 about the possibilities of cutting off vital mineral resources from the western nations. This view was enunciated by subsequent Soviet leaders. Soviet President Leonid Breshnev stated to Somalian President SiadBarre, then an ally of the USSR, that "Our aim is to gain control of the two great treasure houses on which the West depends — the energy treasure house of the Persian Gulf and the mineral treasure house of central and southern Africa."(20)

Termed the "real war" by President Nixon who knew the Soviets well, their resource quest had many aspects. How consistently it was used by the former Soviet Union is a matter of some debate, but the political stirrings and in some cases upheavals in some mineral rich African nations seem to have had at least some Communist involvement. President Nixon made the observation that "The Soviet leaders have their eyes on the economic under-pinnings of modern society. Their aim is to pull the plug on the Western industrial machine. The Western industrial nations' dependence on foreign sources of vital raw materials is one of our chief vulnerabilities. This, as well as the inherent instability of many of the producing nations, dictates Soviet strategy in such areas as the Middle East, Africa, and Latin America."(20) Whether or not the new regime in Russia will continue to pursue economic warfare against the western democracies only time will tell. The record of the USSR in the "Cold War" which followed World War II, and continued until the breakup of the Soviet Union in 1991 was clearly one of economic warfare against the NATO allies, and particularly the United States. This warfare was conducted in several parts of the world. (5,7,9,14,16,20,25,27,30,31,32,33)

Russia and the Persian Gulf

What Russia may do elsewhere in the world regarding mineral resources is not very predictable but it seems certain that Russian interest and involvement will continue in the Persian Gulf, which fact will remain a concern to the western democracies. Russia is now peaking out in its own oil production against a growing domestic demand. The Persian Gulf region has a history of marked and intermittent political upheavals. This pattern is likely to continue, and because it is in Russia's backyard it is a fertile and convenient neighborhood in which to expand Russian influence. Former Soviet foreign minister Molotov stated that the Persian Gulf region is where lies the "center of the aspirations of the Soviet Union." In the 1980's, the USSR made an unfortunate move toward the Persian Gulf by going into Afghanistan which lay in the path. The move was aborted, but Russian interest in the Persian Gulf remains high. Persian Gulf nations will hold the last remaining world oil reserves and Russia is much closer to that region than is the United States.

Russian gas

Russia, with its huge gas reserves, is adjacent to the western European industrialized countries which must import substantial energy supplies. Although the apparent immediate intention of Russia is simply to sell natural gas to obtain foreign exchange, a long term result could be that western Europe becomes increasingly dependent on Russia for energy supplies, and vulnerable to the Russian influence. At present, Russia supplies only about three percent of western Europe's energy, but a recent survey of total world uncommitted (surplus) natural gas supplies showed that the rest of the world, not including Russia, had about 600 trillion cubic feet of surplus, whereas Russia alone, held more than 750 trillion cubic feet. Obviously, over the longer term, Russia will be in an increasingly strong position in the European energy supply situation, and can exert a corresponding influence on this area.(6) For this reason the construction of the large diameter gas pipeline from the gas-rich Russian Yamal Peninsula into western Europe was vigorously opposed by the United States and an embargo was placed by the United States on equipment which would help to build the line. But this was a futile gesture because similar equipment was available elsewhere. The line was built and now carries natural gas to western Europe.

So far the amount of Russian gas exported is modest, but it is growing. The valves controlling the gas supply remain on the Russian side of the border, and little by little western Europe is buying more Russian gas. Production from gas fields which do exist in western Europe is expected to decline after the year 2000, and imports will grow. The sources will have to be Russia, Algeria, Libya, and ultimately the Middle East which now has an estimated one-third of the world's total natural gas reserves. Energy is critical to the survival of any industrialized nation, and this increasing dependence on foreign energy supplies will correspondingly give Europe a lesser control of its own destiny.

Sweden and Russian energy

Sweden has voted to eventually phase out all existing nuclear plants, and to build no new ones. As Sweden has no coal or oil and only a modest amount of hydropower, where future energy supplies will come from is a problem which seems to have been ignored to some extent. Some people in Sweden envision a solar/biomass/wind economy. But whether these combined energy sources can replace steady, predictable, and high quality nuclear power is questionable. For the immediate future, Sweden has begun talks with Russia concerning obtaining Russian oil and gas. In 1988, Sweden's OK Petroleum Company signed a contract with the USSR to purchase two million metric tons of crude oil and oil products. The result of this decision on energy will be to make Sweden economically vulnerable to Russia.

Finland and Russian gas

Finland already gets some of its gas from Russia through a pipeline into southeast Finland completed in 1974. The pipeline network has now been extended into the Helsinki and Tampere districts. Seventy-three percent of the oil Finland uses comes from Russia. Finland also imports substantial Russian coal.

Germany and Russian resources

Fifteen percent of Germany's energy demand is met by natural gas, and of this, 71 percent is imported, 24 percent coming from Russia. When these figures for Germany are sorted out, the amount of energy obtained from Russian gas is not great, but it is likely to grow, because Russia has the resource and will have it in quantity for a long time to come. Germany also imports 95 percent of its oil, about 6 percent of which now comes from Russia.

Turkey

Turkey has long been a thorn in the side of the Russians because it controls Russia's shipping and naval access to the warm waters of the Mediterranean through the Dardanelles. The border between the two countries has been a tense one at times. But the Russians have now penetrated energy-short Turkey with natural gas lines, and in the Turkish capital, Ankara, Russian gas heats homes and factories.

Poland dependent on Russia

Poland is an example of a country which has, in effect, already lost the battle of economic warfare. Nominally it is an independent country, but it is beholden to Russia for very basic needs. Russia can bring Poland to an economic standstill merely by cutting off its oil supply. Russia exports about 80 million barrels of oil and about 19 million barrels of petroleum products to Poland annually. Poland also depends on Russia for large supplies of fertilizer, made from natural gas, and for magnesium, and nickel.

Russian need for foreign exchange

Russia has been obtaining about 60 percent of its hard currencies from the sale of oil and gas. Without this income Russia would have a difficult time supplying its military with needed technological equipment (computers are an important item). Oil and gas earnings also buy grain, something in chronic short supply in Russia. Grain bolsters the civilian economy, but grain also feeds Russia's large standing armies.

Will need for trade be a factor for peace?

For some time to come, Russia will need western technology. Russia will also need grain, particularly from the United States. Recently a number of western oil companies have been making various arrangements to develop or recondition some of Russia's oilfields which, for the most part, have not been well managed. Western oil recovery techniques are far superior to the methods which have been used in Russia. Russia's need for western technology and capital to develop its economy may be a strong factor in keeping peace between Russia and the western nations.

Russia in the future

It is inevitable that Russia, with its generous geological endowment of energy and mineral resources, is geodestined to play an important role in the future of many nations. This may bring the new Russia more fully into the world economy with free trade in vital raw materials without resort to military or economic warfare.

Will the Russia of the future prove to be a friendly neighbor to the western nations, or will it return to its political isolation and antagonism toward the West, using its huge coal, gas, and metal resources as weapons? The "Cold War" was a time when the Soviet Union was active in economic warfare against the NATO allies in many regions. Now that activity appears to have ceased. With its internal economic problems, Russia now shows a somewhat friendlier attitude toward the West, in part at least to obtain loans. In 1996, the International Monetary Fund loaned Russia $10 billion. Previous to that, the United States had loaned Russia considerable sums. The need for more capital, agricultural products, and western technology together with joint petroleum ventures with western companies may replace the economic warfare of the past. This would bring the new Russia more fully into a world economy of free trade in vital raw materials, without resort to military or economic warfare.

Russia, however, is the most natural resource self-sufficient nation in the world, and the least vulnerable to cut-off of a wide variety of mineral and energy mineral supplies. It is in a better position to avoid conflict over mineral resources than is the rest of the world. The reality of Russia's nearly complete self-sufficiency in minerals and energy minerals compared with other industrial nations will be important in future world affairs. Russia is now only marginally a world economic power, but if the current attempted reforms of their economic system succeed, and with the largest and broadest energy and mineral resource base in total of any nation, Russia could do very well economically. In contrast, the western European and North American industrialized nations are increasingly dependent on foreign resources, and at times have to defend their sources by military action as evidenced by the Gulf War with Iraq in 1991. The United States now has permanently stationed naval, army, and air forces in the Persian Gulf area to secure its oil supplies, which adds to the cost of these resources for the U.S.

War or reason?

Struggle for resources, especially oil, will continue as population pressures grow and resources become increasingly scarce.(31) That potential exists not only among the western nations, but is also growing in China and Southeast Asia where rapid industrialization is causing a sharply rising consumption of oil. Will this nearly world-wide increased demand for energy and minerals, compounded by the current exponential growth in population, be resolved by reason, or will the struggle result in war and anarchy as suggested in the very thought-provoking article by Kaplan?(13)

BIBLIOGRAPHY

1 ABBOTT, J. F., 1916, Japanese Expansion and American Policies: The Macmillan Company, New York, 267 p.

2 BAKER, R. L., 1942, Oil, Blood and Sand: D. Appleton-Century Company, New York, 300 p.

3 BILKADI, ZAYN, 1995, The Oil Weapons: Aramco World, v. 46, n. 1, p. 20-27.

4 CENTRAL INTELLIGENCE AGENCY, 1985, USSR Energy Atlas: Central Intelligence Agency, Washington, D.C., 79 p.

5 CONNELLY, PHILIP, and PERLMAN, ROBERT, 1975, The Politics of Scarcity: Resource Conflicts in International Relations: Oxford Univ. Press, London, 162 p.

6 DISPENZA, DOMENICO, 1995, Europe's Need For Gas Imports Destined to Grow: Oil and Gas Journal, March 13, p. 45-49.

7 ECKES, A. E., Jr., 1979, The United States and the Global Struggle for Minerals: Univ. Texas Press, Austin, 353 p.

8 GLEICK, P. H., (ed.), 1993, Water in Crisis. A Guide to the World's Fresh Water Resources: Oxford Univ. Press, New York, 473 p.

9 GORDON, DAVID, and PANGERFIELD, ROYDEN, 1947, The Hidden Weapon: The Story of Economic Warfare: Harper, New York, 238 p.

10 HINDMARSH, A. E., 1936, The Basis of Japanese Foreign Policy: Harvard Univ. Press, Cambridge, Massachusetts, 265 p.

11 HODGES, TONY, 1983, Western Sahara. The Roots of a Desert War: Lawrence Hill & Company, Westport, Connecticut, 388 p.

12 IKE, NOBUTAKA, (ed.), 1976, Japan's Decision for War: Records of the 1941 Policy Conferences: Stanford Univ. Press, Stanford, California, 306 p.

13 KAPLAN, R. D., 1994, The Coming Anarchy: The Atlantic Monthly, v. 273, p. 44-76.

14 KING, A. H., and CAMERON, J. R., 1975, Materials and New Dimensions of Conflict in New Dynamics in National Strategy: Crowell, New York, 293 p.

15 LOVERING, T. S., 1943, Minerals in World Affairs: Prentice-Hall, New York, 394 p.

16 MACKENZIE, RICHARD, 1989, Soviet Motives for Invasion [of Afghanistan] are Starting to Lose Their Veil: Insight, Washington, D.C., June 5, p. 30-31.

17 MARCUS, A. D., 1996, Border Disputes Continue to Roil Mideast: The Wall Street Journal, January 30.

18 METZ, H. C., (ed), 1994, Persian Gulf States. Country studies: The library of Congress, Washington, D. C., 472 p.

19 NEW YORK TIMES NEWS SERVICE, 1995, June 12.

20 NIXON, R. M., 1980, The Real War: Warner books, Inc., New York 341 p.

21 ODELL, P. R., 1983, Oil and World Power: Penguin Books, New York, 288 p.

22 OKITA, SABURO, 1974, Natural Resource Dependence and Japanese Foreign Policy: Foreign Affairs, v. 52, p. 714-724.

23 PARK, C. F., Jr., 1975, Earthbound. Minerals, Energy, and Man's Fate: Freeman, Cooper and Company, San Francisco, 279 p.

24 PATERSON, T. G., et al., 1991, American Foreign Policy: A History Since 1900: D. C. Heath and Company, Lexington, Massachusetts, 723 p.

25 REES, DAVID, 1970, Soviet Strategic Penetration of Africa: Conflict Studies, 77, p. 1-19.

26 RICH, NORMAN, 1973, 1974, Hitler's War Aims (2 vols.): Norton, New York, 584 p.

27 SAWYER, H. L., 1983, Soviet Perceptions of the Oil Factor in U.S. Foreign Policy: The Middle East-Gulf Region: Westview Press, Boulder, Colorado, 183 p.

28 SHABAD, THEODORE, and MOTE, V. L., 1977, Gateway to Siberian Resources (The BAM): John Wiley & Sons, New York, 189 p.

29 SHIMKIN, D. B, 1953, Minerals. A Key to Soviet Power: Harvard Univ. Press, Cambridge, Massachusetts, 452 p.

30 TANZIER, MICHAEL, 1974, The Energy Crisis: World Struggle for Power and Wealth: Monthly Review Press, New York, 171 p.

31 TANZIER, MICHAEL, 1980, The Race for Resources. Continuing Struggles Over Minerals and Fuels: Monthly Review Press, New York, 285 p.

32 TARABRIM, E. A., 1974, The New Scramble for Africa: Progress Press Publishers, Moscow, USSR, 234 p.

33 WU, YUAN-LI, 1952, Economic Warfare: Prentice-Hall, Inc., New York, 403 p.

34 TICKELL, SIR CRISPIN, 1994, the Future and its Consequences: The British Association Lectures 1993, The Geological Society, London, p. 20-24.

35 YERGIN, DANIEL, 1991, The Prize. The Epic Quest for Oil, Money, and Power: Simon & Schuster, New York, 877 p.

CHAPTER 4

The Gulf War of 1990-91: Iraq Invades Kuwait

Why the United States and other major western powers should be concerned about less 7 thousand square miles of desert sand was clearly stated: "...to keep Saddam Hussein from controlling two thirds of global oil reserves and from using that control to blackmail the industrial world and make Iraq a nuclear power."(20)

Kuwait – Great Riches in a Small Package

Kuwait, with an area of 6,880 square miles, is smaller than many counties in the United States. But it has oil reserves four times those of the entire United States, with an area of 3½ million square miles. Kuwait has only a very small domestic oil demand, so almost all of Kuwait's oil production was exported to many countries. These countries were concerned because whoever had control of this oil for export, about two million barrels a day, would have a very large source of valuable foreign exchange, and the income might be used in arming against the West or against other Persian Gulf nations. If Iraq obtained the oil this would be the expected result. But more serious was the eventual threat to Saudi Arabia, the nation with the world's single largest oil reserves. Iraq's despot, Saddam Hussein, had, from time to time, suggested to his fellow Arabs that they cut up Saudi Arabia and divide the spoils.(15)

Why Saddam moved

With finances depleted from a long and inconclusive conflict with Iran, Saddam Hussein saw Kuwait as an exceedingly rich prize in a conveniently adjacent small package. Kuwait had virtually no military forces. Saddam Hussein was deeply in debt from the Iran war which had cost Iraq an estimated $500 billion. At the end of the eight year war, Iraq had debts of more than $80 billion. Half of the debt was owed to Saudi Arabia, Kuwait, and the United Arab Emirates, $30 billion was owed to the other countries including the United States, and was scheduled to be repaid in hard currency, not Iraq money.

There was also a long-standing dispute over the boundary between Kuwait and Iraq which happened to lie over the giant Rumaila oil field, some 50 miles long, with an estimated 30 billion barrels of oil. That field alone had more oil reserves than the entire United States. Iraq held 90 percent of the field but claimed that Kuwait from its 10 percent had pumped $10 billion worth of oil from the side of the field which belonged to Iraq.(3)

The boundary between Iraq and Kuwait had been established by the colonial powers long before oil was discovered there. In 1922, the British High Commissioner for Iraq had designated the modern boundaries of Iraq, Kuwait, and Saudi Arabia. Kuwait was given a coastline of 310 miles, and Iraq got only 36 miles. This was also a matter of contention because it meant that Iraq did not have a good deep sea port. Iraq requested from Kuwait control over the two islands of Warbah and Bubiyuan. Kuwait rejected the proposal to give Iraq Warbah Island, and to give a 99 year lease for half of Bubiyuan Island.

War with Iran had been hard on the Iraqi economy. With the war over, the Iraqi population looked forward to an improved living standard. Money was needed to accomplish this and keep the internal situation under control. With the economic grievances which Iraq had against Kuwait, and Kuwait being such a rich prize with an estimated 90 billion barrels of oil in a small area adjacent to Iraq with virtually no significant defenses, Saddam Hussein saw the chance to both satisfy his grievances and solve his economic problems. Also, because of Saudi Arabia's policy against foreign troops on its soil, Hussein believed that the Saudis would not allow American or other non-Muslim forces to defend either Saudi Arabia or Kuwait.

The invasion

On August 2, 1990, Iraq invaded Kuwait, and in one day overran the entire country. With the possession of Kuwait, Saddam Hussein controlled 22 percent of the world's exportable oil. If he could also take Saudi Arabia, as there were indications that he might, he would control 44 percent of the oil for export. That circumstance which would have made the United States, Europe, Australia, and Japan his hostages. This was clearly an intolerable situation for the industrialized world.

President George Bush of the United States, promptly put together a coalition of nations toward the goal of evicting Iraqi forces from Kuwait. And more important, to prevent Iraq from moving on the much bigger prize of Saudi Arabia. American forces were immediately moved toward the Middle East, and forces of other concerned countries followed. A total of about 660,000 troops were assembled from 28 nations, with the majority from the United States (430,000 troops of which 376,000 were combat ready). Britain and France, with substantial air power supplied by Italy, were major force members. Saudi Arabia committed all the forces it had, and other troops came from the United Arab Emirates. Egypt contributed troops, and in return the United States cancelled Egypt's military debt owed to the United States.

The United States made repeated diplomatic attempts to persuade Saddam Hussein to leave Kuwait, but all fell on deaf ears. Very last minute diplomacy failed to remove Iraq peacefully. The ultimatum for Iraq to leave Kuwait ran out at midnight January 15, 1991. Gulf Coalition aircraft were over Baghdad by 2:30 a.m. January 16. The air war had begun. Control of the air space over Kuwait and Iraq was quickly established by the combined Gulf forces, and Iraqi installations including power plants, and water supply facilities were attacked. The one nuclear facility which was under construction was a special target. The air war was designed to persuade Iraq to leave Kuwait without the need for a ground war.

In the meantime, Saudi Arabia allowed foreign troops on its soil. Heavy military equipment was brought into main bases in Saudi Arabia, with secondary air bases established in neighboring Gulf countries, including Bahrain and the United Arab Emirates. U.S. General Norman Schwartzkopf was put in charge of operations. The Iraqis, in the meantime, were digging into forward positions along the Kuwait/Saudi Arabian border. Iraq had about 550,000 regular army personnel and about 450,000 in reserve.

With the air strikes continuing but no move by Iraq to withdraw, it became apparent that a ground war would have to be waged. The ground attack began at 01:00 Greenwich Mean Time on February 24, 1991. The 16 inch guns of the U.S. battleships Missouri and Wisconsin fired a massive barrage on the Kuwaiti coast in a pretense that there would be amphibious assault.(7) Troops on the ground moved at 04:00 GMT, and ultimately broke through and also outflanked Iraqi defenses.

The "Mother of all retreats"

Saddam Hussein stated on February 21, 1991, that "The Mother of all Battles will be our battle of victory and martyrdom." It is interesting to note that when the Iraqis had moved into Kuwait, an opinion appeared in the U.S. media about the future: "Experts say dislodging Saddam Hussein's troops will be extremely difficult — if not impossible."(18) On February 27, 1991, Richard Cheney, U.S. Secretary of Defense, stated, "The Iraqi forces are conducting the Mother of all retreats." Truce was declared 05:00 GMT on February 28. The ground war lasted less than 100 hours. So much for experts.

Senseless Environmental Warfare

Saddam Hussein had threatened to set fire to all of Kuwait's wells in retaliation for any invasion by coalition Gulf forces. The war moved so fast, however, that not all the wells could be blown up and then set afire, but a large number of them were torched by the retreating Iraqis. The sky turned black from burning oil wells. At one time about 4.6 million barrels of oil were burning per day.(12) Wells not set afire but only blown up, spewed oil over the desert sands to form large pools, and some 11 million barrels of oil entered the Persian Gulf. It was the biggest oil spill in history. The impact on both bird and marine life was devastating.(10)

In this way, as a final gesture of defiance, Saddam Hussein waged environmental warfare on Kuwait with its own oil. It was an environmental atrocity for which there was neither reason nor excuse.(9) Fortunately, the prediction it might take a year or more to bring the fires under control proved untrue, and the last well was capped on November 8, 1991.(10) The effects of the oil in the Persian Gulf, however, will take decades to be obliterated. The Iraqis discharged massive quantities of oil from Kuwait's Sea Island terminal which is 10 miles offshore.(9) The oil coated shorelines all the way to Bahrain and beyond. Because of the narrow entrance to the Gulf at the Strait of Hormuz, it takes five years to flush contaminated water through the strait to the open ocean.(4) However, some two million barrels of oil were recovered from the Gulf waters and the Tigris and Euphrates rivers which empty into the Gulf are nutrient-rich and will be a positive factor in restoring the vitality of the upper reaches of the Gulf which were the most adversely affected.(5) The lakes of oil from the sabotaged wild wells contained about 200 million barrels of oil. Before soaking into the desert sands, they were as much as eight feet deep. They will leave a visible effect for many years.(6)

Political Alignments Modified

One of the major results of the Gulf War was to change political alignments among the Gulf nations. Previously, Qatar, for example, had opposed allowing any foreign troops on its soil, but with the Kuwait invasion, Qatar allowed American, Canadian, and French aircraft to operate out of its capital city, Doha. This put Qatar definitely on the anti-Iraq side, and weakened the strong support it had previously given the Palestine Liberation Organization which was supporting Iraq.

The Israel Issue, and the War

Prior to the Gulf War, there was a fairly solid front among the Arab nations against Israel. But after Iraq invaded Kuwait and threatened Saudi Arabia, things changed. Arafat, chairman of the Palestinian Liberation Organization, threw his support to Iraq, as did Jordan, long an adversary of Israel. In an attempt to enlist Arab support, Saddam Hussein indicated that he was also considering destroying Israel. This idea appealed to both Jordan and the Palestinians, or at least to their heads of state. It was questionable as to whether or not this was the unanimous view of the Palestinians or Jordanians, but it was the official view of these governments. Thus the Palestinians and Jordan were on the side opposite Kuwait, Saudi Arabia, Bahrain, the United Arab Emirates, and Oman.

The Arab oil world had been split.

Iraq attacked Israel with scud missiles, which meant, in effect, that Israel was also placed on the side of the Arab coalition against Iraq. Ultimately, Syria, long in conflict with Israel, threw its support to Kuwait, Saudi Arabia, and the united military Gulf forces. Thus, Syria, a long-time adversary of Israel, was on the same side with Israel.

Jordan, foe of Israel, was active in getting supplies to Iraq during the confrontation, but when Iraq was so thoroughly defeated and humiliated in the brief military action, Jordan found itself on the wrong side, as did Arafat and the Palestinians who he represented. Before the war, Jordan had been a long-time recipient of aid from the U.S. government, and this aid was temporarily suspended. Jordan's King Hussein speedily made effort to make amends and change sides. Arafat did the same thing, and there had been notable dissent among the Palestinians regarding Arafat's support of Iraq. Some 300,000 Palestinians had been employed in various capacities in Kuwait, and much of their earnings had been sent back to the Palestine Gaza strip, and were a welcome addition to that economy. With high unemployment in Gaza, the opportunity to work in Kuwait was very important.

Palestinians Lose

Arafat's support for Iraq infuriated the Kuwaitis, with the result that after the war, the Palestinians were generally regarded as part of the Iraq enemy, and were expelled from Kuwait. Many fled to Jordan taking no resources with them. This put an additional strain on Jordan. Some of the Palestinians in Kuwait supported Kuwait, a few supported Iraq, and many were more or less neutral, but the net result was that the Palestinians lost their jobs in Kuwait in the reorganization of alliances. Jordan got back into reasonably good graces of the West, but Kuwait no longer had a place for the Palestinians.

The War and Religion

Religion is closely bound up with the several Persian Gulf governments. The presence of large numbers of foreign troops tended to modify some of the strict Moslem attitudes which had previously prevailed in the Gulf area. Unable to defend themselves against the military machines either of Iran or Iraq, the other Gulf nations sought closer ties with the West, particularly the United States. Military operations were allowed to be conducted from Saudi Arabian bases, and the presence of female military personnel in the area was a first time occurrence. These increasing military and economic ties which these governments now have with the West tend to strain the religious environment. Some Muslim leaders do not favor close ties with the West, whereas others are more moderate in their views. The result has been dissention among various religious factions. These tensions have the possibility of causing internal strife, and how each nation can deal with these problems will have a considerable bearing on the future of the country. So far the need for both economic ties and military security has prevailed over isolationist religious views. Subsequent to the Gulf War, the United States was allowed to station 20,000 troops in Saudi Arabia.

Religious factions inside Kuwait

Another result of the Iraq invasion of Kuwait and subsequent atrocities committed on the population, was the healing of the division between the Sunni and Shiite Muslims. Iraq was not concerned if the people it mutilated were Shiites or Sunnis, they were all Kuwaitis. The result was that the Shiites and Sunnis banded together in a common cause to help each other to survive.

Alignment of the West

Whereas the Gulf War served to align the Middle East nations (except Iraq and Iran) more firmly with the industrialized West, the war also emphasized to the general population of western nations the importance of Middle Eastern oil to their economies.(19) This was the first time the West had taken combined severe military action to protect access to oil supplies. It set a precedent for a strong potential, if not actual, military presence in the Persian Gulf region for as long as the oil remains. This military concern will only increase as more and more of the world's exportable oil will have to come from this area. This precedent for combined military action by the West to protect and maintain these oil supplies will shape the political future of this region for decades to come.

Western Military Superiority

The long term power position in the Persian Gulf region was impressively defined by the Gulf War. Iraq was armed chiefly with Soviet military equipment which came face to face with western machines and technology. The superiority of the West's armaments, from the combination of infrared television and laser guidance beams to direct "smart bombs," to the pinpoint accuracy of Tomahawk missiles fired from U.S. warships over great distances, was clear for all to see. The subsequent disintegration of the Soviet Union further enhanced the position of the United States and its western allies in the Gulf region. The previous influence in the region by the former Soviet Union, and what is now Russia, has been blunted. However, Russia, now past its peak in oil production and barely able to supply its own needs, will no doubt maintain a strong interest in the Gulf area.

Oil, through the Gulf War, has markedly changed the balance of power and political alignments in the Middle East. The continued presence of military forces in that region will also inevitably affect the culture.

BIBLIOGRAPHY

1 AMOS, DEBORAH, 1992, Lines in the Sand. Desert Storm and the Remaking of the Arab World: Simon & Schuster, New York, 223 p.

2 ANONYMOUS, 1991, War and Change in the Oil World: Oil and Gas Journal, January 21, p. 13.

3 BENNIS, PHYLLIS, and MOUSHABECK, MICHAEL, (eds.), 1991, Beyond the Storm. A Gulf War Reader: Olive Branch Press, New York, 412 p.

4 CANBY, T. Y., 1991, After the Storm: National Geographic, Washington, D. C., August, p. 2-33.

5 EARLE, S. A., 1992, Persian Gulf Pollution. Assessing the Damage One Year Later: National Geographic, February, p. 122-134.

6 EL-BAZ, FAROUK, 1992, The War for Oil: Effects on Land, Air, and Sea: Geotimes, May, p. 13-15.

7 FRIEDMAN, NORMAN, 1991, Desert Victory. The War for Kuwait: Naval Institute Press, Annapolis, Maryland, 453 p.

8 GEYER, ALAN, and GREEN, B. C., 1992, Lines in the Sand: Justice and the Gulf War: Westminister/John Knox Press, Louisville, Kentucky, 187 p.

9 GOLUB, R. S., 1991, "It Looks Like the End of the World:" Harvard Magazine, September/October, p. 29-33.

10 HAWLEY, T. M., 1992, Against the fires of Hell. The Environmental Disaster of the Gulf War: Harcourt Brace Jovanovich, Publishers, New York, 208 p.

11 HIRO, DILIP, 1992, Desert Shield to Desert Storm: Routledge, New York, 591 p.

12 HOBBS, P. V., and RADKE, L. F., 1992, Airborne Studies of the Smoke From the Kuwait Oil Fires: Science, v. 256, May 15, p. 987-991.

13 KHAVARI, F. A., 1990, Oil and Islam. The Ticking Bomb: Roundtable Publishing, Inc., Malibu, California, 277 p.

14 LANE CHARLES, et al., 1990, Requiem for an Oil Kingdom. Kuwait's Wealth Made it a Target of Resentment: Newsweek, August 13.

15 MATTHEWS, TOM, 1991, The Road to War: *Newsweek,* January 28, p. 54-65.

16 MAZARR, M. J., et al., 1993, Desert Storm. The Gulf War and What We Learned: Westview Press, Boulder, Colorado, 207 p.

17 METZ, H. C., (ed.), 1994, Persian Gulf States. Country Studies: The Library of Congress, Washington D. C., 472 p.

18 Portland *Oregonian,* August 3, 1990.

19 RIVA, J. P., Jr., 1991, Persian Gulf Oil: Its critical Importance to World Oil Supplies: The Library of Congress, Congressional Research Service, March 5, 15 p.

20 SAMUELSON, R. J., 1991, Saddam is No Profit Maximizer. The Fight is Over Oil's Power, and not its Price: Newsweek, January 21, p. 36.

21 WATSON, RUSSELL, et al., 1990, Baghdad's Bully: *Newsweek,* August 13.

CHAPTER 5

The Current War Between The States

One of the decisive advantages which the North held over the South in the Civil War was in the possession of a steel industry based on the abundant coal and some small but fairly rich iron ore deposits in the region of Pennsylvania. The South had no significant steel industry and one of the results was that all their railroad equipment had been manufactured in the North. The South had to go the length of the war with only the rail equipment it had when the war started. Replacement rails and other railroad supplies were not available to repair the worn out and war damaged rail system which was important in transporting the materials of war. The South's industry chiefly related to cotton and other agricultural products with little development of minerals. In his farewell address to his troops at the surrender site of Appomattox, General Lee said, "The Army of Northern Virginia has been compelled to yield to overwhelming numbers and resources."

The Economic Wars

Although no subsequent shooting wars have broken out between the states, the unequal distribution of minerals and particularly oil, combined with the great rise in oil prices during the 1970s, has caused some hard feelings. The states, in the United States, and also the provinces in Canada have been sharply divided between what has been called the "energy-producing areas, and everybody else."

The regional tensions and contentions arise chiefly because of revenues which come to a state or provincial treasury from royalties, and various kinds of taxes including severance taxes on minerals. Severance taxes are taxes levied on the minerals and timber when taken from the land. If a state or province exports a substantial part of the oil, gas, coal, or other minerals, then the out-of-state or out-of-province consumer pays a large part of the government revenues derived from that resource.

The Province of Alberta produces about 85 percent of Canada's oil, a much greater percentage than the two largest oil-producing states in the United States, Alaska and Texas. As a result, Alberta gets a very large amount of oil revenue compared with the rest of Canada, from this source. Much of that revenue is paid for by oil-less provinces such as Quebec. Recently the annual per capita income in Alberta from energy production alone was approximately $1,000, while in oil-less Quebec it was $18 (from export of provincially owned hydroelectric power). Quebec, however, has a large population and many votes. The result has been some hard fought inter-provincial political battles.

The Canadian national government also levies high taxes on energy production, so Alberta also pays a large amount of tax on its petroleum and coal output. A headline in the Calgary, Alberta *Herald* read "Alberta pays biggest tab as Quebec benefits."(2) The article noted that " Quebec has received $168 billion more in federal spending than it contributed in tax and other revenues between 1961 and 1992." It goes on: "Meanwhile, Albertans have paid $139 billion more to Ottawa than they got back over the same period, mainly because of high energy taxes imposed under the National Energy Program in the mid 1970s. Quebec has been the largest net beneficiary [among the provinces] in absolute terms over a 32-year period. Alberta, on the other hand, has been the single largest contributor."(2) Obviously, there are strong feelings in Alberta over the exploitation of their energy resources and the distribution of the benefits. It was the geodestiny of Alberta to possess these resources which the other provinces and the national government now want to share.

Severance taxes

Severance taxes levied by the states are usually calculated as a percentage of the value of the product. Increasingly, energy exporters are putting severance taxes on their resources. It is the consumers in energy-poor areas who have to pay this bill. The result is that the energy-rich areas export a substantial part of their tax burden to people in other regions. A study showed that the cost of state and local taxes for a family of four with an assumed annual income of $17,500, ranged from 4.2% in Louisiana and Oklahoma, and 4.5% in Texas, to Michigan where the family paid 9.6% and to New York where the family paid 11.5%.

Geological destiny

Citizens of these various areas had no merit or lack of merit of their own to account for the inequalities in distribution of oil, coal and other Earth resources. It is entirely the result of their geological location — their geodestiny. Why did Louisiana have such a low tax burden? Because of petroleum. For example, in the fiscal year 1980-1981, when oil prices were relatively high, Louisiana took in $1.19 billion in oil and gas lease charges, royalties, and severance taxes. Because most of the petroleum produced was exported to other states, most of these charges went right along with the product to the out-of-state consumers. During that same time Maine, Vermont, Massachusetts, Florida, Wisconsin, Minnesota, and Idaho, fossil fuel energy "have-nots," didn't get a dime of income from oil and gas royalties or severance taxes. They just helped pay the bills for Louisiana, Texas, and Alaska among others of the more fortunate "energy have" states.

Oil import tax

There has been discussion about imposing an oil import tax as a way to obtain revenue for the U.S. government and to reduce demand for oil. This issue also divides the states. Texas would be glad to see such a tax as it would raise the price of the oil they produce. On the other hand, New England, which produces no oil and is especially dependent on oil for winter heating, is totally opposed to such tax.

School taxes

As homeowners know, taxes to support the local schools are a large part of annual real estate tax bills. In the early 1920s the State of Texas set aside one section (square mile) out of each township (36 square miles) to be designated as "school lands," and the income, whatever it might be from such lands, would go to schools. Eventually, many of these lands proved to have petroleum beneath them. Because much of this oil and gas is sold to residents of oil-less Minnesota and other energy-short states, they support the Texas school system.

The citizens of Texas would surely not look favorably on paying for the education of the school children in Minnesota. The fact is, however, that the residents of Minnesota help pay for the education of Texas children. This was determined between about 450 million years ago and about 20 million years ago. Over that long interval of geologic time Minnesota was only briefly invaded by very shallow seas which left virtually no appreciable thickness of sediments in which oil might be generated. In fact, part of that time Minnesota was busy shedding sediments southward toward the Gulf Coast. These are now part of the great thickness of sediments in the Gulf Coast which contain the oil of that region. So, from the upper interior part of the continent came the sediments which were deposited over much of South Texas and Louisiana, and on into the Gulf of Mexico. In these sediments organic material also was deposited that ultimately became oil and gas. From these petroliferous rocks today's citizens of these fortunate areas have received a very nice financial reward paid in part by the residents of less fortunate states to whom the energy resources are sold.

Oil price rises — popular in some places

The State of Texas is really not all that unhappy about higher and higher oil prices. These high prices not only provide good royalty and severance tax income for the state, but, of course, high oil prices promote the oil industry which provides good jobs. Oil-less states, of course, would like to see low oil prices. Historically, Texas has produced about one-third of the oil in the United States. Royalties and severance taxes from that production have supported a friendly and comfortable economic environment over the years which included no corporate or personal income taxes. The absence of these taxes also tends to encourage industry to move in, producing even more jobs and wealth, so the geologic rich get richer and the poor get poorer.

With the great price rise of oil in the 1970s, from around $3 a barrel to more than $35, there was nothing but joy in Houston and Midland, Texas, Tulsa, and Denver. The so-called energy crisis created coal boom towns in Wyoming and Montana, and brought prosperity to many relatively isolated and previously poor communities in North Dakota which were fortunate enough to be located in the oil-producing Williston Basin region. In 1981, with oil prices at more than $30 a barrel, Oklahoma enjoyed oil revenues of $100 million more than expected. Total severance taxes on coal, oil, and gas reached $559 million. New Mexico, that year, passed the largest tax cut in its history and used taxes on coal, oil, and gas to slash income and property taxes. North Dakota used its taxes on oil and gas to take over funding of the schools from the local communities, markedly reducing property taxes. All these benefits were partly paid for by consumers in the energy-poor regions of the country. This represented and continues to represents a very considerable transfer of wealth, very much akin to how OPEC nations get rich while the oil-less or oil-poor countries get poorer.

OPEC's effect

The benefits of the "energy-haves" had actually gone largely unnoticed by the general U.S. public until OPEC raised oil prices very high, with corresponding rises in domestic oil prices. With Texas and other energy-producing states levying taxes as a percentage of the price of the oil, gas or coal, the treasuries of these states began to overflow with money. Meanwhile the energy bills of the other states got higher and higher. From time to time the federal government has tried to limit the amount of severance taxes which could be levied by a given state so as to more fairly distribute the wealth. But, so far the energy-rich states have wrapped themselves in the cloak of "states rights" and each state has remained its own economic kingdom in terms of mineral taxes.

Regional attitudes

Various evidences of this "beggar thy less energy-fortunate neighbor" attitude appear including such things as bumper stickers. On one Texas pick-up truck I noted the statement "LET THE BASTARDS FREEZE IN THE DARK." Another favorite – of some Texans, at least – was "DRIVE 90 AND FREEZE A YANKEE," referring to the fact the reduced highway speeds for cars would save fuel, and higher speeds would use more. But when the price of oil collapsed in the mid-1980s, the Yankees had long memories. In Boston, Milwaukee, St. Louis, and Seattle there wasn't a lot of sympathy for Texas, with a corresponding abundance of Texas oilman jokes. One asked: "How do you contact an oilman in Houston?" Answer: "Call a cab." This joke referring to the fact that some oilmen who did go broke were actually driving taxis. These attitudes are unfortunate, because the United States is a single economy, and all citizens eventually directly or indirectly suffer the hardships of any given area.

The oil price drop in the 1980s did hurt Texas to a considerable extent. In 1987, Texas passed the largest tax increase, mostly in the sales tax, that the state had ever enacted. But even then there was still no personal nor corporate income tax and the average annual tax load per $1,000 in personal income for Texans was $53.76 whereas the average for the United States as a whole was $74.11. Oil revenues still pay a lot of the Texan's taxes, and as most of the oil is shipped out of Texas it is the citizens of the other states who continue to pay those taxes.

Each state unto itself

Under present U.S. laws, each state, in effect, is a nation unto itself, with a balance of payments relationship with every other state. In terms of energy, there are currently 11 states of the United States which are energy self-sufficient and are energy exporters. These are Alaska, Kansas, Kentucky, Louisiana, Montana, New Mexico, North Dakota, Oklahoma, Texas, West Virginia, and Wyoming.

Alaska's natural endowment

There are other examples of how geological good fortune makes for vicious economic inequities. Alaska, because of the Prudhoe Bay oilfield, ultimately had so much oil revenue that in 1981, it repealed its state income tax and did so retroactively back to 1979. Oil revenue for the State of Alaska in 1981 was about $10,000 per citizen. It was then decided that starting in 1982 the state would send each Alaskan citizen — man, woman and child — a check each year instead of having Alaskans send the state any income tax. This oil dividend has been distributed at the rate of about $1,000 a year per person.

In my university geology class the next year, I asked if any student was from Alaska. One student was, and I asked if he had received a check for $1,000. He said he had. Then I then asked who paid for that, whereupon he rather sheepishly looked over the rest of the class and said "you did." The rest of the class was less than happy, because this was at the University of Oregon, and in oil-less Oregon that year the legislature had to meet in special session to raise taxes to meet revenue needs, in part to pay rising government costs due to the higher price of oil. Alaska, however, had simply been raising their taxes not on the citizens, but only on the oil. Since 1955, such taxes had been increased eleven times, first on the oil discovered in the Cook Inlet, and then also on the North Slope oil. Today, Alaska's tax rates on the oil industry are the highest in the United States.(5) In this way, Alaska is able to export its tax burden to the other states, chiefly those on the West coast of the U.S. where most of the Alaskan oil is used.

Alaskans pay lower taxes per capita than do citizens of any other states and the District of Columbia. If there is a drop in revenues, as there was in 1989, due to declining production and weak oil prices, Alaska has simply raised the oil industry taxes, as it did that year moving the oil severance tax to 15 percent.(5) In the case of property taxes, after the opening of the Alaskan pipeline real estate taxes in all the Boroughs (same as counties) were lowered 30 to 40 percent everywhere except in the North Slope Borough formed at the time of the oil discovery. Here, property taxes were raised 50 percent because almost all property taxes are paid by the oil companies which have their installations in the Prudhoe Bay area.(3)

At the terminal end of the Alaska pipeline, the City of Valdez gets the final opportunity to tax Alaskan oil. About 17 miles out of Valdez on the road in the valley through which the oil pipeline goes, and definitely out in the Alaskan woods, one is surprised to see a sign which reads "Valdez City Limits." The pipeline, even though buried in the ground near here, is taxable property. The City of Valdez has extended its city limits rather considerably to take advantage of that fact, and by this means ship out a part of its tax burden with the oil as it leaves the Valdez terminal (which facility is also heavily taxed).

Montana coal

"We in Montana, are wondering why our coal should light the streets of Seattle and Portland," so said the governor of Montana. But Montanans found a good reason "why" by imposing a 30 percent severance tax on Montana coal. That is, a 30 percent tax was added to the price of each ton of coal mined in Montana. Montana has a lot of coal. It does not have many residents and it does not have a state income tax. The people who pay for the city lights of Seattle and Portland as well as their own personal utility bills are helping, in reality, to pay the taxes for the citizens of Montana. And, this will continue for quite awhile, for Montana has very large coal reserves. There are no additional hydropower plant sites in the Pacific Northwest, and the Columbia River Bonneville Power System does not have more water that could be harnessed. More and more Montana coal will be needed in the region to supply the coal-fired plants which may have to be built.

U.S. coal "OPEC"

Montana and Wyoming are potentially the nation's OPEC of coal. Within these two states lies more energy in the form of coal than exists in all the oil of Saudi Arabia. It is energy not in such a convenient form as is oil, however, but sooner or later it will be needed, and, like Saudi Arabia, Montana and Wyoming will rise in relative economic importance. Both Montana and Wyoming are already doing quite well from their mineral resources although with the rather severe price swings in coal and oil, the picture changes from time to time. But with both states having small populations relative to the size of the energy mineral resources which are being produced, the positive economic effect of these resources on each citizen is quite large. As the demand for energy increases, Wyoming and Montana will benefit correspondingly.

Regional resentments

Midwestern utilities which must import this coal have attempted to have the high coal severance taxes reduced. Commonwealth Edison in Chicago estimated that the Montana tax cost its customers annually more than $14 million. With Montana's tax so high, other areas which had to buy Montana coal also complained loudly that Montana was getting rich at the expense of everyone else. Because the complaints became so severe, Montana finally did reduce its severance tax on coal and at present it ranges from 10 to 15 percent, depending on the quality of the coal.(4)

The midwest Governors Conference estimated that residents of its 13 member states in 1979 paid $700 million in severance taxes to other states, and that before the end of the century the cost could be as much as $2.5 billion. Because this increases the costs for all industry in these energy-poor states, it makes them less competitive in attracting new business than the energy-rich states.

The debate between the "have" and "have not" energy states can become quite heated at times. Congressman Sharp of Indiana stated "At some point I am convinced that the Congress will have to address the political, social, and economic issues raised by the specter of a nation balkanized by inter-regional rivalry over energy and natural resource use and production." Congressman Tauke of Iowa noted that Iowa was paying more than $11 million a year to other states in severance taxes, and stated that "Some states have now raised their severance taxes on coal to a level that cannot be justified by any reference to the costs that mining imposes on the state. Energy poor states cannot be expected to stand idly by while other states reap windfalls in state revenues — especially since these windfalls may be used to lure jobs from other states."

An editorial in *Business Week* magazine stated "Congress should certainly pass the bill limiting the tax. If anything 12.5 percent is too high a tax that is levied on consumers all over the country and contributes directly to inflation by raising costs. The Western states are abusing an ancient right and, in effect, taxing the citizens of other states."

"They are also gaining unfair advantage in the continuing contest to attract new industry and jobs. States such as Montana are putting the severance revenue into a trust fund but others are cutting their regular taxes and promoting themselves to business as super-low-cost areas."

"Montana and Wyoming control 70% of the nation's desirable low-sulfur coal. This country does not need a little OPEC astride the Rockies."(1)

Montana trust fund

Recognizing that the coal is a depleting resource, Montana has set up a permanent trust fund, to try to alleviate future problems caused by diminishing revenues from coal. One half of the severance taxes are put into this fund annually. The other half is distributed for current expenses of highways, schools, and various community projects.

Texas revenues

Severance taxes on oil and gas in Texas are considerably lower than either Montana or Wyoming coal severance taxes, but Texas has had such large petroleum production that severance taxes have produced as much as $3 billion annually. Currently, seven states receive more than 20 percent of their revenues from severance taxes levied on the production of minerals, chiefly, coal, oil, and gas.

Whatever the merits of the argument, it is again an example of how the economic destinies of regions, states, or nations caused by the geologic distribution of mineral resources can set one area against another.

Dividing the states: Coast or no Coast

The oil in state-owned offshore waters importantly divides the states into those with coastlines and those without. Beyond the limits of state jurisdiction over adjacent ocean areas (usually out to three miles), the Federal Government has ownership. Offshore oil obliterates political party lines. Where there is offshore oil money lying around, there are no Democrats or Republicans. It becomes a matter of region against region. In one example, in 1985, it was a question of long-standing as to precisely how much money the Federal Government would get from offshore oil, and how much the adjacent state would get. The amount which had accumulated was about $6 billion — a large enough sum to arouse some very strong feelings.

The states with no oily coastal areas, such as Arizona and Wisconsin, were all for giving most of the money to the Federal Government. The states with oily coasts such as Louisiana and Texas wanted most of the money given to them. The fight was based strictly on geography, not on party lines. The states without offshore oil thought, rightly or wrongly, that the other more fortunate states were making out like bandits on royalties. The House Interior Committee reviewed the bill. The Chairman was Morris Udall of Arizona (no coastline at all, much less an oily one) who made an impassioned speech against the bill. Committee members from Texas and Louisiana took the floor to say what a fine piece of legislation it was. The ultimate bill which was passed gave 27 percent of offshore royalties to the states and 73 percent to the Federal Government.

This law still did not give any federal offshore oil money to inland states. In 1988, however, the states with no coastlines figured out a way to get at least some money from federal offshore petroleum lands. The Supreme Court upheld a law which decreed that the individual states could levy a tax on that portion of an oil company's income which came from business within that state, and the oil company could not exclude from its taxable income the income it got from federal offshore oil production. Thus, as a result of this skirmishing among the states, the inland states finally got some offshore oil money, and oil money was what the warfare was all about.

Canada — Alberta

In the case of Canada, the gods of energy clearly have favored Alberta above all other provinces. Alberta is now the chief conventional oil producer, but also is already producing some 270,000 barrels of oil daily by mining the Athabasca oilsands. It also has very large natural gas reserves. Also, as conventional oil and gas production declines, the tremendous Athabasca deposits will gradually come more and more on line, as will the more than 25 billion barrels of oil in the form of heavy oil in the Cold Lake region of eastern Alberta. Alberta also has considerable coal and will remain the Texas (or Saudi Arabia, if you choose) of Canada for a long time to come. Populous Quebec, with no petroleum resources whatsoever, will fight for lower natural gas prices while Alberta will want higher prices, and the national Canadian government in Ottawa will continue to try to referee the argument. If Quebec chooses to separate from Canada, it will have to make special arrangements to continue to obtain Canadian oil and gas.

An Ecological Price

But there are some balancing factors in this war between the "haves" and "have nots" on the benefits of possession of energy and energy mineral resources. Exploiting these resources has some liabilities, mostly environmental. A state such as Louisiana may reap large economic rewards from its petroleum resources for a time, but it also pays a price. The marsh areas where oil drilling has taken place have been invaded in many places by salt water, reducing or destroying the once thriving fisheries of various sorts.

The Future

What will be the future of these economic differences among the states and among the provinces due to the unequal distribution of energy and mineral resources? In the case of petroleum — oil and gas — those areas which have chiefly oil will be depleted first, and those with larger gas reserves will do well for some time longer, as gas reserves appear to be potentially longer-lived in the United States, at least, than do the oil reserves. If the price of oil rises high enough, California may do somewhat better for a time, as there are very large deposits of heavy oil in that state — oil which cannot now be produced by conventional means and takes special and relatively expensive methods to extract. But California's days of oil production are numbered.

Ultimately, as petroleum reserves are exhausted, other energy resources will have to take over, and the once important petroleum producing areas will no longer be a factor on the energy scene. Other energy sources will be more significant, and such economic warfare based on possession of energy and mineral resources which may be practiced will shift to new regions.

The changing scene

In the United States, Texas and Alaska may eventually fall behind Montana and Wyoming as energy producers when oil production declines in the United States, and coal rises in importance as an energy source.

Texas' oil reserves have been declining rather rapidly in recent years, going from 14.8 billion barrels in 1960 to less than seven billion currently, and they are still declining. This trend is not likely to change, for Texas clearly has passed its peak as an oil producer. In the case of the dominating Prudhoe Bay Field in Alaska, half the oil has already been produced. Still, Texas and Alaska will be producing some oil for a long time to come, whereas Maine, Massachusetts, Minnesota, and others will never produce any oil, and will have to continue to pay tribute to Texas and Alaska.

Much as the "have not" areas would like to change the situation, it is probable that legislation can never economically equalize the "have" and "have not" areas of mineral resources which were established by geological processes millions of years ago. Some areas will always be poorer than others in terms of minerals and energy minerals, but the relationships will change as one energy or mineral resource gives way to another in a different area. The Mesabi Iron Range of Minnesota had its day. Ore revenues which that area collected were paid by all the steel users in the United States and wherever else the iron end product was shipped. This revenue paid for many improvements in the Iron Range giving it, for a time at least, a considerable economic advantage over the rest of the State of Minnesota. Other areas of the state were envious of the "rich range towns".

But the envy gradually subsided as the higher grade ores ran out and affluence changed to an economic decline. There was a war, economically, and the Minnesota iron range partially lost it when the high grade hematite ore was depleted, leaving only lower grade taconite. Other places had better ore. This change was inevitable as it is with all mining areas, for sooner or later the ore deposits are depleted.

The same was true of the copper range of Michigan on the Keweenaw Peninsula. This area lost the copper war to Utah and Arizona, which had greater reserves. But, in turn, in 1995, Utah and Arizona were getting very severe competition from Chile, which, with much richer deposits and much lower labor costs, could produce copper for half the cost in the United States. Eventually there may be no copper produced in any state.

Inequities Inevitable

We try to live in harmony with one another, but beneath the very thin veneer which we call civilization there are basic needs of society which will be supplied by whomever has the resources, and more particularly the lowest cost resources. Frictions and economic inequalities arise between those areas which have the resources and those who do not. In the case of the United States and Canada, unequal distribution of energy and mineral resources are destined to continue to cause regional envy and dissention within these countries. This neighborhood economic warfare will probably persist in one form or another for as long as the resources last.

BIBLIOGRAPHY

1 ANONYMOUS, 1981, Stopping a Tax Grab: Business Week, February 16, p. 116.

2 ANONYMOUS, 1995, Alberta Pays Biggest Tab as Quebec Benefits: Calgary (Alberta) Herald, March 28.

3 JACKSTADT, S. L., and LEE, D. R., 1995, Economic Sustainability: The Sad Case of Alaska: Society, March/April, p. 50-54.

4 STATE OF MONTANA, 1995, Department of Revenue, Natural Resource and Corporation Tax Division, Helena, Montana.

5 WILLIAMS, BOB, 1989, Alaska Tax Hikes Cloud Latest Giant's Prospects: Oil & Gas Journal, August 14, p. 26.

CHAPTER 6

Mineral Microcosms

Mineral and energy mineral resources are brought to us from many and varied places although we may be a long distance from sources of these materials. What happens to these places when the resource is exhausted may be of no concern to us. Supplies will be obtained elsewhere. But, for the community or region which produces a given resource from which they derive a substantial portion of their income, the exhaustion of the resource is a very serious — sometimes devastating — situation.

Western American mining towns

One can visualize the ultimate problem of a country or a region with essentially just one resource on which to base its livelihood by examining small scale examples. The American West is dotted with the names of mining towns which no longer exist. If the community could not find another basis for making a living beyond the original mineral economy on which it was established, it died.

The list of once thriving mining towns which are now either gone entirely or exist as mere shadows of their former selves, is very long. If they now have residents they may be a few people who have retired to the solitude of the areas. On the roll call are Tuscarora, Midas, Berlin, and Unionville, Nevada; Bode, Cerro Gordo, and Ballarat, California; Miner's Delight, Wyoming; Mogollon, New Mexico; Leesburg, Idaho; Galena, and Golden, Oregon; and Granite, Montana. The list could go on for hundreds of other names. In some of these towns the population was once as high as 20,000. Hamilton, Nevada, once had such a population, but in 1946, by my count on site, there was just one coyote. Hamilton today has regained a few hardy souls, as some mining has been locally revived.

Tourist attractions and gambling

A few towns have survived as tourist attractions when they are conveniently located near large population centers with more permanent economic bases. Virginia City, Nevada, is an example. Established by the discovery of the rich Comstock silver lode, it now hosts curious visitors from nearby Reno.(12) Leadville, Colorado, is now attempting to build up the tourist trade, and has tried also to bring in some new businesses. But at an elevation of about two miles, the winters are long and the scene is a bit barren being just below timberline. This does not attract a lot of business. Proximity to an interstate highway helps somewhat, but the glory days of Leadville disappeared along with the ore deposits.

Other old, and once virtually abandoned mining towns in Colorado, however, have found a new life as gambling centers. When silver and gold were discovered in 1892, in the walls of what was once a great volcano now eroded into a broad basin, Cripple Creek had a population of only a few hundred. Two years later it was nearly 13,000. By 1990, the silver and gold was nearly gone and the population had dwindled to fewer than 600. Today, Cripple Creek has come back as a gambling mecca, where thousands of tourists clog the streets on weekends. Central City, Colorado, has had a similar history. Residents of greater Denver, a short distance to the east, flock to this old mining town to try their luck at the gambling tables, not the gold diggings. In both of these communities, some owners of either worked out or initially worthless mining claims, are now striking it rich through much less effort than mining by renting parking space which is very much at a premium particularly in the narrow gulch which is Central City.

The Keweenaw Peninsula of Michigan was the site of the richest copper deposits ever found in the United States, and in the 1920s this region produced 80 percent of the world's copper. But most mines were closed by the 1970s. One survived until 1995. A few small communities have remained but with difficulty, and have only modest, seasonal economies. Copper Harbor is one of these. Near the tip of the Peninsula jutting out into Lake Superior, it is picturesque and an interesting place to visit but a difficult area in which to make a living. The rocky land does not lend itself to agriculture and eight to 10 feet and more of snow makes for a hard winter. In this depleted copper mining region, the U.S. Congress created Keweenaw National Historical Park in 1992. A 1995 newspaper report stated, "National Park Service officials say that the 1,870 acre Keewenaw, the site of a copper lode mined from the 1840s to the 1960s, can teach visitors about industrial history, corporate paternalism, architecture and mineral science. But critics contend that the true purpose of the park, championed by Democratic Sen. Carl Levin and former Republican Rep. Bob Davis, is to prop up the economy of a remote area on Michigan's Upper Peninsula by attracting visitors and creating jobs."(8) To the park visitors, the reality of non-renewable resources and how the destiny of this community was controlled by minerals will be very evident.

Across Lake Superior, in the iron range country of northeastern Minnesota, the once rich mining towns of Hibbing, Virginia, and Eveleth, and the ore shipping town of Silver Bay have suffered large economic slides. At one time in Hibbing the economic benefits of being the biggest iron ore mining operations in the United States meant they had money for all sorts of projects. One was a "glass schoolhouse" made of glass bricks. That building and the lavish high school with big beautiful chandeliers hanging in the auditorium, along with velvet covered seats, were the envy of the rest of the State of Minnesota.

But the high grade hematite is gone and iron mining employment is only about half of what it once was. Two World Wars were fought with iron ore from the great Hull-Rust mine at the north edge of Hibbing. It is an ironic twist to the story of Hibbing that with the decline in iron mining, some of the residents formed a company to make chopsticks for the Japanese. The first shipment was sent in 1987. Post-war Japan has made tremendous inroads on the U.S. steel industry, exporting a lot of cheap steel to the U.S. shores, but Japan now gets its ore chiefly from Australia and Brazil. Hibbing, once the iron mining capital of the world, still mines some iron but also began making chopsticks for the Japanese. The promising aspect of the Japanese chopstick venture was that chopsticks are used only once, and with more than 100 million Japanese, market demand would be steady and perpetual. The chopsticks were made from aspen trees which grow well around Hibbing and represent a renewable natural resource, unlike the depletable iron ore.

Butte, Montana, is the site of what is said to have been the "Richest Hill on Earth" according to the sign at the city limits. It may once have been, but the copper ore is largely depleted. The giant Berkeley pit is now half filled with water. From time to time efforts are made to bring other nearby smaller mines back into production, but that production is limited and the smelter at Anaconda, a few miles to the west, stands abandoned. It is doubtful that it will ever smelt another pound of copper, and the Anaconda Copper Company no longer exists.

The same general scenario is apparent in the Coeur d'Alene mining district of northern Idaho. Once the site of the largest zinc mine in the United States, the Bunker Hill and its huge smelter, the valley now survives largely by tourism, some modest mining, some logging, and a little ranching. It is an interesting place to visit to see numerous waste dumps of mines now long abandoned, and see how ingenious people built houses precariously perched on the hillsides of the steep gulches of the area. Perhaps the mines will some day come back —some of them. A few still operate. Here the largest silver mine in the United States, the Sunshine Mine, opens and closes with the rise and fall of the silver market, but the silver mining boom days are quite likely gone forever.

Ajo, Arizona was chiefly a company town, and Phelps Dodge operated the large copper mine there. But, it too, has seen its day, and that day has departed as has Phelps Dodge. The huge abandoned open pit mine remains. The company-built houses are now private dwellings for retired persons who like the sunny, dry climate.

Isolated mineral-based communities of today

One may speculate as to what may happen to some present mining communities which are in terrains or climates, or both, not well suited to agriculture or other enterprises after mining. In Canada, a great belt of nickel deposits is the reason and support for the pleasant but isolated town of Thompson, Manitoba.(4) Farther to the northwest, the towns of Leaf Rapids and Lynn Lake exist because of mines. For now, Leaf Rapids is doing well, but Lynn Lake has seen the decline of mining. In 1994, all the store windows in the downtown area had been smashed and were replaced by plywood. Unemployment and crime were rising. In this area of lakes, swamps, and granite bedrock, an agricultural economy is not possible. Trapping and fishing are only minor sources of income. Lynn Lake is declining like other mining towns when the minerals are gone. When this happens social problems tend to arise among those who cannot readily move and adapt to new and different work and economics.

Farther east in Canada, Timmins, Ontario was once the site of the world's largest gold mine, the Hollinger. The mine is now a tourist attraction offering tours. However, Timmins received a new lease on life when the great Kidd Creek silver, lead, and zinc deposit was discovered west of town. Found by Texas Gulf Sulfur, it is now owned by Noranda Corporation. It is a remarkable deposit, a volcanic plug only a few hundred yards across but containing several billion dollars worth of metals. It illustrates that very large mineral deposits can occupy only a small area, but as the deposits must be mined where they are, they must be accommodated in the environment. However, the environmental impact is small compared to what can be obtained for the general welfare of the economy.

As an interesting sidelight, Timmins is located among a number of mines, and over the years, in order to obtain city revenue, the city limits have been extended to include these properties. As a result, Timmins is reportedly now the largest city in the world in area relative to the size of its population. "Timmins City Limits" are now located way out in moose pasture. Minerals make Timmins possible at present.

Petroleum

Oil and gas tend to be a longer producing resource than are the many small mineral deposits on which the now abandoned western mining towns were built. But oil is also finite. The question increasingly being asked by many smaller communities and even some larger areas is "what do you do when the well runs dry?" Such is the question being asked in Van, Texas. Starting about 1930, when oil was first discovered under the town, more than three-quarters of the city expenses have been paid from oil-related revenues. The income came from about 370 wells pumping at various places in the community, including several in the yard at the Van High School.

But the edge wells of the Van Field are now drying up (more accurately, going to water). The oil revenues are shrinking and will ultimately stop. Who or what pays the approximately 80 percent of city costs then?

In contrast to Van, Giddings, Texas is an example of a more recent oil boom town. In 1976, only five wells were drilled in the vicinity, whereas in 1981, 945 were drilled. The housing shortage in Giddings was so acute as workers flocked in, that one enterprising citizen tried to open a motel using old oil storage tanks as bedrooms. The Giddings population abruptly rose from 3,900 to more than 8,000. More than 110 ribbon cuttings marked the opening of new businesses. In June, 1980, an "Oil Appreciation Week" was held with the theme "Praise God from Whom all oil doth flow." The police force was increased from 4 to 17, and school taxes were cut more than 50 percent.

An acre of farmland at the edge of town which may have earlier produced $200 worth of peanuts per year might bring annually a thousand dollars for a drilling lease. Thirty new millionaires were created as a result. While Van, Texas watched its oil production decline and the town slumping accordingly, Giddings, Texas boomed — for a time. The oil price drop which began in 1982, burst the Giddings oil balloon. Workers were laid off, builders found themselves suddenly with new houses but no buyers. Each of these mineral microcosms had their day. First Van, Texas, and then Giddings.

By definition, microcosms are small and localized. They also tend to have a rather ephemeral existence. A local small but rich mineral deposit is the basis for the boom and also sets the stage for the subsequent "bust." A brief mineral bonanza is the geodestiny of the community.

We shall here go on to stretch the definition of a microcosm to include a territory as large as a state or province and look at Louisiana.

Louisiana

For more than 60 years, the State of Louisiana has been a very rewarding area for oil exploration, with drilling success rates far beyond most other regions. During this time more than 15 billion barrels of oil and 120 trillion cubic feet of natural gas have been produced. On each barrel of oil and on each one thousand cubic feet of natural gas, the State of Louisiana has levied a tax. The tax on oil is 12.5 percent. And the money has rolled in — in huge amounts. The state government of Louisiana for many years obtained about 40 percent of its income from petroleum, and the legislature had so much money it hardly knew what to do with it. Public salaries were raised. All sorts of social programs were started including a denture program for senior citizens which came to be known as "the right to bite." The first $75,000 of value of the homestead was exempted from property taxes which has resulted in 85 percent of the citizens paying no real estate taxes.

To produce the oil in the marshlands, draglines were brought in to cut channels to provide access for the huge drilling equipment. This often changed the ecology of these areas. What were once prime fishing and trapping areas were adversely impacted. Today, Louisiana oil and gas production is in marked decline. One state report estimates that by the year 2000 Louisiana will be 97 percent depleted of oil and 90 percent depleted of gas. Such estimates may be wrong in detail, but the trend is clearly accurate. Louisiana is now experiencing oil and gas withdrawal pains, and this economic pain will persist for some time to come. Unfortunately, in Louisiana's trade for the quick riches of petroleum, many of the once highly productive marshlands which produced renewable natural resources such as fish, furs, and shellfish, have been badly damaged. The Cajuns (the French of southwest Louisiana) went into the oil industry as workers after generations as fishermen and trappers. They now find the oil business is declining but they cannot go back to trapping and fishing. Oil has changed their lives and the economies of the communities for a very long time to come —perhaps, in a practical sense, forever.

Also, during the oil boom, Louisiana spent money on many capital improvements such as bridges, buildings, and roads. But these need to be maintained. With oil revenues declining, money to maintain this infrastructure is not adequate. And, from petroleum at least, it never will be again. Louisiana does have a good agricultural base, and an important petrochemical industry but eventually more and more of the raw materials for the petrochemical industry will have to be imported. Perhaps more important than the probable increase in price of obtaining petrochemical raw materials, is the fact that countries which produce petroleum will try to upgrade their resource before shipping it out as just raw materials as in the past. They are building petrochemical plants of their own, and these are modern state of the art plants which can effectively compete with those of Louisiana. Saudi Arabia, for example, has built an extensive petrochemical complex, and continues to expand it. Kuwait is doing the same.

The rapidity of exploitation of the petroleum resources of the United States is without precedent in the world, and Louisiana is a good example of the rags-to-riches story which accompanied this period. But now the non-renewable riches are diminishing, and if the once highly productive marshlands of that State have suffered long-term (possibly in some areas, irreparable) damage from oil operations, it may be that Louisiana will ultimately be poorer than before the oil industry arrived. Everything has its price. It was a fine economy while it lasted.

As a result of this economic change, there is also a rather sad social change becoming evident. In Morgan City, Louisiana, once a booming area for building and maintaining offshore drilling rigs, authorities have noted an increase in social ills because of the economic pressures from the decline of the petroleum industry. However, because of a lack of large onshore prospects in the United States, and with new technology, the oil industry is presently increasing its activities in the Gulf of Mexico. This has brought a modest temporary revival of the economy of Morgan City and adjacent coastal areas.

Texas

The great State of Texas was once the premier oil-producing state of the United States but now is second to Alaska, and, like Louisiana, has seen the peak of its oil industry. Texas oil reserves in 1960, were nearly 15 billion barrels. By 1993, the oil reserves were estimated to be 6.2 billion barrels, and still declining.(1) Texas, too, is suffering from petroleum withdrawal economic pains.

Houston, which had never really seen a depression before, with the oil price crash of the 1980s saw people simply moving away from houses, giving them to the banks and savings and loan associations. This was true all through the oil states, including Louisiana, Texas, Oklahoma, Colorado, and Alaska. The loans gone bad accounted for a substantial part of the Savings and Loan financial debacle which eventually cost the American Taxpayer in excess of 200 billion dollars to rescue the Federal government's deposit insurance fund guarantee.

Alaska

Alaska is a prime example of a mineral-based microcosm. The State government raises 85 percent of its revenues from oil royalties and taxes, and 89 percent of the gross product of the State is based on the oil industry in its various ramifications. As a result, Alaska's economy is oil-sensitive in the extreme. The oil price depression of 1986-1988, caused Anchorage alone to lose about 12 percent of its residents. One in twelve persons in the state with mortgages on their homes lost them. Statewide, two and a half million square feet of store and office space became vacant. State employment and public services were cut drastically.

The great majority of Alaska's oil revenue comes from the Prudhoe Bay Field which is half gone and continuing to decline, which forecasts a further drop in State revenues. There is, however, a mean average estimate in the "most likely case" of 3.2 billion barrels of economically recoverable oil in the region east of Prudhoe Bay in a small corner of the Arctic National Wildlife Refuge.(2) Development of this area for oil would help Alaska's budget for a time, but cannot replace the ultimate loss of Prudhoe Bay's production. As it has done for other Golden ages, history will record the black-gold era of lavish State spending in Alaska.(6) It will have been a pleasant experience for those who lived at the right time. There is no obvious replacement in sight. Currently Alaska levies a 15 percent tax, the highest in the nation, on all oil produced in the state. Because one-quarter of the U.S. oil production now comes from Alaska, they are able to do this. If they held only a small proportion of the national oil output they would not be able to maintain the tax. Analysis shows that the potential revenue take in severance taxes is small when there are many producing areas. One area cannot greatly increase the tax, or it will drive companies to other sources of production.(11) But, at the moment, Alaska has a dominant position in U.S. oil and they are making the most of it.

Other states: Mineral resources decline; revenue shifts

In the "oil patch" states, public budgets are being revised as crude oil production in the United States has shrunk from some 9.3 million barrels a day at its peak in 1970, to less than seven million barrels today. Metal production is also declining. Montana now gets only a very small amount of income from copper production compared with the past. Coal has become more important. The great metal mining Coeur d'Alene district of Idaho is now a shadow of its former self. All through regions where mineral exploitation once paid many of the bills, sales, liquor, cigarette, and gasoline taxes and other sources of revenue are being raised. The economies must try to shift to new more sustainable and renewable bases and away from the one-crop mineral resources revenues. But with diminished mineral resources the economies will be forever changed. The oil and mining boom days in the U.S. are becoming history.

Alberta, Canada

Among the Canadian provinces, Alberta has by far the largest conventional proven oil reserves, the world's largest oilsands deposits, huge heavy oil reserves, and some coal. As a result, Alberta is the only province which does not levy a sales tax, and its personal and corporate taxes are among the lowest in Canada. This is in marked contrast to such provinces as Newfoundland/Labrador which imposes a 14 percent sales tax, and other provinces where sales taxes range up to 9 percent. Revenues from oil and coal pay most of Alberta's taxes for its citizens. One result of this has been a high rate of migration into the province relative to other parts of Canada.

More Remote Effects

The discovery, development, and eventual decline of the production of petroleum and other mineral resources has a marked effect upon communities and states where these resources exist. But there is also a wider ripple through the economy in areas sometimes far from the site of the resource. Other microcosms, individual enterprises, or projects, or groups of people, which may be far removed from the oil wells, or mines can be affected. When the oil slump hit in the mid-1980s, Floating Point Systems of Beaverton, located in oil-less Oregon, but maker of high-powered computers widely used in the oil industry, saw its orders cut drastically, and layoffs resulted. The nationwide Public Broadcasting System (PBS) of the United States saw the sponsorship of its programs by oil companies markedly reduced. Oil companies had been a significant part of the PBS budget, accounting for more than half of that network's million-dollar or more annual contributors. The helicopter business is markedly affected by the state of the oil industry. At one time more than half the civilian helicopters in the United States were sold for use to ferry workers and supplies to offshore drilling rigs. By recent count, more than 200 of the 800 helicopters in the Gulf of Mexico area were surplus as drilling activity has declined.

The Girls Clubs of America, Inc. felt the decline of the oil industry as ARCO and Exxon cut their donations by more than half. Even TV evangelists felt the effects of America's declining position in the world oil industry because an appreciable part of their income comes from the so-called "Bible-belt" of the South and Southwest, which is also the "oil patch" that is Louisiana, Arkansas, Oklahoma, and Texas. Some of them in asking for more generous donations emphasized in their broadcasts that the oil slump was causing a substantial reduction in the financial support of their ministries.

Thus, in obvious ways, and in more subtle ways, mineral and energy mineral resource exploitation has a large and in some cases an overwhelming impact on many small units of population and some not so small, such as regions and states. In some cases where the depletion of the resource is essentially total, a community may simply disappear. Many have, and more will. But other new communities based on new discoveries or the application of new technologies to waning resources, will spring up overnight. Round Mountain, Nevada, for example, appeared as a result of the application of the modern heap leaching process to low grade gold deposits. There is a huge open pit mine on the east side of the valley and the town which serves the mine is just to the west. The mine has a finite life. When the gold is gone, will Round Mountain survive in some fashion or join the roll of other ghost towns spread across Nevada?

In some cases, further exploration discovers new deposits missed by the early miners using their primitive methods, or improved recovery processes or higher metal prices may make it profitable for mine dumps to be reworked. As a result, a community may stage a comeback, for a time.

Analogies for Countries?

All these mineral microcosms just described are local or regional in nature, but they illustrate in relatively simplified form the problems which some nations will face if they are dependent largely on one or a few depleting resources. The Industrial Revolution, with its demands for huge amounts of mineral and energy resources, has reached far across the world to obtain needed materials. When it was found that some countries had the needed resources, these began to be developed. As this exploitation has been going on for a relatively short time, the resources in many cases have been sufficient to continue to allow their production to the present time. But as these resources are finite, gradually, and at various times they will be exhausted. For some countries this will happen earlier than in other countries but the resource depletion is inevitable. Populations and economies have been built on these resources, but when they are gone, major adjustments will have to be made, and these adjustments should now begin to be planned. For the most part they are not. Some of these countries are discussed in Chapter 25, *Minerals, Social and Political Structures.*

Summary

When mineral resources are discovered and developed, not only the communities and areas where the resource is located benefit, but the places to where these resources are sent benefit also. Mineral resources build our cities, industries, and homes. They supply energy to keep our industrial civilization going. We use these resources every day. If we are to exist with a reasonable standard of living we must find and use more and more of these resources until we have achieved a sustainable economy based on a current renewable resource income.

But, mineral and energy mineral resources are finite. The discovery, development, and then the decline of these resources is the inevitable economic course of events. We cannot fault the resource producers for digging holes in the Earth to obtain these things. By this they live, and by these resources the rest of us live. As long as it is possible for the producers to move to other areas which have the resources after one area is exhausted, the consumers do well. It is the producers and the mineral microcosms built around the mineral production who have the problems.(5) These are now generally either not recognized or else ignored by the consumers in the industrialized world. But as more and more non-renewable resources are depleted, the problems will grow and eventually become painfully apparent to consumers as shortages occur. At present, the fact we are using a diminishing supply of resources is for most people an “out of sight out of mind” situation. But, the problems which mineral microcosms have experienced with the decline of mineral production will eventually be the problems of the world at large.

BIBLIOGRAPHY

1 AMERICAN PETROLEUM INSTITUTE, 1995, Basic Petroleum Data Book: Washington, D. C., (no pagination, large volume).

2 BIRD, K. J., and MAGOON, L. B., 1987, Petroleum Geology of the Northern Part of the Arctic National Wildlife Refuge: U.S. Geological Survey Bulletin 1778, Washington, D. C., 329 p., 5 pls.(maps).

3 FLORIN, LAMBERT, 1971, Ghost Towns of the West: Promontory Press, New York, 872 p.

4 FRASER, H. S., 1985, A Journey North. The Great Thompson Nickel Discovery: Inco Limited, Manitoba Division, Thompson, Manitoba, 388 p.

5 HODEL, D. A., and DEITZ, ROBERT, 1994, Crisis in the Oil Patch: Regnery Publishing, Inc., Washington, D. C., 185 p.

6 JACKSTADT, S. L., and LEE, D. R., 1995, Economic Sustainability. The Sad Case of Alaska: Society, March/April, p. 50-54.

7 MARTINEZ, LIONEL, 1990, Gold Rushes of North America: Wellfleet Press, Secaucus, New Jersey, 192 p.

8 NOAH, TIMOTHY, 1995, Tired of Mountains and Trees? New Park Features Superfund Site, Shopping Mall: The Wall Street Journal, July 28.

9 PAUL, R. W., 1963, Mining Frontiers of the Far West: Holt, Rinehart and Winston, New York, 236 p.

10 POTTER, M. F., 1977, Oregon's Golden Years. Bonanza of the West: The Caxton Printers, Ltd., Caldwell, Idaho, 181 p.

11 REES, JUDITH, 1990, Natural Resources. Allocation, Economics and Policy: Routledge, New York, 499 p.

12 SHINN, C. H., 1992, The Story of the Mine: As Illustrated by the Great Comstock Lode of Nevada: Univ. Nevada Press, Reno, Nevada, 277 p.

13 SMITH, D. A., 1967, Rocky Mountain Mining Camps: Univ. Nebraska Press, Lincoln, Nebraska, 304 p.

14 SMITH, D. A., 1977, Colorado Mining: Univ. New Mexico Press, 176 p.

15 STROHMEYER, JOHN, 1993, Extreme Conditions: Big Oil and the Transformation of Alaska: Simon & Schuster, New York, 287 p.

16 VOYNICK, S. M., 1984, Leadville: A Miner's Epic: Mountain Press Publishing Company, Missoula, Montana, 164 p.

17 WEIS, N. D., 1971, Ghost Towns of the Northwest: The Caxton Printers, Ltd., 319 p.

18 WOLLE, M. S., 1953, The Bonanza Trail. Ghost Towns and Mining Camps of the West: The Swallow Press, Inc., Chicago, 510 p.

CHAPTER 7

The One-Resource Nations

"One-resource nation," as used here, describes a country which does not have a significant industrial base and therefore must depend chiefly on the export of a particular mineral resource in order to obtain foreign exchange to buy manufactured goods from other nations. Some countries with very limited mineral resources, like Japan, do very well in developing foreign exchange by having a successful, competitive industrial complex which can export manufactured goods to pay for imported raw materials. But many countries do not have a commercial or manufacturing base, particularly smaller countries.

Also, to have a variety of minerals there must be a diversity of geology, because different minerals are found in particular and quite different geologic settings. Platinum, tungsten, and oil, for example, each occur in very different geologic associations. The smaller the country, the less likely it is to have a wide spectrum of mineral resources.

However, there are examples where a small country may have just one mineral resource and still be very rich. Four such are Kuwait, Bahrain, Qatar, and Brunei. Each has very large oil reserves relative to their geographic areas and populations.

But even some larger countries, given the irregular nature of the distribution of minerals, may be essentially one-resource nations. Such is the case of Saudi Arabia. It is useful to examine some representative countries which obtain more than half of their foreign exchange from a single mineral. A second group of nations is also reviewed which obtain a significant part of their foreign exchange from a single mineral, or group of minerals.

Algeria

This large North African country has an area of about 919,000 square miles and a population of only 20 million. But only about three percent of the land can produce crops, the rest is inhospitable desert and mountains. The principal resource Algeria has to convert to foreign exchange is petroleum, to some extent upgraded to refined products. Petroleum in one form or another, constitutes 92 percent of exports; a small amount of iron and steel make up the rest. Without petroleum, Algeria would have very little with which to buy the world's goods. Because it has a limited manufacturing base, most items of modern living, including such vital things as medical supplies, have to be imported. Riots in 1988 resulted in more than four hundred deaths. The unrest was caused by food shortages and rising prices brought about by the drop in petroleum prices, which substantially cut Algeria's income. With population still rising, and with petroleum Algeria's main source of foreign exchange, one might wonder what is going to happen in that country when the petroleum deposits are exhausted as will eventually happen.

Angola

This southwest African nation has two principal resources to pay for imports, petroleum and iron ore. The country has been torn by internal strife for a number of years, which has interfered with the shipment of iron ore. However, foreign oil companies have kept the flow of oil going, for the oil is largely from offshore operations, whereas the iron ore comes from the interior of the country, made largely impassable for the transport of the ore because of the civil war. Petroleum and upgraded petroleum products make up about 85 percent of Angola's exports. Like many small countries, there is no significant broad manufacturing base. Most items for life beyond a subsistence level have to be imported. Petroleum pays almost all of that bill.

Bahrain

This is a group of islands in the Persian Gulf with an area of about 225 square miles. The islands are low, with the highest point less than 500 feet above sea level. Oil was discovered in 1931, and since then a large refinery complex has been built which also refines oil from other areas of the Gulf.

Government revenue was derived almost entirely from petroleum until recently when, in part due to the construction of a causeway from Saudi Arabia to Bahrain, various worldly enterprises have brought in some money. Although Bahrain is nominally Moslem, the strict rules of that religion are not rigorously enforced in Bahrain. As a result, gambling and drinking and other aspects of the western world living have come upon the scene. Moslems from other oil-rich but less liberal areas can easily travel the causeway to Bahrain to sample western "culture". Thus Bahrain's economy is supported in two ways by oil, its own oil, and the oil wealth which comes from neighboring nations via the causeway, and also the modern airport.

Without oil, Bahrain would revert to what it was — a fishing and pearl diving archipelago with humidity almost always above 90 percent. Pearl diving and fishing are now greatly reduced, but the humidity remains. Oil and gas, however, come to the rescue again, fueling generating plants producing electricity to run air-conditioners which make Bahrain much more livable than previously.

Brunei

This tiny country on the northwest coast of Borneo, with an area of only about 2,200 square miles and 200,000 people, has one of the highest per capita incomes in the world. Oil is the source of wealth. Brunei's economy depends almost entirely on petroleum. That industry employs seven percent of the working population but accounts for more than 93 percent of Brunei's exports. Without petroleum what will Brunei be like? Can a now more numerous and much more affluent people make an orderly transition back to a more austere non-oil economy?

Ecuador

This delightful and very scenic country is one of the smallest of the South American republics. It might quite honestly and without prejudice be called a "banana republic" for it is the largest exporter of bananas in the western hemisphere. But the chief source of Ecuador's hard currency, that is, foreign exchange which is widely accepted internationally, is oil. Oil provides about 70 percent of what Ecuador earns abroad. Being such a small country, Ecuador has a limited manufacturing base with the result that many things have to be imported from medical equipment to automobiles.

Ecuador's export of oil is usually less than 150,000 barrels a day. Yet this oil brings in nearly three-fourths of Ecuador's export income. Can bananas make up the difference in the future when the oil is gone? Now, even with the oil income, Ecuador has accumulated more than $8 billion in foreign debt. Oil is the chief source of Ecuador's ability to repay the debt. When an oil pipeline was severed because of a huge landslide, Ecuador had to suspend payments on its international debt until the pipeline could be repaired, and oil again flowed. Per capita, each Ecuadorian man, woman, and child has a foreign debt of nearly $1,000. If oil production falls there is little likelihood that the debt can be either serviced or repaid. Eight billion dollars is a lot of bananas!

Indonesia

This is the fourth largest nation in the world in terms of population. It is a group of islands including Sumatra, Java, Bali, part of Borneo, and the western half of New Guinea. Altogether, there are more than 3,000 islands with a land area of about a half a million square miles. Before World War II, it was a Dutch possession, called the Dutch East Indies. Japan invaded these territories immediately after Pearl Harbor to obtain oil. After World War II, Indonesia became an independent nation. The Dutch never returned. Indonesia is the largest oil producer in the Far East and is the tenth largest producer in OPEC. Until recently, oil has constituted about two-thirds of Indonesia's export values. Recognizing this excessive dependency on one resource, the government has been working to strengthen the competitive position of its other exports in world markets, and recently petroleum sales have constituted only about half of the total value of exports. However, government revenues remain heavily dependent on petroleum, and the economy as a whole remains very sensitive to petroleum markets.

Iran

This Persian Gulf nation was frequently in the news in the 1980s because of its conflict with Iraq, and the disruption of oil shipping lanes. Before the outbreak of hostilities between Iran and Iraq, Iran obtained about 94 percent of its export income from the sale of oil and its derivatives. This diminished during the war as Iraq made repeated attacks on Iran's shipping facilities. Iran in turn made great effort to keep their oil exports flowing because oil was the major source of income to pay for both military and civilian supplies.

Iran continues to be dependent on oil for much of its foreign exchange. Faced with a rapidly growing population with rising expectations, income from oil export is very important to maintaining a stable society.

Iraq

This nation on the northwestern end of the Persian Gulf is both Muslim and Arab. For a number of years it has been almost totally dependent on oil for its export revenues. Iraq's oil income was severely disrupted for a time by the war with Iran, but eventually pipelines built to reduce dependency on the Persian Gulf as an outlet for the oil brought Iraq's oil exports almost back to pre-war levels.

Before the war with Iran, Iraq received 99 percent of its foreign exchange from the sale of oil. When Iraq was defeated in the Gulf War of 1991, the United Nations, knowing that Iraq was so dependent on oil for its foreign exchange, imposed a ban on the export sale of Iraq oil. The ban was used as a weapon to insure that Iraq took steps to fully comply with the terms of the surrender. The ban seems to have been only partially successful, as oil was smuggled out of Iraq into Turkey and other areas. Also, the deprivations caused by the ban were inflicted upon the common people whereas Saddam Hussein and the ruling military were able to insulate themselves in various ways from the effects of the ban.

Regardless of when or how Iraq may eventually rejoin the world economy, oil will continue to be virtually its only source of foreign exchange. This situation is likely to continue for many decades. Iraq has very large oil reserves, but no other exportable resources of any significance.

Jamaica

This Caribbean nation has substantial aluminum ore (bauxite) deposits, and has been one of the world's ranking producers. However, the ore deposits are beginning to be depleted. Nonetheless, bauxite and a slightly upgraded product called alumina constitute about 67 percent of Jamaica's export revenues. Jamaica has few basic industries and, like other small island economies (its area is only about 4,400 square miles) has to import virtually everything beyond some agricultural products. Bauxite and its alumina derivative provide the chief money used to buy manufactured goods, and some food supplies.

Kuwait

If one were to compare value of mineral resource per square mile of territory of a whole country, Kuwait is almost in a class by itself. Lying on the upper end of the Persian Gulf between Iraq and Saudi Arabia, this small country with an area of less than 7,000 square miles has more than four times the total oil reserves of the United States in its 3½ million square miles. Kuwait's wealth was its undoing causing the Iraqi invasion, but it also caused it to be rescued by the western industrial nations that needed Kuwait's oil.

The production from Kuwait's approximately 90 billion barrels of oil reserves give it one of the highest annual per capita gross national product figures in the world, more than $30,000. The role which Kuwait plays in the world economy because of its singularly large oil reserves is greatly disproportionate to the size of the country, and its small population.

Management of the vast amounts of money which have poured into the country as a result of its oil wealth has at times created some chaotic financial situations. But Kuwait is trying hard to make well-placed investments abroad so when oil runs out, the oil legacy will survive. Kuwait is keenly aware that it is a one-crop, one-resource country, and good investment decisions are crucial to its future.

Liberia

This west African nation was established by several American philanthropic societies to provide a place where freed American slaves could return to Africa. The state was formally established in 1847. The imprint of America remains on the country in several ways, one being that the unit of currency is called the dollar.

Unknown at the time it became an independent nation, was that Liberia possessed some exceedingly rich iron ore deposits conveniently located near the coast. The result is that Liberia can now lay down high quality iron ore on the East coast of the United States at prices competitive with ore coming from the Mesabi Range of Minnesota. Iron ore concentrates are by far the most important of Liberia's exports. Gold and diamonds in small amounts have also been discovered and are exported.

Libya

Mostly desert, in part mountainous, this nation has an area of about 680,000 square miles, or approximately one-fifth the size of the United States. But Libya has a population of only four million. Only a narrow strip along the coast can be successfully farmed. Another narrow area of marginal grassland grazing area lies immediately to the south, but this thins out rapidly into the Sahara Desert.

Until oil was discovered, Libya was a nation of no great importance with virtually no impact on the rest of the world. The discovery of oil changed that. Oil rapidly became by far the largest source of government revenues, and allowed Libya in various ways to buy into the Twentieth Century. Without oil, Libya could still be what it had been for centuries, a nation of small farms and herding operations. Libya is markedly a one-resource country, with oil consistently making up from 90 to nearly 99 percent of total exports. When the oil is gone, the economy will have to be greatly restructured toward a simpler life. Change is inevitable, and there is nothing in sight to replace the oil.

Mexico

This nation was an oil producer early in this century, but substantial oil production was not developed until fairly recently. Mexico's big oil boom started in the 1970s. Mexico now produces about 2½ million barrels of oil a day, and appears to have oil reserves at least as large and probably somewhat larger than those of the United States.

This somewhat imprecise statement about Mexican oil reserves stems from the fact that as the Mexican oil boom got underway, there was an economic (and political) need to make the situation look as good as possible. It was largely on the basis of oil reserves that foreign banks were willing to loan money to Mexico. Some of the banks apparently believed what now seem rather inflated oil reserve figures, and loaned money accordingly — to their later regret. The most reasonable figure seems to be about 29 billion barrels (compared to U.S. with about 23 billion). Petroleum and its products make up (depending on world prices) between 60 and 75 percent of the value of Mexican exports. Most of this goes to the United States.

Nauru

This is the world's smallest independent nation. It was a German colony until after World War I, when New Zealand, Australia, and Britain governed it. It obtained independence in 1968, and was then left to its own resources which consist of only one thing —phosphate.(4)

Nauru is an island which lies almost in the center of the Pacific Ocean, and has an area, not figured in square miles, but in acres — 5,236 of them. And these acres are made primarily of superphosphate which is easily processed into a very rich agricultural fertilizer. Nauru's export trade consists entirely of phosphate which goes to Japan, New Zealand, and Australia. Much of Nauru's land is devoted to phosphate mining. After the mining is done in a given area, what remains are gray, inhospitable pinnacles of rock. For the approximately 8,100 Nauru citizens only about 1,100 acres (less than two square miles) are available on which to live. This is the world's best example of a nation dependent upon a single mineral resource. Unfortunately, it is almost gone. Australia paid Nauru $75 million in environmental compensation. Australia also did offer to move the entire population to Curtis Island, an Australian island off the Queensland coast. Nauru has rejected the offer.(7) With the loss of almost all of its arable lands to mining, Nauru has essentially abandoned all food production.

Probably as a result of a high fat imported diet, Naurans have a short, 55 year, life expectancy, and high rates of obesity and associated health problems.

Knowing that the phosphate deposits were fast diminishing, Nauru, however, did set up an investment agency to put some of the phosphate income into things which will hopefully sustain them in the future. A few of the investments, however, have been notably unsuccessful. Some of the details are described in Chapter 10: *Mineral Riches and How They Are Spent.*

New Caledonia

Like Nauru, this is an island in the Pacific. It is an independent nation, but remains more or less under the protective wing of its earlier French possessors. It has one of the world's largest nickel deposits. Nickel is a strategic metal, useful in many ways in industry and in war materials. New Caledonia's foreign exchange is derived almost entirely from this metal, and is the only substantial resource the island nation has. As nickel goes, so goes the New Caledonian economy. Being a small island, it must import nearly everything except a few foodstuffs locally grown. Today nickel pays the bills, but when the nickel is gone there is no obvious replacement export to earn equivalent amounts of foreign exchange.

Niger

Several nations in Africa are substantial uranium producers, or have the potential to be. These include Namibia, South Africa, Gabon, and Niger. Of these, Niger is the most vulnerable to the uranium market and when the uranium boom ended in the early 1980s, Niger was dependent on uranium for 84 percent of its foreign earnings. Since that time, the uranium market slump has markedly depressed the Niger economy. Niger has few other resources even in total to take the place of the atomic metal as an income for the country.

Nigeria

This is the most populous African nation with nearly 100 million inhabitants. Nigeria's dependence on oil for its export earnings is almost total. Oil provides at various times between 80 and 95 percent of that country's foreign exchange. For any other resource to replace the oil revenues seems an impossibility. Because of Nigeria's great dependence on oil to support its industry as well as finance the social and political systems, Nigeria is one of the OPEC nations which tends to frequently cheat on its production quota. The rather corrupt political system takes much of the oil revenue.

Recently, Nigeria has been the scene of much strife, lawlessness, and general government mismanagement. The future of that oil-rich nation is very much clouded. Even abundant oil resources do not seem to have made for a better life for most of its citizens. Only a relatively few have benefitted.

Oman

Properly called the Sultanate of Oman, this nation with an area of about 105,000 square miles has a coastline about 1,000 miles on the southeast end of the Arabian Peninsula. It was a country characterized chiefly by date palms and camels until 1964, when oil was discovered in commercial quantities. Now oil provides essentially all government revenue. Copper was subsequently found and a small copper mining and refining industry using the natural gas from the oil fields has developed. Petroleum with a little help from copper, provide more than 90 percent of government income and foreign exchange.

Qatar

Probably not one person in a hundred can identify the location of this nation, but it has a per capita gross national product more than twice as large as that of the United States. Qatar occupies a land area of about 4,000 square miles, the whole of the Qatar Peninsula, which juts out into the southern part of the Persian Gulf adjacent to Saudi Arabia. With the territory claimed in the Persian Gulf, Qatar's total area is about 17,000 square miles. Qatar is a region of sand, gravel, and some limestone ridges. Until the 1960s, it was an exceedingly poor country, dependent chiefly on fishing, pearl diving, and some trading. It had a subsistence existence at best. Now it has a world-class per capita income. Petroleum made the difference, and petroleum and related industries today provide more than 90 percent of the national income.

Qatar has 12 percent or more of the total world gas reserves in the North Field. To use these gas deposits, petrochemical and fertilizer plants have been built. Some of the gas is shipped abroad, chiefly to Japan which takes about six million tons a year in the form of liquid natural gas (LNG). The gas is cooled to minus 260 degrees Fahrenheit, put in huge refrigerated units called trains, and then loaded on ships. At the destination, the liquefied gas is off-loaded, regasified, and piped to the ultimate user. In one form or another, Qatar's economy is based on petroleum. There are almost no other resources except bare subsistence agriculture and a little fishing.

Saudi Arabia

Holding the world's largest oil reserves beneath its sandy soil and below the shallow waters of the adjacent Persian Gulf (Arabian Gulf to the Saudis), Saudi Arabia derives nearly 100 percent of its income from petroleum. Before the discovery of oil, Saudi Arabia was an obscure nation of a few million people, many of them nomadic. The economy was dependent on grazing, small farms, and a few fishing and pearl diving villages. Oil made Saudi Arabia a world economic power, and remains the only substantial resource.

Sierra Leone

This west African nation has an area of about 28,300 square miles, slightly larger than the State of West Virginia. Bauxite and titanium mining supply 90 percent of its exports, but even with this industry the gross national product produces a per capita income of only $200/year. By comparison the U.S. is $21,100. and Britain is $14,500. The economic situation is critical. When the minerals are exhausted it will be a desperate situation, particularly as the rate of population growth is 2.6 percent a year, which will double the present population of 3.5 million in 27 years, and there are no substantial resources available for export to replace minerals by which to obtain foreign exchange.

Suriname

Prior to 1975, this country was called Dutch Guiana. It is located on the north coast of South America, in a region of heavy rainfall and a warm climate. Over the course of geologic time these conditions have combined to weather and leach some of the rocks to form a rich residual deposit of bauxite, the chief ore of aluminum. Today, bauxite, its concentrate, alumina, and some aluminum metal from one refinery, constitute about 70 percent of the value of Suriname's exports. Most of the aluminum in its various forms goes to the United States.

Trinidad

This island lying a few miles northeast of Venezuela was initially famous for its Pitch Lake, an area where tar oozes from the ground and can be mined. Many of the roads of Britain (of which Trinidad was once a colony) and of western Europe were initially paved with asphalt from the Pitch Lake of Trinidad. Later, petroleum was developed by drilling in Trinidad both onshore and in near-offshore areas. From this petroleum base, Trinidad built fertilizer industries, petrochemical complexes, and cement plants. The chief cost of cement production is energy to burn limestone. For this Trinidad uses natural gas because the gas which is produced along with the oil cannot conveniently be exported.

The importance of petroleum to Trinidad is seen on its paper currency which portrays an engraving of an oil derrick. This is appropriate for oil is indeed money to Trinidad, with 90 percent of its export value coming from petroleum in one form or another. Recently, however, oil production and oil-related income has been declining and there are no other apparent resources available large enough to replace it. The result has been close to disastrous.

The declining oil production together with the collapse of oil prices in the mid-1980s, caused Trinidad's gross national product to drop from $4.45 billion in 1982 to $2.82 billion in 1987. Per capita income fell from $7,060 to $3,380 during that time. Government revenues, 70 percent dependent on oil, dropped from $1.67 billion in 1982, to less than $1.18 billion in 1988. Unemployment doubled to 22 percent. Oil production continues to drop. Yet Trinidad's population of one and one-quarter million continues to increase rather rapidly. Harder times have arrived and more may lie ahead. It is an excellent example of the problems of a largely one-resource country, when the resource base diminishes but the population continues to grow.

United Arab Emirates

The UAE is a confederation of seven sheikdoms, the most important of which is Abu Dhabi. It lies on the south end of the Persian Gulf with an area of about 35,000 square miles, and a population of slightly over a million people. Like nearby Qatar, the UAE was a relatively poor area until oil was discovered, principally in Abu Dhabi. Proven reserves in this small nation are now estimated to be more than 61 billion barrels, which is almost three times those of the United States. Oil here, as in all the oil-producing nations bordering the Persian Gulf, is by far the most important export and source of revenue.

Venezuela

Petroleum brought Venezuela into the Twentieth Century largely because of huge oil discoveries in the Lake Maracaibo region of northwestern Venezuela. An affiliate of what is now Exxon Corporation, Creole Petroleum, was the principal developer, but subsequently all oil operations in Venezuela have been nationalized. Venezuela has other mineral resources, including iron ore (a mountain of it called Cerro Bolivar), and some reasonably good bauxite deposits. But even with these other resources, oil contributes 90 percent or more of that nations's foreign exchange income. Venezuela is largely a one-resource country in terms of trading with the rest of the world.

Zaire

Once called the Belgian Congo, Zaire became independent in 1960. Located on part of the great African mineral belt, Zaire obtains most of its foreign exchange from copper. Zinc and cobalt also are important, but copper is by far the largest export item. Without copper, and the lesser amounts of cobalt and zinc, coffee would be the chief exports but the income from those sources would be very small compared with that earned currently by copper. Copper is the coin which allows Zaire to buy at least part way into the modern world.

Zambia

Adjacent to Zaire is Zambia. It also lies over the rich mineral deposits of south-central Africa, and, like Zaire, is a major copper producer. That metal accounts for more than 80 percent of Zambia's foreign exchange earnings. Cobalt and zinc add additional exports to the extent that 98 percent of export income comes from these three metals.

Like many other countries, upon gaining independence from being a colony, Zambia nationalized its mineral industry. Now, however, the largest copper mine needs a considerable capital investment to continue exploration and operation. So the Zambian government states that it, "...is committed to development of private-sector mining and has created a legislative framework to encourage investors in the mining sector — a clear break with the past. Key to the regeneration is the privatization of Zambia Consolidated Copper Mines (ZCCM), the state monopoly. Long starved for investment, ZCCM is having to mine deeper and often uses antiquated equipment, which adds to the overall costs."(1) the World Bank representative in Zambia states, "Ultimately mining may not be the mainstay of the Zambian economy but for the next ten years or so there's nothing else. All other projects are a drop in the bucket compared to what happens to ZCCM."(1)

Zambia's population growth rate is 3.2 percent a year.(3) This means the population will double in about 22 years. The present annual gross national product (GNP) per capita is $390 in U.S. dollars. This compares with the $21,100 for the United States.(5) There is nothing in sight to replace the earnings from metals, once the mines are exhausted. Zambia clearly illustrates the problems many nations will face in now having expanded their economies and populations based on non-renewable mineral resources. These resources have allowed, perhaps encouraged, this growth. Their exhaustion will almost certainly involve some major adjustments.

Russia

In reviewing one-resource nations, in terms of their export earnings, it is important and perhaps surprising to include Russia. Despite possession of perhaps the widest spectrum of mineral resources of any nation, Russia today falls into the category of a one-resource nation in international trade. Substantially more than half of Russia's foreign exchange, which it so urgently needs, comes from the sale of petroleum, chiefly natural gas. This makes about 80 percent of the value of Russian exports.

Russia has the world's largest known natural gas reserves, but it does not have great fields of surplus grain as for example, does Canada which supplements Canada's substantial export of minerals. What grain Russia does grow, it must use domestically. Only natural gas is currently in substantial surplus. Russian oil production is now about in balance with internal consumption, so Russia, in terms of exports, is now a one-resource nation dependent on natural gas. The Russians realize this and are trying to shift their power sources from the current heavy reliance on petroleum more and more to coal and to hydro-electricity, reserving their petroleum (both oil and gas) as much as possible for sale to obtain badly needed foreign exchange.

In the future, however, Russia may be better positioned in terms of a variety of mineral exports. Russia is slowly developing the great mineral deposits of Siberia which heretofore have been locked up by their remoteness and general lack of transportation. Now, an increasing network of roads and the new Siberian railroad, the Baykal-Amur Magistral, will perhaps allow Russia to become a major mineral exporter in the Twenty-first Century. It does have large deposits of nickel and manganese, both essential metals alloyed with iron, to go along with Russia's iron deposits which are the world's largest.

Nations with Several Minerals of Substantial Importance

In addition to these nations in which more than half of their foreign exchange is currently derived from one mineral resource, there are a number of other countries where several minerals combine to be a major source of export earnings. In other countries the mineral or minerals may make up less than half of the total export values, but still are a significant amount. Examples of such nations are considered here.

Australia

This nation, which is a whole continent by itself, is commonly thought of as exporting chiefly wheat, beef, lamb, and wool. Actually, the largest single export in value is coal. Australia is also now the world's largest producer of bauxite, after discovery of tremendous deposits in the northwestern part of the country. And, it has the world's second largest iron reserves, for which Japan is a major customer.

Bolivia

Mining is the single most important industry in Bolivia with tin making up nearly half the value of production. Minerals in total make up about half of Bolivia's exports. Bolivia does not have a diverse industrial base, and must import most manufactured products. Without minerals, imports would be cut in half.

Chile

Copper for Chile accounts for about 48 percent of its exports, and when other minerals are included, the total mineral value is more than one half of Chile's exports. Copper will continue to bulk large in Chile's economy, and perhaps even more so in the future than at present, for Chile holds an estimated one-fifth of the world's copper reserves, some of which are high grade.

Egypt

This is the largest of the Arab nations in population, now about 60 million. It is also among the poorest. Imports in recent years have been more than twice the value of exports, and Egypt is sustained to a considerable extent by the generosity of its oil-rich neighbors, and by the United States. Egypt does have some oil production, about 800,000 barrels a day. Crude oil and some upgraded petroleum products make up about 42 percent of Egypt's export values, which is more than the value of cotton for which Egypt has long been famous. Cotton is only about 34 percent of total exports. Egypt could probably use all the oil it produces domestically to replace the energy now supplied by animals and humans. But with a large population, there are many hands to do the work, and oil is urgently needed to provide foreign exchange.

Morocco

This interesting and colorful country has a good agricultural base. Situated near Europe and enjoying a mild climate, it sells much of its agricultural produce to Europe, especially in the winter. Morocco has a unique position among world mineral producers because it has the world's largest phosphate deposits. Every living cell must have phosphorus, and phosphate resources are very unevenly distributed in the world. Morocco's phosphate and its derivative, phosphoric acid, make up about 40 percent of Morocco's exports. Because phosphorous is so important for life itself, this very large deposit of this critical mineral is indeed good geological fortune for Morocco! It will provide Morocco with substantial foreign exchange for many years to come.

Namibia

This nation, previously known as Southwest Africa, gets 40 percent of its government revenue from royalties on diamonds recovered from beaches on the coast. Namibia also has a very large uranium mine. Together, diamonds and uranium provide substantially more than half of Namibia's income.

Norway

This delightful and picturesque nation has only four percent of its land which is arable. For centuries Norway has had to turn to the sea for its livelihood, first to fishing. More recently it has found another reason to go to the sea — the petroleum production from the large oil and gas discoveries in the Norwegian segment of the North Sea. Phillips Petroleum was one of the leaders of this exploration work. Oil and gas are now a substantial source of foreign exchange for Norway. Although Norway's petroleum production will peak by the year 2000, along with the rest of North Sea production, fortunately these resources relative to the current rate or production are quite large, and are likely to easily carry Norway safely into the Twenty-first Century with considerable petroleum to sell. By that time, further development of hydro-electric facilities may provide Norway with another important energy export. In view of the world-wide decline in ocean fisheries, energy production from both petroleum and hydroelectric projects will be a very important part of Norway's economy. As petroleum production ultimately is exhausted the question will then remain as to what might fill its place. At present, nothing is known which can do this.

Peru

This nation has a fairly broad spectrum of minerals including petroleum, copper, lead, zinc, silver, and gold. Indeed, gold and silver were the undoing of the Inca Empire because Pizarro and his troops came to plunder the Inca treasury. Peru today is a poor country, and much of the population lives at a subsistence level. Exports are chiefly minerals which make up more than half the foreign exchange, but no one mineral makes up the majority of the tctal. Of the minerals, copper is the most important. Peru's economy has often been in difficult straits, with a sizable foreign debt. Peru's poor population continues to increase and without mineral exports, its economy would face disaster.

South Africa

This nation has been described as a mineral treasure storehouse. The description is valid. Most of South Africa's export income comes from the sale of a wide variety of metals. No one mineral accounts for as much as half of that country's exports although gold amounts to 40 percent. Collectively, however, the metals and diamonds of the Republic of South Africa earn nearly three-quarters of the country's foreign exchange. The important metals include gold, chrome, copper, manganese, tin, iron ore, silver, and platinum.

Zimbabwe

On the 18th of April, 1980, Rhodesia (Southern Rhodesia) became the Republic of Zimbabwe. Like Zambia and Zaire, Zimbabwe is one of the geologically fortunate African nations whose territory includes some of the rich mineral deposits of southern Africa. Within its borders Zimbabwe has what is called the Great Dike of Rhodesia, a unique geological feature which is a huge wall or dike of rock rich in chromite. The dike is three to four miles wide, extends vertically to an unknown depth, and is more than 300 miles long. Nothing comparable as a mineral deposit is known anywhere else in the world.

Zimbabwe also has substantial deposits of gold, copper, nickel, and iron, which together make up about 54 percent of the country's exports. It is also a rich agricultural country and exports tobacco, sugar, and other plant products. Nonetheless, for hard currency, mineral exports have been Zimbabwe's mainstay and will continue to be for some time to come.

Other mineral-dependent nations

A number of other countries also receive half or more of their foreign exchange from mineral exports. These include Botswana, 76 percent (nickel, copper); Mauritania, 58 percent (iron); Papau New Guinea, 51 percent (copper, gold); and Togo, 52 percent (iron). The average value of minerals in the export total of all lesser developed countries which export minerals, not including oil, is 51 percent.(5)

What of the Future?

From these examples, it is clear that many nations are largely, and in some cases almost entirely dependent, beyond a mere subsistence existence, upon one or a combination of several mineral resources for their livelihood. But these minerals are a one-crop, non-renewable national resource. When they are depleted and are no longer a significant source of national income, what will happen? In many, if not most instances, there is no apparent other resource to replace the lost export earnings.

The experiences of many smaller communities and regions dependent on a single mineral resource have been described in the previous chapter, *Mineral Microcosms*. The rise and fall of such communities eventually may be re-enacted on a larger scale by these single resource nations. In the case of smaller mineral resource based communities, the population could move away to other parts of the country and do other things. But this is not possible for a whole nation. There are large problems ahead for those which obtain a significant part of their national income from minerals.

Philosophy of Resource Use

Nauru has been cited as a nation which used up its single mineral resource in a wholly predictable and relatively brief time, and also with predictable negative effects on the environment. But it would be an oversimplification to fault Nauru for shortsighted environmental policies. The extraction of phosphates has fully supported Nauruan people since the beginning of the century. For each generation, the mining and sale of phosphate was a reasonable choice. The recognition that the resource income must and could be replaced by investment income was also a reasonable national policy.

Each nation tends to deal with its physical environment and the resources which it offers, based on current need. Serving the population that exists at any given moment is every government's primary mission. Indeed, if it does not do so, it is usually soon replaced by a government which will respond to current needs. Decisions whether or not to extract resources cannot always be balanced with future idealist goals. This is particularly true of the poorer nations dependent on one or only a few resources, and which frequently have a rapidly growing population. Do people today deprive themselves of a better standard of living for the benefit of generations yet to come? It is a philosophical question that has largely been ignored. In a few countries it is being considered, but even in the United States, finite, non-renewable resources are being used for the present generation. Farmland is being paved over, groundwater is being mined in many areas, and petroleum resources are in a markedly declining trend. Technology is looked upon as the agent which will secure the future, but for some resources there are no replacements. The one-resource nations are simply now more obviously facing the same problem and philosophical question which other more generously endowed nations are able to largely ignore for the moment.

BIBLIOGRAPHY

1 ANONYMOUS, 1995, Mining. The Business of Copper: (advertisement) Fortune, July 24, p. 144-145.

2 DICKSON, DAVID, 1988, Norway: Boosting R&D for a Post-oil Economy: Science, v. 240, p. 1140-1141.

3 ERBSEN, C. E., 1988, Associated Press World Atlas: New York, 184 p.

4 HOWELLS, WILLIAM, 1981, The Good Fortune of Nauru: Harvard Magazine, p. 40-48.

5 HUNTER, BRIAN, (ed.), 1992, The Statesman's Year-book. 129th Edition: St. Martin's Press, New York, 1702 p.

6 METZ, H. C., (ed.), 1994, Persian Gulf States. Country Studies: Federal Research Division, The Library of Congress, Washington, D. C., 472 p.

7 SHENON, PHILIP, 1995, Pacific Island Nation is Stripped of Everything: New York Times, December 10.

8 TILTON, J. E., (ed.), 1992, Mineral Wealth and Economic Development: Resources for the Future, Washington, D. C., 121 p.

CHAPTER 8

The Good Geo-Fortune of the USA

Mineral resources have shaped the course of history and development of many nations, but two examples stand out, the United States and Saudi Arabia. Each is a phenomenon which will never be repeated. Each is worthy of special note. Saudi Arabia is discussed in the next chapter.

Wilderness to World Power

The United States changed from a three million square mile wilderness to the most powerful and affluent nation in the world in about 200 years. In terms of the total energy minerals and minerals spectrum, the United States was without equal among nations at the time the Declaration of Independence was signed. However, the citizens did not know then what riches were in this relatively unexplored country. As the pioneers moved westward and exploration proceeded, word spread of the great natural resources of this region both in minerals and fertile land. A flood of immigrants swept into this undeveloped territory.

Right time

The good fortune of the United States was that the country was established at the right time, with motivated people, and at the right place. In terms of the right time, the USA emerged as a nation shortly after the Industrial Revolution began. It began in Great Britain and promptly spread to Europe. New inventions and new technologies developed rapidly. The technologies enabled people to extract and process important raw materials such as iron in great quantities. The invention of the steam engine fostered the development of the railroad which then was able to haul the raw materials cheaply and in great quantities to the factories and then distribute finished products across the country.

Motivated people

The people who came to North America for the most part were from lands where they had been under kings and oppressive landlords. Most who came were not the nobility. Why would anyone already well off in Europe come to the wilds and the primitive living of North America? A few did, but the great majority were the oppressed and those with little material wealth. But the majority were united by one thing — a burning desire to establish a nation where all persons would have equal rights regardless of status at birth, and where wealth gained by hard work could be kept. This was written into the Constitution and the Bill of Rights, and democracy and the free enterprise American capitalistic system emerged.

Right place

Great Britain and Europe had fewer mineral and energy resources than did the new nation across the sea. When the immigrants of the time which were largely from these northern European lands came to North America, they found a much greater and richer spectrum of mineral resources than what they had left. They had seen the beginnings of the Industrial Revolution, and now they had tremendous natural resources by which to build industries and cities on the North American continent.

This combination of the right time (during the spread of the Industrial Revolution), together with a poor but ambitious free people, and the right place (three million square miles of virgin land with a tremendous variety and quantity of mineral resources), produced the great economic and military might of the United States in just the first two hundred years of its existence.

Of the three factors, the great variety and abundance of mineral and energy resources was probably the most important. Without these , even a free people at that time would have seen the industrial age largely bypass them or arrive much later. But, the lavish geological endowment of the United States shaped its destiny from then to the present.

Boyer has expressed it very well: "Fertile farmland, vast forests, open ranges, coal, hydro-power, and abundant petroleum provided more opportunity for each person to gain material abundance through hard work and initiative than people had ever experienced before."(1) Kennedy notes that the Industrial Revolution arrived just at the time the United States was able to settle the matter of the Civil War, and could combine the new technology with its great natural resource wealth:

> "Of all the changes which were taking place in the global power balances during the late nineteenth and early twentieth centuries, there can be no doubt that the most decisive one for the future was the growth of the United States. With the Civil War over, the United States was able to exploit its many advantages — rich agricultural land, vast raw materials, and the marvelously convenient arrival of modern technology (railways, the steam engine, mining equipment) to develop such resources..."(8)

Canada?

One might suggest that Canada also had the same potential as did the United States. But Canada has somewhat less conveniently arranged mineral resources. It does not have high grade iron and coal adjacent to the inexpensive Great Lakes transportation system. There is very little oil in eastern Canada unlike the numerous oil fields in Pennsylvania and Ohio where people first settled and industry was established in the United States. There is no area in Canada comparable to the prolific Gulf Coast region of the United States where the famous Spindletop oil gusher discovery was made in 1901. Canada's major oil industry really dates only from post-World War II, and, although important, it does not rival the size and wide geographic distribution of oil fields all across the United States. Also, Canada's northern position with its more hostile climate and the difficult terrain of lakes, bogs, swamps, and large areas of tundra underlain by permafrost has delayed its development.

Canada did not attract the size of population necessary to form the basis for a big industrial complex with sizable internal markets needed to foster large scale manufacturing such as occurred early in the history of the United States. It may be, however, that because of this delay, the best is yet to come for Canada. Their world's largest deposits of oilsands, for example, will be an asset for many decades to come, and large high grade iron ore deposits still remain. The United States has already depleted its high quality iron deposits as part of the price for its phenomenal economic growth.

Russia?

Russia had the resource potential, except for agricultural lands, during the phenomenal rise of the United States to accomplish similar development. But it lacked a political system which encouraged or, most importantly, economically rewarded individual initiative. Wealth could not be accumulated by enterprising individuals, and it could therefore not be re-invested in more enterprises to build a better life for the average citizen.

The early immigrants to the United States came to a land with no established government or political system. This allowed them to establish a political and economic environment where individuals were rewarded for their initiative, and money could be retained and invested. Although Russia's agricultural base is not as large as that of the United States, Russia now has the richest mineral and energy spectrum of any nation. If the new Russia can establish and retain a political and economic society which allows individuals to realize and be rewarded for their full potential, Russia could significantly raise its standard of living and substantially increase its position as a world economic power. However, Russia would have missed one advantage which the United States had in its rapid economic rise — a relatively small population and large mineral wealth. Today Russia would have to spread its geological wealth over several times the number of people the United States had when it rose to its affluent world position. In 1880, as the United States began to enjoy the advantages of the Industrial Revolution, the population was approximately 50 million. The new Russia's population today is about 150 million. Large mineral wealth spread over a small population creates a rising standard of living. This has been clearly illustrated in such countries as Saudi Arabia and Kuwait. Russia missed a great opportunity. Now, although it does have mineral wealth, it has a large population. Also, Russian oil production is now peaking before the benefits of its oil riches have been enjoyed to any large degree by the average citizen. The United States combined its oil wealth with its world class motor vehicle industry to bring a degree of affluence and life-style to the average citizen which would be difficult if not impossible now for Russia to duplicate.

U.S. Resources

Oil and rapid rise to the top

The United States has had an incredibly fast ride to the top of the world economic power and affluence. Oil was discovered in Pennsylvania in 1859, and large very high grade coal deposits had been known before then also in Pennsylvania. What were then the world's largest known iron ore deposits came into production in Minnesota in 1884. The United States soon became the world's largest producer of coal, iron, and oil — three basic ingredients of the dawning industrial age, and for building a higher standard of living.

Following the 1859 discovery of oil, the United States was completely self-sufficient in petroleum for more than 100 years. Ultimately it was the possession of these large oil resources and the self-sufficiency thereon, which brought about the reversal of strength between Great Britain and the United States. Until World War I, coal was the dominant energy source, and British coal mines had been a major source. After World War I, oil became the major fuel on which the world depended. Britain at that time had no oil production. With the arrival of the age of oil, economic power went to the United States. One might note that the current increasing dependence of the United States on foreign oil has substantially decreased the relative world economic strength of the United States — a matter taken up later in this volume.

Abundant raw materials

For many years the United States was the world's dominant producer of most vital raw materials. Until 1992, the United States was the world's largest copper producer, and is still the world's largest cement producer. Until about 1950, it produced half the world's oil. It has long been the leader in molybdenum and lead output, and it has the largest recoverable coal reserves in the world, far more than Britain.

Helpful geography

The United States not only benefited from a uniquely favorable sequence of events combining mineral discovery with the developing technology of the Industrial Revolution to use those resources, but the geographic arrangement of some of the resources was also very fortunate. This was especially the case with iron ore and the raw materials needed to smelt the ore to produce iron, the single most important metal in our industrial civilization. The richest iron ore deposits then known in the world were discovered in the Mesabi Range of northeastern Minnesota. The large lower grade taconite deposits had locally been fractured, weathered, and leached of worthless rock material leaving behind the mineral hematite, which is 60 percent iron.

These rich iron ores were easily and economically brought together with the two other ingredients of steel-making, high grade coal and limestone, by the fortunate geography of the northern Great Lakes region. Iron ore could be brought down first by rail (and it was not uphill, an economically important fact for the transport of heavy iron ore) to Lake Superior where cheap water transport moved the ore to steel mills in Chicago. Then, later the Pittsburgh area adjacent to the rich Pennsylvania coal fields also became a steel producing center. Both areas had abundant coal and limestone to combine with iron ore to produce iron and steel.

Timely Discoveries and Inventions

Timely discoveries

The rich iron ore discovery came just when it was needed — at the time the railroads began to dominate transportation. The engines, the cars, and the rails all demanded great quantities of steel. The blast furnaces around Chicago, Cleveland, and Pittsburgh produced it. American steel production was only 20,000 tons in 1867. But, by 1895, it had passed the British production of six million tons, and reached ten million tons before 1900. Ultimately, a large steel network of rails stretched from coast to coast, an impossible task for that time had it not been for the great iron ore deposits which had been so timely discovered and developed.

Steel also built factories and machines by which more goods were produced. The railroads efficiently distributed the manufactured products such as steel farm implements for the pioneers breaking sod in the Midwest and the Great Plains. The railroad brought needed equipment and supplies to miners and ranchers of the mountain regions, and to the growing settlements on the west coast, previously supplied in part by ship which had to go way around South America.

Steel made the world's first sky-scraper possible. After the great Chicago fire of 1871, large areas of the city needed to be rebuilt. An architect named William Jenney demonstrated that walls of buildings no longer had to be used for bearing the weight of the structure. Rather, with the abundant and relatively cheap steel now available, he could build a steel frame to act as the skeleton of the building. Using lighter weight materials, the structure could be walled in. Thus was born the first sky-scraper, the 10-story Home Insurance Building finished in 1885. It was such a success that two more stories were added later. The giant steel mills came into being because of the rich iron ore deposits of the Mesabi Range which built the great railroad network, and then provided the structural steel to build the huge complexes of office buildings and factories we know today.

Copper and the electric age

About the time the steel business was booming, the electrical age was dawning. The electric motor had been invented about 1854. In 1879, Thomas Edison produced the first usable electric light, and visualized lighting cities. But how could electric current be transmitted to these lamps for use in the home, offices, and factories, and to the motors which could replace so much of the hand labor in the factory?

Again, geology had favored the U.S. with the presence of huge native copper deposits, some of the richest known in the world, on the Keweenaw Peninsula of Upper Michigan. These deposits were brought into production to meet the demands of the electric age. Copper became the workhorse for the electrical industry. Upper Michigan, located not far from the industrial East and Midwest where much of the copper was being used, produced huge amounts of this most useful metal. And it was inexpensive copper — native copper. A tunnel in one mine struck a deposit of pure solid copper about 50 feet long with an average thickness of about 14 feet, and weighing more than 500 tons. The copper, being so malleable, could not be blasted out, but instead had to be cut into small pieces. This procedure was economic because the mass was pure copper requiring little smelting and refining.

Michigan copper was made into thousands of miles of wire which carried electric power to homes and factories. It made the workday more pleasant and efficient, and domestic life brighter. Copper wire carrying electricity allowed factories to efficiently operate three shifts a day instead of one, if the production of the factories was needed. Copper helped to greatly increase the productivity of the American economy.

In the United States, the Rural Electrification Administration (REA) begun in the 1930s, brought light and power to rural America and substantially improved the living standards of people in rural areas. Copper wires carried the power and still do.

Copper and communication

In the 1830s, Samuel Morse established his telegraph line from Washington to Baltimore. Copper telegraph wires soon spanned large areas of the nation, first running along railroad tracks, and then spreading out and connecting many otherwise isolated communities with the outside world. Telephones began to appear, and copper wires were available to put this most useful instrument into many places. Business and industry were greatly helped by this communication system. All this was facilitated by the abundant rich copper deposits in Michigan which could be developed at just the right time to promote the electrical age in the United States in all its many and varied useful forms. It should be noted that the Michigan copper deposits fed far more money into the American economy than did the gold of the California gold rush.

Cheap steel, oil, and cars

Then came the development and mass production of motor vehicles made possible by the abundance of cheap steel combined with the discovery of oil in increasing amounts in many parts of the United States. After first finding oil in Pennsylvania, drillers soon discovered oil in New York, West Virginia, Ohio, Texas, Louisiana, Oklahoma, Kentucky, Kansas, Colorado, Wyoming, and California. Oil was available coast to coast. By 1909, the United States was producing more oil than all the rest of the world combined. With oil found all across the U.S., and with the development of trucks and automobiles, soon a nationwide network of roads and service stations was established. Travel came into vogue, and oil was inexpensive. The average citizen could afford it.

In 1930, the great East Texas Oil Field was discovered, the biggest ever found in the 48-adjacent states, and oil prices dropped briefly to as low as 4 cents a barrel. The United States found itself with more and more and cheaper and cheaper oil, and ultimately led the world from coal as the major fuel into the present dominance of oil. In myriad ways, oil has powered the United States to its dominant world position, including being the world's largest motor vehicle manufacturer.

Minerals and Two Wars

Ample mineral and energy minerals won two wars

In the United States in the early and middle decades of the Twentieth Century, mineral and energy mineral resources were in seemingly endless supply. The United States successfully provided its allies with vital energy and mineral resources first to win World War I, when it was said that the "Allies floated to victory on a sea of oil." U.S. oil supplies again played a vital role in World War II. Japan's and Germany's Achilles Heel was lack of oil. In terms of metals it has been said that both wars were fought out of the great hole in the ground which is the Hull-Rust iron mine on the north side of Hibbing, Minnesota.

Minerals: Standard of Living and World Power

With cheap and abundant mineral and energy mineral resources, the United States enjoyed a phenomenally rapid rise to the world's highest material standard of living. It is generally agreed that the physical standard of living involves how much energy each individual is able to draw upon and mineral resources which result in all the various things which processes using energy can produce. This high living standard includes cars, airplanes, ships, and trains, and energy to move them, comfortable dwellings, household appliances, and myriad other things enjoyed in industrial societies, including light, heat, and air-conditioning on demand. In these matters the citizens of the United States have done very well to the present, and mineral resources and energy to process them are a fundamental part of achieving this affluence, now enjoyed by more than a quarter of a billion people. Along with this developed the strongest military position the world has ever seen.

Abundant cheap energy, high grade metal deposits

In viewing the future, as compared with the past, it is important to note that the United States rose to its pre-eminent industrial position and its high standard of living on abundant, cheap energy, and rich mineral resources. It took much energy to mine and smelt the ores to produce the metals vital to industrial development. It took vast amounts of energy to conquer the frontier and do the work needed to convert a raw wilderness into the world's largest and most affluent society. As recently as the period of 1940 to the 1960s, much of that time the United States enjoyed $3 a barrel oil, natural gas at around 15 cents a thousand cubic feet, and coal at about $4 a ton, all available within the United States. Abundant and inexpensive energy sources and high grade iron and copper deposits were exceedingly helpful to a young and rapidly growing nation. High grade metal deposits take less energy than do low grade deposits to mine and smelt. The combination of high grade ores and inexpensive energy compounded to provide very inexpensive finished products to foster economic growth. Conversely, as ore grade decreases it takes more energy to produce the same amount of metal as previously. When this is combined with higher energy costs, the result is substantially higher end product costs. The earlier more favorable economic circumstances of high grade mineral deposits and low energy costs (coal and oil) will never again return to the United States.

The zenith

The peak of power of the United States may have been symbolized by its use of the ultimate energy weapon, the atomic bomb, to end World War II in 1945. At that time, the United States was the sole owner of this fearsome form of energy. It was the possession of a particular metal, uranium, within its borders which had allowed the United States to arrive at this zenith of world power.

Foreign Oil Dependency and Some Loss of Economic Control

With the break-up of the Soviet Union in the early 1990s, one might argue that the United States then stood alone with no other comparable world power on the scene. However, unlike 1945, in the 1990s the United States was no longer self-sufficient in its principal energy need — oil. In fact, it was importing more than half of its supplies, with no ability to reverse this trend. The United States no longer had complete control of its economic destiny. Part of it was now in the hands of foreign oil producers. And the continuing imbalance of foreign trade, in which imported oil was the largest single (and growing) component, was hurting the prestige and value of the U.S. dollar in world markets.

Future "Warfare"

After the explosion of the atomic bomb, the United States and the world may have entered into a time when it may be that wars of the future will not be fought by the violent methods of the wars of the past. Possession of atomic weapons now by a number of countries and the probable destructive ramifications of atomic warfare for all sides of a conflict are such that it is unlikely anyone would really win. Therefore, the military might of the United States may not be so important in the future as in the past.

China's Mao Tse-Tung said that, "political power comes out of the barrel of a gun." But, a more recent view is that, "political power comes out of a barrel of oil." The international influence which the oil-rich Persian Gulf countries, Saudi Arabia in particular, now have, coming from being relatively unimportant largely desert nations, would seem to bear out this observation.

The new battlefield may be on economic and industrial fronts, using mineral and energy mineral resources chiefly for non-military purposes. The United States already is in that conflict with its former military adversary, Japan, which has been doing very well on the economic front. However, without access to energy and mineral resources Japan would be almost totally crippled. It has a vitally exposed natural resource Achilles heel of which it is very much aware. Only its ability to purchase its mineral and energy supplies allows it to survive as an industrial power.

Resource Depletion

The United States rose to international economic dominance in record time, but in the process it used up many of its own and highest grade resources. The rich ores of the Mesabi Iron Range are now gone. All the high grade native copper mines of Upper Michigan are closed. The United States must now search for oil off the frozen north coast of Alaska and in the deep waters of the Gulf of Mexico. The U.S. is no longer nor will it ever again be self-sufficient in oil. Its oil reserves, once the largest known in the world, are now dwarfed by those of several other countries. Although it uses about 28 percent of the world's oil, the U.S. now holds only about four percent of the estimated proved world oil reserves.

The United States has changed from being an exporter of energy and mineral resources to a net importer on an increasingly large scale. In the process, the United States has also gone from being the world's largest creditor nation to being the world's largest debtor nation in less than 20 years. Oil imports now are the single largest item in an annual balance of trade deficit. These costs are certain to go up in the future.

In 1920, the United States produced 80 percent of the entire world's oil. In 1970, the graphed curves of U.S. oil consumption and production crossed. That year, the United States was able to produce just enough oil for its own needs. By 1994, more than half its oil was imported. The downtrend of domestic oil production and the increase in domestic consumption continues.

A Unique World Event

The saga of the astonishing rise of the United States in affluence and power will never again be duplicated anywhere in the world. There are no more virgin continents to exploit. The story of the growth of the United States has been a phenomenon beyond comparison. The question now is: where does it go from here?

Jones has drawn an interesting analogy between the past and present position of the United States with respect to mineral resources, and oil in particular:

> "For its first 150 years, the United States was a boy with so much candy that no matter how much he ate he always had some to give away. For the next 25 years, he had as much as he wanted provided he gave very little away. Today he cannot supply enough to meet his very large, and increasing appetite, and must, for the first time, go to the world's candy store and stand in line like every one else to buy it."(7)

BIBLIOGRAPHY

1 BOYER, W. H., 1984, America's Future. Transition to the 21st Century: Praeger Publishers, Westport, Connecticut, 168 p.

2 BROBST, D. A., and PRATT, W. P., (eds.), 1973, United States Mineral Resources: U.S. Geological Survey Prof. Paper 820, Washington, D. C., 722 p.

3 CAMERON, E. N., (ed.), 1972, The Mineral Position of the United States 1975-2000: Published for Soc. of Econ. Geologists Foundation, Inc., by Univ. Wisconsin Press, Madison, Wisconsin, 159 p.

4 CAMERON, E. N., 1986, At the Crossroads. The Mineral Problems of the United States: John Wiley and Sons, New York, 320 p.

5 GRONER, ALEX, 1971, The American Heritage History of American Business & Industry: American Heritage Publishing Company, New York, 384 p.

6 HESSION, C. H., and SARDY, HYMAN, 1969, Ascent to Affluence: A History of American Economic Development: Allyn and Bacon, Inc., Boston, Massachusetts, 896 p.

7 JONES, ARTHUR, 1976, The Decline of Capital: Thomas Y. Crowell Company, New York, 202 p.

8 KENNEDY, PAUL, 1987, The Rise and Fall of the Great Powers: Random House, New York, 677 p.

9 POSS, J. R., 1975, Stones of Destiny. Keystones of Civilization: Michigan Technological Univ., Houghton, Michigan, 253 p.

CHAPTER 9

The Extraordinary Geodestiny of Saudi Arabia and the other Persian (Arabian) Gulf Nations

The nations of the Persian Gulf are Muslim countries, and all are Arab, except Iran. The national language of Iran is Persian, with Kurdish, Turkic and several other languages also spoken. Over several centuries Iran (named Persia until 1935) has been a traditional enemy of the peoples of the Arabian Peninsula.(13) For that reason, Saudi Arabia does not recognize the name Persian Gulf, but calls it the Arabian Gulf. Regardless of these differences, each nation which includes in its borders any part of the Persian Gulf geologic province shares the extraordinary destiny given it by geological events of the past.

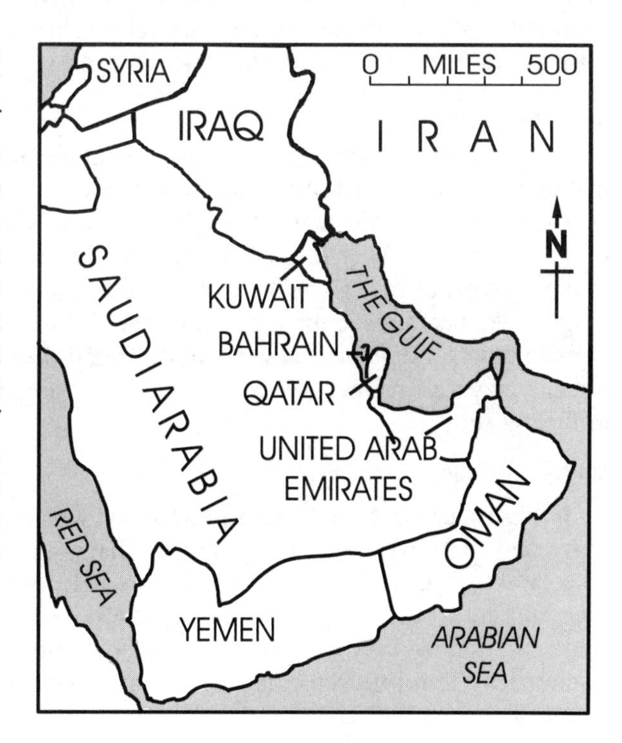

Figure 3. Geography of the Persian Gulf region.

(Source: From *The Persian Gulf Unveiled*, by permission of John Bulloch)

To understand the reason for the term "extraordinary" being applied to the Gulf nations, one must know something about how such a huge accumulation of oil as exists today in that region could be formed. Oil in the great quantities which can be so relatively easily and therefore cheaply produced as in the Gulf took a special combination and sequence of geological events. These have occurred elsewhere where oil is found, but not on the scale as happened in the Gulf. The oil accumulation in the Gulf region has no other world equal. It is extraordinary.

Forming Commercial Oil Deposits

Here is a simple description of how economic deposits of oil are formed (with apologies to my petroleum geologist colleagues for omitting many details). Consider what geologic processes had to do over millions of years to enable people to say "fill 'er up" at the gasoline station.(7,10)

Source beds

Oil is formed from the accumulated remains of myriad very small plants and animals. When these organisms die, their remains must quickly get into a place where there is no oxygen, for if they are oxidized they are destroyed, and their energy potential lost. That energy must be preserved so that the organic material can later be oxidized — burned as fuel in an engine or as a boiler fuel. Or its organic material must be preserved so that it can be used as a raw material for petrochemicals (plastics, etc.). The rocks in which oil is formed are termed "source beds" and are commonly a mud into which in a quiet water environment the organic material was deposited. It has to be a quiet water environment or else the relatively light organic material will be swept away. Sediments which are deposited in these quiet waters along with the organic material tend to be clays which, with the organic material, form a dark colored organic rich mud.

The "oil-window"

After entering an oxygen-free environment (technically termed anaerobic), the organic material must be buried to a depth which provides enough heat and pressure for it to be "cooked" into oil. If the temperature is too low, this will not happen, or if it is too high, the oil which may have been formed is subsequently destroyed, or will remain only as gas. (This is why, in general, at depths greater than about 16,000 feet, only gas is found). This optimum range of temperatures in the Earth in which oil forms is called by petroleum geologists the "oil-window." (Some basins which have been drilled were found to not yet have had their organic materials reach the temperature of the "oil window," — come back in another 20 million years!)

Reservoir rocks

If this organic rich mud enters into the "oil window" and oil is formed, then the oil must move into rocks which are permeable — that is rocks in which the oil fluid can move. It must be able to move if it is to ever get into a well bore drilled into the rocks. This permeable rock may be a bed of well sorted and not too well cemented sandstone, a limestone with some solution cavities in it, or any sandstone, limestone, or other rock which has been fractured by Earth movements so there are some spaces in which to hold the oil. This permeable rock which contains the oil is termed the "reservoir rock". It should be noted that the reservoir rock will have water in it, and the oil will migrate out of the mud source bed into the water-filled reservoir rock and displace the water, floating on top of the water. But the process is not reversible. That is, once the oil gets into the more permeable medium which is the reservoir rock, it will not go back into an adjacent less permeable rock. This relates to the importance of a "cap rock," and the fact that a fine grained tightly compacted rock such as a shale will not allow oil to enter.

Cap rock

But the oil, once it gets into a reservoir rock, cannot be allowed to continue moving upward until it reaches the Earth's surface. If it does, it will escape and be lost. So oil must encounter a "cap rock" which will prevent further vertical migration. This rock has to be relatively impermeable. Commonly this may be a stratum of shale, or a bed of gypsum, anhydrite, or salt. Many otherwise favorable situations for oil occurrence have been drilled only to find that the oil escaped long ago. There was no cap rock to seal in the oil, or by uplift the cap rock reached the Earth surface and was eroded away allowing the oil to escape.

Trap

A cap rock, however, does not prevent oil from migrating laterally. A barrier to its lateral migration must be present to stop the oil, much like a dam in a stream will form a pool of water. The barrier is termed a "trap," and some sort of trap is necessary by which to pool the oil in sufficient quantities to justify drilling. This barrier may be a fault which has moved impermeable rock against the permeable oil reservoir rock preventing further migration. It may be a plug of salt which has risen up from an underlying stratum of salt (the "salt dome" common in parts of the Gulf Coast of the United States) which has cut off the migration of the oil. It can be an elongate up-fold in the Earth's crust called an anticline, or it may be a dome-shaped fold. In either case the oil will migrate to the top of the folded structure as it is lighter than the water initially in the reservoir rock. But it cannot migrate back down the other side of the fold as the heavier fluid, the water, prevents this from happening. So the oil is trapped at the top of the fold. Commonly a gas cap will lie beneath the cap rock and above the oil. The driller tries to drill the upper part of these folds. Drilling too far down dip will find only water. The trap may be a tilted reservoir rock in which the upper end is less and less permeable to the point where the oil can no longer migrate. This is a lens or "pinch-out" type of trap. In oil traps, there usually are what are termed a "cap" and two "legs." There is a gas cap at the top, then an oil leg, and below that a water leg. The driller tries to drill into the oil leg, leaving the gas cap and the water leg to provide pressure to drive the oil to the well bore.

Best traps are big traps

Just as the bigger the dam in the stream, the bigger the pool of water, the bigger the trap the bigger the pool of oil is likely to be, assuming there was enough oil in the source bed to fill the trap. The biggest and most easily found traps are huge anticlines or domes. Some anticlines may be tens of miles in length and several miles across. The amount of oil which can be stored in such traps is enormous.

Some traps may be simply lenses of sand buried at depths of 10,000 feet or more. They are not visible at the surface, and can be hard to detect by subsurface methods. But anticlines frequently have a surface expression. The up-folded rocks stand out clearly. In early days of prospecting for oil, for example in Pennsylvania, Oklahoma, and California numerous anticlines were easily located by people with no geological training, and were quickly drilled up. If the anticlines did not have a surface expression, later the development of the reflection seismograph technology made it possible to detect buried folds and other types of traps. The first oil field to be located by this method was in Texas in 1924.

Silled basin

All these factors in forming commercial oil deposits are important, but the most important one is to have the oil-forming organisms quickly deposited in an environment where they will be preserved, otherwise there will be no oil. With no oil, reservoir rocks, cap rocks and traps are of no use.

The efficient preservation of large quantities of organic material can best occur in a slowly sinking "silled basin" — that is, a downwarp in the Earth's crust which will continue to sink and at the same time have a very shallow sill over which the basin is connected to the ocean proper. Water from the ocean can flow in and out of the basin and support the plankton (microscopic plants and animals — the source of the oil), but, because of the sill, these moving waters are surface water. The bulk of the water in the basin is very little disturbed. There is little turnover of the deeper water, and the bottom of the basin has an anaerobic environment — without oxygen. Here the dead organic material can accumulate and be preserved to be buried by sediments brought into the basin by adjacent streams, and ultimately may enter into the "oil window." It is the distilled concentrated energy of this preserved organic material which is recovered in the form of petroleum.

The Extraordinary Persian Gulf Oil Geologic History

Given these several factors which make for commercial oil deposits, what makes the Persian Gulf region special? Other places in the world have commercial oil deposits.

Classic silled basin

To have a lot of oil there must first be preserved a lot of organic material. The Persian Gulf is the classic geologic example where this has occurred. It was for millions of years a slowly sinking elongate sag in the Earth's crust with a very narrow, shallow passage to the Indian Ocean at the Strait of Hormuz. Over geologic time, organisms flourished in the warm surface waters of the Gulf. When they died their remains sank to the bottom of this silled basin where there was little or no turn-over of water. It is estimated that it takes at least 60 years for all the water in the Gulf to be changed.

There are other geological situations in which oil may be formed (they all need an anaerobic environment), but a big, silled basin is best, and the Persian Gulf is a classic silled-basin, oil-forming environment. And, indeed, it is classic for in this basin, because of its optimum oil-forming conditions during the geologic past, has accumulated more than half of the world's known oil reserves. There are other somewhat similar basins which also have become great oil producers, such as the Lake Maracaibo Basin in northwestern Venezuela, and the Permian Basin of west Texas. But the largest and best of all such basinal situations is the Persian Gulf. It formed great source beds for oil.

Reservoir rocks

Some of the reservoir rocks in the Gulf region are well sorted sandstones. Some of them are highly porous and permeable limestones. Both of these are exceptionally good storage strata for the huge quantities of oil which the excellent Gulf source beds produced. These permeable beds not only store large quantities of oil but allow oil to flow long distances to well bores.

Cap rocks

Both relatively impermeable beds of shale and of anhydrite are present in the Persian Gulf area. Each is an excellent seal, and some geologists believe this has been one of the most important factors in making the Persian Gulf region the oil reservoir it is today.

Traps

Eventually, as is the geologic history of many basins, the Persian Gulf region was compressed and folded. This resulted in some very large broadly arched structures, anticlines, and domes. An important feature of these structures in the Persian Gulf basin is that there are very few smaller folds on their flanks. They are huge broad folds. When this fact is combined with the excellent permeable reservoir rocks filled with oil, it means that one well can drain oil from a large area. If there were numerous small folds on the broad larger structures, or if the anticlines and domes were broken up by numerous faults, many wells would have to be drilled to obtain the oil. Fewer wells mean lower costs. Also, the huge folds are easy to find. Exploration costs are low. They are low also because the desert terrains and the shallow offshore areas of the Persian Gulf offer easy access to exploration, and drilling equipment. Building production facilities such as storage tanks and pipelines likewise encounter few difficulties. This can be contrasted with drilling in the stormy North Sea, or in the jungles of Colombia and Peru, where pipelines have to be built over the Andes by which to get out the oil. Or on the north slope of Alaska, where temperatures drop to 60 below zero F., and the oil must be piped 800 miles over three mountain ranges to a port.

The Persian Gulf basin was bigger in the past than it is now, with the result that there are oil accumulations beneath the land margins of the Gulf today, as well as beneath the shallow Gulf waters. Both environments are easy areas in which to work.

The total geological setting

In summary, the Persian Gulf region is a place where vast quantities of oil source beds were deposited. The accumulated sediments over the source beds provided enough heat and pressure to cause the organic rich muds to enter into the temperature of the "oil window." Cap rocks are present to prevent the oil from escaping upward. The region was subjected to broad folding which formed huge oil traps, relatively unbroken by faults or smaller folds. It is easy to find these big oil traps, and it takes relatively few wells to drain them. This results in very low exploration and production costs. These costs are variously estimated to be less than one dollar a barrel of oil compared with other oil producing regions where the costs may be several times that amount.

No other region of the world enjoys this exceptional combination of geological circumstances which has made the Persian Gulf region the holder of more than half the known world oil reserves. The countries which by good geologic fortune possess parts of this oil-rich basin are Iran, Iraq, Kuwait, Bahrain, Qatar, the United Arab Emirates, Oman, and Saudi Arabia — which has the largest part of the basin.

The Petroleum Geodestinies of the Persian Gulf Nations

It is obvious that petroleum has profoundly influenced the course of all nations who share this treasure in the Persian Gulf. How possession of petroleum has affected each nation differs in detail, but the broad blueprint which has been followed has been that of bringing relatively obscure and poor nations into the modern world in a phenomenally short time.

Saudi Arabia

This is largely a desert country, with a land area about a third that of the 48 adjacent United States. Ninety percent of it is too dry to be cultivated. It had been a rather loose organization of tribes, many of which were desert nomads, together with some fishermen along the coast of the Gulf. However, a very able desert chieftain, Ibn Saud, unified these peoples into a kingdom which was recognized by the British by treaty in 1927. Ibn Saud became King of the Hijaz and Najd and its Dependencies. The country was renamed the Kingdom of Saudi Arabia on September 22, 1932.

The discovery oil well was completed on March 3, 1938, a well which through 1979, had produced more than 27 million barrels of oil and is still producing today.(20) At that time the population of Saudi Arabia was approximately three million.

In the less than sixty years since oil discovery, Saudi Arabia has changed from the position of an undeveloped third world country to being an economic giant. It has constructed the world's largest and most expensive airport, and the world's most modern communication system with its own satellites. It has built railroads, schools through the university level, advanced medical facilities, miles of modern highways, and its own fairly sizable airline.

Early oil production was not very large, and therefore economic development proceeded rather slowly at first. Also, the years of World War II, 1939-1945, intervened to delay progress. The United States until 1970, was self-sufficient in oil, and had been a major exporter of oil so that there was no great demand for Saudi oil. As late as 1954, Saudi Arabia had only 147 miles of paved roads.

Oil and the development boom

With only 147 miles of road in 1954, by 1986 Saudi Arabia had more than 50,000 miles of pavement. The number of vehicles using these roads increased from 60,000 as late as 1970, to nearly two and one-half million in 1990. The Saudi Public Transport Company now has 900 buses providing low cost transportation between all sizable cities and many villages.

Saudi Arabia has come almost as far in less than 60 years in the rise in standard of living, and the use of modern technology and equipment, as the United States did in 200 years, or what European nations did over many thousands of years. In terms of coming into the modern world, the Saudis arrived almost overnight — making the trip from a simple agricultural economy to the technological Twentieth Century. Oil did it.

It is truly said that money cannot buy happiness, but it can buy almost everything else. That is the story of Saudi Arabia. It was able to make the leap into the Twentieth Century and obtain all the material things which characterize this industrial age, by simply buying it.

The Saudis did not have to wait for the development of the telegraph, the telephone, automobiles, the electric light, the radio, antibiotics, television, and jet airplanes. All these had been invented and were ready on the shelf waiting to be purchased when Saudi Arabia began to receive its oil money. The United States contributed much ingenuity and many inventions to the Industrial and Technological Revolution. Saudi Arabia simply bought it all, to some extent with money from the United States which paid for Saudi oil.

Saudi Arabia, along with its great oil deposits, also had another advantage which allowed it to come this far and so fast. It had a small population. Saudi Arabia has the world's largest oil reserve of any single nation, and the production profits have been spread over relatively few people. As a result, raising the standard of living in terms of material things has been a comparatively easy task. It could not have been so easy if the oil wealth had to be spread over, say, 100 million people. But the oil income has been substantial, and it was able to make a big difference in the lives of all Saudi Arabian citizens.

There is still no lack of Saudi oil to be produced, and it will be available well into the Twenty-first Century. But there have been some ups and downs, even for the Saudis with their oil wealth. In the middle of the 1980s, a temporary oil surplus developed. Saudi Arabia chose to support the price of oil by cutting back on production to less than four million barrels of oil a day, far below the potential flow, and the 10 million barrels a day they once briefly produced.

The welfare state and budget deficits

Cutting back on production, however, caused a domestic problem in that Saudi Arabia had begun a number of long term projects which had to be suspended. Also, social programs of various kinds were implemented with regard to health care, education, and other matters. Saudi Arabia has been called by some observers the world's largest welfare state. The problem is how to maintain it indefinitely, and with a growing population to serve.

As a result of production cutbacks, the Saudi government found itself short of funds. They announced on December 30, 1987 plans to borrow up to $8 billion to help finance the 1988 budget. To help cut the deficit, the 1988 budget was almost 17 percent less than the previous year's program. The new budget projected expenditures of $45.3 billion and income of $28.1 billion. By 1996, this deficit had been somewhat reduced, but Saudi Arabia was still spending more than it was earning.

Thus Saudi Arabia now has completely caught up with the United States and the other advanced nations by entering into the era of government deficit financing.(18,22) Welcome to the modern world!

Beyond oil?

The final chapter on the geological destiny which petroleum has given to Saudi Arabia has yet to be written. It depends on how the reality that petroleum is a finite resource can be accommodated. There is a Saudi Arabian saying: "My father rode a camel, I drive a car, my son rides in a jet airplane — his son will ride a camel." With at least 160 billion barrels of proved oil reserves and probable additions of 142 billion barrels(19) it may be that it may take an additional generation or two before the Saudis are back to camels, but someday the oil will run out. The present scene may then be looked back upon as the Golden Age, or more properly the Black Golden Age of oil.

Nevertheless, the rise of Saudi Arabia floating up on a sea of oil into the modern world, has been an unmatched phenomenon. They have made good use of most of their wealth. The problem for them now is to invest it in such a manner that they can build an economy which will last beyond oil. That is a very great challenge.

Other Persian Gulf Nations

The story is somewhat more mixed with regard to how petroleum has influenced events in the other nations which have shared in the Gulf's oil wealth.

Iran

Oil in the Persian Gulf region was first discovered in Iran, and for a long time a combination of British and Iranian interests, the Anglo-Persian Oil Company, produced the oil. But in 1979, all foreign interests were nationalized. Iran's oil reserves are very large, variously estimated to almost three times those of the United States. And, Iran is not so thoroughly drilled up as is the United States. There are still large areas to be explored. Iran's population is growing quite rapidly and oil continues to play a very important part in supporting the economy. With more oil still to be discovered and produced, Iran's geological fortune has not yet run out, but it may not be enough to keep the very rapidly growing population complacent, particularly as Iran has in the last few decades put a large amount of oil revenues into military operations.

Indeed, Iran's population growth appears to be outstripping what rise there is in oil income. The eight year war with Iraq (1980-1988) cost Iran $1 trillion, from which the economy has not yet recovered. Before the Islamic Revolution the annual per capita income was $3,500. It is now less than $1,200. Utilities and food prices have increased 150 percent which has caused some riots and even deaths. Of the 70 million population, 45 percent are younger than 15 years of age which portends a further rapid growth of population. With a growth rate of 2.29 percent the population will double in 31 years. The unemployment rate is 30 percent and the inflation rate is 40 percent. Iran is likely to be the first oil-rich Gulf nation, which, despite its oil riches, within ten years will be poorer on an individual basis instead of wealthier. This trend will cause internal social and economic problems which could spill over internationally in the vital Gulf region.

Iraq

This country, with an area of 169,190 square miles, slightly larger than the State of California, has had a turbulent history. It was part of the Ottoman Empire from the Sixteenth Century until it was taken from the Turks by British forces in 1916. It was made a kingdom under the League of Nations mandate in 1921. Iraq became independent in 1932, under the Hashemite Dynasty which was overthrown in 1958 by a military coup which formed a Republic. In 1968, the Ba'ath Party seized control and established the Revolutionary Command Council.

In 1980, Iraq invaded Iran and a full-scale war continued for eight years. The war eventually became a stalemate, and Iraq eventually gave it up but was left with a huge war debt. On August 2, 1990, Iraq, with no warning, invaded Kuwait. Even in the face of a large assembled military force from the western nations, Iraq refused to withdraw from Kuwait. At 12:01 the night of January 17, 1991, a coalition of forces led by the United States began an air attack on Iraq, and on February 24th a land offensive began wherein Iraq sustained massive military losses and was forced to withdraw from Kuwait. Over many years, Iraq has waged an internal war including poison gas attacks against its Kurdish population.

Iraq's oil fields are only partially related to the Persian Gulf. Considerable oil is produced in the northern part of the country from structures in a different geologic province. However, there are very large oil deposits in southern Iraq geologically in the Persian Gulf basin. Much of the oil of this region has only recently begun to be developed. Estimates are that several fields in this area are probably large enough to support production to the extent of as much as 300,000 to 500,000 barrels a day.(6) The stakes here are large, and one reason given for Iraq's invasion of Kuwait was a dispute between the two countries over their boundary which lies over a major oil field.

Iraq's oil reserves are estimated to be about 100 billion barrels, some four times those of the United States. Unfortunately, most of Iraq's excellent oil inheritance to date seems to have been squandered in military mis-adventures. However, there are still very large oil reserves to be developed, probably as much as 35 billion barrels.(19) If Iraq can alter its course from militarism toward a broadly based beneficial civilian economy, the average Iraqi may yet benefit considerably from that country's good geological fortune.

Kuwait

This country owes its entire modern existence as something not much more than a terrain of sand, to the oil of the Persian Gulf. What Kuwait has done with its oil riches is described in Chapter 10, *Mineral Riches and How They are Spent*. Holding about 10 percent of total world oil reserves, the Kuwaitis are keenly aware of how their destiny is very much under the influence of their disproportionately large participation (relative to their 6,880 square mile area) in the oil wealth of the Persian Gulf.

Bahrain

This country is a group of 33 low-lying islands with a total land area of about 340 square miles. It lies in the lower Persian Gulf about 18 miles off the coast of Saudi Arabia. The largest island, Bahrain, is 30 miles long and eight to 15 miles wide. Bahrain had its oil developed rather early among the Gulf nations. The discovery well was drilled by the Standard Oil Company of California (now Chevron) and came into production on June 1, 1930.

Bahrain's oil deposits are small, but because of its small area and population these have been sufficient to allow the country to also participate in the fortunes of the Persian Gulf region. Petroleum has brought Bahrain from being a largely fishing-based subsistence economy into much improved standard of living. With its strategic offshore position in the shipping lanes Bahrain has used its oil money to capitalize on this situation, becoming a Gulf trading and financial center.

Qatar

In land area Qatar is the second smallest country of the Persian Gulf nations covering slightly less than 4,000 square miles. Additional area making a total of about 17,000 square miles is claimed in the Gulf Waters. It is controlled by the Gulf's largest ruling family, Al Thani. Qatar is a barren peninsula scorched by extreme summer heat. Although oil was discovered in 1939, little was developed until much later. In the early 1960s it was a poor British Protectorate whose chief industry was pearling. Oil has since the middle 1960s changed it to an independent state with modern infrastructure, services, and industries. It was built mostly with foreign labor and technology, all paid for by oil.

In addition to oil, Qatar sits atop the world's largest known gas field, some 2,300 square miles in area.(3) In the early 1970s, Qatar flared about 80 percent of the gas produced with the oil. In 1974, it was still nearly 66 percent flared, but by 1979, less than five percent was burned off. Now some of the gas is converted to liquid natural gas (LNG) which is shipped out. It is also used by Qatar to manufacture petrochemicals which now total about one half million tons annually. The LNG goes mostly to the Chubu Electrical Power Company of Japan which has agreed to buy four million tons of LNG a year for 25 years. This represents about two-thirds of Qatar's expected annual gas production of about six million tons. With oil reserve depletion expected by about 2023, Qatar has also been building industries which can make use of its natural gas and provide domestic employment.

United Arab Emirates (UAE)

This nation is the consolidation of seven emirates (emirate defined as a state or jurisdiction ruled by an Emir, a native ruler) which were former British dependencies. The merger took place in the winter of 1971-1972. The total UAE area is somewhat uncertain due to disputed claims concerning some islands, but the land area is about 30,000 square miles. It stretches for about 300 miles along the southeastern end of the Persian Gulf. The UAE has oil reserves estimated to be about 61 billion barrels, with a probable 41 billion barrels yet to be discovered.(19) Oil income is huge, and the economy is almost entirely based on petroleum.

Before the discovery of oil, the products of these separate emirates were fish and pearls. Soil and freshwater resources were limited. The income used for foreign trade was based on slaves who dived for pearls. The slave trade continued until 1945. Other occupations were mostly family or small enterprises which hammered metals into pots, livestock herding, and limited date palm cultivation. A substantial part of the population were nomads.

Oil markedly changed these ways of life. People began to work in the oil industry in various ways. But the population was so small and unskilled that in order to take care of the rapidly developing petroleum economy, foreign workers were brought in. The total population was estimated in 1993 to be about 2 million, but of these only about 12 percent were actually UAE citizens, and they constituted only about seven percent of the labor force. What good fortune oil has brought to the UAE is briefly described in Chapter 10. The transformation has been striking.

Oman

Oman has an area of about 81,000 square miles, approximately the size of the State of Kansas. Desert makes up approximately 82 percent, mountains 15 percent, and coastal plain about 3 percent of the land area. Oman lies on the southeastern end of the Arabian Peninsula, and its northern coast borders the southwest side of the Gulf of Oman, which is an extension of the Arabian Sea. The rest of Oman borders the Arabian Sea which is the northern portion of the Indian Ocean (see Figure 3). Much of Oman is not in the Persian Gulf basin, but the northwestern portion of Oman which borders the UAE does include a part of the oil productive Persian Gulf geologic province.

Until the discovery of oil, Oman was the poorest country on the Arabian Peninsula. As recently as 1970, it had only six miles of paved road.(16) A complete census has never been taken, but the population is estimated at about 1.5 million. Oman's growth rate of 3.5 percent annually which means the population will double in 20 years.

Oil was discovered in 1962, but until the export of oil began in 1967, Oman's national budget was supported entirely by religious taxes, customs duties, and loans and subsidies from the British. Its economy now is dominated by petroleum and the service sectors chiefly related to petroleum. Petroleum contributes about 82 percent of government revenues. The government is using its petroleum wealth to try to develop a sustained economy based on renewable resources, chiefly fishing and agriculture. It also is building small manufacturing operations. Some of these are based on other mineral resources which include limestone (for a cement plant), and iron ore, copper, lead, manganese, nickel, silver, and zinc. Some of the deposits are fairly large. Coal also exists and is being developed against the time when oil will no longer be available for electric power generation. However, the coal reserves are small, only about 22 million tons, which will be sufficient for domestic use for a time, but not enough for export. Tourism is also being encouraged, and Oman now has seven international hotels, and several golf courses. It also has an electric power grid. In 1969, all of Oman had only one generating plant which produced just one megawatt of power, used only in the capital city, Muscat. Now several other communities are lighted, and an electric grid is planned for the interior. The power is generated with petroleum as the fuel.

Oman's development has been entirely petroleum-based, and its immediate future must rely almost solely on that resource. In the somewhat longer term, Oman's metal deposits, now being developed with oil revenues, may help to maintain the economy. Whether a truly sustainable economy can be built from these resources, especially against a rapidly expanding population, is a question. For the present the good fortune of possessing petroleum has given Oman an entrance ticket to the modern world.

A Brief Fortunate Interval

No other countries have been so profoundly changed in such a short time as have the Persian Gulf nations by the discovery of the single resource, petroleum.(4) Their destinies have been forever altered by geological events which no one surmised a century ago.

Appropriately, these Muslim nations have come to call their petroleum riches "a soft loan from Allah." With the discovery of huge oil reserves in the United Arab Emirates, the President, Sheik Zayid ibn Sultan al-Nahayan, said, "the problem is not how to get money, but how to spend it.(15) The same could be said for all the other oil-rich Gulf nations. They know that this is a brief fortunate event in their countries's histories. One of the things which has come of this is the recognition of the need for what has been termed "Arab Future Studies."(1) These studies are now underway to make practical workable plans for the time when the oil is gone. It was the geodestiny of these countries to become rich from petroleum. But as oil is finite, it is also the geodestiny that these countries must eventually exist without the resource which made them rich. Beyond any other countries, the Persian Gulf nations clearly demonstrate the vital role which Earth resources play in determining the course of nations.

BIBLIOGRAPHY

1 ABDULLA, ISMAIL-SABRI, et al., 1993, Images of the Arab Future: St. Martin's Press, New York, 242 p.

2 ABERCROMBIE, T. J., 1979, Bahrain: Hub of the Persian Gulf: National Geographic, v. 156, n. 3, p.300-328.

3 ANONYMOUS, 1989, Qatar's Huge North Field is Biggest Middle East Gas Project: Oil and Gas Journal, May 22, p. 50-51.

4 BULLOCH, JOHN, 1984, The Persian Gulf Unveiled: Congdon & Weed, Inc., New York, 224 p.

5 FINDLAY, A. M., 1994, The Arab World: Routledge Press, New York, 224 p.

6 GROVE, NOEL, 1974, Oil, the Dwindling Treasure: National Geographic, p. 792-825.

7 HAUN, J. D., (ed.), 1971, Origin of Petroleum: American Association of Petroleum Geologists, reprint series n. l, Tulsa Oklahoma, 192 p.

8 HUNTER, BRIAN, (ed.), 1992, The Statesman's Year-book, 129th Edition: St. Martin's Press, New York, 1702 p.

9 KIPPER, JUDITH, and SAUNDERS, H. H., 1991, The Middle East in Global Perspective: Westview Press, Boulder, Colorado, 359 p.

10 LEVORSEN, A. I., 1967, Geology of Petroleum: W. H. Freeman and Company, San Francisco, 724 p.

11 LINDSEY, GENE, 1991, Saudi Arabia: Hippocrane Books, New York, 368 p.

12 MANN, C. C., 1969, Abu Dhabi: Birth of an Oil Sheikdom: Khayats, Beirut, Lebanon, 141 p.

13 MANSFIELD, PETER, 1991, A History of the Middle East: Viking, New York, 373 p.

14 METZ, H. C., (ed.), 1994, Persian Gulf States. County Studies: Federal Research Division, the Library of Congress, Washington, D. C., 472 p.

15 NATIONAL GEOGRAPHIC, 1972, Peoples of the Middle East (map): National Geographic, Washington, D. C.

16 RANGE, P. R., 1995, Oman: National Geographic, v. 187, n. 5, p. 112-138.

17 REIFENBERG, ANNE, and TANNER, JAMES, 1995, U.S. Oil Companies Fret Over Losing Out on Any Jobs in Iraq: The Wall Street Journal, April 17.

18 REIFENBERG, ANNE, and PEARL, DANIEL, 1996, Oil-Rich Gulf Nations Fall Deeper in Red. Despite Rise in Crude Prices, Spending tops Revenue: The Wall Street Journal, June 4.

19 RIVA, J. P., Jr., 1995, World Oil Production After Year 2000: Business as Usual or Crises?: Congressional Research Service, The Library of Congress, Washington, D. C., 20 p.

20 TIME-LIFE BOOKS EDITORS, 1986, Arabian Peninsula: Time-Life Books, Alexandria, Virginia, 160 p.

21 WORMSER, M. D., (ed.), 1981, The Middle East, 5th Edition: Congressional Quarterly Inc., Washington D. C., 275 p.

22 WALDMAN, PETER, PEARL, DANIEL, and GREENBERGER, R. S., 1996, Terrorist Bombing is Only the Latest Crisis Facing Saudi Arabia: The Wall Street Journal, June 27.

CHAPTER 10

Mineral Riches and How they are Spent

Eons to form

Mineral resources have taken hundreds of thousands or in many cases millions of years to be formed in nature. The banded iron deposits of Minnesota called taconite, took several million years to be deposited. Then by the process of uplift and subsequent eons of weathering, these deposits were locally further enriched to form the extra high grade iron deposits called hematite. These deposits then remained in the Earth for more millions of years before being mined.

In the case of oil, untold myriad organisms lived and died, and sank into an unoxygenated watery environment to be slowly buried by accumulating sediment. Ultimately, when buried to sufficient depth and with proper increase in temperature, some of these deposits became oil which had to further migrate into reservoir rocks which in some cases themselves had to be arched, or pierced by salt domes, or faulted to form traps for oil. Economic oil formation is a long and complex process.

Used in an instant

This is the great heritage of geological events upon which we now build and maintain our civilization. But, in a geological fraction of a second these resources are now discovered, developed, and consumed. Then these gifts from the past in the case of mineral energy resources, petroleum, coal, uranium, are gone forever. Some metals can be reclaimed and recycled. Still there is an inevitable percentage dissipated which cannot be recovered.

How the Wealth Has Been Preserved

One of the interesting and important aspects of the discovery and development of mineral deposits has been the way in which individuals, communities, and nations have used this great wealth which can be produced only once. Although the resources are consumed by the generation which discovers and develops them, there is an argument that in some fashion at least some of these riches should be passed on in some form to future generations who did not have the good fortune of living during the time of great mineral wealth.

To be sure, some of this wealth will be there in the future in the form of the buildings, factories, houses, roads, bridges, railroads, locks, canals, power stations, water lines, sewers, and many other structures for which mineral resources are used. But, in the case of fossil fuels, although they help to construct these physical features of civilization, their use leaves no legacy as much is lost to recreational use and for supplying immediate transient energy needs of the current population such as space heating.

Thus, the legacy is mixed in that in some cases the resource will be of great and lasting benefit to people and to regions, but in other instances it may literally mean that more people in the long run may suffer hardships as populations expand on a resource base which will eventually be depleted.

Individuals and Their Wealth

It should be noted that although it is the miner or the oil wildcatter who discovers the mineral, they are not necessarily the ones who will make the largest profit. Frequently it is the developer or entrepreneur who survives and reaps the greatest rewards. Miners come and go but the general store and the storekeeper selling the potatoes, flour, nails, and shovels stays on. It has been said that "those who mined the miners were the ones who got rich."

Leland Stanford — gold

Leland Stanford supplied the California '49-ers with their mining and household needs. The miners are lost in history, but Leland Stanford survives as the founder of a great university which he named The Leland Stanford Junior University (still the legal name), in honor of his son. The gold of the Sierra Nevada has largely been mined out, but its legacy, through Stanford University, in the form of educated human minds, is the most important resource of all. A nation can benefit greatly from the discovery of mineral resources long after that resource is gone.

Levis

An interesting sidelight to the California gold rush was the origin of the now famous blue jeans called Levis. Levi Strauss was a poor immigrant in New York. He made tents out of canvas material. His brother went to California during the gold rush and enthusiastically wrote back to Levi that there was a great demand for tents for the miners. But by the time Levi arrived in California the demand for tents had fallen off. Instead, there was a great need for durable work pants which could also be made from heavy tent-like material, denim, with which Levi worked. Levi Strauss set up his factory in San Francisco which now supplies the famous Levis to the world — in a sense a legacy from the California gold rush.

John D. Rockefeller

It is reported that students going to the chapel at the University of Chicago, when they sang the doxology, would sing, "praise John from whom oil blessings flow." It was John D. Rockefeller, with his oil money, who established the University of Chicago. At this same prestigious university, from a laboratory underneath the stadium, emerged the atomic bomb which ended World War II.

Rockefeller became an object of much hostility because of his single-minded development of the great oil complex which became the Standard Oil Company. In 1911, it was broken into many parts by anti-trust legislation. But, even before he had any wealth, Rockefeller was a generous person. During the first year he worked, he gave frequent small contributions to charities, missions, and Sunday schools. Later, as his wealth greatly increased from his oil interests, his gifts became very large, and he ultimately donated $530 million to various causes.

Rockefeller established the Rockefeller Sanitary Commission, later incorporated into the Rockefeller Foundation which promotes world-wide health. Its research determined that the hookworm disease could be eliminated, it conducted research on tuberculosis, yellow fever, pellagra, malaria, and viruses. The Foundation made and continues to make numerous grants to medical schools, not to increase their number, but to improve their quality.

One of Rockefeller's less known grants was to Spelman College, a school founded in Atlanta in 1881 which was then and still is devoted to the education of African-American women. The school, known as the Atlanta Baptist Female Seminary when the grant was made, was renamed Spelman in honor of the abolitionist parents of Rockefeller's wife, Laura Spelman Rockefeller.

Other bequests from the Rockefeller oil legacy include land given to create Acadia National Park. In 1901 Harvard University President Charles Eliot and George Dorr organized a board of trustees to acquire as much land as possible in the scenic Bar Harbor, Maine area, to set it aside for public use. The surprising single largest bequest was 11,000 acres given by John D. Rockefeller, nearly a third of Acadia National Park. Now, more than 3 million people visit the area each year. Rockefeller money also purchased the land for the site of the United Nations in New York. Rockefeller funds established Rockefeller University concerned with the causes and cures of diseases, population studies, and related disciplines. Land contributed for the site of Memorial Sloan-Kettering Cancer Center, land contributed for the establishment of the Jackson Hole Preserve (Wyoming) "to preserve the primitive grandeur of the area," and a total of 33,000 acres of land for national parks in the Grand Tetons, the Virgin Islands, and parks in New York State were also Rockefeller gifts. Rockefeller oil money was among major contributions to the New York Zoological Society, Lincoln Center for the Performing Arts, the United Negro College Fund, and the Metropolitan Museum of Art. Rockefeller money restored the capital of colonial Virginia, Williamsburg, visited annually by more than a million people.

Altogether, Rockefeller established or was a major contributor to forty-seven philanthropies. Most, if not all will survive even beyond the entire time of the world Petroleum Interval. In the course of their existence, these philanthropies will touch millions of people who will never have heard of Rockefeller and his oil company.

Henry Flagler — oil

Not nearly so well known was Rockefeller's partner in Standard Oil, Henry Flagler. He used his oil money to open up Florida, pushing a railroad all the way through the state to Key West, building hotels along the way. He also bought huge tracts of land and lured people to Florida both to live and as tourists. Oil money helped to put Florida on the map.

Andrew Carnegie — iron ore

Carnegie built U.S. Steel Corporation ultimately drawing upon its high-grade iron mines in Minnesota and Michigan and became a multimillionaire by the age of 40. He was born into a poor weaver's cottage in Scotland in 1835. By the age of twelve he was the family bread winner, and that year he and the family emigrated to the United States where he became a telegraph operator at the age of 17. By savings and careful investments he eventually ran his own business and ultimately got into the steel industry and in 1873, established the first fully modern steel mill. The country was growing rapidly and steel was in great demand, especially for railroads. Carnegie made a fortune.

But always remembering the poverty of his past, Carnegie said that "it would be a disgrace to die rich." However, with this view he did not simply give money away. In fact, he was contemptuous of alms-giving. Instead, he gave his money, as he said, "to assist, but rarely or never do it all."

The route he favored for his money was to build "ladders upon which the aspiring can rise." To this end, in small towns all across America there remains a solidly built familiar structure, the Carnegie Library. He gave a library building to nearly every community which offered a site for it and would agree to maintain the building.(2) Ultimately, 2,500 libraries were built in English-speaking countries. Forty million dollars was spent for 1,679 libraries in 1,412 communities in the United States. These libraries originated from the iron ore deposits of the Upper Great Lakes region.

True, these libraries are small and for the sake of economy, most of them are of only one design, but for many towns this was the start of bigger libraries as communities outgrew the original structures. In many communities this was the first, and for many years, the only library for miles around. Who knows how many youngsters got their first introduction to science and literature in these places, and how many went on to distinguished careers which grew from this seed. These libraries continue to be places where students in relatively remote areas have a library. In the State of Oregon, Carnegie built 32 libraries, of which 16 are still in use in such places as Union, an isolated community of fewer than 2,000 people in the Blue Mountains of central Oregon. Without Carnegie money there would probably be no library there, but Carnegie brought one in 1912, where it has continued in service ever since.

The Carnegie Foundation, true to Carnegie's philosophy that his money should be used to "build ladders" on which people could climb, was the initial underwriter of the award-winning educational children's program *Sesame Street*. Untold children who have never heard of the man who made his money from iron ore have benefitted from this TV presentation.

The Carnegie Endowment for International Peace is another legacy from iron ore. This fund, in a variety of ways, seeks to promote world peace, and has been active in the cause from the time it was set up to the present day.

The Carnegie Museum in Pittsburgh, the Carnegie Institutes in both Pittsburgh and New York, and Carnegie Hall in New York are Carnegie's ideas of "ladders" upon which people can aspire to greater things. Carnegie also generously gave to Cooper Union in New York which was established as an evening school to train engineers, and was later expanded into daytime operations. Because it is endowed by Carnegie and others, it has no tuition charges. The Carnegie Foundation for the Advancement of Teaching based in Princeton, New Jersey, recently announced a comprehensive plan to revamp elementary education, and gave $1.5 million in grants to 13 schools across the United States to embark on this project.

Carnegie also gave generously to black industrial schools including Hampton and Tuskegee, which promote industrial arts and character building. He also established an interesting endowment, the Carnegie Hero Fund, which annually awards medals and financial grants to persons who have exhibited outstanding acts of bravery on behalf of fellow citizens. Carnegie gave money to Pratt Institute which was established as a co-educational school for "practical training." These are prominent examples of what American iron ore did for society at large. Although the high grade ore which made these things possible is now depleted, the legacy lives on.

Riches from iron ore even reached the Old World. Andrew Carnegie was a loyal Scot, and finally retired to Scotland, where, among other things, he set up the Carnegie Trust for

Universities of Scotland. Carnegie's first library grant was to his birthplace, Dunfermline, Scotland. In total, Carnegie gave away some $350 million dollars at a time when the dollar bought much more than it does now. Although he died without being poor as he had started, he gave away far more than he retained on his deathbed. The money which was left at this death, was given to various causes which he had designated. His children inherited none of his money, because Carnegie believed that people should earn their way in life.

Carnegie could be a generous Scot because the geological processes which took place more than a billion years ago in the Upper Great Lakes iron ore region of the United States had determined that he could become wealthy.

Cecil Rhodes — gold

A Rhodes Scholarship is one of the highest honors and most valuable awards a student can receive. By means of this fund some of our most distinguished leaders have obtained their higher educations. These scholarships were established by Cecil Rhodes who conceived the idea of re-working old gold mining areas in what ultimately became known as Northern and Southern Rhodesia, now Zambia and Zimbabwe. These regions were probably the greatest gold fields of the ancient world. The gold occurred in an area 500 miles long and 400 miles wide. The mines had earlier produced large quantities of gold, but sufficient gold remained so that about 90 percent of the approximately 130,000 new registered claims were staked on the sites of ancient workings. Rhodes, through his British South African Company, brought in settlers, chiefly miners, in the decade of the 1890s, and by 1905 this area was once again a world-class gold-producer.

Rhodes made a modest fortune from his gold mining interests. In his will he provided money for about 200 scholarships for a term of three years each on a perpetual basis. The British South African Company went out of business in 1923, when the gold deposits were depleted, but the intellectual legacy from the gold will survive as long as Rhodes Scholarships exist.

Guggenheim — gold, silver, copper, diamonds

The Guggenheim family derived their several individual fortunes from the silver of the Leadville, Colorado area; gold in the Yukon; copper in Alaska, Utah, and Montana; and diamonds in Africa among other mineral resources. The several branches of this family set up a number of foundations, and made a variety of charitable contributions including $12 million to the Mayo Clinic, and $22 million to the Mount Sinai Hospital. The Guggenheims established art museums and a free dental clinic.

The most famous of the foundations is that set up by Simon Guggenheim in 1924, in memory of his son. This, the John Simon Guggenheim Memorial Foundation, sponsors Guggenheim Fellowships which go to artists, scholars in the liberal arts, and to scientists. More than 10,000 individuals have benefited from this endowment, giving encouragement particularly to promising young talents. One beneficiary was the late Linus Pauling, who often gave credit to his Guggenheim Fellowship for helping him out at an important time in his life. Pauling later won two unshared Nobel Prizes, the only person ever to do so.

Hearst publications — silver and gold

The Hearst publishing empire is another legacy that came from mineral wealth. A substantial part of the Hearst family millions was derived from the silver-rich Comstock Lode discovered in 1859 at the present site of Virginia City, Nevada. Here, George Hearst of San Francisco made his first fortune. He did the same thing again in the gold strike in

South Dakota of 1877. With this money he established and expanded the Hearst publishing empire.

Rhodes, Carnegie, Rockefeller, Hearst, Stanford, and the Guggenheims stand out as individuals who more or less single-handedly left large cultural monuments. But thousands of people whose identities have been lost to history have also built lasting monuments by their collective contributions. Mining and oil boom towns soon brought families and the desire for something besides saloons as cultural centers. The now famous San Francisco Opera came to San Francisco as a result of the gold rush. The Alaska gold rush also brought the opera briefly to that northern land, but the opera did not stay.

Numerous Other Benefits

In many small grants, and some not so small, minerals have contributed to institutions of higher learning. It is a rare university or college in the United States and Canada which has not received money, equipment, or some scholarship from a mining or oil company. With an endowment fund of nearly $3 billion, chiefly from oil revenues, the University of Texas is the world's richest educational institution. To fill the very fine campus buildings, the University has hired outstanding faculty from great universities world-wide. But with oil revenues now declining in Texas, the institution is challenged to see if it can be maintained as it was in the more opulent oil-supported past. The trustees of the fund are well aware that oil is a one-time crop and have been working to invest the money wisely for the long term.

The Province of Alberta — Oil

With the major discovery of oil in Alberta after World War II when LeDuc Number 1 was brought in, both Edmonton and Calgary greatly expanded. The Calgary Philharmonic, now one of the premier orchestras of Canada, is housed in a beautiful facility paid for largely through the prosperity that oil brought. An examination of the back of any Calgary Philharmonic Orchestra program where the benefactors are listed is a quick way to find out which oil companies have offices in Calgary.

In Canada, most mineral resources belong to the government (the "Crown"). With their rich oil and gas fields, Alberta has enjoyed prosperity considerably beyond that of the other Provinces. With the oil revenues the Alberta fund has been established from which bequests are made each year to a variety of cultural and artistic organizations. It also provides various financial advantages to Alberta residents with respect to taxes and loans. Presumably, if carefully managed, this fund will survive long after oil.

Dinosaurs benefit

In 1985 Alberta completed a magnificent structure to house some of its more spectacular ancient residents. In the scenic badland topography on the northwestern edge of the old coal mining town of Drumheller, about 70 miles from both Calgary and Edmonton, now stands the world-class Tyrell Museum of Palaeontology. There is none better anywhere, and the star attractions, the dinosaurs, are wonderfully well displayed. In asking about the funding for this facility, a reply from the museum's director, Dr. Emyln H. Koster, described it this way: "As one of the more costly cultural facilities of the Alberta Government, the Tyrell Museum capital project coincided with an affluent early part of this decade due to a very buoyant oil and gas industry." It was money very well spent on a museum second to none.

Alaska

This state has become the richest state in the Union on a per capita basis, because it has a small population and the largest oil field in the United States. However, being so dependent on the fluctuating price of oil, Alaska has seen some briefly depressed times. But for the moment, Alaska is rich.

Since the 1967 discovery of oil on the North Slope (chiefly the Prudhoe Bay Field, but there are other smaller fields nearby), the State of Alaska has received to date about $35 billion in taxes and royalties from oil. The Prudhoe Bay Field provided most of it. How have these monies been spent, or invested? The Alaskan government, even before the 1967 Prudhoe Bay discovery, was spending 1.6 times the national average, with some of the money coming from petroleum produced in the Cook Inlet area. But, with the completion of the Alaska pipeline, oil and money from Prudhoe Bay began to flow in quantity.

Strohmeyer describes what happened then:

> "The state did not know what to do with the gusher of money that poured in after oil began to flow from Prudhoe Bay in 1977. Alaskans had expected the royalties would ease the lean state budget and flesh out capital expenditure. But even the most optimistic projections did not foresee the world events that would send oil prices into orbit. By 1981, oil had reached a high of $34 a barrel, an increase of more than 1,000 percent in thirteen years. As owner of the Prudhoe Bay oil fields, the state collected about 12 percent royalties from the sale of every barrel piped to Valdez. During the Iranian crisis alone, Alaska's revenue tripled to $.5 billion."(16)

Alaska increased its state spending 20-fold. As a result, Alaska currently has the lowest personal tax burden and highest per capita state government spending in the United States. With the new wealth, a number of state programs were set up including a loan subsidy program for farmers growing barley, which has resulted in $70 million of unpaid loans. As an interesting aspect of Alaska and barley it is reported that, "many of the farmers who were taking money from the state to grow barley were double-dipping —simultaneously taking money from the federal government to *not* grow barley."(7) Several hundred million dollars have also been lost in loans to other economically doubtful enterprises including a venture to produce a dog-powered washing machine. A performing arts center in Anchorage cost $70 million and loses one million dollars a year. High school students have been sent on European field trips.(7)

A state trust fund was set up with investments now about $16 billion. But in 1982, Alaskans voted themselves an annual dividend from the fund which takes most of its earnings. That year each Alaskan, man, woman, and child, received $1000. In 1996 each Alaskan received $1,130.68, including any baby born up to the last minute of 1995. A family of four got $4,522.72. A University of Alaska report says that the state is now spending about $1.2 billion more than it can continue to spend in the years ahead. It has been calculated that had Alaska increased its spending at just the average rate of other states over the past 25 years and earned a very attainable 5.15 percent on its excess oil income, that it would have had a $75 billion fund at the end of 1993. This would earn about one billion dollars more in annual interest than Alaska spent in 1993. Thus, Alaska appears to have missed the opportunity to enjoy permanent financial security.(7)

But the citizens have demanded the annual dividend payment in spite of the declining revenues from the Prudhoe Bay Field. By this law, the state had to give out some $423 million in 1988. It is hard to wean citizens off of this oil welfare and make them think of the future. (This is proving to be true in some oil-rich countries which also have started very generous welfare programs in various forms — free utilities in Kuwait, for example).

The present generation of Alaskans are living up their oil inheritance in several ways. Alaska has no sales tax, no income tax, and the lowest property taxes in the United States. A survey of the 50 states by *Kiplinger's Personal Finance Magazine,* August, 1995, reported that for a retired couple with a $50,000 year income, living in a 2,000 square foot house, the annual tax burden in Alaska would be $253. The highest tax burden among the states was in oil-less Wisconsin at $9,528. Having oil makes a difference.

No other oil fields of the size of the approximately 12 billion barrel Prudhoe Bay Field are expected to be discovered in Alaska. About 60 miles to the east of Prudhoe Bay, the geology suggests that perhaps about 3.2 billion barrels might be found in a small portion of the coastal plain of the Arctic National Wildlife Refuge. The amount of oil there will be uncertain, however, until the area is drilled. But, the idea of drilling at all has split Alaskans and others over environmental effects. The local native populations support the project as it would mean substantial income for them.

Alaska's oil revenues are certain to decline in the long run, to the point where there will be little or no such income. If future generations of Alaskans are to be considered, then the abundant but transient oil revenues of today must be used differently. In many aspects of daily life, humanity tends to live for the moment. Jackstadt and Lee examined the myriad political pressures which have prevailed upon the Alaskan politicians to spend money for local interest projects, and for the vote to pass out annual dividends to all the citizens: "This political process has allowed the current generation of Alaskans to capture benefits at the expense of future generations by plundering much of the wealth of a nonrenewable natural resource."(7)

The Prudhoe Bay Field initially held estimated total recoverable oil of about 12 billion barrels. Seven billion have already been produced. Production has peaked, and is now declining. Estimates are that even if new oil fields are put into production on the North Slope, sometime within the next ten years, oil production from that region will decline to only half of what it currently is. The economic impact of this circumstance, given Alaska's current spending habits, will be severe.

Countries

Mineral wealth not only funds institutions, and endows states and provinces, but in a number of instances it has transformed entire countries. In total, however, the use which oil-rich countries have made of their wealth is a very mixed bag. Pope has commented, "If oil wealth is like winning the lottery, nations act like lottery winners: They tend to blow the money."(13) Some do, and some don't. Examples of both follow.

Brunei — oil

This country not many years ago was a hot, moist, small country (2,226 square miles) on the northwest coast of Borneo. The population lived largely at a subsistence level. Then oil was discovered. Now Brunei is a rich, hot, moist, small country where citizens enjoy a per capita income among the world's highest, at more than $20,000 a year. However, this is somewhat misleading as a considerable amount of the oil income apparently goes to the

Sultan, the national ruler. Nevertheless, because of employment opportunities in the oil fields and increased secondary business from the presence of the oil industry, oil wealth has reached the general public who now enjoy more than one car per family. The government collects no taxes, and provides almost totally free education, medical expenses, and old age pensions. For those without financial means in this Muslim country, there is government assistance in making Hajj, the important pilgrimage to Makkah (Mecca).(6) There is also a generous welfare system, cheap subsidized food, and even subsidized television sets. For the 46 percent of the 85,000 workforce who work for the government, there is a free pension system and very low interest loans by which to buy cars and houses. How much of the oil income the Sultan receives is apparently a state secret. *Fortune* magazine has stated that the Sultan was the world's richest person, with assets of $25 billion. Another independent valuation puts the figure at $33 billion (1994 estimate). Those who manage the Sultan's money dispute these figures, but admit that the Sultan Sir Hassanal Bolkiah's personal fortune does put him in the multi-billion class. The Sultan and his large family reside in a palace with 1,788 rooms, costing $500 million. The throne room seats 2,000 people and is hung with twelve two-ton chandeliers.

The monetary reserves of Brunei are estimated to be between $25 and $30 billion, or more than $100,000 for each Brunei citizen. This money is held for the good of the country and is presumably being put into sound investments. The Brunei Investment Agency officials say they avoid common stock, but buy money market instruments, currencies, and property. The Sultan himself bought the Dorchester Hotel in London and the Beverly Hills Hotel in California. In 1994, he bought the Palace Hotel in Manhattan for a reported price of "more than $200 million."

Saudi Arabia — petroleum

The discovery of oil dramatically changed Saudi Arabia and the life of its citizens. The enormous oil wealth has presented some formidable investment challenges.

In 1932, the Saud family united the various tribes of the Arabian Peninsula into a kingdom, becoming the Kingdom of Saudi Arabia. The Port city of Jiddah as recently as 1940, was a walled city of about 50,000 inhabitants. Jiddah is now a thoroughly modern city of 1.5 million citizens with plazas decorated with sculptures, high-rise apartments, wide boulevards, and the tallest building in the Middle East, the 44-story National Commerce Bank. At one time during the development of all this, there were 355 ships waiting to unload cargo at Saudi ports.

As recently as 1962, there was only one radio station in all of Saudi Arabia. The first railroad did not reach the Capital, Riyadh, until 1951. In 1950, there were no paved roads anywhere in Saudi Arabia, but by 1982, there were more than 7,000 miles of main highways, and more than 6,000 miles of paved secondary roads. More than 15,000 miles of earth-surfaced rural roads had been constructed to some 7,000 villages, which, for the most part, had never previously seen a road.

The International Airport at Jiddah, constructed at a cost of four billion dollars, covers more than 40 square miles, and is 50 percent larger than Kennedy, La Guardia, O'Hare, and Los Angeles airports combined. The Jiddah airport, at the time of completion was the world's largest in area, and is capable of receiving as many as 100 aircraft an hour. There are four grades of terminals, the most lavish of which is the Royal Pavilion designed for the royal family. It has a copper roof, white marble walls, and its main reception hall has Thai silk wall paper and gold embroidered tapestries. Adjacent is an ultra-modern press room with complete radio and television equipment.

But this airport at Jiddah was soon overshadowed by the airport built 22 miles north of Riyadh. This facility is more than twice the size of Jiddah airport, covering an area of 94 square miles.

In 1978, there were only about 125,000 telephones in all of Saudi Arabia. By 1985, just seven years later, there were more than a million phones in operation, serviced by equipment supplied by major communications companies in the Netherlands, Sweden, and Canada. Saudi Arabia today has the most advanced computerized telephone exchange system overall of any country in the world. Because the entire system has been so recently built, there is no old obsolete equipment in service. As part of the communication system, Saudi Arabia has a domestic satellite system of 11 mobile and three fixed Earth stations, which allow almost anyone with a telephone anywhere in the Kingdom to dial directly to anywhere in the world. For a villager who 40 years ago had no roads, no lights, no sewer, no telephone available anywhere in the village, and spent the day herding sheep or camels, this is quite a change.

Schools through the university level, and hospitals have been constructed. The Saudis have spent vast sums of money to speedily bring themselves into the Twentieth Century. Although some inefficiencies are inherent in such a mammoth and rapid undertaking, most of the oil money has been employed wisely for the Saudis know that even the world's largest oil reserves are finite.

Along with the production of oil and development of large shipping facilities for the resource, Saudi Arabia has realized that, rather than being just a raw material supplier for foreign interests, it should try to process more of its petroleum into added value finished products. To do this Saudi Arabia has developed and is expanding a large petrochemical complex located in the newly established cities of Jubail and Yanbo. The cost is about $45 billion, and uses the talents of more than 40,000 people from 39 countries. These facilities may ultimately rival those for which the Houston, Texas area has long been known. The shift in emphasis from being a raw material producer to upgrading the resource to the finished petrochemical end products will have a large economic impact on other such facilities around the world. Saudi Arabia has the oil and gas reserves to support these plants for many decades, perhaps a century or more to come, or at least long after the reserves of the United States and of the North Sea (Britain and Norway) have been exhausted.

The shift of a raw material source to another country tends also to mean the loss of the plants which upgrade the raw materials. With these plants go jobs, as well as the peripheral smaller suppliers and related jobs supporting these major processing facilities. Losing the capability to produce raw energy materials or metals is much more important to a domestic economy than the simple percentage the raw material itself represents in the gross national product. This point is largely unrecognized by the politicians and the public at large in the United States where there has not been a great deal of general concern and support for the oil and mining industries. Indeed, these enterprises, fundamental to the economic well-being of the country, have been treated rather unkindly at times. But, in the meantime, Saudi Arabia, Qatar, and other petroleum producers are putting up refineries and petrochemical plants in their lands. Some metal producers are also doing the same thing by building plants to upgrade their raw materials before shipment abroad (Chile for smelting copper, Venezuela is using its gas to set up aluminum smelting facilities for its own ore and that of neighboring Suriname).

Water has been a limiting factor in the economic life of Saudi Arabia, but petroleum has come to the rescue there too. Using the abundant natural gas supplies produced along with the oil, the Saudis have developed desalinization plants now producing more than a billion gallons of fresh water a day from the sea. Electric power is generated using waste heat from these plants and other facilities all based on petroleum.

The story of the electrification of Saudi Arabia is perhaps the most spectacular of all. The first public electric generating plant was opened in Taif in the late 1940s. Since then, nearly the entire country has been electrified, and all major buildings are now air-conditioned.

One of the interesting items which oil brought to Saudi Arabia is a 60-ton solid granite bathtub. In 1983, King Faud ended his world-wide hunt for a perfect bathtub in a granite quarry in Manitoba. A flawless piece of granite was cut and hauled to Montreal and then shipped to Italy where some of the world's finest stone workers cut, sculptured, and polished it, including giving it the final touch, the crest of the royal family carved on the tub. It was then shipped to Riyadh and installed in the palace.

Mineral bonanzas can cause some quite unexpected things to happen in a national economy. In Saudi Arabia its huge petroleum deposits ultimately produced an unlikely result — a glut of wheat. In the late 1970s when the Saudis thought that the wheat producing countries might retaliate for oil embargoes by cutting off wheat shipments, Saudi Arabia began a program to encourage domestic wheat production. This was done by means of a subsidy which guaranteed farmers a price of $26 a bushel.

The program was so successful that in a few years the local production of wheat had reached two million tons whereas domestic consumption was only one million tons. The Saudis cut the subsidy price to $14 a bushel, still five times the world price at the time, so the wheat continued to pour in. When the government tried to cancel its commitment to buy the wheat, the farmers raised such a protest that the government continued to pay up.

The problem continues, and what is perhaps more serious is that Saudi agriculture, including wheat in particular, uses 84 percent of the water used in the country. Seventy percent of the water comes from underground aquifers which are being replenished very slowly. The water table is dropping. They have been mining their groundwater to produce a crop they do not need and paying five times the world price for it. Striking oil can cause some odd complications.

Another result of the Saudi oil bonanza has been the training of Saudi young people. At one time, virtually any male (women were generally excluded then) who could qualify academically could study abroad at government expense, in almost any subject he wished. Educated people will be the country's ultimate greatest asset — paid for by petroleum.

However, in spite of the large oil income, by 1985 the Saudi budget was in deficit, as expenses exceeded oil revenues, and they began to live off their accumulated capital. As of 1996, it was still running a deficit of several billion dollars. The affluent society so rapidly and spectacularly built by oil income faces more austere times, the long term effects of which will profoundly affect Saudi Arabian political and social structure.

Regardless of problems which have arisen from the sudden great wealth of oil money for Saudi Arabia, the country has energetically transformed itself to a thoroughly modern nation to the immense benefit of its citizens. Probably no other government and country has done so much to spread the benefits of the oil wealth to all its people. In some relatively oil rich countries the oil wealth seems to somehow "disappear" before it becomes evident in projects and services to the general public. But in Saudi Arabia, the benefits of the oil riches are widely evident although the royal family has had some advantages in this regard, which has caused some discontent.(14,19)

Kuwait — oil

On a smaller but similarly impressive scale, the story of Saudi Arabia and what oil did for it has been repeated in Kuwait. Kuwait is a very small nation which lies at the upper end of the Persian Gulf. It was a British Protectorate until 1961 when Kuwait was given full independence. Prior to the discovery of oil, it was an insignificant land of sand whose principal exports, such as existed, were skins, wool, and some pearls. The discovery of oil changed all that. Beneath that tiny desert area lies more than 90 billion barrels of oil yet to be produced.

With an area of 6,880 square miles, Kuwait is smaller than some counties in the United States, for example, Harney County, Oregon (10,166 square miles), but the proved reserves, that is, oil already discovered, are four times the reserves of the entire United States including Alaska, with a total area of $3\frac{1}{2}$ million square miles.

The discovery and development of oil has meant free education for all Kuwaiti citizens as far as they want to go. It means no income taxes, subsidized housing and utilities, free medical care, and the government gives more than $7,000 to every couple upon their marriage. The county has its own airline, Kuwait Airways, which flies Boeing 747s on regularly scheduled service from New York and London to Kuwait. The Kuwait government and private interests share ownership in Mobile Telecommunications Company which now has 110,000 mobile telephone subscribers, and 150,000 pagers. This is per capita the highest ratio of mobile telephones anywhere in the world. Among numerous minor facilities built by the Kuwait government, is an Olympic-size ice-skating rink, a novelty luxury in this hot dry land.

To develop Kuwait as rapidly as has been done, a great number of workers were imported. At the peak of activity, 75 percent of the work force were foreigners, and they out-numbered the Kuwaitis. There is obviously something of an internal threat in this situation, but it has merit also in that when times are slow economically or for security reasons the country can simply decree that foreign workers return to their homelands. More recently, that is what is being done. After the Gulf War, the number of foreigners in Kuwait has been greatly reduced. Nearly all the Palestinians have been forced to leave.

There are only about 600,000 Kuwait citizens among whom to divide the income from the world's third largest oil reserves. Understandably, Kuwait makes it difficult for any one to become a Kuwait citizen.

Rich as it is today, Kuwait has a keen eye on the future. Since gaining independence in 1961, Kuwait has invested its large oil revenues into a great variety of assets including $2 billion in the most respected stocks listed on the New York Stock Exchange.(18) Kuwait has set up what is called the Kuwait Reserve Fund for Future Generations. This organization and the Kuwait Petroleum Corporation have bought 14 percent of the West German car-maker, Daimler-Benz, 20 percent of the German mining complex Metallgesellschaft, and about a quarter of the giant German chemical company, Hoechst.

Perhaps endorsing the saying "there will always be an England" even after oil, Kuwait bought 5.1 percent of the British Midland Bank and has also invested heavily in English real estate. They own a one million square-foot complex of restaurants, shops, and offices along the Thames in London. They own about nine percent of British Petroleum Corporation. Kuwait has put more than $2 billion into Madrid, Spain, and is looking at possible investments in Australia, Singapore, Malaysia, and Hong Kong. In the United States, Kuwait has purchased Santa Fe International, a California-based drilling company, and they also have a very large portfolio of United States real estate including 100 percent ownership of the Kiawah Island Resort in South Carolina. They own several New York skyscrapers but have kept this fact obscure in deference to the Jewish tenants of some of these buildings. And they have invested in the stock of many of the top 500 firms (the *Fortune* 500) in the United States.

In order to assure a good market for its oil, Kuwait has bought most of the European market and refining operations of the former Gulf Oil Corporation (now part of Chevron Corporation), and they are adding to this system by building more gasoline stations of their own. The Kuwaitis have a brand of gasoline they market as "Q8" which is distributed through more than 4,700 retail stations in a number of countries including Italy, Denmark, Sweden, Belgium, the Netherlands, and Luxembourg.

Recently, Kuwait has been earning more money from these enterprises than from the export of its oil. As early as 1987, foreign investments earned $6.3 billion, and oil earned $5.4 billion.(11) When the oil is gone Kuwait hopes and believes that these investments will continue to allow them to pursue an affluent lifestyle like they now enjoy. Clearly Kuwait is very conscious of the finite nature of its oil riches, and has been working hard, and quite successfully to put itself in a position to survive beyond petroleum. A Kuwait government economist states, "Our investments will be the main source of income for generations after the oil runs out." In late 1989, Kuwait announced it was drawing up plans for multi-billion dollar petrochemical and petroleum refining industries to upgrade its end product rather than just shipping out raw crude oil. By 1995, this program was well established. Of the two million barrels of oil produced a day, 1.2 million was exported as crude. Of the other 800,000 barrels, part was refined and shipped as finished product, and part was further upgraded to petrochemical specialty products by Kuwait's Petrochemical Industries Company. The gas is used to produce fertilizer.

After the invasion of Kuwait by Iraq and the subsequent defeat of Iraq, Kuwait is even more conscious of the great wealth it holds and the need to carefully preserve it in some fashion for future generations. The Gulf War temporarily modified some of Kuwait's financial plans. The cost of the war to Kuwait used most of the cash it had accumulated (approximately $20 billion) for further investments. Kuwait also spent $11.7 billion on military equipment, including 200 U.S. made tanks, 40 fighter planes, and 200 British armored personnel carriers — a regrettable tribute cost to the instability of the Persian Gulf region, caused chiefly by Iraq and Iran. Mostly, however, Kuwait has spent its oil money on butter, not guns. In view of the vital interest which the western nations have in protecting Kuwait's oil demonstrated by the Gulf War, Kuwait's major emphasis in using its oil money for long term investments seems a wise course. The West will look after Kuwait militarily.

Bahrain

Oil production in this nation of Gulf islands is now declining. Recognizing this problem early, the government of Bahrain built a huge aluminum smelter which uses the energy from

Bahrain's gas deposits, which are a longer term reserve than its oil. The large number of oil tankers and other vessels related to the development of the Persian Gulf region made Bahrain's island location an ideal place at which to build large drydock facilities. Bahrain's location was also strategic for an international offshore banking center. This was developed and has thrived.(1) Also thriving has been the recently introduced more western style entertainment attractions, facilitated by a relaxation of some of the strict social rules of the adjacent Muslim states. The 22 mile causeway to Bahrain from the mainland allows for easy access to Bahrain's offerings.(3)

United Arab Emirates —oil

Dubai, of the United Arab Emirates, in six-page ads in American business magazines, quotes Sultan Ahmed in Sulayem as saying "We knew the oil boom would not last forever, and that a more stable and diversified source of capital was required." Dubai's answer to this has been to create what they call "the Hong Kong of the Middle East." This is a mammoth free-trade zone which first required they dig a 2,500 acre harbor at the edge of the desert. Around it on 25,000 acres an industrial and warehouse complex has been built.

The concept is that the United Arab Emirates location at the lower end of the Persian Gulf gives them a geographic advantage in being the trans-shipper of goods from the Middle East into Africa and Asia. The free-trade zone imposes no inconvenient tariffs, and offers a number of other financial advantages. Also, Dubai now has a first-class airport, its own international airline, four hospitals, and a golf course. A $14 million cricket and hockey stadium, tennis and squash courts, and a bowling green have been built. They also advertise that beyond golf, tennis, and the like, tourists can watch camel racing from a special viewing building. They further note that "One popular pastime pursued in Dubai that is found nowhere else in the world is 'wadi-bashing'. Its' rules are deceptively simple. It involves nothing more elaborate than driving in wadis, which are dry river beds that flow from the Hajar Mountains into the desert, or, on the East Coast, toward sea." So far, this "free-trade zone" project seems to have been a success.

Along with these quite useful developments, however, the United Arab Emirates (UAE) in 1995, began to spend money on armaments, and embarked on a large military purchase program which included up to 12 ship-based helicopters, at a cost of about $350 million. The UAE also negotiated contracts for the purchase of up to 80 long-strike warplanes, patrol aircraft, helicopters, and frigates. How effective this small country would be in defending itself in any major military engagement even with this equipment is questionable.

Oman — oil

This is the only nation in the Persian Gulf region where the ruler bears the title *sultan,* which denotes a Muslim sovereign combining political and religious authority. The present ruler, Sultan Gaboos, came to power in 1970. Commercial oil production in Oman began in 1967. At 1992 production rates, the reserves were estimated to last only about 17 years. With this in mind, the emphasis has been on diversifying the economy as quickly as possible by developing agriculture, fishing, and tourism.

Oman has also seen a large shift in population from rural areas to the urban areas. Severe housing shortages have developed. Providing the infrastructure of electric utilities, water and sewer in these increasingly densely populated urban areas has been a priority and taken a great deal of money. Oman, like the other Gulf countries, has also been spending its oil money to buy military equipment.

Iraq

This is a nation with tremendous oil resources but otherwise a rather poor country. Iraq's population has grown to the point where much of its food must be imported. Yet, under Saddam Hussein, Iraq spent huge amounts of money on weapons which were first used in a futile eight-year war with Iran, the total cost of which came to about $500 billion. This money went for massive amounts of military hardware which at the time of the Gulf War, included 4,000 tanks and 2,500 heavy artillery pieces. The use of Iraq's diminishing oil income to finance failed military adventures will go down in history as one of the greatest mistakes ever made by a national leader, and a tragedy for his people.

Iran — oil

This country with a rapidly growing population, also has used much of its oil wealth for military equipment. Initially it was the Shah who began the military built-up because he was determined that Iran would be the dominant nation in the Persian Gulf. He also wanted to be able to protect Iran from its long-time enemy, Iraq. He bought planes and tanks to the extent that it was estimated at one time there were three planes for every qualified pilot. Iran's military build-up continued after the Shah's departure. War with Iraq did come, and during its eight years Iran used virtually all its oil income for military expenses. Nothing was set aside for the future. Today, with the large increase in population during the past decade, oil income now available should be used for the daily necessities. The rapidly rising population (doubling in 24 years at the present growth rate), is now increasingly hard pressed to maintain a modest standard of living. Iran is spending some money to upgrade its existing oil refineries and to develop a modest petrochemical complex, but the government continues to spend considerable of its oil income on armaments. Iran has obtained two submarines and will get a third. It has purchased five Chinese-made patrol boats which can be armed with missiles, and, has obtained surface-to-air and surface-to-surface missile systems. These armaments have been placed at the mouth of the Persian Gulf, the strategic Strait of Hormuz, through which 50 percent of the world's oil is now transported. To counter this, the United States stationed 18 naval vessels in the region. Like Iraq, Iran's use of declining oil revenues for materials of war seem a very poor use of finite resources.

Libya — oil

Libya is another example of an initially very poor nation catapulted into the Twentieth Century by oil, and oil alone. Libya has an area of about 680,000 square miles, about one-fifth the size of the United States. Almost all of it is desert. Without oil, Libya would still be living on the fringes of the Twentieth Century, looking in from the outside, and it would continue to be mainly a country of nomads and small farms.

At the time oil was discovered in 1950, Libya had a population of about two million people. With only a narrow strip of greenery along the coast suitable for conventional agriculture, the land, in an agricultural economy, could not support many more people. Libya's impact on the world economy prior to the discovery of oil was slight indeed. But now it has oil revenues from production of about two million barrels a day, and its population has more than doubled to about four and one half million.

Libya has had a mixed spending program for this wealth. Early after the discovery of oil, under the leadership of the wise old desert chieftain, Idris, who became king, money was spent on roads, sewers, wells, and other generally useful public enterprises for the benefit of many people. But King Idris had to go to Greece for some medical attention, and while he was there, a Colonel led a coup to depose Idris. Thus, Col. Qaddafi took power, and the government spending emphasis shifted from civilian projects to military hardware. The ground equipment of 20 tank battalions, modern missiles, trucks, and artillery came mostly from the Soviet Union. For the air, French Mirage fighters were purchased. Six ex-Soviet submarines and numerous small surface craft make up the Libyan navy. In 1989, Libya was found to have built a chemical weapons plant. Some public works were continued, but the planning became a little haphazard. In part of a major road system, where clover-leaf interchanges were constructed, apparently the project was less than well thought out. At several places it was necessary to make a U-turn against on-coming traffic in order to get on another road.

But these things are of less concern than Libya's use of oil revenues in exporting terrorism, and invading neighboring Chad to the south. The terrorist inclinations were discouraged by a U.S. Air Force bombing attack, and by a U.S. Naval presence off the Libyan coast, which resulted in the shooting down of several Libyan fighter planes, and the sinking of some menacing torpedo boats. The Libyan invasion of Chad was deterred by French military forces which were dispatched to that inland country.

However, Libya has made a few constructive investments to keep it going beyond the time of oil. These include a one-half interest in an oil refinery in Hamburg, Germany. Libya also owns a small airline, Libya Arab Airlines (LAA). What may or may not be a good investment is a grandiose scheme to pipe water from deep below the sands of the Sahara to the coastal areas. Col. Qaddafi stated that his $25 billion Great Man-made River Project would convert the country, which is now 95 percent desert, into a "garden of Eden." The plan called for 12,500 miles of pipeline to carry water to all areas of the coast. However, as the wells were drilled in a desert area where there is no appreciable recharge, it is likely that the water there has accumulated over a long period of time. Any heavy pumping of the aquifer will easily exceed the recharge and water levels will drop markedly. There is, in fact, geologic evidence to indicate that the water to be used is what is sometimes termed "fossil" water — water preserved from the geologic past which cannot be duplicated again.(5)

The water must make a nine day, 750 kilometer trip through a pipe four meters in diameter. The announced expectations are that it will make the desert bloom and new lands will be opened for the rapidly increasing Libyan population. With considerable flourish Col. Qaddafi opened the valve of the initial stage of the project. However, just as the money spent on the current military which will eventually be of little or no value, this project also is likely to have a limited life. Probably contrary to the expectations of Moammar Qaddafi, rather than being a monument to great thinking, it will become a monument to folly. The groundwater that has been sparingly used in isolated oases could last for many thousands of years for the scattered peoples of the desert. But under the proposed heavy pumping, it will be a short-lived resource. It will provide the illusion of a great sustainable future which will not be realized. Ultimately the economy built on that false base will come to grief.

It is sadly apparent that tanks, guns, planes, and second-hand Soviet submarines will be a large part of the legacy which future Libyans will inherit from their once-affluent oil era. But continuing to pursue this course of spending, in early 1989, Libya purchased several Su-24 Soviet bombers. Again one has to wonder about the logic behind a desert country of about four million people, basically poor except for the temporary oil riches, spending its money on used submarines and bomber planes. These hardly seem to be good use of money obtained from essentially the only resource Libya has for making investments of more lasting value. The judgment of history on the present way in which these transitory oil riches are being spent is likely to be severe.

Oil buys arms

Military spending by the oil-rich countries, principally in the Middle East, and including Libya, has over-shadowed all other investments. In the 12 years following the first Arab oil embargo in 1973, with the subsequent huge rise in oil prices and concurrent great increase in revenues to oil producers, the Persian Gulf nations spent more than $640 billion for military purposes. Iraq and Iran spent billions between them, which served only to finance an eight-year war of attrition which ended in a stalemate. An estimated one million people were killed and probably twice that many were wounded. Without the oil money to finance the advanced weapons of war, it is probable that at least the casualty figures would have been somewhat less. Oil has been a mixed blessing to these countries. In the case of Iran and Iraq, the billions spent accomplished nothing but death for a million people and untold misery for many more. What a tragic way to waste forever the proceeds from a non-renewable resource.

Mexico — oil

Mexico is another nation which geology destined to have rich mineral resources. However, initially this may have actually been a curse for it brought in the Spaniards and others who plundered and ultimately destroyed much of the native civilizations, as well as bringing smallpox and other diseases which decimated many communities.

The mineral resources which Cortez and his cohorts sought were silver and gold. More recently the black gold, oil, has become the most important mineral resource of Mexico. Oil was discovered quite early in Mexico but only in modest quantities although what was probably the world's largest single producing oil well was drilled in Mexico making some two to three million barrels of oil a day for a time.

But it wasn't until the second half of the Twentieth Century that the great oil fields of the Tampico-Vera Cruz area, especially offshore, were discovered. As a result, Mexico has larger proven oil reserves now than does the United States. But, unlike Saudi Arabia, the oil of Mexico has not done a great deal to transform the country. The government-owned oil industry has been ineffective in translating oil wealth into general social good. True, it has built the tallest building in Mexico City, the Pemex building, which houses the Mexican national oil company, Petroleos Mexicana (Pemex). The company is not efficiently run, and is grossly overstaffed, with more than 108,000 employees. This may be compared with Exxon which has five times the revenues, yet employs fewer people.(10) Pemex has brought local prosperity to some of the coastal communities where the oil fields are located, but for the vast majority of Mexicans, the oil has had little effect except perhaps to keep the price of gasoline from being as high as it might be if Mexico did not have the oil resources. How much of the oil money has gone into increasing the Mexican standard of living is questionable.

An ironic result of the big Mexican oil discoveries may be that Mexico will actually be deeper in debt because of them. On the basis of the oil finds, foreign banks, most of them American, were persuaded to loan large amounts of money, now totaling more than 100 billion dollars, to Mexico. Where much of this money has gone is somewhat unclear, but what visibly remains is a huge foreign debt. The peso which collapsed was subsequently replaced by the "new peso". It, too, collapsed and reached an all-time low against the U.S. dollar in February 1995, creating an economic crisis. In borrowing against their oil reserves, the oil riches, in effect, have been spent before the oil is pumped. There is also some question as to whether or not the oil reserves actually are as large as claimed by the government for collateral on the debt.

Nauru — phosphate

This little Pacific island nation has a single mineral deposit, phosphate. This phosphate deposit was nearly depleted by 1996, and the last will be mined by the year 2000. Nauru's challenge has been to place investments somewhere which would continue to support the population after the resource is depleted. To handle these investments, the country formed the Nauru Phosphate Royalty Trust. One such investment is Nauru House, the 52-story tallest office building in Melbourne, Australia. The building is big enough to accommodate the entire population of Nauru if they chose to convert it into a hotel.

Nauru also has its own airline, consisting of five Boeing planes. Air Nauru touches all major air terminals in the Far East, and also is the only line to run a north-south route through the mid-Pacific region. Nauru has a shipping line, the Nauru Pacific, and two excellent fishing vessels. Looking toward both immediate and future tourist trade, Nauru bought a square-mile tract of land in the West Hills district of Portland, Oregon, with the intention of making it a luxury home development. Other investments include hotels and golf courses in Guam and Hawaii.(15)

A few of the investments have not done well, including $2 million in a failed London musical, and some $60 million placed on the advice of an Australian law firm into what proved to be bogus letters of credit and bank notes. Nauru has recovered about $48 million of this so far.(15)

Whether the good investments can become the permanent source of income for Nauru before the phosphate deposits are gone remains to be seen. One plan as a solution for the future has been to use some revenue from the phosphate deposits to flatten out the coral pinnacles which now remain in the mined out area and then haul in topsoil and try to grow pandanus, mango, and breadfruit trees. But, environmentalists say that it is doubtful that the land can ever be made to produce sufficient food to support the population.(15) Another suggested possibility is to buy an island from a neighboring Pacific island country and move there. In any case, Nauru will be the first nation in the world to see its mineral resources on which it is mostly or wholly dependent, become totally exhausted, and it will have to go on into the future on the basis of its investments alone. It will be an interesting situation to watch. Nauru does issue very colorful postage stamps which collectors seem to like. How long this source of income will last one can only guess.

Nigeria — oil

This, the largest African nation in population, recently has become a major oil producer and is a member of OPEC. When the Arab nations cut off the oil supplies to the United States in the 1970s, it was fortunate that Nigeria at that time rather than Saudi Arabia was the principal foreign supplier of U.S. oil from the eastern hemisphere. Nigeria then had an oil surplus to make up in part for the oil which the Arabs did not supply and thus there was no great shortage. Nigeria continued to send regular oil shipments to the U.S., without which the U.S.' situation would have been much worse.

With its considerable oil income, Nigeria has embarked on a variety of projects, ranging from fine new government office buildings to steel mills. A bold plan of national industrialization brought hope to many Nigerians who flocked to newly established industrial centers in search of jobs.

When President Shagari, who was from the northern part of the country, came to power in 1979, he had barely enough votes from the other regions to capture the office. In order to establish a political consensus for his regime he tried to spread the oil bonanza money throughout the country in the form of large public-works projects. The effect of these projects was that the percentage of the Nigerian population living on farms dropped from 85 percent to 65 percent, as people moved to the cities. They also changed their tastes from home-grown millet, yams, casaba, and sorghum to imported foods, so that some of the advantages of the oil income in terms of providing foreign exchange were lost simply to fill the new demand to buy foreign foods.

International salesmen, seeing large amounts of oil money, entered the scene and sold projects to Nigeria which in some cases were not very well suited to the economy, with the result that some have become economic disasters.

Nigeria has made some efforts to invest its petroleum revenues in enterprises for the long-range national good. An example was the completion of a world-class fertilizer complex at Onne, Rivers State, near Port Harcourt. The $800 million project serves domestic as well as export markets. With over-cropped, nutrient depleted soils in many parts of Nigeria, this fertilizer will be most useful. But the problem with this project for the very long term is that the feedstock for this plant is natural gas (from which the ammonia fertilizer is made), so it is still dependent on petroleum and cannot last beyond the life of Nigerian petroleum. However, the project which now obtains the gas from the Aalakiri Field about nine miles away, will save Nigeria about $100 million a year in fertilizer imports. This is a substantial help in the near future, and for a number of years to come. But if agriculture and population expand on this non-renewable resource, what is the ultimate outcome? Can the greater population continue to be supported after the petroleum is gone?

Unfortunately, overall, Nigeria has not managed its petroleum wealth carefully. In 1995, *The Wall Street Journal* reported, "And Nigeria, after squandering its petrodollars, is now bankrupt. It needs $20 billion of oil company investment to keep its industry running, the World Bank estimates. Production from its huge fields discovered in the 1960s is now slackening."(17) Kahn has made essentially the same observations reporting that the political economy of Nigeria remains one of gross indebtedness, inefficiency, and mismanagement. An irreplaceable resource which could have done much more for the country is being largely squandered.(8)

During the last decade and to the present, much of the oil money seems to have been diverted into the pockets of a relatively few military leaders in charge of the country. In 1995, against the outcry of many nations, a leading dissident and several others were executed. One of the charges the dissidents had been making was that the oil money was not being properly distributed. The military controlled government apparently could not stand for these statements.

In 1995, Johnathan Powers in the *International Herald Tribune* wrote: "Above all, Nigeria compels notice because its political repression shows how, even when you have world-class oil reserves, wealth can easily turn to dross."

Norway

This is a land of mountains, beautiful fiords, and lots of rock. Less than four percent of Norway is arable. Consequently, Norway has always had to live in considerable part from the sea. It continues to do so from oil and gas discovered in the Norwegian sector of the North Sea. As a result. the Norwegian government revenues are now derived almost half from petroleum. Marianne Andreassen, a Norwegian finance ministry official, says of the oil income, "The money is dropping from the heavens down into our hands. It's a gift."(13) Norway now enjoys the second highest standard of living in Europe, after Switzerland. Nine of the 10 biggest corporate taxpayers in Norway are oil companies.(13)

How have these monies been spent? Norway has used much of it to keep unemployment low. Billions of dollars of petroleum revenues have been poured into what turned out to be money-losing projects in agriculture, iron mining, smelters, and fishing to keep people employed. Although this solves employment problems temporarily, it does not provide a long term sound economic base.

Norway has begun to change some of its priorities in spending its oil money, and is putting more into research. Hallvard Bakke, Minister of Cultural and Scientific Affairs, has stated, "Our main objective is to expand the possibilities of the Norwegian economy and to give it more feet to stand on so that it does not have to rely on oil and oil-related industries." An example of this is the decision to pursue genetic engineering and apply it to one of Norway's new and rapidly growing enterprises, aquaculture. Money is being invested to study the Norwegian fiords as a place to raise a variety of fish such as halibut, salmon, and cod. Perhaps by genetic engineering some fast growing fish species can be adapted especially to fiord ecology. Roads and bridges have also been built so that these are the best in Scandinavia. However, most of the money seems to have been spent in subsidizing a very generous social security system. All health care and higher education are free, and farmers enjoy very high crop subsidies. In 1991, the Norwegian parliament established a petroleum fund that was to get billions in oil money. But so far, the fund has received nothing, and the social security system faces problems. Even if Norway began immediately to put all the oil income into the fund, the fund will be broke five years before the social security demands reach their peak.

The Overall View

In review, we have seen mineral dependent countries such as Nauru which almost immediately faces the fact of its one and only mineral being totally gone. In contrast, Saudi Arabia, has a resource in petroleum which will last for many decades. But all these countries have built their economies on a depletable resource. It is evident that some have invested their income wisely for the future. These are chiefly the countries which have a small population and a large resource (for example, Kuwait, Saudi Arabia). They have had a surplus of cash from their oil income which they could invest either within their own countries or abroad.

Other countries, generally with larger populations, have had to use their mineral income to meet current expenses, and they have not been able to put away any great amount of investment money for the future. Brazil has been in this situation to some extent. It has opened up large, rich iron mines but the iron ore it exports unfortunately has not been able to pay for the oil it has had to import.

In some countries graft and corruption have siphoned off substantial oil income for the benefit of only a few.

Thus the legacy of how these mineral monies are used will run varied courses. Some run their course as soon as they are received — they are spent immediately in one fashion or another, or simply looted by the reigning political groups. But the mineral income money which is invested wisely by some of the governments today will markedly affect the lives of their citizens for the better probably for many generations to come. In retrospect, such leaders will be remembered and honored for their integrity.

Individuals seem to do it better

On a smaller scale, the wisdom and foresight of those individuals who have gained their wealth from mineral bonanzas and wanted to share them will also affect future generations through the universities, museums, scholarships, research foundations, libraries, and other beneficiaries of their gifts. These carefully and specifically designated uses of mineral riches may be the most efficient employment of these funds, rather than those monies which are subject to political hazards by being processed through governmental agencies.

Mineral riches can survive

But if money from minerals has been or is carefully invested either by governments or individuals, the high grade iron ore of Minnesota, now gone, and the oil of Kuwait and Saudi Arabia, after it has all been produced, can help many people in the future. Most of them will never realize from whence their benefits came. Many of us today have already had our lives touched at least in a small way by these endowed situations.

"We have all drunk from wells which others have dug."

BIBLIOGRAPHY

1 ABERCROMBIE, T. J., 1979, Bahrain: Hub of the Persian Gulf: National Geographic, v. 156, n. 3, p. 300-328.

2 BEEMNER, R. H., 1988, American Philanthropy, 2nd. ed.: Univ. Chicago Press, 291 p.

3 BULLOCH, JOHN, 1984, The Persian Gulf Unveiled: Congdon & Weed, Inc., New York, 224 p.

4 DEMPSEY, W. M., 1983, Atlas of the Arab World: Facts on File Publishers, New York, 38 maps, 42 p. text.

5 GARDNER, GARY, 1995, From Oasis to Mirage: The Aquifers That Won't Replenish: World Watch, May/June, p. 30-36.

6 HANSEN, ERIC, 1995, The Water Village of Brunei: Aramco World, May/June, Aramco Services Company, Houston, p. 32-39.

7 JACKSTADT, S. L., and LEE, D. R., 1995, Economic Sustainability: The Sad Case of Alaska: Society, March/April, p. 50-54.

8 KAHN, S. A., 1994, Nigeria. The Political Economy of Oil: Oxford Univ. Press, Oxford, 248 p.

9 KORETZ, GENE, 1995, Alaska's Lost Opportunity: Business Week, May 1, p. 30.

10 MACK, TONI, and MILLMAN, JOEL, 1995, Petroleum Machismo. Mexico is in Crisis, But at the Powerful Nationalized Oil Company it is Busines as Usual: Forbes, April 10, p. 46-49.

11 METZ, H. C., 1994, Persian Gulf States. Country Studies: The Library of Congress, Washington, D. C., 472 p.

12 POLLACK, J. R., and DAVIS, J. M., Alaska at the Crossroads. With Prudhoe Bay in Decline, What's Next for Alaska?: Oil and Gas Journal, August 3, p. 35.

13 POPE, KYLE, 1995, Norway's Oil Bonanza Stirs Fears of a future When Wells Run Dry: The Wall Street Journal, October 3.

14 ROBERS, JIM, 1995, The Tent of Saud: Worth Magazine, p. 35-40.

15 SHENON, PHILIP, 1995, Pacific Island Nation is Stripped of Everything: The New York Times, December 10.

16 STROHMEYER, JOHN, 1993, Extreme Conditions: Big Oil and the Transformation of Alaska: Simon & Schuster, New York, 287 p.

17 SULLIVAN, ALLANA, 1995, Western Oil Giants Return to Countries That Threw Them Out: The Wall Street Journal, March 9.

18 SYMONDS, W. C., et al., 1988, Kuwait's Money Machine Comes Out Buying: Business Week, New York, March 7, p. 94-95, 98.

19 WALDMAN, PETER, PEARL, DANIEL, and GREENBERGER, R. S., 1996, Terrorist Bombing is Only the Latest Crisis Facing Saudi Arabia: The Wall Street Journal, June 27.

CHAPTER 11

Minerals, Money, and the "Petro-currencies"

This chapter might well have been titled "Minerals as Money" for during many centuries the two were one and the same. They still are in a few places. Pieces of gold and silver, and sometimes copper, lead, tin, and zinc were literally used as money. When people first began to travel relatively long distances from their homeland, they encountered barter systems different from what they had known and there was a need for some common medium of exchange. Copper ingots were used, as well as smaller pieces of gold, silver, and bronze.(11) Gradually, gold and silver became the preferred metals for exchange even before coins were invented. Pieces of these metals were passed back and forth among merchants who kept scales for weighing them.(14) In the eyes of many people, pieces of gold and silver still remain the only valid forms of money, and are called the "currencies of last resort." In India and China they are still used as that. And, as long as politicians have the records they do in managing nations' finances there will remain good reasons for that view. Recent peso and ruble devaluations reinforce the perception.

Honest Money = Store of Value

Money, to be truly money in the classic sense, has to be a store of value whereby one person's goods or services can be traded for money and that money later can be given for goods or services of equal value. If inflation intervenes between these two transactions, paper money does not retain its initial value. On the other hand, gold and silver have proven to be much more dependable. Paper money, at times, has actually had its value decrease by half or more in just a matter of days, as in the German inflation of the early 1920s. In Brazil, Peru, Chile, Mexico, and a number of other nations, inflation has at times exceeded 100 percent a year. The degree to which paper money can become virtually valueless is striking. At the time this is being written, there are 40,917 Turkish lira, and about 5,000 Russian rubles to the dollar. When I visited the Soviet Union a few years before its breakup, the ruble's official exchange rate was at U.S. $1.60.

Printing paper versus mining metals

It takes energy and materials to wring gold, silver, copper, and zinc from the Earth, and smelt and refine them. It takes much less energy to cut down a tree, process it into numerous pieces of paper, and then print some number on the bits of paper and call it money. Also, unlike a gold coin which has a fixed gold content, one ounce, for example, a piece of paper can be printed in any denomination. You cannot legitimately stamp any other figure than "one ounce" on a one ounce gold coin, but governments can and do print any number they wish on a piece of paper and call it money.

Johannes Gutenberg, with his invention of the printing press, in effect, invented inflation via paper and every politician who votes for deficit spending should have a statue and a shrine to Gutenberg in his office. His invention, in the hands of fiscally irresponsible governments, has destroyed the life savings of many people around the world, and continues to do so.

Metal coins

The first use of metal for coins is lost in antiquity, but it may have started in China as near as history can tell (a likely possibility as the Chinese seem to have been first in a great many things). Interestingly enough, very shortly after the Chinese began issuing precious metal coins, other Chinese began to make them wholly or partially out of lead, so it may be also that the Chinese can be given credit for the lead nickel, and counterfeiting in general. This counterfeiting became so rampant that Chinese officials ultimately decreed that anyone operating a lead mine without government sanction would be executed.

The use of metal and metal coins as the only money continued for many centuries. The capture of state treasuries such as the Persian, the Babylonian, and the Greek were major military objectives. Along with slaves, the gold and silver booty was paraded through the streets of the victors. The parade of the precious metals and the slaves vividly illustrated the importance of two things which are still vital for civilization — mineral and energy resources, the slaves representing at that time energy in useful form.

Sal-ary

After Greece, Rome gradually became the dominant power in the Mediterranean area. Rome did this by developing a large and well disciplined army. The soldiers were given a ration of salt as part of the monthly payment for their services, and from this we have the expression "is he worth his salt?" The fact that payment was part in salt in regular installments also gave rise to the term "salary" derived from the Latin word *sal,* for salt.

Rome, Precious Metals, and Money

The Italian Peninsula has a fine climate but it is deficient in precious metal. To pay the army with something in addition to salt, Rome needed sources of gold and silver. Some had been obtained from the looting of the Greek treasuries, but that supply ultimately dwindled. However, when Rome defeated Carthage, among the spoils of the campaign was an area known as Spania, now called Spain. From Spain, the Roman military returned with much gold and silver. Rome also took possession of the mines which produced these metals. At one time, the Romans had 40,000 miners working the silver deposits of Spain, sending all the metal to Rome.(4) Silver ultimately became the main metal of Roman coinage, and the unit of exchange was the denarius. Rome, as noted, had very little in the way of precious metals, so it was the mines of Spain as well as the earlier captured Greek gold and silver which financed Rome's armies, and made Rome the dominant power in the then known civilized western world.

The Romans, on their delightful peninsula, had luxurious villas overlooking the various bays on the Mediterranean. They also had beautiful homes across from Italy on the south shore of the Mediterranean, for they had taken over parts of North Africa including what are now Libya and Tunisia. The Romans loved luxury, and the gold and silver captured by their armies, and that produced from the mines worked by slaves in the conquered lands, chiefly Spain, provided the money for imports which Rome needed. Paper money had not yet been invented, so Rome could not run a deficit in balance of trade as do some nations today. Precious metals were the only medium of exchange. Ultimately, the precious metals from the mines of Spain became the chief source of money for Rome, and this situation continued for many years.

But, gradually a difficulty arose. The mines got deeper and deeper, and the problem of water flooding in the mines became increasingly hard to manage. The Romans were quite ingenious for a time with regard to this matter, but ultimately they did not have adequate systems or pumps by which to remove the water. Finally, the ore which could be reached in the mines was exhausted.

There were many causes of the fall of the Roman Empire, but one surely was the demise of the Roman currency. The great gold mines of Spain for three centuries had produced more than 300,000 ounces annually. The silver mines of Spain produced even more metal. With the depletion of the mines, Rome's treasury became empty. But Rome had to continue to issue coins to pay the army, for the army had become, in its later stages, an entirely mercenary force. The sons of Rome no longer served in the military as in earlier times. They chose to pay others to take their places rather than leave the luxuries of Roman living for the spartan military routine.

Debasing the coin of the realm

The loyalty of Rome's mercenary army was purchased, and not inborn as it would be with native sons, and the army definitely wanted to be paid in something of value. Replacement mercenaries could not be hired with debased currency. In 9 A. D. the Emperor Augustus was unable to replace the army of his General Varus which was destroyed by the Germans.(6) The Roman answer to the lack of silver and gold for their coins was to begin to debase the coins by adding other metals — chiefly lead and copper. In a period of only 50 years, the dinarius went from being about 70 percent silver to only about 10 percent silver. As the dinarius was losing its standing, a new coin, the "Antonianus" was issued and also the "double dinarius" but this had only 50 percent silver. When civil wars broke out, the expenses brought on a monetary crisis, and the silver coin was more rapidly debased and dropped to a 5 percent silver content. In fifty years it had fallen to less than a tenth of its original value.(4)

But the citizenry and the army were not deceived. If you visit the museum at Sabratha, the ancient Roman city a short distance west of modern Tripoli in Libya, you will see an interesting display of Roman coins. The earlier coins were pure silver and gold. But later coins, although carrying the same stamped face value on them, were debased. However, the citizens, the merchants, and the army knew about the cheapening of the coins. To compensate for this debasement of their money, they clipped the earlier, more valuable coins to, in some cases, merely narrow pie-slice shaped pieces, to equal the value of the later issued debased coins. On the wall where these coins are mounted, you see narrow slices of earlier pure precious metal coins which remained in circulation and which were equal in value in the contained silver to the value of the later still round but debased coins.

The earlier Roman coinage, with its pure precious metal content, was the most highly prized and valued currency of its time in the Mediterranean region. However, as precious metal supplies dwindled, and debasement of the coins proceeded, but with the government continuing to try to pass these debased coins as full value, the Roman currency became a despised currency and with this change of view, the prestige of the Roman government also declined.

By 218 A. D., the Roman Emperor Elagabalus made the basic Roman coin, the denarius, wholly of copper. This coin was not accepted by the eastern traders, which was a great blow to Rome for Rome had been the empire's trading center and the basis of its economic life.(8) In the year 220 A. D., silver was so scarce that the debased currency could no longer be supported, and the government repudiated its debts. This resulted in Rome's fall as a trader of consequence, and the inability of the government to pay for itself or the military.(6)

Eventually, by the fourth century A. D., the situation was so bad with Roman coins of various degrees of debasement in circulation, that it was necessary to weigh and assay each coin. In these circumstances the Roman currency became virtually worthless. The mercenary Roman legions could not be properly paid, and their loyalty to Rome disappeared with the disappearance of the silver and gold from the coins. With a debased currency no longer accepted by the traders, the Roman citizens could not import the luxuries which they had once enjoyed. The chief export of Rome had long been money accumulated from previous times when by conquest they had access to precious metal wealth. Rome had depended on plunder instead of creating wealth and when this policy began to fail, Rome began to decline.

Poss makes an interesting observation, perhaps with modern parallels:

> "This imbalance in trade contributed greatly to Rome's ultimate ruin. Paying for lavish quantities of imports, in amounts that exceeded exports, with a currency that had been depreciated since the Punic Wars and further debased by such emperors as Nero, Aurelius, Commodus, Caracalla and Severus, Rome's currency was becoming unacceptable. The Empire's trading capital, the basis for its economic life, was being destroyed.
>
> "With Rome thus facing economic disintegration, nothing could have added greater distress than the withering of her metallic supplies. Germanic hordes everywhere were rising in constant revolt. They began to overrun provinces of the Empire, pillaging Gaul, Spain, Macedonia, Cyprus, and Asia Minor, long the storehouses of Rome's invaluable mineral deposits."(8)

Finally, with the northern hordes beating on its doors, and with no money of value to pay for defense, the Roman Empire collapsed. The debasement of the currency was only one of the reasons for the fall of Rome, but it was an important one. It may have been the single most important one, for the Roman historian, Antonius Augustus, wrote "Money had more to do with the distemper of the Roman Empire than the Huns or the Vandals." Recently Vicker writes, "Romans could do many things: they couldn't control their money, and this may have played a part in their downfall."(14)

The Gold Standard

Allowing money to be freely converted into gold is what is termed the gold standard. This prevents governments from printing paper money which does not have a gold backing and restrains politicians from excessive spending. By the 1870s most major nations of the world had adopted the gold standard. The United States was one of the last, adopting it in 1900.

Until the Great Depression of the 1930s in the United States, the dollar was tied to the price of gold. That is, the dollar could be converted to gold at a fixed gold price. As long as the price of gold remained fixed (which it was for many years, first at 20 dollars an ounce and then briefly at 35 dollars an ounce), there was confidence in the U.S. dollar because it could at any time be redeemed in gold. At the same time, because of this, the government could not simply print money because the money had to have gold behind it.

But the great many government programs which the Roosevelt Administration in the United States put in during the 1930s required that more dollars be printed. The Gold Standard Act in the United States, passed in 1900 and officially repealed in 1971, was for all practical purposes, invalidated when President Roosevelt signed the Gold Reserve Act of 1934. It provided that the dollar could no longer be converted into gold, and the government prohibited the manufacture of gold coins.(10) This meant that a private money system still based on gold could not be established. Silver continued to back some currency with the presence in circulation of silver certificates.

However, in the 1960s the United States did offer to redeem all the silver certificates with silver, and then took all the silver out of the coinage. The silver coins were becoming more valuable, in terms of the inflated paper currency, than their stamped face value. Gresham's Law took over — "bad money drives out good money." Such silver coins existing at the time speedily were taken out of circulation as people searched their change for these bits of real money. The new coinage was nickel and copper —and later even the copper in pennies was replaced with the much cheaper zinc. Try to find a silver coin in circulation in the United States today! With copper coated zinc pennies, how far down can the debasement of a metal currency be carried? Ultimately it may be that even zinc is too valuable for pennies. Then what? Now the paper dollar no longer has any precious metal restraints attached to it as the record of inflation since that time clearly shows.

The United States' federal debt is now in the vicinity of $5 trillion. There is little prospect that the gold standard which could cause some financial discipline will ever be instituted again. The official government link between precious metals and money is gone forever. No other nation maintains this tie either. One might suggest that a gold standard could have avoided the several debacles of the Mexican peso, but the political environment there would make such monetary restraint impossible.

Alan Greenspan, who later became chairman of the U.S. Federal Reserve Board, stated with regard to political establishments and the gold standard:

> "Under a gold standard, the amount of credit that an economy can support is determined by the economy's tangible assets, since every credit instrument is ultimately a claim on some tangible asset, but government bonds are not backed by tangible wealth, only by the government's promise to pay out of future tax revenues...Thus, government deficit spending under a gold standard is severely limited.
>
> "The abandonment of the gold standard made it possible for the welfare statists to use the banking system as a means to an unlimited expansion of credit...there are now more claims outstanding than real assets.

> "In the absence of the gold standard, there is no way to protect savings from confiscation through inflation. There is no safe store of value...The financial policy of the welfare state requires that there be no way for owners of wealth to protect themselves.
>
> "Deficit spending is simply a scheme for the 'hidden' confiscation of wealth. Gold stands in the way of this insidious process...If one grasps this, one has no difficulty understanding the statists' antagonism toward the gold standard."(9)

If countries were forced back to the gold standard it would speedily show what was happening to their currencies. Unfortunately, there is not enough gold in the world now to allow a world-wide establishment of the gold standard. This is fine with politicians for they cannot tolerate the honesty of gold and silver as money. With no such restraints now, and by voting for generous public spending, politicians have been able to use the public treasury as a campaign fund. This gets them re-elected but it also ensures continued inflation — the depreciation of the currency. Politicians hand out money but the citizens ultimately pay for it with the declining value of their savings and of their money in general.

Paper as "Money"

The end form of "money" is paper; there is nothing cheaper. In Hong Kong, when you exchange your money for the local currency, if the final exchange figure has some pennies on it, you will get slips of paper which are Hong Kong pennies. How long will pennies in the United States be made of metal? Only to the time when the currency has depreciated to the point where there is no metal which can be produced the size of a penny and still be worth less than a penny as metal. Copper did not survive, and zinc is the next to go. There are already suggestions that the concept of a penny be abandoned, and that things in the United States be priced to the nearest nickle.

The record of paper money has been miserable, as witness the 1923 50 million mark German banknote I have before me. In late 1923, near the end of the wild German inflation, workers were paid twice a day, and their wives met them at the plant gate at noon to get the morning pay so it could be immediately spent because prices would be higher by evening.

Worldwide use of paper

Gradually, around the world, paper money has taken the place of gold and silver. Some countries still cling in various ways to precious metals for their currencies, chiefly by issuing special coins for various occasions. But now all countries including the major nations issue paper money which is not backed by gold or silver. Unfortunately, for the general public, paper money as a store of value has a very poor history. Not only is paper money subject to government printing press inflation, but also it can be more easily counterfeited, and involved in other sorts of fraud than can precious metals currency.(2)

How Other Minerals Back Currencies.

But other things besides precious metals can and do, in effect, back a nation's currency to some extent. The ability to produce food, cars, refrigerators, medicines, medical equipment, and many other items tends to support a nation's money and therefore the materials from which these things are made including soil, iron, and copper, and energy resources in the ground (coal, oil, gas) are valuable. In the case of Japan, which has very few natural resources, the yen has been strong because of what the Japanese can do by processing and up-grading the natural resources they import. But their economy is a precarious one, for without basic raw materials within their own borders, they are vulnerable, and they know it.

Before World War II, Japan moved into Manchuria, China, and Southeast Asia to establish a colonial empire from which to obtain raw materials. They were defeated in this endeavor. Now they are trying to compensate for this by buying into mineral resources abroad, either by 100 percent ownership or, more often, through joint mineral resource developments in other countries.

Taking note of the current strength of the Japanese yen in international markets, in 1994, the U.S. Bureau of Mines reported:

> "Japan has intensified its overseas mineral exploration and development projects for copper, gold, lead, silver, and zinc since 1992. Significant joint exploration and development projects were undertaken in Australia, Chile, China, Mexico, Mongolia, and the Republic of South Africa, and in the United States. In 1992, Japan successfully launched its first geological observation satellite with the world's most advanced sensor and radar equipment to aid in the worldwide search for copper, gold, iron, lead, silver, zinc, and other metals. The satellite began transmitting data in 1993 covering 10 portions of the earth's surface, including 2 targeted areas in the United States."(13)

In this way, Japan is endeavoring to establish, through legitimate economic means, a worldwide network of mineral supply sites to take the place of the colonies which it could not establish by military force. The strong yen, in part due to the weakness of the dollar caused by the U.S. deficit in balance of payments (with oil imports the largest single item), and inability to deal with the domestic deficit, has become Japan's method of conquest. To a degree, the United States has given this to Japan.

"Petro-currencies"

There is no better example of how the possession of mineral and energy mineral resources affects a nation's money than the case of oil. As oil became more and more important in world economies and clearly was the only thing making the money of some countries valuable in world trade to any degree (Nigeria, Qatar, United Arab Emirates, Kuwait, Libya, and others), the concept of "petro-currencies" developed. The term has been in use only recently, but when Kuwait's chief resource lies beneath its desert sands in the form of several times the amount of oil in the United States, there can hardly be a better description of the Kuwait dinar.

Among the Persian Gulf states, oil provides about 80 percent of government revenues, and the sale of oil is the chief thing which supports these currencies. In North Africa, Libya has virtually no other resources than oil, and the Libyan pound floats on the surface of the ocean of world currencies because it floats on oil. When the oil is gone the Libyan pound will sink.

Oil and the British pound

The same is true to some degree for the British pound since the discovery of oil in the British sector of the North Sea. With that discovery and the subsequent self-sufficiency of Britain in oil, and actually, for a brief time the possession of an export surplus of oil, the British pound has done a great deal better than in times past. It has had its problems, however, which were clearly tied to oil. In the 1980s, when the price of oil declined markedly for a time, the British pound dropped to the point where it was almost "a dollar a pound." Then, as oil prices rose, so did the pound, a clear example of the importance of the possession of that energy mineral resource to the value of the British currency.

Former British Prime Minister Margaret Thatcher's greatest stroke of good luck was probably the coincidence of her early tenure with the surplus foreign exchange account which the government ran on the basis of oil production and export. This helped Mrs. Thatcher to become the Prime Minister with the longest tenure during recent British history. Without North Sea oil revenues, the government's budget deficit would have been much higher, and if the North Sea oil income had not been available to reduce the budget deficit, it would have required income tax increases of between $30 and $35 billion (pounds converted to dollar equivalent). There would have been another million people added to Britain's jobless lines, which circumstance would have probably contributed to a considerable amount of civil strife.

Over the longer term, the North Sea oil has given Britain a lasting legacy in the form of some 80 billion pounds of foreign investments which Britain made at the time it had the oil money with which to do it. This was earned as new oil fields were discovered and oil prices went up and up. As oil prices slumped in the mid-1980s, combined with a leveling off of production, the income from oil declined to the point when, in 1986, the income from investments made earlier from oil money exceeded the income directly received from oil. This again tended to support the British currency and economy. Thus the value of the British pound has been helped substantially by the oil deposits in the North Sea.

Britain, however, has come close to killing the goose which has been laying the golden, or in this case, the oil-filled egg. The average tax rate, when adjusted for inflation, is about 78 percent of the profits of the oil companies, and for some oil fields, the rate is over 90 percent. With such a high tax, companies declined to develop the many small field discoveries which have been made, with the result that Britain will have passed the peak of its oil production in the late 1990s. The hope for future development lies in the more than 40 smaller fields, some with only one-twentieth the recoverable oil in the larger oil fields. As one observer said, "The government milked hell out of this industry, and the cow nearly fell down." But more recently the British Government has reduced its tax rate on all new fields and on some marginal fields with the result that some of the smaller discoveries are now being developed.

Oil Supports Many Currencies

Looking abroad again, other countries aside from Britain and the Persian Gulf nations are heavily dependent on oil for maintaining the value of their currencies. In general, among the OPEC nations (some of which are not in the Persian Gulf region), between 50 and 90 percent of government budgets have been financed recently by petroleum revenues. Nigeria, Africa's most populous nation, has its economy now heavily built on oil. Without oil, Nigerian money would be of considerably less value in the international market place.

Oil and U.S. Balance of Payments

In the late 1980s, and into the 1990s, the value of the U.S. dollar plunged to a record post-war low against other major currencies. The chief reason for this was the large deficit in balance of payments combined with a huge budget deficit at home. A significant part of the trade deficit resulted from the cost of importing oil, and other raw materials. Hodel and Deitz have pointed out that, "Our purchases of foreign oil have contributed more to the growth of the trade deficit than any other single commodity...In the twenty-two year period from 1970 to 1992, petroleum imports have totaled $924.5 billion. This is more than 73 percent of the cumulative trade gap of $1.26 trillion during that twenty-one year period."(5)

This situation continues to get worse for by 1994, the United States began to import more oil than it produced. Petroleum is now the largest single item in the U.S. deficit in international payments. It is likely to continue so as more and more oil has to be imported to take up the gap caused by declining domestic production.

How sensitive the United States dollar is to oil was illustrated when the rumor came out that a large oil discovery was made in the Baltimore Canyon area off the east coast of the United States. This had long been a major exploration hope and target. On the discovery rumor, the dollar immediately jumped in value. Subsequently, when the rumor proved false (and the Baltimore Canyon area has now been one of the great disappointments in east coast off-shore drilling), the dollar dropped.

Money and international prestige

The U.S. dollar continued to drop, and politicians have applauded this situation for presumably it has made exports more competitive. But the low value dollar is a two-edged sword. The weakening of the British pound (propped up temporarily by the oil from the North Sea) coincides with the decline of Britain's position as a world economic power. The pound, once the world's premier currency, and the international unit of exchange, is no longer that. If, in turn, the United States' dollar becomes a greatly weakened currency, the prestige of the United States is also weakened. No one respects a country where it takes a handful of money to buy a loaf of bread. The disaster which overtook the Russian ruble at the time the Soviet Union disintegrated certainly did not enhance the international prestige of Russia.

The value of its currency, and how it is respected, is to a considerable extent a measure of that country's political strength and economic influence around the world. It has been said that the world market valuation of a nation's currency is an "international report card" on how well that country is managed. Testifying before the United States Congress in 1995, Felix Rohatyn, a senior partner of the investment firm Lazard Freres stated, "We are gradually losing control of our own destiny; the dollar's decline undercuts American economic leadership and prestige. It is perhaps the single most dangerous economic threat we will face in the long term because it puts us at the mercy of other countries."

The dollar since World War II, has been the reserve currency of the world. Most international accounts including oil and all other commodities are now settled in dollars, but there is an ominous trend emerging. In 1965, the world's foreign-currency reserve holdings of dollars were 80 percent, but in 1995, they had declined to about 58 percent.(3)

Oil Priced in Dollars

Oil is currently priced in the international market in U.S. dollars, but because of the continued weakness in the dollar, the oil producing countries have suggested that perhaps oil should be priced in terms of a "basket of currencies — that is by using the average value of several combined currencies. If oil is no longer priced in U.S. dollars, it will be a blow to U.S. prestige. The fact that the most important commodity in world trade, oil, is presently priced in U.S. dollars goes far toward making that currency the most widely accepted around the world. Whereas the huge oil import bill for the United States is not the only culprit in destroying the image of the dollar as a strong currency, it is an important one. This factor is still growing, and there is no early end in sight for this trend as U.S. oil production continues to decline and imports correspondingly rise. If, at some point, a major oil exporting country may chose to price its product in some other currency than the dollar, the international repercussions would be large, and it could set off an unfortunate train of events for the dollar.

In 1994, when the United States for the first time imported more oil than was produced domestically — a very serious event — it unfortunately was given little public notice. With the huge U.S. appetite for oil, and now being in the permanent position of having to import increasing amounts of oil as domestic production declines, the impact of this on the U.S. balance of payments deficit will remain a major problem. Also, because of substantially expanding world oil demand, particularly in Asia, it is probable that oil prices will go up faster than inflation, which will exacerbate the difficulties of the U.S. international trade account, and further imperil the dollar. Oil will almost certainly continue to be the largest single item in the U.S. import bill for the foreseeable future, and it will grow relatively larger.

In October, 1995, Secretary of the U.S. Treasury, Robert Rubin, stated, "It is in the interest of the United States to have a strong dollar." However, this will be a challenge for at that time the U.S. was daily importing more than 6.8 million barrels of crude oil and 1.6 million barrels of refined oil products with a total cost of more than $150 million daily, meaning that just to pay for the oil bill that much value in products would have to be exported each day to balance the trade. And, that year U.S. oil production reached a 40 year low, with the trend continuing downward to the present.

Oil backs regional credit

In some states of the United States, oil also is money in the form of how bonds are rated and how easily they can be sold. A brokerage house titles one of their investment publications "General Obligation Credits in the Oil Patch," which indicates oil production remains a strong factor in the economies and budgets of states such as Louisiana, Texas, Oklahoma, New Mexico, and most notably, Alaska. In the first three states mentioned, bond ratings were lowered when the price of oil dropped and oil production and exploration activity were reduced.

Alaska's Good Fortune — For the Moment

In the case of Alaska, however, because of Prudhoe Bay's strong production, Alaska's bond ratings were raised in 1980, and have not been reduced since. But, when Prudhoe Bay production begins to dry up and if other production is not found in Alaska, no doubt that state's bond ratings will be reviewed. Alaska, more than any other state, has its budget tied directly to the petroleum industry. Revenues from oil and gas make up about 86 percent of the total unrestricted state income. Bond ratings are important for they determine what interest has to be paid on the bonds. Therefore a good bond rating is worth money to the organization which issues them, and if it is oil which gives the bond a good rating then it is oil which is worth money to the bond issuer.

Oil Via Dollars = An International Currency

Oil, as already noted, is currently priced and paid for internationally in dollars. As long as the dollar is strong enough to be wanted around the world, the presence of the dollar in a nation's accounts makes the budget easier to handle by providing universally acceptable foreign exchange and reduces the pressure on the local currency to carry the load. If the oil money were not there it would almost certainly would take much more of a local currency to buy anything on the international market. Oil, being priced and sold in dollars which are accepted internationally, gives each oil-producing country, despite the condition of its local currency, money which is recognized as money around the world.

Oil Money Into the Banking System

Minerals, oil in particular, have had a much more profound effect upon the world's economic system and its money and its structure than simply influencing the value of a particular currency or providing a foreign exchange medium. When the price of oil was raised from less than $3 a barrel to more than $35, the world saw the biggest transfer of wealth in all history, and it continues. Being unable to locally absorb this flood of money (denominated in U.S. dollars as the price of oil is, world-wide), the money from the oil-rich Arab world had to be deposited where it would earn income until the time it could perhaps be effectively used internally in the country in various projects. The money had to be placed somewhere and the chief place it was put was in the major banks of Europe, and particularly the large banks of the United States.

Faced with this huge influx of money, these banks in turn needed to put the money out so that it could earn the interest the banks had to pay on it, plus a profit for the bank. The places which needed money the most and apparently were willing to pay for it, or so they said, were the developing countries which included chiefly Mexico, Brazil, Peru, Argentina, and several African nations.

The Arab money poured into the banks, and the banks in turn poured it out to needy undeveloped regions. It seemed to be a happy arrangement all around.

But apparently the bank managers did not take time to realistically look at how these debtor countries might pay back the money, given their social, economic, and political situations along with (locally, if not nationally) the dishonesty of some government officials — almost a tradition in certain areas. Some of these loaned dollars were subsequently siphoned off by graft and corruption. Some money escaped the plunder of the times and did get into projects of various sorts, many of which, however, were poorly managed at best. Some of them were bad ideas to start with. Peru was one of the first countries to begin to delay debt repayment, one of their reasons being "these projects were so bad you should have known better than to lend us the money we asked for on them." (This is quite literally what was said!)

Mexico, with a population of some 80 million people, was also early to borrow from this pile of oil money in U.S. banks. Mexico subsequently built up a debt of more than $100 billion which meant that each Mexican man, woman, and child owed a foreign debt of more than $1000 each. In a country where the annual income is about $700, clearly this debt is really not manageable, and the prospects that this debt would ever be paid were almost nil. But to prevent the debt from appearing as a loss on the balance sheets of U.S. banks, the banks continued to loan out more money in order for Mexico to pay the interest on the debt.

The ultimate result, of course, was simply to make the debt even larger. The Mexican peso gradually declined from an earlier value of 12 to the U.S. dollar to more than 2500 to the dollar. A "new peso" was instituted, but in 1995, the situation collapsed again. The Mexican peso hit an all-time low in terms of its value against the U.S. dollar, and Mexico was having to offer 35 percent interest on its bonds in order to sell them — a situation which could not possibly continue. A hastily (literally overnight) arrangement was made led by the United States in conjunction with some other nations and international monetary organizations whereby Mexico was given a $47 billion dollar commitment to maintain its credit and currency. This was stated politically as assistance to Mexico, but it should be noted that it also was for the benefit of international bankers to whom Mexico already owed substantial debts. Whether these maneuvers will achieve a successful end to the financial follies of the past remains to be seen. History is not encouraging. Interestingly enough, the only collateral which the lenders recognized for this arrangement was that of Mexico's oil revenues, a clear case in which oil was being used essentially as money.

Oil Money Benefits Other Countries

Oil money is not only aiding the Middle East countries which have the oil, but substantial populations in that general region are also becoming dependent on oil money although those people may not have much oil themselves. This increasing dependency on an exhaustible resource by expanding populations is a potentially explosive situation.

Egypt has some oil production, but with its 60 million people, and growing rapidly, the oil revenues are insufficient to pay for all the needed imports, and Egypt is increasingly dependent for its survival on foreign aid. The United States, for political purposes to balance in the Arab world to some degree the aid it gives to Israel, also gives aid to Egypt, but significant amounts also come from neighboring friendly oil-rich Arab countries, chief of which are Saudi Arabia and Kuwait. But when the oil revenues of its rich neighbors begin to decline, Egypt will no longer be able to be supported as it is now.

With Egypt's population increasing by about one million persons every nine months, then what? There is no apparent happy solution. Egypt is building out on an oil money limb which is not even its own, and which will ultimately be cut off. In nature, the over-extension of a population upon a resource which diminishes is well known, and the results tend to be disastrous. Perhaps this situation will not be played out for some time, however, as both Kuwait and Saudi Arabia have very large oil reserves. But as the oil begins to run out in these two countries, they will no doubt look to their own interests first.

Another oil-less Arab state benefiting from its oily neighbors is Jordan. It gets a little export income from one mineral, some small phosphate deposits, but it has become a gold trading center for some of the Arab world. On the streets of downtown Amman, the Jordanian capital, may be seen robed Arab men and women frequenting the gold shops. Arabs have long placed their faith in gold. Many have gotten wealthy on oil, but perhaps looking to the longer view they continue to invest in gold as something which will last beyond the oil, and also perhaps rightly not trusting any of the currencies of the world as a reliable store of value. The gold trade in Amman is brisk.

Minerals Provide Foreign Exchange for Small Economies

Foreign exchange is urgently needed by many smaller countries whose economies are not big enough to justify the presence of an automobile plant, an electronics complex, laboratory facilities for the production of vital medical supplies, or other sorts of manufacturing which must have a reasonably large domestic market to survive. Without a diversified industrial base, foreign exchange has to take up the gap and pay for the imported items. Where oil is not available as money, metals and other minerals may provide the foreign exchange so urgently needed. Peru, Bolivia, Chile, Zaire, and Zambia, for example, have no greatly diversified manufacturing base. Many things have to be imported. These countries are not highly developed technologically, and therefore cannot sell technology or the things technology produces. Accordingly, these lesser-developed countries must fall back upon production of raw materials, commonly minerals, for much of their foreign exchange. These minerals then provide the only generally accepted money which they have for use in the world market.

Bolivian tin accounts for between 40 and 70 percent of that country's foreign exchange, the exact amount depending to a considerable extend on the rather widely fluctuating tin market. Peru gets most of its foreign exchange from exports of gold, silver, copper and a little oil. Zambia's copper provides more than 80 percent of its foreign exchange. Copper is a major export of Chile, which has some of the world's largest and richest deposits that will provide valuable foreign exchange for Chile for many years to come.

The dire need to export "something" for foreign exchange tends also to disrupt mineral resource markets. Chile, with its relatively rich copper ores and low cost labor, can produce that red metal at an average current cost of less than 60 U.S. cents a pound, substantially below production cost in the United States. Even if Chile had to sell its copper at a loss it would tend to do so because it needs foreign exchange at almost any price. Chile at times has dumped large amounts of copper on the market despite low prices and demand. The result in the United States has been that many copper mines had to close. But on the other hand the money Chile earns goes in part to pay the interest on the $20 billion foreign debt owed by Chile, much of it held by U.S. banks. It is a complicated financial world which mineral resource economics creates.

Another aspect of this dependence by some countries on the export of minerals or energy minerals is the fact that as the price of these materials drops, there is need to export more to maintain the same level of income. In turn, by exporting more, still lower prices result, and so on. These countries need to export minerals not only to obtain foreign exchange, but also to provide internal employment. However, in having to produce more because of declining prices to obtain the same amount of foreign exchange, there is a compensatory factor in that more employment is created. This has the advantage of keeping the local population reasonably contented and is good for the political machine in power. Even if the price of the product falls below the cost of production in a given country, that country is still likely to continue production. This, at first glance, does not seem logical, but the economics of this were neatly explained by a Minister of Mines of one African nation, who said in regard to a gold mining project which was by most standards uneconomic, "You do not understand. As long as the metal brings in more dollars than we have to spend for supplies from the outside, it is economic. The labor, power, and other supplies are obtained locally and paid for in our currency, the value of which we control. We need the thousands of jobs to keep the people quiet. Therefore the project is economic."

Thus it is not just the matter of earning foreign exchange which causes over-production and decline in price of minerals and energy minerals, but it is politics. The amount of the resource produced is increased in order to "keep things going," the "things" in many cases being the current political organization which may involve considerable channels of graft keeping the present politicians in power. If these channels are not kept properly filled with money there are likely to be changes. It might even come to the point where heads may literally roll. Some of the cheating on OPEC production quotas may have its origin in these circumstances.

Domestic Oil Saves Foreign Exchange

Being self-sufficient in oil does make a difference — a lot of difference. If the United States were still self-sufficient in oil and had not sent billions of dollars abroad to pay for the oil, it would not be such a huge international debtor as it is now. It was the increasing dependence of the United States on foreign oil which enabled the OPEC countries to raise the price of oil as high as they did, and thus produce such a great transfer of the western world's money to them, with the resulting tidal waves in international banking.

No doubt the price of oil would have been raised anyway but the extent to which it was raised would have been less if the United States, the world's largest single consumer of oil, had been self-sufficient in oil. This is an example of oil as money, and the problems which come to a highly industrialized, high oil-consuming nation which is running out of oil.

The Longer Term Balance of Payments Problem

Unfortunately, unlike Japan which long ago knew it had to export to survive, the United States was content with its own large internal market and sufficient domestic mineral and energy mineral resources (oil, coal, and uranium) for its own use, and did not need foreign exchange. But as the oil self-sufficiency disappeared, and at the same time U.S. demand greatly increased for the gasoline-efficient cars which the Japanese already made and could export to the U.S., the United States in the decade of the 1980s went from the world's largest creditor nation to the world's largest debtor nation, with a huge deficit in the balance of foreign trade. This is an almost unbelievable turn of circumstance. With the balance of payments problem, and the ever-rising oil bill, getting itself out of this economic hole will be a very large task for the U.S. As oil production in the United States continues to decline, the huge and growing foreign oil bill will further worsen the balance of payments problem, and, in turn, put the value of the U.S. dollar increasingly in jeopardy.

The U.S. Department of Energy states that within the next several years the oil import bill to the United States will reach $100 billion in today's dollars. Domestic production keeps this figure from being even larger. The Alaskan Prudhoe Bay Field alone, from only about half its total oil produced to date, has saved the United States about $135 billion in foreign oil import costs, but unfortunately that field is already into its inevitable production decline.

Find more U.S. oil?

Relaxation of drilling restrictions in the offshore areas of the United States, and the development of the single potentially largest new oil area onshore in the United States, the approximately one percent of the area of the Arctic National Refuge 60 miles east of Prudhoe Bay, would help in this oil balance of payments problem, but would not eliminate the need to import oil. However, there is currently great opposition to such developments. Regarding offshore drilling on most coastal areas of the United States, the Congress in 1995, reaffirmed the 13 year-old ban. Decisions on these matters, in a democracy, are in the public realm, and the public should know the effects of whatever decisions are made. Balance of payments and the value of the dollar affect everyone, and oil imports will continue to be the largest single factor in the U.S. foreign trade deficit for the forseeable future.

Continuing Transfer of Wealth

In various ways the worldwide recycling of the huge amounts of oil money which causes these large transfers of wealth is still in progress, and it will continue for decades to come. When or how it ends no one can say, but the world's money and the world's total economic structure will never be the same again. Oil turned it upside down. The locations of the world's great oil deposits have had and will continue to have profound effects on the economic position of the United States and every citizen therein.

This continuing transfer of wealth to the oil producing countries affects all industrialized nations. It will ultimately be the citizens of these nations who will have to pick up all the bills in various ways — the higher oil costs, and perhaps problems in the banking systems. Higher taxes, a depreciated currency, higher prices for things of daily living, and a lower standard of living may also be some of these costs. As more and more basic raw materials have to be imported, a depreciating domestic currency means these imported materials cost more. Oil, who has it, and who does not, was and remains a major factor in these situations.

It is striking to consider these profound effects upon the world of a single energy mineral — oil — which less than 200 years ago was of no great concern to anyone. One of its chief uses at that time in the Middle East, where numerous oil seeps existed, was still in the treatment of camel mange as it had been for several thousand years.

In Summary

Minerals in the form of precious metals have been literally money during much of modern civilization. In the form of oil and coal, iron, copper, uranium and the many other minerals which modern economic society needs, these materials now cause a great transfer of wealth. Minerals are money in a very real sense, and who has them, and who does not will continue to markedly influence the various national monetary systems, and in turn the destinies of nations. The increasing need of the United States to import oil and the transfer of money out of the United States to pay for this affects the U.S. balance of payments. In turn this is having a markedly negative effect on the value of the dollar, and its prestige in the world economy.

BIBLIOGRAPHY

1 BERESINER, YASHA, and NARBETH, COLIN, 1973, The Story of Paper Money: Arco Publishing Company, Inc., New York, 112 p.

2 BERESINER, YASHA, 1977, A Collector's Guide to Paper Money: Stein and Day, Publishers, New York, 255 p.

3 CURTOPELLE, CHARLES, 1995, The Long and Winding Decline of the U.S. Dollar: Bull & Bear Financial Report, v. 13, n. 3, Longwood, Florida, p. 6-7.

4 FRANK, TENNY, 1927, An Economic History of Rome: The Johns Hopkins Press, Baltimore, 519 p.

5 HODEL, D. P., and DEITZ, ROBERT, 1994, Crisis in the Oil Patch: Regnery Publishing Inc., Washington, D. C., 185 p.

6 LOVERING, T. S., 1943, Minerals in World Affairs: Prentice-Hall, New York, 394 p.

7 OIL AND GAS JOURNAL, 1996, OGJ Newsletter, March 4.

8 POSS, J. R., 1975, Stones of Destiny: Michigan Technological Univ., Houghton, Michigan, 252 p.

9 RAND, AYN, with additional articles by NATHANIEL BRANDEN, ALAN GREENSPAN, and ROBERT HESSEN, 1966, Capitalism: The Unknown Ideal: The New American Library, New York, 309 p.

10 REED, P. B., (ed.), 1990, Coin World Almanac. Sixth Edition: Amos Press Inc., Sidney, Ohio, 743 p.

11 SCHWARZ, TED, 1976, Coins as Living History: Arco Publishing Company, Inc., New York, 224 p.

12 STRAUSS, S.D., 1983, The Quest for Gold and Silver: Including a History of the Interaction of Metals and Currency: Mineral and Energy Resources, Colorado School of Mines, Golden, Colorado, v. 26, n. 6, p. 1-10.

13 U.S. BUREAU OF MINES, 1994, Mineral Commodity Summaries 1994: U.S. Bureau of Mines, Washington, D.C., 201 p.

14 VICKER, RAY, 1975, The Realm of Gold: Charles Scribner's Sons, New York, 244 p.

CHAPTER 12

The Petroleum Interval

As Col. Drake watched oil slowly flow from his 69½-foot deep well at Oil Creek near Titusville, Pennsylvania in August of 1859, he would not have imagined in his wildest dreams, nor could anyone else at the time, have predicted that this black liquid would be the basis for huge industries not yet even thought of — the automobile, and aircraft industries ("what is an automobile or an airplane?"). No one there thought that this gooey dark stuff would provide material by which millions of miles of roads around the world would be paved, or that an entirely new way of life and a new economy would emerge for the world through the use of this substance. No one would have predicted that it would be the cause of terrorist bombing and war.

If anyone had told the people gathered at Oil Creek that this dark liquid would eventually propel millions of people in cigar-shaped containers with metal wings across the United States and around the world at an altitude of 35,000 feet and at a speed of 600 miles an hour he would have been regarded as totally insane. Or, if he had suggested that thousands of products including clothing, medicines, insecticides and plastics ("what is plastic") would be made from that dark liquid, he would have been considered equally mad.

Or perhaps someone would have predicted that some of the nomadic tribes of remote desert regions of the Arabian Peninsula had vast quantities of this material beneath their desert sands, and because of this before the end of the Twentieth Century they would become a nation which would be able to greatly influence the economies and futures of almost all nations on the Earth. And would anyone have anticipated that this circumstance would result in the greatest transfer of wealth the world had ever seen, and ultimately make shambles of parts of the international banking system? Such events would have been regarded as totally wild and unbelievable. Yet these are only some of the things which oil has done to and for the world since it began to be used in quantity. The British Statesman, Ernest Bevin, in 1948 stated, "The kingdom of heaven may run on righteousness, but the kingdom of Earth runs on oil."(4) His observation remains valid today, and will remain so for several decades to come.

Start of petroleum industry

Previous to the Drake well, petroleum had been produced from wells in other areas of the world including one in Ontario, Canada. China records the use of petroleum as far back as 3000 B. C. Wells were drilled to as deep as 3,000 feet using strings of bamboo rods much like modern cable tool rigs.(2) Gas which they discovered was distributed by bamboo pipes.(37) But the potential of drilling for oil was not generally understood until Col. Drake's discovery. This, combined with the American entrepreneurial spirit, and the need for energy to fuel the Industrial Revolution launched the oil industry on its remarkable rise, first in the United States, and then around the world.

It is difficult to overstate the changes which have taken place by the fact that in 1859 world crude oil production was only a few barrels a day, and it is now over 60 million. No other material has so profoundly and universally changed the world in as short a time as has oil. Iron might be the nearest candidate for such a claim, but the influence and use of iron gradually spread over the world during many centuries, whereas in less than 150 years oil accomplished its changes. This is almost instantaneous even in the perspective of the relatively short human history.

Gradually diminishing resource

It is also difficult to visualize the changes which will take place in tomorrow's world as petroleum supplies gradually diminish toward the point of exhaustion. The changes which the use of oil brought have, for the most part, been pleasant. Widespread use of oil had done many good things for many people. The decline of oil supplies may have the opposite and quite unsettling effect. The decline of the "Petroleum Interval" will be gradual so there may be time to make adjustments, but as decline of oil production is predictable, accommodation to that event should begin to be made now. Unfortunately, little is being done.

The "Petroleum Interval"

The Stone Age lasted for hundreds of thousands of years, perhaps more than a million. The Copper Age was shorter but still many centuries long. The Bronze Age was a bit shorter, but still several hundred years, and the Iron Age has been with us for many centuries.

The Petroleum Interval which we are now in, began on August 27, 1859 when Col. Drake demonstrated that oil could be readily obtained by drilling. It has now lasted about 140 years. The term "interval" has been chosen rather than "age" because the time of the use of petroleum in great quantities as we are doing now, will be much shorter than the "ages" of the past. It will be but a brief bright blip on the screen of human history, lasting in significant form for no more than 300 years. How fortunate we are to live during this interval — that is, those of us who are able to enjoy its benefits. Present indications are that more than half the world's oil is now being produced within the span of a single lifetime.

The increase of petroleum production and the many and varied uses discovered for it is a singular event in the history of the exploitation of Earth resources. Petroleum is doubly valuable for it can be used not only as a very convenient, high energy content per unit weight source, but also as a raw material base for many products which other forms of energy such as uranium, sunlight, wind, and electricity cannot be used. Myriad petrochemicals are an example. For the sake of clarity, the term "petroleum" includes both oil and natural gas, but because oil is the more important of the two, the emphasis in this chapter is on oil.

The Petroleum Industry

Although oil, mostly in the form of gasoline, enters into the life of nearly every American every day, the public knows amazingly little about oil consumption and production in the United States, and about U.S. oil and gas reserves. It is important in a democracy that citizens know the facts relating to a given situation so that rational decisions can be made about the future. Here are some of the basics about the U.S. petroleum industry, along with examination of the trends which suggest what may happen in the future.

Petroleum industry: huge, complex, and vital

It did not exist 200 years ago, yet today the petroleum industry, in all its many phases, is by far the largest single industry in the United States, and in the world. No other industry employs so many people, many of them highly skilled, in so many different ways, and pays so many taxes to various levels of so many governments. The scope of the petroleum industry is enormous. It ranges from the local service station, to huge offshore platforms drilling for oil in thousands of feet of water, to vast petrochemical complexes producing literally thousands of products including plastics, pesticides, paint, fertilizer, and medicines. And the petroleum industry is the chief energy supplier for our industrial civilization.

Technology

No other industry utilizes so many different technologies and technical personnel to apply them. Just to determine where to drill is an exceeding complex matter. How do you locate an oil reservoir which lies 10,000 feet below the ocean floor, and where the ocean is more than a mile deep? Oil exploration involves everything from satellites, to geochemistry and geophysics. Using reflection and refraction seismograph data involves the processing of billions of bits of information by computer which create 3-D seismic imaging to visualize the Earth's interior.

Drilling and completing wells to depths of 20,000 feet and more requires huge and complex mechanical equipment and a great variety of electronic instruments. In deep water, computers control electric motors and propellers on the ship or drilling platform to keep it in place over the drill hole. Drilling platforms in the open ocean from which 40 or more wells may be directionally drilled are small cities in themselves. They must be built to withstand storms in the North Sea, hurricanes off the Gulf Coast of the United States, and typhoons in the South China Sea. Sometimes the storms win and destroy or disable the mobile drilling rigs. It is a hazardous business, and insurance costs are high.

Logging the well (studying the well bore and the rock cuttings from it), involves many technologies, from the use of gamma ray and neutron ray instruments to the basic electric logging device which records the self-potential and resistivity curves. These provide the well-site geologist with some of his most useful information. Well logging is what the very successful worldwide French company, Schlumberger, was founded upon, and is still its basic business.

Special drilling mud must be used to lubricate the drill bit and carry the rock cuttings to the surface. The mud also protects the producing formation from being blocked off permanently, while at the same time assuring that the well will not blow out and possibly cause a highly destructive fire — which has happened. It is said that drilling mud is "mud with a college education." Mud engineers are employed to see that the drilling mud is designed to fit the conditions of the particular well being drilled. The conditions change as the well is drilled through water zones and various kinds of rocks. It is not a simple matter. The mud must block out water zones, and keep the hole from caving in, but must not plug off production oil zones.

Once a well is completed and oil is brought to the surface, another host of technologies come into use. These may range from laying a five hundred mile pipeline across the Andes, to all complexities of the refining process during which the hydrocarbon complex of crude oil is broken into numerous components. These are then sorted out and reconstructed by sophisticated chemical and physical processes into myriad end products from gasoline to plastics to medicines. Organic chemistry, an exceedingly large and complex study, finds one of its broadest applications in the oil industry.

The oil industry makes use of some of the world's brightest minds, and pays excellent wages. In the United States the oil industry in exploration, drilling, production, oil field maintenance, refining, petrochemicals, and marketing, directly employs about a million and a half people.(1) But the effect of the oil industry does not stop with a tanker truck bringing gasoline to the neighborhood service station. The huge world automobile industry is based on oil. This industry in the United States, in its various ramifications, is said to account for 25 percent of the national economy. The tourist industry also, which is so important to some countries, and many remote areas (and some not so remote such as Disneyland) is also based on oil. And that most fundamental of all enterprises, agriculture, is now largely dependent on petroleum to plow and fertilize the fields, to destroy the weeds and injurious insects, to harvest the crop, and to deliver it to market.

In the United States, by petroleum-based agriculture, two percent of the population feeds the entire nation and has surpluses to export. Natural gas is the basis for manufacture of important plant fertilizer, and oil allows one man with a tractor and equipment to plow, plant, cultivate, and harvest large areas. Petroleum is a remarkable substance.

Oil — the convenient, high energy source

Oil in its various derivatives such as gasoline and diesel fuel, is by far the most convenient to handle and transport, high energy per unit weight energy source we have. It can be pumped, carried in cans, or put in fuel tanks to power mobile machinery. It can be stored easily and conveniently, and it burns with little or no ash which needs disposal. It can be hauled great distances, indeed, around the world in huge ocean-going vessels, and it can be loaded or unloaded easily and quickly by pipes and pumps. There is no other energy form as versatile in its uses, which can be transported so far so easily, and, except for uranium, deliver such a large net energy return at the far end of the trip as oil. And uranium is by no means so versatile in its end uses.

Coal can also be transported great distances, but the amount of coal it takes to transport a cargo of coal requires more energy used compared with the amount received at the destination than in the case of oil. Also, coal is dirty to handle, bulky, and when burned leaves considerable ash and other pollutants such as sulfur. Coal contributes much more to air pollution than does the burning of oil, although oil cannot be held harmless in this regard as any resident of Los Angeles can attest. Oil is clearly the world's premier energy source, and also the basis for the huge petrochemical industry. There is nothing in sight to replace oil in the volume and diverse ways in which it is now used. Even if solar energy in various technologies could replace the energy petroleum now provides, solar energy cannot be the raw material for petrochemicals. Paints, plastics, medicines, insecticides, and fertilizer cannot be made out of sunlight. This is a most important fact commonly overlooked when people visualize a post-petroleum, solar-energy based economy. Petroleum is much more to the economy than just energy. Our modern agriculture is based on petrochemicals. And plastics which are made from oil are vital to many industries.

USA is where it started

The United States is the birthplace of the modern oil industry, and the United States produced the first group of oil-finders, who quickly spread across the country discovering oil from coast to coast. Early oil was fairly easy to find and it did not take much training to recognize an anticline — an upfold in the Earth's crust — which in oil provinces frequently contain oil. But soon the search had to seek less obvious oil occurrences. The need arose for technologies which could find the hidden oil deposits, and also to get more oil out of existing oil fields. A number of universities established petroleum geology and petroleum engineering departments. Geologists and petroleum engineers from these schools ranged first over the United States, and then more widely, to Mexico, Canada, and South America. They went to the Middle East and North Africa, and finally around the globe from the arctic to Australia and the Indonesian Archipelago. Eventually foreign governments began sending their own students to the United States for petroleum studies. When petroleum nationalizations took place widely across the world foreign students trained in the U.S. (and in Britain and France) moved in to staff what became their new national oil companies.

U.S. oil dominance — for a time

In 1920, more than two-thirds of the world's oil came from wells in the United States. During the period 1859 to 1939, 64 percent of all the world's oil produced came from the United States, and the United States had used that oil to help it to become the world's economic leader, with the world's highest standard of living. Even as late as 1950, the United States still produced half the world's oil. It was far more than self-sufficient, and was a major oil exporter.

Growth and dispersal of the oil industry

But the importance of U.S. oil production in terms of world production was dropping. From producing more than two-thirds of the world's oil in 1920, the United States in 1996 was the source of only about 11 percent of world crude oil supplies. The center of oil production had moved, and become less concentrated. Sixty-six countries now are oil producers. In the 1980s the Soviet Union was the world's largest single oil producer, with the United States second. Saudi Arabia had the potential at that time to be the world's largest producer, but wanting to stabilize oil prices, the Saudis were only producing about four and one-half million barrels of oil a day. But the oil situation changes. Saudi Arabia is now first in oil production, the United States second, and with the disintegration of the Soviet Union, Russia proper is now third.

World Oil Reserves

To understand who has the oil now and in the future, it is important to note that there are a variety of terms used by various people concerning "reserves." Also, the term "resources" is commonly confused with "reserves." "Resources" as compared with "reserves" as used in the oil industry are all the oil (and/or gas) theoretically in an area.(28) "Reserves" are those resources which are "producible, within a known time and with known techniques, at known costs and in known fields."(28) The term "proved reserves" is even more definitive in that it is commonly used to designate petroleum which can be economically produced (that is, with a reasonable profit) by known technology. Other terms applied to "reserves" include "probable," "possible," "inferred," "identified," and "undiscovered." To these terms, the percentage probability of their being realized is frequently added. One might make the observation that these various terms obviously give individuals, organizations, and governments a wide latitude as to what to report — for various purposes. The figures given "undiscovered" reserves are particularly "flexible."

Laherrere has written about this and makes the general observation that "The main reason why it is so difficult to determine the world's reserves is that there is no agreement on how to define them. There are many vested interests which make use of lax definitions to propose numbers that meet their political objectives: oil is money and reserves are, so to speak, oil in the bank — a bank far underground where no auditors can check the account. We must approach the subject with caution and read the reports with great care if we are to succeed in decoding their often-hidden messages."(30)

Ivanhoe has discussed the "use" and "abuse" of the terms "reserves" and "resources" stating, "Well-intentioned, but irresponsible scientists who continue to discuss resources instead of reserves may be a significant cause of our government's lack of realistic energy policies."(28) This point is well illustrated in discussions about oil shale in the United States (see Chapter 13), where it is frequently stated that the "resources" there are as much as two trillion barrels of oil. But to date no oil has been commercially produced, and there is no present assurance that obtaining oil from shale will ever be economic. However, the up to two trillion figure has at times been cited as reason not to be concerned about future supplies of oil in the United States. However, with the recent failures by major companies to economically produce oil from oil shale, the difference between "reserves" and "resources" is now becoming more clearly and painfully apparent.

"Political reserves"

Ivanhoe further states that "Government petroleum ministries have an inherent interest in announcing the 'good news' of large national hydrocarbon reserves inasmuch as large political reserves are useful for national prestige and in negotiations for OPEC production quotas, World Bank loans and grants, etc. Sudden unsubstantiated reserve increases announced by any government ministry should be viewed with considerable skepticism. They may be mostly the puffery of political reserves which will increase a nation's paper reserves, but have no effect on near-term oil production." He adds, "Natural gas is commonly converted to BOE [barrels of oil equivalent)...to increase a company's or nations BOE reserves. However, gas is not the economic or social equivalent of crude [oil] due to the inherent convenience, safety, and flexibility of oil. Natural gas's main global use is still as a boiler fuel for electric power plants to which a pipeline or LNG [liquid natural gas] tankers must provide an umbilical from gas field to generator."(28) Therefore, the conversion of gas to oil in a country's "oil estimates" gives a somewhat false impression of the true oil reserves. It may be noted also, that the figures which are published in various estimates of oil reserves of various countries are usually those supplied by the country, without any outside audit.

Reasonable estimates

Based on what are believed to be reasonable realistic estimates, statistics have been compiled by Riva for the significant oil producing nations as to how much has been produced to date (cumulative production), how much is now being produced (current production), the proved reserves (those which can be produced economically now), and how much might be found in the future (probable additions) and what the grand total might be. These figures are based on existing production technology and the current price of crude oil. It should be noted that if the price of oil doubled, based on today's dollar, that more oil might be credited to probable additions. The reserve figures, therefore, are in part subject to economic conditions, and can never be a precise estimate of the future. However, the figures do give a picture of the relative oil reserve situation of the various nations. Representative selected data are in Table 1.

Salient points of Table 1 are that United States proved oil reserves are now 23.0 billion barrels and rank 10th in world oil reserves. Saudi Arabia is first with 261.2 billion barrels

Table 1. World Oil Statistics 1994

(Billions of barrels)

Listed in order of probable total oil production. Some minor producing countries not included.

Country	Production to date	Current Production Rate/Year	Proved Reserves	Probable Additions	Total Oil
Saudi Arabia*	71.5	2.98	261.2[1]	41[1]	374
United States	165.8	2.45	23.0	76	266
Russia	92.6	2.55	49.0	119	261
Iraq*	22.8	0.16[2]	100.0	45	168
Iran*	42.9	1.33	63.0	52	158
Venezuela*	47.3	0.85	63.3	37	148
Kuwait*	27.6	0.68	96.5	4	128
Mexico	20.5	0.97	27.4	60	108
U.A.E.*	15.1	0.80	61.0[3]	37[3]	98[3]
China	18.8	1.06	24.0	48[4]	91
Canada	16.1	0.61	5.1	33	54
Libya*	19.0	0.50	22.8	8	50
Kazakhstan	3.2	0.18	3.3	40	46
Nigeria*	15.5	0.69	17.9	9	42
Indonesia*	15.2	0.48	5.8	10	31
Norway	6.3	0.82	9.3	15	31
United Kingdom	12.3	0.68	4.6	11	28
Algeria*	9.1	0.27	9.2	2	20
Egypt	6.2	0.33	6.3	5	18
Brazil	3.6	0.23	3.6	9	16
India	3.6	0.19	5.9	3	12
Qatar*	4.8	0.16	3.8	2	11
Australia	3.9	0.19	1.6	5	10
Colombia	3.6	0.16	1.9	3	9

*OPEC members

(Source: Riva, 1994a; American Petroleum Institute 1995). [1]Riva 1995d lists proved reserves of 160 billion barrels and probable additions of 142 billion barrels. Difference from other published figures may be reflection of official overestimation of reserves which occurs occasionally for various purposes. [2]Small production currently because of United Nations embargo on oil shipments following Gulf War. Potential is 0.73 billion yearly. [3]Riva, personal communication 1995. [4]China has recently estimated that the Tarim Basin of western China has undiscovered reserves of 74 billion barrels. An earlier U.S. Geological Survey estimate puts the figure at about 10 billion barrels, within a range of two to 20 billion barrels which seems more realistic.

Table 2. Estimated World Oil Distribution in Selected Countries 1995

(Billions of barrels)

Arranged in order of original oil endowment

Country	Estimated Original Oil Endowment	Remaining[1] Oil	Percent of Original Oil Remaining
Saudi Arabia*	377	302	80
Russia	262	168	56
United States	260	92	35
Iran*	152	108	71
Iraq*	149	126	85
Venezuela*	130	82	63
Kuwait*	128	100	78
U.A.E.*	118	102	86
Mexico	96	74	77
China	87	67	77
Libya*	56	37	66
Canada	49	33	67
Kazakhstan	47	43	91
Nigeria*	41	25	61
United Kingdom	38	25	66
Indonesia*	34	19	56
Norway	31	23	85
Brazil	27	23	85
Algeria*	19	9	47
Malaysia	14	11	79
Egypt	14	7	50
Azerbaijan	14	7	50
Colombia	13	9	69
India	12	8	67
Argentina	12	6	50
Australia	10	6	60

*OPEC member

(Source: Adapted and modified from Riva 1995d, U.S. Geological Survey, and *Oil and Gas Journal*). [1]Includes both discovered and estimated remaining oil to be discovered. Original oil endowment figures differ slightly from total oil figures in Table 1, as figures are drawn from different years (1994, 1995).

(this is the official Saudi figure; outside estimates put the figure at about 160 billion), and Iraq is second with 100 billion barrels.

Perhaps the most significant figure, however, is the great amount of this oil which the United States has already produced compared with any other country, 166 billion barrels. This huge volume of oil energy resource was one of the main forces which so rapidly propelled the United States to its present affluent position.

Further significant statistics appear in Table 2. This shows the estimated original natural endowment of oil among the more important oil producing countries, the estimated amount of oil still remaining, both discovered and undiscovered, and the percentage that the remaining oil represents of the original oil endowment.

Of the 26 countries listed, only two have produced more than half their oil. Of its original estimated oil endowment, the United States has 35 percent remaining, and Algeria 47 percent. All other countries still have more than half of their oil remaining to be produced. The countries with the largest amounts of remaining oil of the original endowments are all in the Persian Gulf area except Kazakhstan.

That the United States has the smallest percentage of oil left of its original natural supply reflects how intensively it has been explored. To supply its rapidly growing economy with oil as it has done, the U.S. is now the most thoroughly drilled country in the world, and is fast running out of places in which to make further major oil discoveries.

The chances of finding any significant number of major oil fields, especially the size of Prudhoe Bay are slight.(44) The balance of power in terms of oil production has definitely left the United States, never to return. In contrast to this limited future potential for U.S. oil discoveries, Saudi Arabia, already with the world's largest oil reserves, continues to extend its oil fields. In 1989, about 190 kilometers south of Riyadh, and far outside the old Aramco concession area, Saudi Aramco drilled a well in the Al Hawtah region with a production potential of 8,000 barrels a day of high quality sweet (low sulfur) crude oil. This discovery was made in rocks of Paleozoic age which lie beneath the strata from which the present large Saudi production comes. This strike opens up large new territories and stratigraphic zones which can be brought into production later. These will supplement still further Saudi ability to sustain oil production long term.

Where the oil is

We have come a long way in oil exploration from when Col. Drake simply drilled near some oil seeps. Oil occurs in basins in the Earth's crust, which slowly sank and into which organic materials were deposited. Most of these basins were in shallow areas of the ocean along the continental margins or actually on the continents themselves, which have from time to time been invaded by the sea. Accordingly, to determine where the oil is, one looks for these basins around the world. So far some 600 have been identified, and from rather thorough mapping the broad geological features of the world we know this number is close to the total number of basins which exist. Of these, about 400 have been explored by the drill to a greater or lesser extent. The other 200 are in hostile working environments such as northwest of Greenland and have been explored only in a minor way. Significant oil production has been found in about 125 basins.

Making world oil estimates

In the approximately 400 basins which have been drilled, two basic figures can be determined: how much oil has been found, and in how many cubic miles of sediments this oil occurred. Dividing one figure into the other gives a world average of how much oil can be expected to be found per cubic mile of sediment in these basins. The next step is to calculate the volume of sediments in all the potential oil basins, in cubic miles or whatever unit is selected, and then multiply that figure by the amount of oil which, on the average, is found per cubic unit. Thanks chiefly to reflection and refraction seismic technologies, and advanced gravity and magnetic studies, sufficient information is available to reasonably accurately estimate the volume of sediments in all the basins of the world, drilled and undrilled. Geophysical ships, helicopters, airplanes, trucks, and strong backs have hauled technical equipment to the near and far corners of the Earth. Our knowledge now of basin size and sediments is quite good. The world is no longer unexplored. It has been measured and we know it quite well.

From these calculations, and known drilled resources, it has been estimated that when, in the 19th Century, humans began to exploit oil deposits in quantity, the Earth originally contained about 2,330 billion barrels of recoverable oil and that approximately 31 percent or 723 billion barrels have already been consumed.(49,53) A theoretical figure of 75 years of future sustained production at the present rate of 22 billion barrels a year can be estimated, but this is subject to a number of variables, including the different rates at which various countries develop their oil resources, and the likely fact that the rate of production will increase for a time from the present rate as demand for oil continues to rise, particularly in Asia. This may reduce the total life of world oil reserves, unless the price of oil rises to the extent that world oil demand is stabilized or even decreases. But regardless of these uncertainties, it is clear that the Petroleum Interval will be very brief. The world is finite and we have a good estimate of what oil it holds.

There is apparently somewhat more oil yet to be produced than has been used so far. One estimate is that there are 10 barrels of recoverable oil yet in the ground to seven and one half barrels which have been produced. Dating the present Oil Interval back to Drake's well in 1859, means that we have been enjoying oil for about 140 years. However, the consumption of oil now is so much greater than in the past, that it is erroneous to assume that we are less than half through the Oil Interval. Oil in the volumes we use it today will not be available even for 140 more years. We are closer to the end of the Oil Interval than to the beginning. The second fact is that the areas which produced most of the oil previously are not necessarily the areas which will produce most of the future oil. Some regions got into significant oil production much earlier than others, and the United States was the very first. Some places had less or more oil than other places, and production life and peaks will differ accordingly. Combining these factors means that some countries are just now beginning to be producers, for example, Yemen. Others are now in essentially peak production like Britain, and some nations are past their production peak, as is the U.S.

The United States is the world's largest consumer of oil. With five percent of the world's population, the U.S. has recently been using about 26 percent of the world's oil. It produces only about 11 percent, and has less than three percent of total world oil reserves. How long can this situation last?

Because the United States is such a large user of petroleum, a short history of U.S. oil production is useful to put the present situation in perspective.

History of U.S. Oil Production

From the time the Drake well was drilled in 1859 to 1909, U.S. oil production grew to about 500,000 barrels a day. At that time, this was more oil than all the rest of the world combined was producing. The United States continued to produce half or more of the world's oil until as recently as the early 1950s. U.S. oil production reached its peak in 1970, at about 9.3 million barrels a day to which can be added the liquids which come from the production of natural gas, giving a total liquid production of about 11.3 million barrels a day.

"Allowables"

In 1930, the great East Texas Oil Field was discovered, the largest field ever found in the 48-adjacent states. At that time, there were no regulations on oil well spacing for efficient production, or limits on rates of production to prevent wasteful practices. The wells were allowed to run wide open and the price of oil dropped as low as four cents a barrel! On an emergency basis, the Texas Railroad Commission was given authority to limit oil production of each well — an "allowable" it was called.

The Texas Railroad Commission retained this authority permanently. Each month until 1970, when the United States could no longer supply its own needs, the Texas Railroad Commission and similar regulatory agencies set up in other oil producing states, issued an "allowable" for each well. By this method, production and price were controlled. Just enough oil was allowed to be produced to take care of the needs of the United States at a relatively stable price, and to sell to foreign markets the U.S. supplied at that time. Increase the amount of oil produced and the price would drop. Decrease the amount and the price would rise. Because the United States was still producing half or more of the world's oil until 1950, this system, which was eventually adopted by all oil producing states, in effect controlled world oil prices.

This operational procedure gave stability and predictability to the oil industry for many years. Under these conditions, companies could plan ahead, raise capital from investors, and embark on long range exploration programs. The industry flourished, and it provided American consumers with the greatest quantity, the lowest cost, and broadest spectrum of petroleum products of any nation in the world. There were no shortages. But, as production peaked, demand continued to rise. The wells were eventually allowed to run up to what is called their MER —the maximum efficient rate of production.

After about 1970, the "allowable" system was not used so much to control the price of oil but as a conservation measure to see that the wells were produced efficiently. Running a well wide open is wasteful and does not allow the maximum amount of oil to be recovered from a reservoir. This rate of production differs from well to well, and is set by state regulatory authorities.

U.S. production peaks — dependence on foreign oil begins

Late in 1970, even with wells running open to their MER, the oil needs of the United States were not completely satisfied. The curves of production and consumption crossed. Production began to decline but consumption continued to rise. At that point the United States became increasingly and permanently dependent on foreign oil, and lost control of the price of oil. Saudi Arabia became the "swing producer" meaning it had "shut in production" by means of which it could increase or decrease the world oil supply and therefore influence if not control its price. This marked an economic milestone. From then on, the United States would never again control oil prices. It lost an economic weapon which

Table 3. Comparative Oil Well Production by Country 1994

Some minor producing countries omitted

Country	Number of Wells	Barrels daily	Average per Well
Abu Dhabi	993	1,835,000	1,848
Algeria	1,273	1,132,269	889
Angola	403	542,205	1345
Argentina	11,709	666,000	57
Australia	1,209	535,000	467
Brunei	858	171,781	200
Canada	42,679	1,743,200	41
China	49,700	2,961,000	60
Colombia	5,819	445,799	77
Commonwealth of Independent States (old USSR)	122,820	7,306,400	59
Egypt	1,228	908,065	739
France	465	56,911	122
Germany	1,605	56,701	35
India	3,344	611,365	183
Indonesia	8,622	1,487,061	172
Iran	751	3,603,000	4,798
Kuwait	350	1,811,000	5,174
Libya	1,087	1,390,000	1,279
Mexico	4,740	2,768,766	584
Nigeria	1,936	1,883,000	973
Norway	436	2,602,117	5,968
Oman	2,099	806,380	384
Papua New Guinea	29	120,569	4,158
Qatar	288	407,000	1,414
Saudi Arabia	1,400	7,811,000	5,579
United Kingdom	839	2,552,693	3,043
USA	582,768	6,661,573	11.4
Venezuela	14,798	2,364,084	160

(Source: *Oil & Gas Journal,* December 25, 1995)

passed to other countries. By this fact, determined by its geological inheritance, the U.S. lost control of an important part of its economic destiny, as the oil crises of 1973 and 1979 showed. It was an unpleasant surprise to the American public.

Comparative productivity of wells by country

To get an international perspective on U.S. oil production it is instructive to compare oil well production in selected countries. (Table 3)

Note that the large per well daily production for Norway, the United Kingdom (Britain), and Papua New Guinea are all offshore wells. There is no onshore production. It is not economic to complete low production wells in these offshore operations. Therefore, only large wells are produced, which tends to distort the comparisons with the Persian Gulf producing nations where both onshore and offshore wells are produced and enter into the statistics. In the United States both onshore and offshore wells are produced, with an average of only 11.4 barrels/day, as the onshore wells are now chiefly very low producing and bring down the average. Note also that the United States, as of 1994, was producing oil from about 583,000 wells. This is far more than the country with the next largest number of producing wells, the former USSR with approximately 123,000 producing wells in an area more than twice the size of the United States. This emphasizes how thoroughly drilled the United States is, and why U.S. oil companies are having to go abroad to find exploration opportunities.

Oil well production by states

A comparison of oil well production by States in the U.S. is also interesting, including the current production of Pennsylvania where oil production began. (Table 4)

Table 4. Comparison of Average Oil Well Production in Selected States 1994

State	Number of Wells	Barrels oil daily	Average/well
Alaska	1,641	1,558,762	949.9
California	40,528	941,274	23.2
Kansas	43,149	128,036	3.0
Louisiana	28,250	1,127,784	39.9
New Mexico	18,085	180,400	10.0
Oklahoma	92,785	249,244	2.7
Pennsylvania	19,439	6,904	0.4
Texas	180,943	1,696,984	9.4
Wyoming	10,634	217,885	20.5

(Source: *Oil & Gas Journal,* December 25, 1995)

Alaska helps the average considerably with its 949.9 barrels per well per day to bring up the U.S. daily average to 11.4 barrels. Without Alaska's wells, the average production per well in the 48-adjacent states is 8.6 barrels a day, a sharp contrast to the more than 5,500 barrels a day from the average Saudi Arabian well.

In the two year period from year ends 1992 to 1994, average daily U.S. oil production declined from 7,170,969 barrels a day to 6,661,573, or 7.1 percent. In the same two year period in the three largest oil producing states which account for approximately 66 percent of total U.S. production, Alaska production dropped 9 percent, Texas dropped 7.8 percent, and Louisiana dropped 1.8 percent. The reason for Louisiana's relative better showing was that oil companies there were still allowed to drill in some offshore waters.

Of the 31 states which produce oil, 30 showed a greater or lesser decline. If there are large oil reserves yet to be discovered in the United States, they are not currently being found. There are probably two reasons for this. First, it is doubtful that very large undrilled oil reservoirs do exist, and, second, such substantial deposits as may exist are largely in areas now off limits to drilling such as off-shore California and the 8 percent of the Arctic National Wildlife Refuge which is the coastal plain.

The record from 1970 through 1993 of U.S. oil production, proved reserves, the number of wells drilled for oil, and the amount of oil found per drilled well (reserve additions per well) is shown in abstracted form in Table 5.

Table 5. United States Oil Statistics 1970-1993

(Millions of barrels)

Year	Production	Proved Reserves	Wells Drilled	Found per well
1970	3,326	39,001	21,522	.590
1975	2,888	32,682	26,253	.051
1980	2,971	29,805	45,316	.066
1983	3,016	27,735	52,577	.055
1986	2,973	26,889	26,523	.055
1989	2,586	26,501	13,896	.163
1992	2,446	23,745	12,389	.122
1993	2,339	22,957	11,643	.133

(Source: Adapted from Riva, 1995b)

What the numbers show

From 1970 through 1993, U.S. production has dropped from 3.326 billion barrels a year to 2.339 billion. In order to meet U.S. oil demand, more oil is now being imported than is produced. Proved reserves have declined from approximately 39 billion barrels to about 23 billion barrels (41 percent). The number of wells drilled annually for oil has dropped from 21,522 to 11,643. The amount of oil found per well drilled has declined from 590,000 barrels to 133,000 (77 percent).

The history of U.S. proved oil reserves from 1947 to 1994 is shown graphically in Figure 4. At the end of 1947, the proved reserves were 23.3 billion barrels. Peak of reserves was reached in 1970, with the Prudhoe Bay Field oil. Reserves have been in a steep decline since then. U.S. crude oil production in 1996 was approximately 6.4 million barrels a day. This was the lowest production rate since 1954.(1)

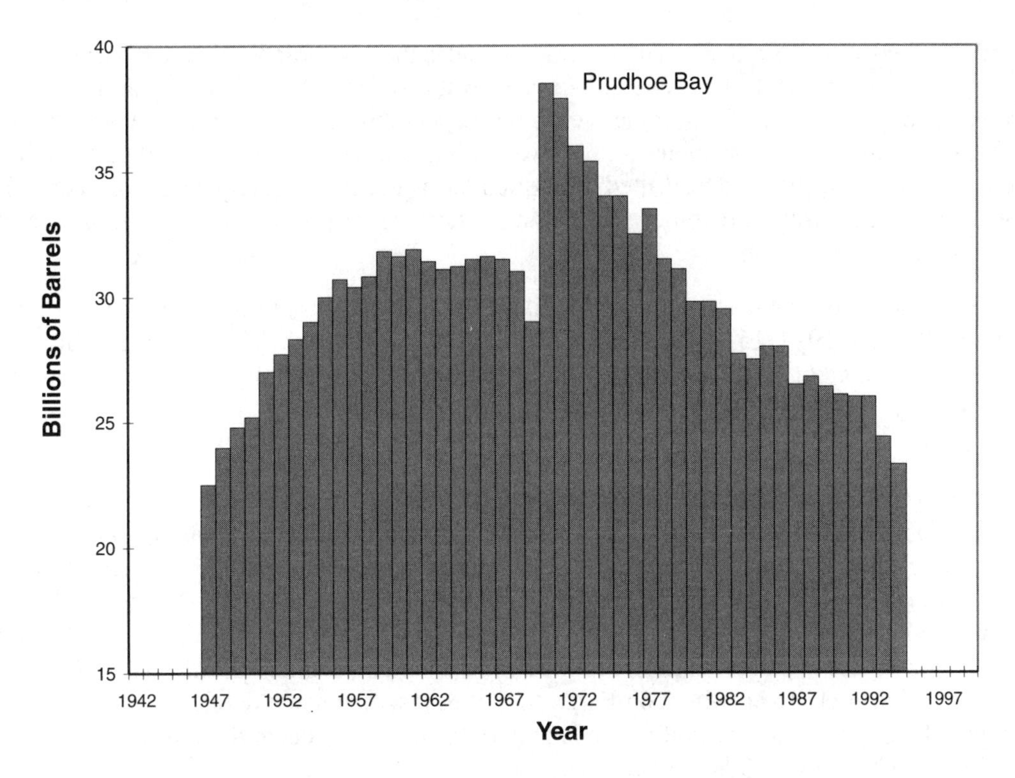

Figure 4. U.S. crude oil reserves, 1947-1994.

(Source: American Petroleum Institute, 1995)

The drop in both reserves and the amount of oil found per well drilled is particularly significant. The U.S. is clearly losing its oil reserve base, and the very marked drop of 77 percent in the amount of oil found per well drilled from 1970 to 1993 indicates that the best prospects have been drilled. What is left is much more marginal production. The U.S. has become a fully mature and now aging oil province. The best prospects left are almost all offshore.

Stripper wells

Three-quarters of U.S. oil wells are what are called "stripper wells." These are defined as wells which produce less than 10 barrels a day. Of the 589,034 producing wells in the U.S. as of January, 1994, 77 percent or 452,248 were in the stripper well category, with an average production of 2.2 barrels a day.(1)

The inevitable decline of an oil producing area is illustrated by the present production of 0.3 barrels a day per well in Pennsylvania where the U.S. oil industry began. Texas, the U.S. premier oil state, produced 21.7 barrels a day per well in 1974, and 9.7 barrels in 1993.(1)

Abandoned wells

Many wells each year are abandoned not because there is no more oil in the reservoir, but because the value of the oil which can be produced each day does not equal the cost of the daily operation of the well. It is not a matter of energy recovered versus energy expended, but rather the cost of maintaining the pump, the cost of the energy to run the pump (usually electrical), and the cost of paying the personnel to manage the operation. In this situation, a higher price for the oil would prolong the life of many wells. Inevitably, however, it becomes a matter of having to put more energy into lifting the oil than is obtained from the oil. Many wells gradually “go to water” — that is water invades the reservoir so that each barrel of oil pumped from the well brings several barrels of water with it. Water is heavy and requires considerable energy to lift it up the well bore, depending on the depth of the well, from several hundred to several thousand feet. Also, oil field waters are almost always highly mineralized and wells must be drilled into which this water can be safely disposed. It cannot be dumped into streams or other surface areas. Drilling disposal wells costs money and energy.

Eventually, for one reason or another, all oil and gas wells have to be abandoned. In the United States in 1994, 16,914 stripper oil wells were abandoned, whereas only 10,804 new wells were drilled.(1,51) If more oil wells each year are abandoned than are drilled, a nation is gradually going out of the oil business.

Drill deeper?

Perhaps in the United States we could simply drill deeper to get more oil in already proven areas? Sometimes this can be done, but there are limitations here also. Kansas, for example, has a long history of shallow oil production. It has historically been a great place where a small operator with limited capital could explore for oil because the wells are almost all less than 6,000 feet deep. “Post-hole” drilling this is sometimes called. But, reality is that 10,000 feet down, or less in places, in most of Kansas you would be in granite, which is notably not oil-producing rock. Everywhere on Earth, the oil drill will at some depth reach rocks with no petroleum prospects. Even if the sedimentary basin is very deep, there is still a limit as to the depth below which no oil occurs.

Deep wells — gas only

In regions where sedimentary rocks, the kind in which oil occurs, are very thick, there is a natural limit to the depth oil exists because of the geothermal gradient. That is, the deeper you go into the Earth the hotter it gets. Generally below 16,000 feet, with some minor exceptions, oil cannot survive, and only gas exists.

There is a further problem in drilling deeper for oil. The deeper one goes, the greater the pressure from the overlying rock, the pore space available to hold the oil is reduced by compression, and the individual pores tend to be smaller. So deeper wells may not be as good producers as are the more moderate depth wells, although, again, exceptions occur. Drilling deeper incurs higher costs but may not produce any more oil than from shallower wells.

Different kinds of reservoirs make a large difference

The kinds and details of the rocks which hold the oil, the oil reservoirs, make a great deal of difference in the amount of oil which the rocks can hold, and how much oil can be produced from them. Most of the oil reservoirs are either sandstone or limestone. Sandstones may be clean with little material other than rounded sand grains in which case there is considerable space for the oil to be held, and it can move easily through the formation. If the sandstone is not clean but has clay or limy material in it the amount of pore space is less and the oil flows less easily to the well bore.

The ability of an oil reservoir rock to transmit a fluid — in this case oil — is called permeability. It is measured in terms of a unit called the darcy. In Saudi Arabia, some formations have permeabilities which are measured in the range of one darcy or more. In the United States, permeabilities are commonly found in the range of millidarcies — that is, thousandths of a darcy. Partly as a result of this difference in permeabilities, the average production per well per day in the United States is 11.4 barrels. The daily average production per well in some of the better Saudi Arabia oil fields is in excess of 10,000 barrels, and in some cases exceeds 15,000 barrels a day because the reservoirs are nice clean sands or fractured limestones, This high rate of production could only exist with good permeability in the producing formations.

Good permeability also means that a given well can more efficiently drain a larger area than can a well which is producing from a "tight" formation — one with low permeability. Therefore, fewer wells have to be drilled in rocks of good permeability which reduces the cost of the oil per barrel. The Mid-East reservoirs and many other reservoirs abroad are much better reservoirs than what are now left in the United States. Consider the fact that a hole no more than a foot in diameter (usually less) is drilled through an oil-bearing formation but this hole has to drain from 10 to 40 or more acres of area. It is a great help to have an oil-bearing formation through which oil can move relatively freely to the well bore. In most oil fields of the world today on the average only about 20 percent of the oil in the reservoir is recovered by what are called primary methods, flowing or pumping.(51) Other more expensive methods are used to help recover additional oil, but it is impossible to get all the oil out of a reservoir.

Record Mexican well

Mexico apparently holds the world record for daily oil production from a single well. It apparently was drilled into a highly porous, almost cavernous limestone reservoir with very high permeability. In 1910, the famous Potrero del Llano well number 4 of the Mexican Eagle Oil Company was brought in with an initial flow of 100,000 barrels a day. But it soon got out of control and flowed about two million barrels, most of which was lost before the well could again be controlled. That single well ultimately produced more than 100 million barrels of oil. No such well, or one even close to it, has ever been drilled in the United States.

The contrast in reservoir production capabilities is strikingly apparent by noting that in 1992, the United States produced 2.6 billion barrels of oil from 602,000 wells. That year Saudi Arabia produced 2.9 billion barrels of oil from only 1,400 wells.(1)

USA: Most of the world's wells

The United States is now the most thoroughly oil-explored and drilled nation in the world. Approximate calculations (worldwide records are not complete enough to be precise) indicate that to date about 4,600,000 wells for oil and gas have been drilled in the world. Of this number approximately 3,400,000 or about 74 percent have been drilled in the United States. A major oil field covers a number of square miles. In the United States onshore there now are very few reasonably prospective areas where there is enough space between dry holes, producing wells, or depleted wells to accommodate a major oil field. Except for certain offshore areas, chiefly in the deeper waters of the Gulf of Mexico, in arctic and sub-arctic waters, and a portion of the Alaska North Slope which includes a national wildlife refuge, there is hardly an undrilled area large enough to hold a major oil field. Many of the prospective but unexplored areas which do exist are now off limits to drilling by environmental regulations.

Some "hostile" areas open

There are a few areas in United States territory which are still relatively unexplored and have the potential of some substantial finds. But as the more accessible places have already been drilled, what remains are areas not very friendly in terms of access and working conditions. These are termed in the oil industry "hostile" territories. Such areas include the Beaufort Sea, the Chukchi Sea, and Norton Sound, all located off northern and northwestern Alaska. These are regions of severe storms and difficult working conditions. Some drilling has been done in Norton Sound with disappointing results to date. The cost of drilling in these areas is very high. The most expensive oil well ever drilled in the world was drilled by Sohio Oil Company, their Mukluk Number 1, on an artificial island off the Alaska north coast. It cost $1.6 billion including regional exploration costs and was a dry hole.(11) The geologic structure which was drilled was gigantic, but the oil had escaped the trap some time in the past.

The most prospective onshore area yet undrilled lies between Prudhoe Bay and east along the Alaskan coast to the Canadian border, and includes a part of the Arctic National Wildlife Refuge. Also, the offshore areas of the 48-adjacent states have the possibility of holding some major oil fields. However, in 1995, the U.S. Congress reaffirmed its 13 year ban on most coastal drilling in Federal waters from Maine around the entire United States to Washington State. Some of the outer continental shelf areas are open for drilling but exploration and drilling here is in water depths in some places greater than 7,000 feet. In 1996, in a joint endeavor because it is so costly, Shell, Amoco, Mobil, and Texaco drilled a well in a water depth of 7,800 feet in the Gulf of Mexico.

U.S. Oil and "Giant" and "Supergiant" Fields

In the U.S. a "giant" field has meant one which originally contained at least 100 million barrels of economically recoverable oil.(44) (For world oil outside the U.S. it has been defined as having at least 500 million barrels of recoverable oil. The 100-million definition of "giant" is used here.) The rare "supergiant" field is universally defined as having at least five billion barrels of recoverable oil. The U.S. has only two.

There have been approximately 40,000 oil fields discovered in the world. Of these, according to the U.S. Energy Information Administration, 34,067 have been found in the United States as of the end of 1994. Of these 34,067 fields, only 288 are "giants." Yet, these 288 fields have contributed nearly 60 percent of total U.S. production and still hold slightly more than 61 percent of the remaining U.S. oil reserves. It is important to the future of the U.S. oil industry that areas where such "giants" might be found are made available for exploration.

The average U.S. giant oil field originally contained 430 million barrels of recoverable oil, and had an areal extent of about 10 square miles.(44) Recently, in the United States, very few giant fields have been found. In the decade of the 1960s, with the aid of the newly developed digital seismology, and with still some large areas to explore, 22 giant fields were discovered, including the supergiant Prudhoe Bay. But in the 1970s only 14 giant fields were discovered and in the 1980 decade only three were found. Since that time none has been confirmed.(44)

Where are the "Supergiant" Fields?

Of the over 40,000 oil fields found in the world, the great majority of them have had little impact on world oil production. There are only about 40 so-called "super-giant" fields (those with more than five billion barrels of recoverable oil). But these 40 fields originally held more than half of the oil discovered in the world to date. Significantly, 26 of these 40 super-giants are in the Persian Gulf area. Libya, Russia, Kazakhstan, Mexico, the United States, China, and Venezuela have the others.

Importance of giants

Riva, in a survey of the world oil status, cites the importance of giant oil fields in concluding that the Persian Gulf OPEC countries which have the great majority of the big oil fields, will eventually control the oil industry. He states:

> "This conclusion is based on a fundamental geological fact — that most of the world's oil is reservoired in a few very large fields. There are only about 40 known super-giant oil fields, but they originally contained more than half the world's known recoverable oil. In fact, less than 5 percent of the world's oil fields originally contained more than 95 percent of the world's known oil. Giant oil fields are immense concentrations of natural wealth upon which modern civilization and major oil companies are built. A single well in one giant oil field can deliver more than 10,000 B/D at a trivial cost in money and involving little labor. Such large fields, due to their substantial areal extent and anomalous geology, are normally found early in an exploration cycle."(49)

Riva goes on to suggest that because nearly all the world's sedimentary basins are known to some degree, and most have seen some drilling, it is unlikely that anything comparable to the present producing oil basins can be found in the future. The dominant position of the Persian Gulf oil producers seems assured. This increasing concentration of known world oil reserves in the Middle East has great significance for the future of the high energy consuming nations.

U.S. forced to move abroad to find more oil

With the remaining few good prospects onshore in the 48-adjacent States, and the current drilling restrictions on offshore areas, major oil companies now have to go overseas to find the large fields they need to survive. Such prospects as still exist onshore in the 48-adjacent states are for the most part pursued by smaller independent operators. The U.S. oil companies, long-time world leaders in exploration and production, have begun to run out of domestic acreage to explore.

Now, the best oil prospects are abroad. In 1988, ARCO, which depended heavily on domestic oil in the past, and was the major company in the Prudhoe Bay Field, spent 60 percent of its exploration budget overseas. Texaco spent half its exploration budget that year outside the United States. Other companies are similarly finding that the "oil patch abroad" is better hunting ground than is the USA. In the past five years, U.S. oil companies have spent $30 billion more on foreign oil exploration and development than in the United States.(57) This trend will continue for the duration of the Petroleum Interval for U.S. companies.

An example of this comes recently from California, for many years the second largest oil producing state, surpassed only by Texas. In 1995, Unocal Corporation, which originated in California and for many years based most of its production there, sold all its oil and natural gas production facilities in California. The chairman of the company stated that "Our strategic focus today is on natural gas development in the expanding markets of Asia and exploiting our gas reserve inventory in the Gulf of Mexico...proceeds from this sale will be directed to projects that offer the highest growth potential." Unocal's exploration efforts now are largely directed toward Indonesia and the Gulf of Thailand. Oil production in the State of California has been in decline since 1985.(1)

Even the premier professional petroleum geology association, The American Association of Petroleum Geologists, established in Tulsa, Oklahoma in 1917, now has an increasing number of members who are not American. To reflect that, it has added to its masthead "An International Geological Organization." It is a clear statement that the oil industry is no longer centered in the United States as it was for nearly 100 years.

No longer under American control

The significance of this is that oil resources, so vital to American industry and the national economy and way of life, are now chiefly in foreign hands. The oil companies may be American, but they are subject to the rules, regulations, taxes, politics, and whims of foreign governments. This is a vastly different situation from before 1970 when the United States was still self-sufficient in oil.

General Oil Exploration Economics, and the U.S.

To those unfamiliar with the oil industry, the monetary cost of finding oil might simply seem to be the cost of drilling the well. But that may be only a small part of the total expenses. There is the cost of acquiring leases, sometimes paying a competitive bonus for them even though there is no assurance that any oil will be found. The federal government and state governments get money from lease sales of their lands, and they get the lease money regardless of drilling results. For example, in the period from 1968-1978, oil companies paid the U.S. government more than $18 billion for bonuses and rentals on leases, before a single drop of oil was produced.(2) Of course, saying that the "oil companies" paid this money is incorrect. This cost is part of the cost of the oil and is ultimately paid by the public. The only money oil companies get to work with is what they get by selling their product to the consumer. If any oil is found on these lands leased from federal or state governments, these governmental bodies also get a royalty from each barrel produced. They cannot lose, and this cost also is ultimately passed on to the consumer.

After leases are obtained, the areas must be explored in detail. This exploration is done not only by geological studies but also by the use of expensive geophysical equipment designed to "see" into the Earth and find what are hoped to be oil-bearing strata and structures. Satellites are also used in the search for oil. Processing the geophysical and satellite data is a huge task involving billions of bits of information which can only be handled by computers. Then the data have to be interpreted by highly trained people. Rocks collected on the ground or by exploratory drill holes are sent to laboratories for analyses of their organic content to see if they might really have been a source to generate oil.

If drilling is conducted in the ocean, the ocean floor and what lies beneath it must be studied by means of a ship, fully manned with highly trained technical personnel and with a great deal of expensive equipment. Finally there is the cumulative cost of all the materials and equipment, beginning with the cost of mining and processing the ore which ultimately went into producing the steel tools with which to drill for oil, steel for making the drilling derrick, and to also provide the steel casing for the completion of the well. In some instances, two to four miles of steel casing goes into wells.

Energy cost

All these expenses should also be calculated in terms of energy costs. There also is the energy cost of building the ship and the energy cost of the materials used to build the ship if a vessel is involved in oil exploration. There is the energy cost of running the ship, or the energy it takes to haul the geophysical equipment over the tundra of the arctic regions, or through jungles, or perhaps through the air to remote sites by helicopter. There are the energy costs of doing the basic geological field work, if the prospect is on land. The gasoline to run the Jeep I used in Peru, and in western United States as a petroleum geologist had to eventually be accounted for by getting it out of the oil which this activity helped to discover. If this exploration work did not result in oil discoveries, then these energy costs must be paid for by oil from wells somewhere which do produce.

There is energy expended in drilling for oil. Rotating 10,000 feet or more of drill pipe with a heavy bit on the end against rock, and pumping mud through the drill pipe to carry up the rock cuttings takes lots of energy. Not all oil drilling operations are successful so the energy costs of drilling dry holes have to be charged against producing wells.

Energy profit ratio

All this energy expended in thousands of ways used to finally discover oil and produce it has to be added up and compared with the amount of energy in the oil which these efforts produce. This ratio — of energy produced compared to the energy used — is the all-important energy profit ratio. As we have to drill deeper to find oil, and as we have to move into more difficult and expensive areas in which to operate, the ratio of profit to energy expended declines. Already, in some situations energy in the oil found is not equal to the total energy expended. Also, although some wells flow initially, all wells eventually must be pumped. Pumping oil is expensive, particularly if it is being pumped from a considerable depth. It takes energy to move the steel pumping rods up and down, in some cases as much as three miles of them.

The rods also wear out from the sand or other abrasive materials which may come into the well, or be "eaten" out by the sulfur compounds and other corrosive chemicals which may be in the oil. The rods then have to be replaced, so there is the energy it took to make the rods, and the energy it takes to make new rods to replace the old rods.

The most significant trend in the U.S. oil industry has been the decline in the amount of energy recovered compared to energy expended. In 1916 the ratio was about 28 to 1, a very handsome energy return. By 1985, the ratio had dropped to 2 to 1, and is still dropping. The Complex Research Center at the University of New Hampshire made a study of this trend and concluded that by 2005 at the latest, it will take more energy, on the average, in the United States to explore for, and drill for, and produce oil from the wells than the wells will produce in energy.(19)

Exploration trends and costs

The easy oil to find in the more obvious traps has been found. In the United States, and other mature to aging oil producing regions, the search for oil has had to turn to oil traps not visible on the Earth's surface. These include buried anticlines, buried sand bars, shore lines, and stream channels. These oil traps may lie at depths of many thousands of feet. Finding them involves costly exploration and a lot of drilling. The shallower oil has been found and wells are going deeper. The average well depth in the United States in 1953 was 4,035 feet. In 1993 it was 5,668 feet.(1) It is costing more and more to find oil.

Oil per foot of drilling

We are drilling deeper but are we finding more oil? In the United States during the early 1930s, about 250 barrels of recoverable oil were found per foot drilled. By the 1950s, this figure had decreased to about 40 barrels a foot, and by 1981, it was down to 6.9 barrels per foot drilled. Oil producers in the United States are drilling deeper and finding less oil per foot drilled.

To exacerbate the matter, the cost of each foot drilled goes up with depth. Because the wells were getting deeper the footage was getting more expensive. In 1963, the average cost per foot of well drilled in the U.S. was $10.58. In 1993, the average cost was $75.30. Although there is an inflation factor involved here, the rise in cost of well drilling has been far higher than the inflation rate because the wells have to be drilled deeper. It costs not just twice as much to drill a 10,000 foot well than a 5,000 foot well, but three to four times as much. Wells drilled offshore from the expensive drilling platforms which have to be constructed, are much more costly than wells drilled on dry land.

"Drilling on Wall Street," but no new oil

When oil prices tumbled in the middle and late 1980s, in the United States, the oil industry, faced with the highest exploration and production costs in the world, drastically cut back domestic exploration. The number of operating drilling rigs dropped from more than 4,000 to fewer than 800. In 1995, the number averaged less than 700. Many operators went out of business. Other oil people decided to go "drilling on Wall Street." That is, they found it was cheaper to buy oil already discovered, by buying oil companies, than it was to go out and drill for it. There were corporate raiders such as T. Boone Pickens going through the industry, the result of which was a "restructuring" of the industry by mergers and acquisitions. But none of this activity found a barrel of oil. It simply made lawyers, accountants, and some entrepreneurs some quick money.

New oil discoveries vs. field extensions

Another trend in oil exploration in the United States is a low discovery rate of new oil fields. From 1977 to 1986, it was reported that 20 billion barrels of oil were added to U.S. reserves. But of this, less than two billion barrels were discovered in new fields. Most of the additional oil was found around the edges of already known fields, and by drilling inside known fields. Clearly oil hunting has become much more marginal than in the past. This is the part of the inevitable destiny of a finite Earth resource. Geology finally prevails. There is only so much oil.

U.S. Oil Reserve Decline

What is happening generally to United States' oil production is illustrated by Texas which was the major oil producer among the states until the Alaska Prudhoe Bay discovery. In 1960, Texas had oil reserves of about 14.8 billion barrels. By 1975, that figure had dropped to about 10 billion barrels, and by 1985 the figure was only about 7.8 billion barrels, only about half the reserve figure of 1960. By 1993, the figure had dropped to 6.17 billion barrels.(1) Clearly in the great oil producing State of Texas, more oil is being produced than is being found, which, in longer term, means that Texas is going out of the oil business.

Well production decline

During the same period of 1960 to 1993, the average production per well in Texas dropped from 13.3 to 9.7 barrels a day, a decline of 27 percent. Oklahoma saw a drop in daily well production from 6.9 to 2.8 barrels, a drop of almost 60 percent. As a historical note, the daily average production per well in the State of Pennsylvania, the birthplace of the U.S. oil industry, was just $\frac{4}{10}$ th of a barrel in 1993.(1)

In 1960, each Texas oil well on the average produced about 4,630 barrels of oil a year. In 1985, each well annually produced about 3,950 barrels, and in 1993 the average was 3,540. In Oklahoma, the 1960 total yearly oil production per well was 2,518 barrels. In 1993, it was 1,022 barrels.(1)

General long term decline

Oil reserves in the United States have continually declined ever since the year 1970 when additional drilling had proved up the Prudhoe Bay Field discovered in 1967. There is little likelihood that this trend can be reversed for very long. However, every discovery of oil, large or small, adds to the longevity of U.S. domestic oil supplies, and reduces the amount of oil which has to be imported.

It is worthwhile, however, to take note of a study headed by William L. Fisher, Director of the Texas Bureau of Economic Geology, made public in 1989. This study has an optimistic figure of 247 billion barrels of U.S. remaining oil. This amount is reached by assuming both the use of existing and technology yet to be devised, and an ultimate price for the final barrels of $50, based on a 1989 dollar value.

In 1995, the U.S. Geological Survey concluded that "Assuming existing technology, there are approximately 110 billion barrels of technically recoverable oil onshore and in State waters. This includes measured (proved) reserves, future additions to reserves in existing fields, and undiscovered resources."(56) Of this 110 billion barrels, the Survey includes 20 billion barrels of oil already discovered, which leaves 90 billion to be added in some fashion. Of this 90 billion, the survey estimates that 60 billion barrels can be added by oil found in existing fields. There is some precedent for this as "...the reestimation of reserves in old fields each year has added far more to measured (proved) reserves than have new discoveries."(25) At some point, however, new fields of some consequence have to be found. In this regard, the Survey estimates that undiscovered small fields (less than one million barrels) will provide 6.3 billion barrels, and undiscovered large fields (greater than one million barrels) will account for 24 billion barrels of future oil.

The foregoing estimates are moderately encouraging. However, if there is that much oil left to be discovered, critics ask why is it not being discovered? Instead, oil reserves have been declining since 1970.

There is no doubt considerable oil left to be produced in the United States, but the economics of the matter are uncertain. As already noted, most oil fields at present recover much less than half of the oil in place by the primary production methods of flowing and pumping. Much more remains potentially to be recovered. How much can be recovered is the multi-billion dollar question. At a price and with a time lag for the putting in place of existing technology and the development of new recovery techniques, more oil can be obtained. However, trying to recover oil beyond the initial flow (if the new well flows, some do not) and then simple pumping, involves costly processes, and the amount of energy in the oil recovered relative to the amount of energy used in the recovery process becomes a declining situation. It is never possible to recover all the oil in a reservoir. This statement applies to all oil fields around the world.

It seems certain that relative to current U.S. oil consumption of about 18 million barrels a day, the U.S. will never again be able to produce enough oil to supply its own oil needs. Imported oil has become a permanent material and economic fact of life, and the U.S. no longer controls the price.

Present U.S. oil position and trend

There are about 23 billion barrels of conventional oil reserves now identified in the United States. Daily U.S. production is about 6.4 million barrels of oil and about 1.9 million barrels of natural gas liquids for a total of 8.3 million barrels. At the same time, crude oil imports are 6.7 million barrels a day, and product imports are 1.7 million barrels a day for a total import volume of 8.4 million barrels. The United States is now importing more oil than it produces.

The supergiant Prudhoe Bay Field had initial reserves of about 9½ billion barrels of oil. With extensions and enhanced recovery technology it may have ultimate reserves of 12 billion barrels. This field alone now produces about one-quarter of U.S. oil. It is half empty, or half full if you wish, but half the oil is gone. The Prudhoe Bay Field is the largest field ever found in the United States, and no comparable U.S. field has been discovered since the Prudhoe Bay find in 1967.

But, as all oilfields eventually do, Prudhoe Bay is beginning to show its age. North Slope Alaska's oil production, which is chiefly from Prudhoe Bay, has been dropping since 1988. At its peak, 2.1 million barrels of oil moved daily down the Alaska Pipeline to Valdez. In 1996, this volume was down to 1.5 million barrels. That year the Alyeska Pipeline Service company which runs the 800-mile pipeline announced plans to close five of the 11 pumping stations in the following four years. This was in part due to the addition of a drag-reducing additive to the crude oil which makes it flow more easily, but the main reason for the shutdown of pumping stations is the decline in production. Since 1977, the Prudhoe Bay area has been the centerpiece of U.S. oil production, providing about 25 percent of the domestically produced oil. But it is fading, and unless additional oil supplies are found in the region (the best prospects are in a small portion of the Arctic National Wildlife Refuge about 60 miles east of Prudhoe Bay), Alaska's oil production will continue to decline. In the next 15 years it is projected to drop to 500,000 barrels a day, at which time it may not be economical to operate the pipeline.

As its conventional oil reserves decline, the United States is clearly living off its capital.(18) Some of the most promising prospects that remained in recent years have been disappointing. On the East Coast, the Baltimore Canyon area had been regarded as an excellent prospect. Shell Oil Company drilled four wells in what at that time was the deepest water in which any wells had ever been drilled, 7,000 feet, and all four wells, very expensive wells, were dry. In total, several tens of millions of dollars were spent by oil companies on this Baltimore Canyon "play." All the money was lost. The largest undrilled geologic structure in the eastern Gulf of Mexico, the Destin Dome, also was found dry when finally drilled.

About 600 miles offshore of southwest Alaska, in very rough seas, is the Narvin Basin, as big as the Gulf of Mexico. This was a major exploration target, and during the 1990s Amoco, ARCO, and Exxon, with the aid of the most sophisticated state of the art geophysical equipment to help them locate the best drill sites, drilled nine holes, all dry. Discovery of the large fields which would be significant to U.S. oil reserves are unfortunately not being made.(45)

Future U.S. Major Oil Discoveries

With almost three and a half million wells now drilled in the United States, making it the most thoroughly drilled up country in the world, where can significant oil discoveries still be made? An analysis of remaining U.S. oil exploration ares was made by Riva in 1994:

> "The discovery of new large fields from which large amounts of oil can be produced from relatively few wells could help shore up declining oil production. However, such fields are unlikely to be found in the mature, onshore oil producing regions in which several million wells already have been drilled and where the average well is producing less than 10 barrels a day. Rather, geologic studies indicate that almost all of the remaining undiscovered large oil accumulations in the United States are likely to be site-specific to a relatively few areas in Alaska and on the outer continental shelves. The Arctic National Wildlife Refuge on the North Slope of Alaska is especially prospective, containing the largest undrilled onshore geologic structures known in the United States. Most of these areas remain off limits to exploration, a questionable policy that has engendered much debate given that there is no oil-based substitute for 80 percent of the oil consumed domestically. The results of looking for oil where it is unlikely to be found have not been rewarding. In 1992, new oil field discoveries were exceptionally low, scarcely more than one day's production. U.S. oil production will continue to fall and imports, currently the equivalent of 15 supertankers offloading each day, will continue to rise."(48)

A perspective on the problem the U.S. now has in finding large oil fields can be gained from noting the discovery by the Shell Oil Company in 1996 of a field in deep water in the Gulf of Mexico. This field, called Mars, reportedly holds 700 million barrels of oil, and was enthusiastically and correctly reported by the media to be the largest domestic oil find since Alaska's Prudhoe Bay nearly 30 years ago. How much of this oil is economically recoverable is not now known. But even if all 700 million barrels are recoverable, this field is less than one-twelfth the size of Prudhoe Bay. It raises U.S. proved reserves by approximately three percent. The field will be a producer for many years to come, and will help to slightly slow, but not reverse, the decline of U.S. oil production. But the important fact to consider is the difficulty of drilling to a depth of 14,000 feet, in water more than one half mile deep, and the many years between this discovery and the previous notable oil strike. Also, relative to U.S. oil consumption, of nearly 6.6 billion barrels a year, this field is not large. But, even fields of this size are becoming increasingly rare discoveries in the heavily drilled U.S.

Russian Oil Production Peak

The history of the oil industry in the former USSR somewhat parallels that of the United States, but at later dates. USSR production passed that of the U.S. in 1974, and reached peak production in 1988, exceeding the best yearly production the United States ever had. The USSR then produced about 12 million barrels a day and was at that time the world's largest oil producer. But when the USSR broke apart the production declined. The northern areas in which Russia is now exploring tend to have more gas rather than oil compared to the older, more southern fields. Russia may increase its production to some extent, but it will almost certainly never have the volume of oil production it had, even adjusting proportionately for the loss of the southern, now independent republics. In 1993, Saudi Arabia became the world's single largest oil source, a position it holds to the present and is likely to hold for the indefinite future.

Russia, like the United States, is finding it increasingly difficult to find good oil fields. In the present most important oil province in Russia, the Western Siberian Basin, Russia reports that in 1970 it was necessary to drill about 865,000 feet of hole to get an increase in production of 20,000 barrels of oil per day. By 1985, the footage drilling figure had reached 7,000,000 needed to add that 20,000 barrels a day. Again, like the United States, Russia is finding less and less oil per foot drilled — characteristic of a mature to aging oil producing province.

Also, new wells being drilled were obviously having to be located in less productive areas. In 1976, the average Russian new well initially produced about 730 barrels a day. In 1985, the average yield of each new well was down to about 300 barrels a day. Russia, like the United States, is clearly running out of its more prospective acreage. In addition, in 1985, General Secretary Gorbachev reported that the costs of producing an additional ton of oil (about six barrels) had risen 70 percent the past 10 years. To find new areas to drill, Russia, like the United States, has had to move northward into more hostile areas with thicker permafrost, colder weather, more remote locations, and more difficult and longer road-building situations. These problems have substantially increased the costs of conducting oil exploration.

Some Countries Doing Very Well

First the United States, and later the USSR were the world's largest oil producers and oil production was rising. But in both countries, production now has passed its peak and reserves have fallen. The U.S. reserves dropped from 39 billion barrels in 1970 to about 23 billion now. The USSR reserves declined from 83 billion barrels in 1975 to about 57 billion now (figure for 1994 includes all the Republics once in the USSR and including Russia, so data are comparable to those of 1975).(1)

The U.S. and Russia are now in a declining oil producing situation, but some countries have been doing well. A comparison of the oil reserve positions of selected countries illustrates what has happened between 1970 and 1994. (Table 6)

These data show the big gainers in oil reserves during the past 20 years were the Persian Gulf nations. Mexico and Venezuela also did very well. The U.S. and Russia were the big losers, along with the United Kingdom, and Indonesia. Canada also declined.

At the beginning of 1995, total world oil proven reserves were estimated to be 999.7 billion barrels, of which 660.3 billion (66 percent) were in the Middle East. Of the total Middle East Reserves, Saudi Arabia officially claimed to hold 261.1 billion barrels (39 percent), which is 26 percent of the total world reserves.

It is also significant that the Middle East countries are the world's lowest cost oil producers with an average cost of less than $2 a barrel. Saudi Arabia probably has a cost of 50 cents or less. To replace the oil being produced in the United States, the cost is now somewhere between $6 and $20 a barrel, some areas are not now economic to drill.

Oil production in countries outside the Middle East is generally much more costly, particularly in the harsh environment of the North Sea where Norway and Britain have nearly all their production. Northern Alaska and the Canadian arctic likewise are expensive areas in which to search for and produce oil. Offshore areas elsewhere involve exploration and drilling in relatively deep waters. The conveniently located oil in the world has mostly been discovered.

Table 6. Oil Reserves in Selected Countries 1974 & 1994

(Billions of barrels)

Country	1974	1994
United States	34.2	24.0
Commonwealth of Independent States (includes all former USSR)	83.4	57.0
Saudi Arabia	164.5	258.7[1]
United Arab Emirates (UAE)	33.9	61.0[2]
Kuwait	72.8	94.0
Iran	66	92.9
Iraq	35.0	100
Indonesia	15	5.8
China	25	24
Venezuela	15	63.6
United Kingdom	15.7	4.5
Mexico	13.5	27.4[3]
Norway	7.3	9.2
Canada	7.1	5.1

(Source: American Petroleum Institute, and Oil & Gas Journal 1995; Riva 1994b). [1]This is official Saudi Arabian government figure. Other estimates are lower. [2]Riva, personal communication 1995. [3]This figure is substantially lower than the 50.9 figure which the API and the OGJ carry for Mexico. There is some indication the larger figure was inflated, perhaps for financial purposes. Figure of 27.4 billion barrels appears to be more realistic. API and *Oil and Gas Journal* use figures given by the official agencies of the various countries.

Saudi Arabia and other Persian Gulf nations also have very large fields which can produce for a long time so that additional vigorous exploration is not really needed for some time to come. They can live very well on the oil already found. This is not true of the United States which is using more than twice as much oil as it can produce.

Saudi Arabia is currently producing about eight million barrels a day to stabilize prices, but it can easily produce ten million barrels a day or more if it wished. The United States has no reserve production capacity. The wells are running as wide open as they can be efficiently run. This is also true of Russian oil fields. Clearly, oil production more and more will center in the Middle East. The Persian Gulf Muslim nations are geodestined to have the last word on oil. OPEC will survive and eventually emerge with fewer but stronger members. OPEC nations outside the Middle East will gradually drop by the wayside as their oil reserves are exhausted.

Reserve to Production Ratio

Some countries have produced their oil much earlier and faster than other areas. Some have passed their peak; some are about at their peak; and some are still coming up to their full potential. It is also important to note that oil is now being produced by various countries at markedly different rates relative to their reserves. The number of years reserves of oil will last at the current rate of production is called the reserve production ratio (R/P). (Current yearly production divided into proved reserves).

However, this will not be an accurate prediction of how long the reserves will last because it does not take into consideration the effect of additional discoveries which might be made, nor the expansion of reserves by further drilling in existing fields. And it assumes that the present rate of production will remain fixed. In the Middle East countries, for example, Saudi Arabia and Kuwait, there is excess production capacity which may be brought on stream should world demand increase to the extent that additional production would not depress prices. OPEC now tries to maintain a production quota among its members to hold the price of oil where they want it. However, some countries want to increase their share of the market. Kuwait plans to ask OPEC to increase its production quota from the present two million barrels a day to three million a day within the next five years. If this would come at the expense of another country's quota, it might produce some strains within OPEC as has happened in the past.

The reserve production ratio (R/P) does give some general indication of the probable relative life of oil production in various nations. The United States, for example, has already probably produced more than 60 percent of what was its original amount of conventionally recoverable oil, and has a current reserve to annual production ratio (R/P) of 10/1. Canada has 8/1, Norway has 10/1, and Britain has 13/1. In marked contrast, Kuwait's R/P is 116/1, and Saudi Arabia has 55/1 (based on conservative figures of proved reserves of 160 billion barrels). If the Saudi government's reserve figures of 261 billion barrels are used, the R/P ratio is 89/1. These two Gulf nations have long oil-producing futures, even at substantially increased rates of oil extraction. They will have an important influence on world economies for many years to come. Iran, chronic trouble-maker in the Persian Gulf region, for better or for worse can also look ahead to many years of oil production at current rates, with a R/P of 53/1. Libya, with somewhat the same reputation as Iran, will have oil to produce at current rates of exploitation for at least 46 more years.

The United Arab Emirates with a R/P of 75/1 will also be in the oil producing business for a long time to come. Iraq, with the current United Nations embargo on exporting oil has at the moment an R/P of 526/1. However, Iraq is very much in need of oil income to finance basic imports. When oil exports can be resumed, the daily production will probably be about two to three million barrels, giving Iraq a R/P of about 50/1 — still a long time of oil production.(53) It may be hoped that Iraq will use this money for better purposes than in the recent past.

A generalized summary of the time to which current production can be sustained by various countries is given in Table 7.

Like the reserve/production ratio statistics, these are relative projections because discoveries and changes in rates of production will alter these projections. But Table 7 does give some perspective on who has the oil and how long they might have it.

Table 7. Time to which Current Oil Production can be Sustained

(by Country)

Less than 10 years	Less than 50 years
United States	China
Canada	Nigeria*
United Kingdom	Algeria*
Indonesia*	Malaysia
Norway	Colombia
Egypt	Oman
Argentina	India
Australia	Qatar*
Ecuador	Angola
	Romania
	Yemen
	Brunei

Less than 100 years	More than 100 years
Saudi Arabia*	Iraq*
Russia	United Arab Emirates (UAE)*
Iran*	Kuwait*
Venezuela*	Kazakhstan
Mexico	Turkmenistan
Libya*	Tunisia
Brazil	Uzbekistan
Azerbaijan	
Trinidad	
The World[1]	

*OPEC member

[1]Total world crude oil production in 1997 is approximately 62 million barrels a day.

(Source: Riva, 1995b)

A measure of oil wealth per capita in selected countries has been calculated in Table 8. These figures are probably more significant for those countries which depend chiefly on oil for their national income than for countries which also have other sources of wealth, as, for example, the U.S. with its huge agricultural base. Japan's vulnerability clearly stands out for it has no significant oil wealth nor significant other mineral or exportable agricultural wealth. It is simply an island factory strictly dependent on imports and exports.

Natural Gas

Natural gas is chiefly methane (CH_4) with minor amounts of ethane. It may also contain heavier hydrocarbon gases including butane, pentane, and propane and it may have the

Table 8. Barrels of Oil Reserves Per Capita in Selected Countries

(Based on 1990 population: World Almanac, and 1990 proved crude oil reserves: American Petroleum Institute)

Country	Barrels of proved crude oil per capita
Canada	233
China	23
Commonwealth of Independent States (former USSR	218
Great Britain	70
India	7
Indonesia	45
Iran	955
Iraq	5,525
Japan	0.4
Kuwait	44,762
Libya	5,700
Mexico[1]	650
Norway	2,749
Saudi Arabia[1]	17,344
United Arab Emirates (UAE)	57,474
United States[1]	106
Venezuela	3,063

[1]If the probable more conservative reserve figures are used for Mexico the barrels/capita are 334; for Saudi Arabia 11,497. If 1995 population and oil reserve figures are used for the USA, there were 86 barrels/capita.

non-combustible gases, nitrogen and carbon dioxide. Like oil, gas is derived from buried remains of plants and animals which, over time, were subjected to heat and pressure. Some sediments which contain relatively more woody and other plant materials tend to be "gas prone", and contain little or no oil. Also, generally at depths greater than 16,000 feet the temperature of the Earth is so hot that oil cannot exist and only gas survives.

Gas has a more widespread geological occurrence than does oil, although in general it is found in oil provinces. However, whereas most oil fields have associated gas, many gas deposits have no associated oil. Large areas of black shale in eastern and central United States contain considerable gas, as do other dark shales in Montana and Wyoming. These rocks, however, have low permeability, and the rate of gas flow from wells drilled is small. Locally, wells are drilled for farm houses and ranches for domestic gas supply.

Natural gas is also associated with coal deposits. Underground coal miners are subject to the hazards of asphyxiation from deadly "coal gas" (methane), and gas explosions. Some wells are drilled into coal seams to produce this gas. This technique has a particular economic value in areas which, because of complex geology (folding, faulting), cannot be mined satisfactorily for coal. They can be drilled to produce gas. In total, however, the amount of natural gas which can be produced from coal seams is small compared with gas from conventional deposits.

Originally, most gas was found as a result of drilling for oil. When oil comes to the surface and the pressure is reduced, some light oils vaporize and become gas. Also gas which is in solution in the oil will come out. For many years, gas was regarded as of little value, and was flared (burned off). This wasteful practice has largely stopped. Now, the gas is either piped away to be used for home heating, industrial heat, or electric power generation, or it is pumped back into the oil reservoir to help maintain the reservoir pressure to produce the oil. Pipelines are an efficient way of transporting gas for great distances, and gas pipelines now extend across many countries and across many borders. The natural gas industry has shown steady growth over many years.

Gas can be used to replace oil in space heating and as fuel for boilers, both in industrial heating, and for electric power generation. Use of gas as fuel for electric power plants is the most rapidly growing use for natural gas today. Gas is also the basic ingredient for the production of ammonium nitrate, a fertilizer extensively used in agriculture, and also in explosives. Gas is the least polluting of all fossil fuels, and has a lower particulate emissions that does oil or coal.(23) In equivalent energy, about 5,600 cubic feet of gas equals one barrel of oil.

Who has the natural gas?

With the common geologic association of gas with oil, the countries with large oil supplies also have large gas deposits. However, because gas is also found where oil does not occur, there are countries which may have more gas than oil. For example, the tiny Sheikdom of Qatar which does not rank among the 20 leading nations in crude oil reserves, ranks third in the world in gas reserves, holding 12 percent of the total. In its approximately 17,000 square miles (land and claimed sea area), Qatar holds 36 percent more gas than does the United States in its three and one half million square miles.(1) Table 9 lists countries holding major world natural gas reserves in 1975 and in 1993.

Among the major countries holding gas reserves, only two have had a decline since 1975, the United States and Algeria. Algeria sends gas across the Mediterranean by pipeline to France. In 1995, the U.S. Geological Survey estimated that the proved U.S. gas reserves were 135 trillion cubic feet and that reserve growth by improved production techniques and expanding existing fields would result in an additional 322 trillion cubic feet, more than twice the known reserves.(56) However, infield drilling in North America's largest gas field, the Hugoton, which lies astride the Oklahoma-Kansas border area, did not result in increased gas reserves.(34) In 1992, the total amount of gas discovered in new fields amounted to two weeks of domestic production and in 1993 new field gas discoveries equaled two had a half weeks of production. Riva reports: "In the past decade only about 15 trillion cubic feet of gas was discovered in new lower-48 State fields by exploratory drilling. This is less than one year's output. Gas exploration has been hampered by a drilling moratoria off much of the east and west coasts and Florida. It is ironic that environmentally inspired drilling moratoria will inhibit environmentally favored natural gas."(50)

The increase in U.S. gas supplies projected by the U.S. Geological Survey may or may not materialize. The projections are also based on a large amount of gas to be obtained from what are termed "continuous-type accumulations (unconventional)," which include shales

Table 9. Countries Holding Major Proved Gas Reserves, 1975 & 1995

(Trillions of cubic feet)
Arranged in order of 1995 reserves

Country	1975	1995	1995 R/P[1]
USSR (as of 1975)	810	(C.I.S.)[2] 2,057	82/1[3]
Iran	330	741.6	742/1
Qatar	none significant	250	1250/1
United Arab Emirates	200	204.6	256/1
Saudi Arabia	55	185.9	169/1
United States	237.1	163.8	9/1
Venezuela	43.5	130.4	163/1
Algeria	229	128	71/1
Nigeria	45.5	120	600/1
Iraq	27.5	109.5	1095/1
Canada	56.7	79.2	13/1
Norway	24.7	70.9	79/1
Mexico	15	69.7	54/1
Malaysia	15	68	97/1
Netherlands	94.8	66.2	24/1
Indonesia	15	64.4	34/1
China	25	59	98/1
Kuwait	32	52.9	264/1
Libya	26.5	45.8	229/1
United Kingdom	50	22.2	9/1
Australia	38	19.6	25/1

(Source: American Petroleum Institute for 1975, Riva 1995e) [1]R/P is proved reserves divided by annual production rate. [2] C.I.S. is the Commonwealth of Independent States, the former U.S.S.R. [3]This R/P figure is for Russia proper 1995. R/Ps of now independent republics in 1995 are Azerbaijan 63; Kazakhstan 277; Turkmenistan 68; Ukraine 36; Uzbekistan 68. Other republics have only minor natural gas resources.

and tight sandstones. These were credited in the 1995 assessment with 358 trillion cubic feet and termed technically recoverable.(23) However, producing this gas usually requires special treatment and the recovery rate is low.

The results of drilling for natural gas during the decade of the 1980s is reviewed by Riva:

"During the 1980's about two wells (exploratory, development, and dry holes) had to be drilled for gas to net one additional producing gas well and an average of almost 18,000 wells per year were drilled. If lower-48 State per-well gas production continues to decline at the same rate as in the past, a yearly average of more than 20,000 wells will have to be drilled for gas to sustain the current 17 trillion cubic feet during the next decade, and over 30,000 wells per year would be necessary to increase it to the level (20 trillion cubic feet) projected by the Department of Energy. Gas drilling in the 1990's has averaged less than 13,000 wells per year and continues to decline."(50)

What has happened to natural gas resources in the United States is exemplified by the State of Texas. In 1960, Texas had reserves of about 119 trillion cubic feet. By 1975, this figure was down to about 71 trillion cubic feet, in 1985, the figure was 42 trillion cubic feet, and in 1991, it was 36 trillion cubic feet. Overall in the United States, the natural gas reserves were 262 trillion cubic feet in 1960 and 167 trillion cubic feet in 1991.(1) During the past ten years the United States' gas reserves in the lower 48 states have been increased almost entirely by drilling in existing fields. This has been only moderately successful and in some fields not at all. The 15 trillion cubic feet which have been discovered in new fields by exploratory drilling compares with the approximately 19 trillion cubic feet used annually.(1) If gas production levels are to be maintained in the United States, new fields must be discovered and this is not happening to any significant extent.

The United States at present is living off its natural gas capital. Considerable increased drilling and technical effort will have to be made if the trend of declining reserves is to be stopped or reversed.

World Distribution of Natural Gas

During the next several decades, natural gas can be expected to provide increased amounts of energy around the world, particularly because it is more widespread in its occurrence than is oil. Worldwide, natural gas consumption is expected to continue to grow at the rate of 2.5 to 3 percent annually compared with oil's 1.5 to 2 percent growth. The regional occurrence of natural gas deposits and their estimated volumes in 1975 and 1995 are shown listed in Table 10.

Table 10. World Regional Natural Gas Distribution 1995

(Trillions of cubic feet)

Region	Estimated Undiscovered Resources	Original Gas Resources	Cumulative Production	Remaining Gas
North America	856.5	2118.3	949.1	1169.2
South America	291.1	523.9	43.7	480.2
Europe	299.9	736.4	220.2	516.2
Former USSR	1840	4358.9	461.4	3897.5
Africa	411.4	788.7	35.7	753.0
Middle East	1013.7	2665.7	57.7	2608.0
Asia/Oceania	561.4	998.1	86.1	912.0
Total World	5274.0	12190.0	1853.9	10336.1

Approximate percentage of estimated total world original gas resources produced to 1995: 15.2%

(Source: Slightly modified from Riva 1995e)

Most natural gas is now transported to the point of use by pipeline. The United States currently is importing gas by this method both from Mexico and Canada. However, gas can also be cooled to a liquid state and shipped in cryogenic tankers great distances by sea. This adds significant additional expense over crude oil shipment because of the cost of refrigeration equipment, and the energy involved in keeping the gas cold. Nevertheless, it is an economical method of gas transport, and natural gas from Algeria has been landed by tanker on the East coast of the U.S. Japan now gets considerable gas from both Alaska (Cook Inlet area in southern Alaska) and the Middle East, by cryogenic tanker, and has recently entered into a long term gas purchase agreement with Qatar.

Jobs and Products from the Petroleum Industry

The petroleum industry, from exploration, to production, to refining, to marketing, and to myriad products from the petrochemical industry is the largest single industry in both the United States and the world. However, with the decline of the domestic oil industry, the United States, during the decade of the 1980s lost an estimated 450,000 jobs.(22) The decline is continuing, and the greatest loss has been in exploration and production where, from 1982 to 1996 jobs declined from 754,500 to 305,000, a drop of more than 50 percent. (Source: U.S. Bureau of Labor Statistics). This is a simple but most significant indication of the fact that oil exploration prospects in the U.S. are very much fewer than in earlier times. Jobs were not only lost directly in the oil industry, but those businesses which manufacture equipment and provide supplies and services to the oil industry lost jobs. If drill pipe and steel casing are not used for well drilling in Texas, jobs are lost in the steel mills of Pittsburgh. There is a great economic ripple effect from the oil industry through many parts of the economy. The more oil we import, the more jobs we export to other countries. With every barrel of oil the U.S. imports, an equivalent "barrel of job" is exported. With continuing rising oil imports, more U.S. jobs will be lost. The trend is continuing. An estimated additional 50,000 jobs were lost in the first half of the 1990s. And it is estimated that for each job lost directly in the oil industry, that two jobs are lost in peripheral related industries.

Oil — much more than just energy

Crude oil is a remarkable material. Not only can it produce high quality, convenient fuels, gasoline, kerosene, and diesel, but it also is a versatile raw material from which to make myriad useful products. Petrochemicals are used one way or another by nearly everyone in the industrial world every day. Plastics in particular are used very widely in automobile parts, plastic cabinets for television sets and radios, in products including food packaging, television and radio parts, in paints and inks, dyes, videotapes, pharmaceuticals, and in many agricultural chemicals. There are thousands of different petrochemical products in daily use with an annual value of hundreds of billions of dollars. When solar energy or nuclear power is suggested as a replacement for oil, frequently the multitude of uses to which oil can be put beyond simply being an energy source are overlooked. There is no evident replacement for oil in the volumes and the myriad ways in which it is now used.

Petrochemicals, once a U.S. domain

The petrochemical industry was born in 1920 in an experimental plant near Bayway, New Jersey, built by what is now Exxon corporation. But with the discovery of large oil deposits in Texas the petrochemical industry moved to Houston, and until quite recently the undisputed world center of the petrochemical industry was Houston, Texas, and vicinity. First, with ample domestic crude oil available, and then gradually supplemented by rising imports of crude oil, the U.S. petrochemical industry flourished. Now, however, more and more foreign producers are upgrading their crude oil and natural gas at home and exporting the finished product. Saudi Arabia has constructed a huge petrochemical complex, and other Persian Gulf nations are beginning to do the same. Qatar is moving forward with a $700 million petrochemical plant using gas from its North field, the largest single natural gas deposit in the world. Kuwait's refinery and its Petrochemical Company now retain about 40 percent of Kuwait's petroleum production within the country to process it into finished products. In 1996 a Middle East Refining and Petrochemics Conference and Exhibition was held in Bahrain, emphasizing the rapidly growing importance of Persian Gulf countries' interest in upgrading their petroleum rather than simply shipping out unprocessed oil and gas.

Instead of sending crude oil to the United States for petrochemical raw material, the petroleum producers are exporting the end product not only to the United States, but to other nations of the world which were formerly the petrochemical customers of the United States. During the 1981-1990 period, the United States petrochemical industry share of global exports declined 26 percent.(29)

When the oil industry declines in the United States and expands abroad, not only are oil industry jobs lost, but related sources of income are lost to the United States. This trend is far-reaching, and now extends beyond the Persian Gulf. The headline reads "Thailand rapidly developing into world class petrochemical producer."(6)

The oil industry in general, and the petrochemical industry particularly produce a phenomenally wide array of products. Unocal (formerly Union Oil Company of California) researchers broke down a typical barrel of U.S. crude oil and listed what just one barrel can produce:

- Enough gasoline to drive a medium-sized car (17-miles-per gallon) over 200 miles
- Enough distillate fuel to drive a large truck (five-miles-per gallon) almost 40 miles or, if the jet fuel fraction is not used by a plane, it can run that same truck nearly 50 miles
- Enough liquefied gases (such as propane) to fill 12 small (14.1 ounce) cylinders for home, camping, or workshop use
- Nearly 70 kilowatt hours of electricity at a power plant generated by residual fuel oil
- Asphalt to make about one gallon of tar for patching roofs or streets
- About four pounds of charcoal briquets
- Wax for 170 small birthday candles, or 27 wax crayons
- Lubricants to make about a quart of motor oil

The petrochemicals in a barrel of crude oil can make a great variety of products, literally thousands. There are enough petrochemicals still left in that same barrel of oil to provide the base for one of the following:

- 39 polyester shirts
- 750 pocket combs
- 540 toothbrushes
- 65 plastic dustpans
- 23 hula hoops

- 65 plastic drinking glasses
- 195 one-cup measuring cups
- 11 plastic telephone housings
- 135 four-inch rubber balls

The special naphthas in a barrel of oil are used mainly for paint thinners and dry-cleaning solvents, and they could make nearly a quart of one of these products.

The miscellaneous fraction of what is left still contains enough by-products to be used in medicinal oils, absorption oils, road oil, and fractionating oils. (Source: *Seventy Six*. Unocal publication, September/October 1978)

Cars and trucks and airplanes run on oil in two ways. Not only do oil products give them the energy to move, but since the discovery during World War II that rubber could be made from styrene and butadiene, two chemicals derived from oil, the world's tires, in effect, come out of oil wells. The Japanese stimulated the U.S. research which led to the development of synthetic rubber. They had advanced into southeast Asia during World War II and cut off almost all natural rubber supplies to the United States. But petrochemistry and oil solved the problem.(2)

Vehicles other than aircraft, also run on oil in a third way. Asphalt roads are a mixture of about 10 percent very heavy ("bottoms") remains of oil refining mixed with 90 percent gravel or crushed rock.

A Permanent Trend to Move Abroad

For all the employment and useful products which oil brings to a nation, unfortunately, there is not much the United States can do about the present general oil industry trend to expand overseas at the expense of the U.S. For a time the United States was pre-eminent. Now, other oil areas in the world far outrank the United States in oil reserves and that is where the oil industry is going. Where the oil was put and the influence it has on the United States, and now has on the world is one of the most significant of the geodestinies imposed on all nations by the Earth's geological history. The influences are profound and inevitable.

Those Who Will Miss the Petroleum Interval

One aspect of the Petroleum Interval largely ignored by the people fortunate enough to have the generous use of oil supplies, is that major segments of world population will never enter into and enjoy this time of oil as the developed countries know it. The figures are convincing.

Total world crude oil production is now about 62 million barrels a day. (Liquids condensed from natural gas add another approximately nine million barrels of liquid fuel a day). Consumption of oil is a very good index of standard of living as we view it today. The oil standard of living presently enjoyed by the United States is achieved by using approximately 18 million barrels a day. But it would not be possible to produce enough oil to establish and maintain such a standard of living for the entire world population. If China used oil on a per capita basis at the same rate as does the United States, the Chinese alone would use approximately 81 million barrels of oil a day, which is 10 million barrels more than the entire present world oil production. The pleasant Petroleum Interval will also bypass most of the more than three-quarters of a billion people in India, as well as many people in

Africa and South America. These "oil-less" people will only get a passing distant glimpse of the benefits which oil bestows on the fortunate people who have substantial access to it.

For China and India combined, at the U.S. rate of oil consumption, more than 140 million barrels of oil would be needed each day. Such production cannot be achieved. And these calculations have left out the oil needs of all the rest of the world. Human labor, and elementary energy supplies such as biomass (wood, crop wastes, animal manure), will remain for many years, perhaps centuries, the important energy sources for many regions.

If oil, however, were devoted chiefly to industrial use by China and India, the volumes of oil used might be reduced to one-half or even one-third of the amount required to achieve an overall U.S. standard of living. But even at that reduced figure, the amount needed would be in the vicinity of 50 million barrels of oil a day, leaving very little for the rest of the world. Indeed, what would be left from world oil production would be less than the United States alone uses today.

It is clear that in terms of oil there are large populations which cannot become industrialized like the western world today. For China, however, there are huge deposits of coal which can be further developed. These are already used extensively, with an attendant severe air pollution problem. But it is not feasible to convert this coal to great volumes of oil. China's current oil production of some 3 million barrels of oil a day, even if doubled or tripled, will not give them a western style oil-economy. The Chinese are said by some observers to be "essentially self-sufficient" in oil. But less than three million barrels of oil per day divided by more than one billion Chinese, means they are self-sufficient in oil only in the way they use it today. They are not self-sufficient if they would use oil to replace hand labor and to appreciably raise their standard of living.

Oil provides many amenities for those who have it. In the United States, it may be a weekend journey to the beach, a drive to the ski slopes, a trip with an outboard motor or perhaps even a yacht over interesting waters. These are things the average Chinese will never enjoy.

A comparison of per capita consumption of gasoline among several industrial nations shows how much more the United States makes use of the Petroleum Interval, even compared with other advanced nations. (Statistics from the World Resources Institute).

Per Capita Consumption of Gasoline 1995 (Gallons per person per year)	
United States	484
Sweden	221
France	181
Britain	176
Italy	148
Japan	133

The Future

The near-term outlook

What is the probable scenario for the Petroleum Interval in the fairly immediate future? Conoco Oil Company has recently made an estimate.(3) Some of their general conclusions are that within the next decade or less, that world crude oil demand will rise to about 80 million barrels a day from the current 62 million. Production from OPEC countries will rise to about 33 million barrels a day from a base of 24 million in 1990. Production of countries outside of OPEC will remain flat. U.S. oil demand will grow to nearly 20 million barrels a day from the current 18 million. U.S. crude oil production will fall to about 5.6 million barrels a day from its present 6.4 million. Imports will have to make up the difference at a cost which may be as much as $150 billion a year.

Conoco also states "It is unlikely the U.S. will be able to reduce its dependence on imported energy supplies, let alone eliminate it. Therefore it is imperative to manage that dependence." For solutions, Conoco recommends conservation and fuel diversification particularly for transportation. Conoco adds, "Since price is the best incentive for conservation, one step that would lead to lower oil dependence and improved air quality is raising gasoline taxes. A 50 cent per gallon increase would reduce demand and encourage purchase of more efficient automobiles." Finally, Conoco suggests that areas now off limits to drilling be opened to substantially contribute to U.S. efforts to manage its growing oil import dependency, and that the U.S. government, by tax incentives, encourage U.S. companies to increase their worldwide exploration efforts. This last recommendation is a clear cut confirmation that the U.S. itself is rather thoroughly drilled.

Long view

Finally, when one looks at figures for oil reserves, and sees the wide differences country by country and region by region, the question arises as to just when and how the Petroleum Interval will end. For individual nations the end of its oil reserves may come soon or it may be a hundred years from now. But we have a global economy. Can one country enjoy abundant oil for decades after other countries, previously used to ample oil supplies and with economies built on that base, run dry of oil?

Because of the very uneven distribution of world oil resources, there will be increasing concern as to how to manage these finite resources. Will resource distribution occur peacefully or will there be attempts to solve it by military action? A revision of lifestyles, and implementation of strong energy planning and policies are part of the not too distant future for the western oil-short industrial nations. With the time-lag required in planning and making major changes, the time to begin is immediately at hand, but human nature being what it is, the problem may be faced only on an eventual crisis basis. We may hope even under those circumstances, actions will be reasonable and not violent. The behavior of the American public in the long gasoline lines of the 1970s, however, is not reassuring. Even murder was committed.

Most likely the end of the Petroleum Interval will be gradual wherein no crisis point is reached, just slow change. But, especially with continually rising populations, and no sufficient substitutes for oil at hand, there is the possibility of a chaotic breakdown of society. But some adjustments can be made to extend the time available to make changes. For example, the complex of hydrocarbons which oil is, has long been thought by some to be too valuable to simply burn for space heating such as is now done in home oil burners. Because of this biochemist and former Harvard University President, James Bryant Conant, believed that by 1980 there would be a law against such a use of oil. His guess on time was not correct, although his premise was. It is probably not a matter of if, but when such legislation occurs. Space heating can be accomplished by electricity directly, or in some regions by using electric heat pumps. Some of this power can be generated by coal sources and much more by nuclear plants.

The last large general use of oil will probably be in motor vehicles which will, at the end, be more efficient than are those which presently exist. For that day, a new generation of cars is being planned, but these too ultimately will have to be replaced by vehicles using other forms of energy than oil. It may be noted that the use of oil today to produce plastics may actually be an economical use insofar as the plastics are used in vehicles to reduce weight. The reduction in weight over the years of the life of a vehicle may save more oil than was used in making the plastic. More plastics in vehicles of the future seem certain.

One cannot make petrochemicals — medicines, plastics, paints, synthetic rubber, insecticides, and myriad other very useful products which come from oil from sunlight or from uranium. As oil becomes scarce, it will gradually be used more and more for its highest value end product which will be petrochemicals rather than energy.

Economic interdependence may promote stability

Because of the increased economic inter-dependence of nations, particularly the recent substantial financial investments of the Middle East oil producing nations in the western industrialized world, there may not be any more sudden cut-offs of oil supplies. The only prospect that might change this would be actions by religious or political extremists who in their zeal would ignore the economic consequences of oil disruption which would injure even their own countries.

With economies of so many nations built on the use of petroleum, especially oil, the destinies of the producing and the consuming nations are already markedly intertwined. The decline of the Petroleum Interval will be a time of stress for both producers and consumers, but the producers will be in the better position until their oil runs out. How the world economy, and particularly the energy supplies for the economy will be structured after oil, is difficult to visualize, but the economic alliances and balances of power will undoubtedly be considerably different than today's world economy based on oil.

Competition for exported oil

Phase one of the saga of oil was the rise of the industry in the United States. Phase two was the move to find oil abroad. Phase three was the nationalization of oil companies abroad and the shift of the world center of oil production from the United States to the Persian Gulf. Phase four, here now, is the increased widespread use of oil particularly by the countries which have not enjoyed the Petroleum Interval to date. There is clearly not enough oil to allow all the world to enjoy the Petroleum Interval as do the nations of the industrialized world. But the other nations are now demanding and obtaining relatively more oil than in the past. Avoiding conflict over oil will be a continual and increasing challenge.

An interesting fact is that the United States, and to a lesser extent other industrialized nations, will finance this ability of the previously have-not oil countries to expand their oil consumption and ability to bid against the western industrialized for oil supplies. The United States buys many things from China now, and in turn pays for them in dollars by means of which the Chinese can buy oil which is priced in dollars. The same applies to southeast Asia and India. This most populous part of the world will be increasingly bidding for what world supplies of oil are available, in considerable part with money earned by exports to western countries. In 1996, the U.S. had a deficit in balance of payments with China of $40 billion.

"China's energy demand will jump by almost 160% during 1993-2015...While coal will continue to dominate China's energy sector, market shares for oil and natural gas are expected to rise sharply."(7) China is looking toward a large increase in automobile production for domestic use. The World Bank has predicted that China's economy will outstrip that of the U.S. by 2020. This is part of what has been called "Asia's coming to age in the 21st Century." As one oil executive has put it: "All over the world, millions and millions of people are beginning to learn the benefits that petroleum can bring. Their lives are improving, and their children's prospects are even brighter."(5)

Edward Carr, writing in *The Economist*, states:

"The shape and size of world energy demand is increasingly being determined not by rich countries but by the fast-growing developing countries of Latin America and Asia. Just as the oil producers in OPEC achieved sudden prominence after more than a decade of obscurity, so the full effect of the shift among consuming countries will not be felt for many years. But it is equally inexorable; and no less far-reaching.

"By 2010 the share of total energy consumption accounted for by the rich countries will have fallen below 50% for the first time in the industrial era. Eastern Europe and the former Soviet Union will consume a sixth. The share of developing countries will have climbed from 27% now to 40% and be rushing upwards faster than ever. The growth in energy consumption in developing countries between 2000 and 2010 will be greater than today's consumption in Western Europe...Energy consumption in India has more than tripled since 1970. In China, consumption has increased 22-fold since 1952...And there is still a huge suppressed demand in both countries...In 1955 all the power on Taiwanese farms was proved by humans or animals; by 1975 one half was mechanical, and the use of oil-based fertilizers was common...A planner at one large company believes that there could be 70 million motorcycles, 30 million lorries and 100 million cars in China by 2015...Two billion people are on the move; the price of oil will follow."(15)

In 1995, General Motors signed a preliminary agreement with China — finalized in 1997 — to build a joint-investment automobile plant in China on which GM will spend $1.6 billion. However, this will be only a small part of China's trend to motorization. Ford is investing $95 million in a Jiangling Motors plant near Shanghai that will build Ford-based vans. The major move appears to be replacing the omnipresent bicycle with mopeds, and the increased demand for oil to fuel these vehicles is expected to be very large. Oil companies are now markedly expanding their retail sales operations in China and Southeast Asia in anticipation of this event.

Neither China, India, nor southeast Asia have the prospects of supplying more than a very small fraction of their anticipated oil demand, resulting from their present and near future large industrialization. Ivanhoe notes, "The industrializing LDCs [lesser developed countries] will soon become hard competitors with western nations for world crude exports."(28) They will be bidding against the western industrialized world for oil, and they will have more money than in the past to do so. China and all of southeast Asia currently have a very large positive balance of payments with the United States and other industrialized nations. International political and economic strains are almost certain to occur.

Length of the Petroleum Interval

There will always be some oil left in the ground. It is impossible to get it all out. How much is gotten out depends on how much people are willing to pay for it, and how much energy is gotten from the oil compared with the amount of energy used to obtain it. Hubbert noted that it took from 1859 to 1969 to produce 227 billion barrels of oil. The first half of this took from 1859 to 1959 to produce. The second half was produced in just the 10 years from 1959-1969.(24) He further estimated that the middle 80 percent of world oil production will be during a time period occupying probably "some 58 to 64 years."(24) This means that most of the world's oil will have been consumed in a period of less than one lifetime — a very, very brief time even in human history.

World oil reserves trend — down

The world has been and continues to consume its oil reserves faster than they are being replaced. Petroconsultants SA of Geneva, Switzerland estimated that as of 1994, world oil and natural gas condensate reserves amounted to about 904 billion barrels. In the period 1980-1994, 311 billion barrels of oil were produced, whereas only 204 billion barrels were replaced by discoveries and revisions of reserve estimates in existing fields. OPEC in 1980-1994 produced 107 billion barrels of oil, and added only 24 billion barrels.(8)

As of 1996, Petroconsultants further estimate that about 23 billion barrels of oil are produced yearly with output rising, but less than seven billion barrels are found each year. They state that the U.S. is definitely past its peak production but most countries have not reached the half-way point. However, 20 or so countries are a year or so from their peak and will soon enter a decline. As a glimpse of the future for the Persian Gulf region, Petroconsultants state that by 2010 the five Middle East oil producers, Saudi Arabia, Kuwait, Iran, and Abu Dhabi, all members of OPEC, will produce 60 percent of the world's oil, up from 27 percent today.

Peak of world oil production

Currently the increased production from non-OPEC nations is taking care of the rise in world-wide oil demand, and OPEC is not increasing its share. But this is a temporary situation. The Middle East OPEC nations are clearly destined to have control of the last of the world's oil. Colin J. Campbell of Petroconsultants expects the five Middle East countries to produce 60 percent of the world's oil by the year 2010, up from 27 percent in 1996. Campbell in 1996 stated, "the share of worldwide production from the Middle East is set to rise. How these countries use this position is anybody's guess, but they won't sell cheaply." He added, "The five Middle East countries [Saudi Arabia, Kuwait, Iraq, Iran, and Abu Dhabi] will peak [in production] about 2015, while the rest of the world will peak about 2000." With world oil production now exceeding discoveries by 16 billion barrels a year, and production rising to meet demand, the exhaustion of global oil reserves is clearly in sight.

MacKenzie has written a concise and well documented estimate of the peak of global oil production, stating:

> "Over the past fifty years, many oil companies, geologists, governments, and private corporations have performed scores of studies of Estimated Ultimately Recoverable (EUR) global oil. (EUR is the total amount of oil that will eventually be pumped from the earth). Taken together, the great majority of these studies reflect a consensus among oil experts that EUR oil reserves lie within the range of 1,800 to 2,200 billion barrels. As of the end of 1995, the world had consumed about 765 billion barrels of these ultimately recoverable reserves.
>
> "Given these estimates of recoverable oil, and plausible assumptions of moderate growth in demand (about 2 percent per year), we can use a simple model to calculate when world oil production might begin to decline driven by resource constraints. At the low end, for EUR oil equal to 1,800 billion barrels, peaking could occur as early as 2007; at the high end (2,300 billion barrels), peaking could occur around 2014. (An implausibly high 2,600 billion barrels for EUR would postpone peaking only another five years —to 2019."(31)

For the world as a whole, one, or at the most two generations will see the end of the Petroleum Interval as we know petroleum in its volume and varied use today. Ivanhoe, in an excellent summary article on future world oil supplies, states:

> "Unscientific reserve claims for political reasons may obscure the fact that most large, economic oil fields have been found, and permanent oil shock is inevitable early in the next century...the world's total EUR [EUR is the total amount of oil that will eventually be recovered from the Earth] may first peak about the year 2000...after which it may fluctuate along a horizontal production line (restricted by Saudi Arabia/OPEC) before inevitable decline towards a low baseline after year 2050...Thus, the question is not whether, but when, the foreseeable permanent oil crunch will occur. The next paralyzing and permanent oil shock will not be solved by any redistribution patterns or by economic cleverness, because it will be a consequence of pending and inexorable depletion of the world's inexpensive conventional crude oil supply."(28)

Expectations for alternative energy sources

The great hope is that alternative energy sources will be found to gradually fill the void left by diminishing oil supplies. Carr writes: "Eventually industrial economies may give up burning fossils. The alternative could be solar or nuclear, fission or fusion, hot or cold; but some combination of technology and scarcity will liberate industry and consumers from the arbitrary dictates of geological deposits."(15)

However, the world at present is a long way from becoming independent of fossil fuels, particularly petroleum. In the area of transportation oil provides 97 percent of the world's energy. Changing this significantly will take a long time, and the world peak of oil production is already in sight. The present global annual oil discovery rate of seven billion barrels is part of a declining trend, while at the same time, the world consumption rate of 23 billion barrels per year is trending upward. Within the next five years the world demand for oil is expected to rise another seven billion barrels a year.

Until the time of alternative renewable energy sources is completely upon us, and we are at present far from it, nations and individuals will continue to be subject to the geodestinies imposed by use of mineral fuels. And these various destinies awarded to various nations will surely come into sharper focus relatively soon — within the next decade or two.

Further reading

For anyone wanting to read further about petroleum, and the industry, two outstanding works are Yergin's *The Prize* (58) which relates in great and remarkable detail the history of this industry, and Anderson's *Fundamentals of the Petroleum Industry* (2) which includes history with special reference to the United States, and also describes the technology. For those with no geological background, an excellent description of how the search for petroleum is conducted is that by Robert Stoneley: *Introduction to Petroleum Exploration for Non-Geologists:* Oxford University Press, 1995.

Historical Note: The "Oilman's Barrel"

The standard measurement used for crude oil is the "barrel" which contains 42 U.S. gallons at 60 degrees F. No such barrel now exists. The origin of this measurement has been widely researched and debated. The most comprehensive study of the origin of the "barrel" is the 122 page volume, *The Oilman's Barrel*, by Robert B. Hardwicke, 1958, University of Oklahoma Press. A brief summary of some ideas on the origin of the "barrel" is given in a note on page 788 of Yergin's *The Prize*, 1991. The exact unquestionable origin of this measurement appears to be lost in history, but the more commonly accepted ideas are cited here.

When crude oil was first produced commercially in Pennsylvania, a variety of barrels were used — whatever containers were handy, like those used for wine, beer, whiskey, cider and other liquids. But the need for a standard measurement soon became apparent to establish an oil trade. The "tierce" was a defined wine measurement in early use in Britain, as was the herring barrel, and a barrel for spruce beer. These were 42 gallon containers. These barrels may have become the standard oil field measurement because they were also in common use in the United States. They were very carefully made tight containers so they would also hold oil very well.

The *Oil and Gas Journal* of April 15, 1921 states very firmly that "In 1864-65 the first standard barrel was made by Samuel Van Syckle at Miller Farm, near Titusville, PA. It was 42 gallons capacity, the size fixed in the year 1461 in England for the herring barrel, during the reign of Edward IV. Van Syckle specified the size of the staves to be used and made an honest 42-gallon barrel. Almost immediately he had practically a monopoly of the business and the odd size barrels gradually disappeared."

Another reported origin relates that the lack of a standard measure caused confusion, friction, and short tempers. In 1866 a group of oil producers got together in West Virginia and issued a proclamation that stated they would sell oil only by the gallon but that an allowance of two gallons (for leakage and impurities) would be made for every 40 gallons, thus making a total of 42 gallons. From this the 42 gallon barrel emerged, and ten years later a Council of Producers formally adopted the 42 gallon barrel as the standard measure.

These are among the more likely possibilities for the origin of the "oilman's barrel." There are several others. There is also a related question of how the abbreviation "bbl" came to be used for the oil barrel, because there is no second "b" in the word "barrel." This seems to have been explained to most everyone's satisfaction. Oil historian, Paul Giddings, in his book *Standard Oil Co.,* 1955, writes that kerosine was shipped in blue painted barrels and gasoline was shipped in red painted barrels. The term "rbl"" was used to designate the red gasoline barrel and "bbl" indicated the kerosine blue barrel "bbl" symbol was the designation most commonly seen in refinery shipment records. Apparently this ultimately became the general symbol for all oil in barrels. (NOTE: "Kerosine" is the petroleum industry spelling. "Kerosene" is the more common spelling.)

Conversion equivalents

That theoretical 42-gallon barrel of crude oil is equal in energy to 5,800,000 British thermal units (Btus), 5,614 cubic feet of natural gas, or 0.22 short ton (short ton = 2000 pounds) of bituminous coal. The various crude oils differ in density, but the average barrel of crude oil weighs about 310 pounds.

BIBLIOGRAPHY

1 AMERICAN PETROLEUM INSTITUTE, 1995, Basic Petroleum Data Book: American Petroleum Institute, v. 15, n. 1, Washington, D. C., (no pagination, large volume).

2 ANDERSON, R. O., 1984, Fundamentals of the Petroleum Industry: Univ. Oklahoma Press, Norman, 390 p.

3 ANONYMOUS, 1992, Conoco Details World Energy Outlook: Oil & Gas Journal, June 22, p. 28-30.

4 ANONYMOUS, 1993, The Dramatic Story of Oil's Influence on the World: Oregon Focus, January, p. 10-11.

5 ANONYMOUS, 1994, ARCO's Mike Bowlin Predicts 'New Prosperity" in Years Ahead: Oil & Gas Journal, November 21, p. 35-36.

6 ANONYMOUS, 1995, Thailand Rapidly Developing into World Class Petrochemical Producer: Oil & Gas Journal, April 3, p. 23-27.

7 ANONYMOUS, 1995, China's Energy Demand, Investment Needs to Soar: Oil & Gas Journal, April 10, p. 30-32.

8 ANONYMOUS, 1995, World Production Out Paces Reserves Replacement: Oil & Gas Journal, September 11, p. 22.

9 ATTANASI, E. D., and ROOT, D. H., 1994, The Enigma of Oil and Gas Field Growth: American Association of Petroleum Geologists Bull., v. 78, n. 3 (March), p. 321-331.

10 BIRD, K. J., and MAGOON, L. B., 1987, Petroleum Geology of the Northern Part of the Arctic National Wildlife Refuge, Northeastern Alaska: U.S. Geological Survey Bull. 1778, Washington, D. C., 329 p., 5 pls.(maps).

11 BRAY, RICHARD, 1995, Personal Communication.

12 CAMPBELL, C. J., 1991, The Golden Century of Oil 1950-2050: The Depletion of a Resource: Kluwer Academic Publishers, Dordrecht, The Netherlands, 345 p.

13 CAMPBELL, C. J., 1993, The Depletion of the World's Oil: Petrole et Technique, n. 383, Paris, p. 5-12.

14 CARR, EDWARD, 1988, Future Crude Oil Supply and Prices: Oil & Gas Journal, July 25, p. 111-112.

15 CARR, EDWARD, 1994, Energy: The Economist, June 18, p. 3-18.

16 FISHER, W. L., and GALLOWAY, W. E., 1983, Potential for Additional Oil Recovery in Texas: Univ. Texas Bureau of Economic Geology Circular 83-2, Austin, Texas, 20 p.

17 FLAVIN, CHRISTOPHER, and LENSSEN, NICHOLAS, 1990, Beyond the Petroleum Age: Designing a Solar Economy: Worldwatch Paper 100, Worldwatch Institute, Washington, D. C., 65 p.

18 GALL, NORMAN, 1986, We Are Living Off Our Capital: (Interview with JOSEPH P. RIVA, Jr., Earth Science Specialist, The Library of Congress), Forbes, New York, September 22, p. 62, 64-66.

19 GEVER, JOHN, et al., 1991, Beyond Oil. The Threat to Food and Fuel in the Coming Decades: 3rd Ed. University Press of Colorado, Niwot, Colorado, 304 p.

20 GROVE, NOEL, 1974, Oil, The Dwindling Treasure: National Geographic, June, p. 792-825.

21 HALBOUTY, M. T., 1981, Energy Resources of the Pacific Rim: American Association of Petroleum Geologists, Tulsa, Oklahoma, 578 p.

22 HODEL, D. P., and DEITZ, ROBERT, 1994, Crisis in the Oil Patch. How America's Energy Industry is Being Destroyed and What Must be Done to Save it: Regnery Publishing, Inc., Washington, D. C., 185 p.

23 HOWELL, D. G., (ed.), 1993, The Future of Energy Gases: U.S. Geological Survey Professional Paper 1570, 890 p.

24 HUBBERT, M. K., 1971, The Energy Resources of the Earth: Scientific American, September, p. 60-70.

25 ISMAIL, I. A. H., 1994, Shifting Production Trends Point to More Oil From OPEC: Oil & Gas Journal, December 26, p. 37-40.

26 IVANHOE, L. F., 1988, Future Global Oil Supply: American Association of Petroleum Geologists Bull., v. 72, n. 3, p. 384.

27 IVANHOE, L. F., and LECKIE, G. G., 1993, Global Oil, Gas Fields, Sizes Tallied, Analyzed: Oil & Gas Journal, February 15, p. 87-91.

28 IVANHOE, L. F., 1995, Future World Oil Supplies: There is a Finite Limit: World Oil, October, p. 77-88.

29 KRENEK, M. R., 1995, Latin America Second Only to Asia in Petrochemical Prospects: Oil & Gas Journal, April 17, p. 44-47.

30 LAHERRERE, JEAN, 1995, World Oil Reserves —Which Number to Believe?: OPEC Bulletin, February, p. 9-13.

31 MacKENZIE, J. J., 1996, Oil as a Finite Resource: When is Global Production Likely to Peak?: World Resources Institute, Washington, D. C., 22 p.

32 MASTERS, C. D., 1981, Assessment of Conventionally Recoverable Petroleum Resources of Persian Gulf Basin and Zagros Fold Belt (Arabian-Iranian basin): U.S. Geological Survey open-file report, 81-986, Washington, D. C., 7 p.

33 MASTERS, C. D., 1985, World Petroleum Resources — A Perspective: U.S. Geological Survey open-file report 85-248, Washington, D. C., 25 p.

34 McCOY, T. F., et al., 1994, Infill Wells Contradict Claims of New Gas in Huge Hugoton Field: Oil & Gas Journal, April 25, p. 56-61.

35 MEYER, R. F., et al., 1984, Preliminary Estimate of World Heavy Crude and Bitumen Resources: in The Future of Heavy Crude and Tar Sands: McGraw-Hill, Inc., New York, p. 97-158.

36 MEYERHOFF, A. A., 1976, Economic Impact and Geopolitical Implications of Giant Petroleum Fields: American Scientist, v. 64, p. 536-541.

37 MEYERHOFF, A. A., and WILLIAMS, J. O., 1978, Oil Broadens Chinese Development Role: Oil & Gas Journal, v. 76, n. 29, p. 91-98.

38 MEYERHOFF, A. A., 1981, Oil and Gas Potential of Soviet Far East: Scientific Press, Ltd., Beaconsfield, Bucks, England, 176 p.

39 NUSSBAUM, BRUCE, 1983, The World After Oil. The Shifting Axis of Power and Wealth: Simon and Schuster, New York, 319 p.

40 ODELL, P. R., 1983, Oil and World Power (7th Edition): Penguin Books, New York, 288 p.

41 RIVA, J. P., Jr., 1987, Enhanced Oil Recovery Methods: Congressional Research Service, The Library of Congress, Washington, D. C., 14 p.

42 RIVA, J. P., Jr., 1988, Domestic Oil Production Under Conditions of Continued Low Drilling: Congressional Research Service, The Library of Congress, Washington, D. C., 12 p.

43 RIVA, J. P., Jr., 1991, Persian Gulf Oil: Its Critical Importance to World Oil Supplies: Congressional Research Service, The Library of Congress, Washington, D. C., 15 p.

44 RIVA, J. P., Jr., 1992a, Giant Oil Fields and Domestic Oil Production: Congressional Research Service, The Library of Congress, Washington, D. C., 7 p.

45 RIVA, J. P., Jr., 1992b, Contribution of Giant Oil Fields: Disappearing Search for Elephants in the U.S.: Oil & Gas Journal, April 27, p. 54-56.

46 RIVA, J. P., Jr., 1993a, Domestic Natural Gas: Part of the Solution or Part of the Problem? Congressional Research Service, The Library of Congress, Washington, D. C., 18 p.

47 RIVA, J. P., Jr., 1993b, Oil and Natural Gas in the Russian Federation: Congressional Research Service, The Library of Congress, Washington, D. C., 44 p.

48 RIVA, J. P., Jr., 1994a, Current World Oil Status: Geopolitics of Energy, v. 16, n. 5, p. 1-5.

49 RIVA, J. P., Jr., 1994b, Domestic Oil: Past, Present, and future: Congressional Research Service, The Library of Congress, Washington, D. C., 15 p.

50 RIVA, J. P., Jr., 1995a, Domestic Natural Gas: A Fuel For the Future?: The Professional Geologist, March, p. 9-10.

51 RIVA, J. P., Jr., 1995b, Domestic U.S. Crude Oil: A Declining Asset: Congressional Research Service, the Library of Congress, Washington, D. C., 22 p.

52 RIVA, J. P., Jr., 1995c, The Domestic Natural Gas Status: Congressional Research Service, The Library of Congress, Washington, D. C., 20 p.

53 RIVA, J. P., Jr., 1995d, World Oil Production After Year 2000: Business as Usual or Crises?: Congressional Research Service, The Library of Congress, Washington, D. C., 20 p.

54 RIVA, J. P., Jr., 1995e, The Distribution of the World's Natural Gas Reserves and Resources: Congressional Research Service, The Library of Congress, Washington, D. C., 13 p.

(Note: Single copies of Congressional Research Service publications can be obtained from The Library of congress, Washington, D. C., 20540. These are public documents and may be freely reproduced, and are an excellent source of concise, basic information.)

55 TIPPEE, BOB and BECK, R. J., 1995, Oil Demand Continues to Grow in the U.S. and Worldwide: Oil & Gas Journal, July 31, p. 45-65.

56 U.S. GEOLOGICAL SURVEY NATIONAL OIL AND GAS RESOURCE ASSESSMENT TEAM, 1995, 1995 National Assessment of United States Oil and Gas Resources: U.S. Geological Survey Circular 1118, Washington, D. C., 20 p.

57 YERGIN, DANIEL, (ed.), 1980, The Dependence Dilemma. Gasoline Consumption and America's Security: Center for International Affairs, Harvard Univ., Cambridge, Massachusetts, 167 p.

58 YERGIN, DANIEL, 1991, The Prize. The Epic Quest for Oil, Money, and Power: Simon & Schuster, New York, 877 p.

CHAPTER 13

Alternative Energy Sources: Non-renewable

Energy, it is worth repeating, is the key which unlocks all other natural resources. Nations which can continue to obtain substantial energy resources are those which are going to be best able to survive and compete.

Petroleum alternatives

Petroleum (oil and gas) is the most important world-wide energy source today. When people refer to "alternative" energy sources, generally they mean alternatives to the conventional supplies of oil and gas. The discussion here of alternative fuels is in that context.

Petroleum, like all fossil fuels (fuels obtained from the remains of organisms, both plants and animals of the past) is finite. It is a remarkably versatile and high quality energy on which much of our civilization now depends. When it is gone what are the possible replacements?

Transition not simple

A fact which can hardly be over-emphasized is that the transition from a petroleum-based economy (especially the oil phase of petroleum) to an alternative energy based economy, whatever those alternatives are, will not be simple and easy. There is a vast difference between drilling for oil and having it either flow or be pumped out of the ground, and any other energy source. The ease of handling and transport of oil and its various derivatives to near and distant places, to be used in all sorts of motors, or for heating are characteristics of oil that are unmatched by any other energy form.

The transition to other energy supplies cannot easily be done. "Go Solar" with its radiant Sun symbol makes a nice bumper sticker, but running that 3,000 pound car 60 miles an hour for 400 miles without refueling is difficult with solar energy, particularly on a cloudy day or at night, or even when the solar energy is converted into electricity in storage batteries. The public seems to have adopted a "don't worry, alternative energy will be available" placebo, which is more comforting than real in terms of the problem of replacing such a huge high quality and easily obtained energy source which is oil.

Energy mix

In coming decades there will be large differences from today in what is termed the "energy mix," that is, the amount that the various energy sources contribute to the total energy supply. Until 1880, wood was the principal fuel used in the United States. From about 1880 to about 1945, coal became the largest single energy source. Since that time, petroleum (oil and natural gas) has been the most important energy source and now constitutes about 65 percent of United States' energy supply. Nuclear energy has had low acceptance in the U.S., and recently development of nuclear power has stopped. No new plants have been started in the U.S. since 1976. Solar energy, geothermal energy, wind power, and certain other minor energy sources are very slowly being developed in the United States. In some other countries, however, nuclear energy is being more vigorously pursued, as, for example, in France, Japan, and Korea.

The generalized history of the energy mix in the United States is shown graphically in Figure 5.

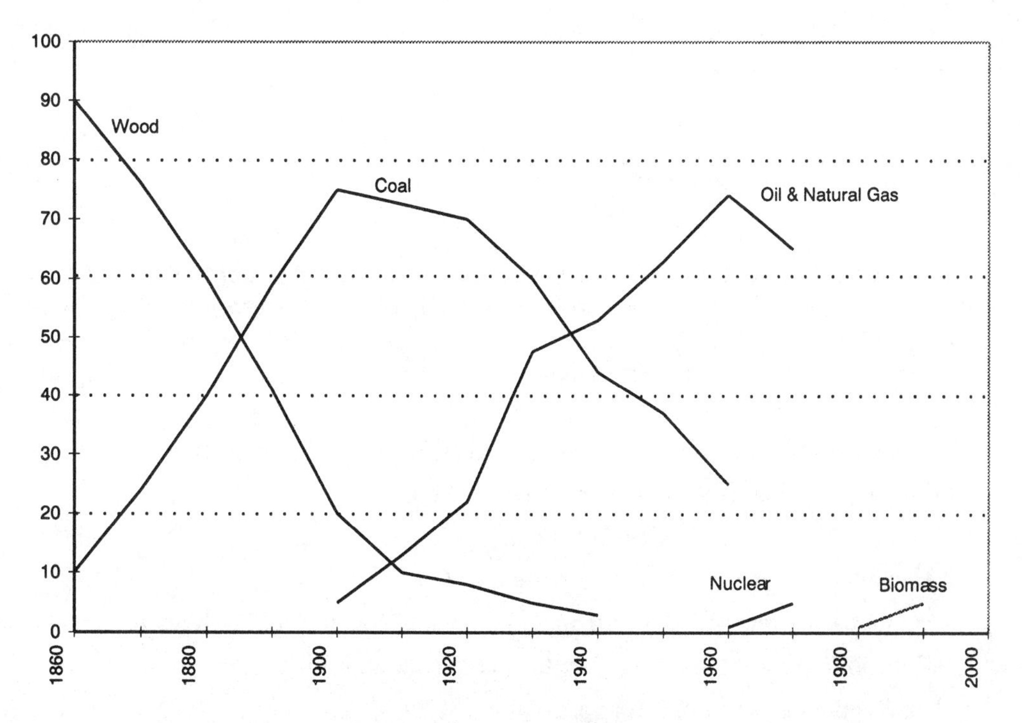

Figure 5. Graphic history of changes in energy mix in the United States.

(Source: Modified from Energy Research and Development Administration, U.S. Department of Energy, 1990)

As of 1996, the percentage energy mix in the United States was oil 39.88, natural gas 21.31, coal 26.05, hydroelectric power 3.79, and nuclear power 8.77. This mix is continually changing, but at different rates at different times. In more recent times the changes have been more rapid. Wood was the only energy source for more than 200 years after the arrival of the first European settlers in North America. Then coal came on the scene followed very shortly by petroleum, and then nuclear power. How fast will the next shift occur to other sources of energy?

Chevron Corporation has documented the energy mix in the United States as of 1986, and made an estimate of the mix in the year 2000. (Table 11)

The lesson here is that energy sources wax and wane, and also that estimates can be wrong. For example, the larger role ascribed to nuclear will not happen given the anti-nuclear sentiment now current in the United States, and solar energy may be given more emphasis.

Table 11. U.S. Energy Mix 1987 & 2000

Percent Share

	1987	2000 (estimate)
Oil	44	41
Gas	21	20
Coal	22	24
Hydro/Misc.	4	5
Nuclear	5	7
Solar/biomass	4	3
	100	100

(Source: The Chevron Corporation, 1987)

The major current energy supplies (oil, gas, coal) are finite and in terms of human history to date, destined to be relatively short lived. It is inevitable that there will be major changes in sources within the next century. Then what alternative energy sources exist to replace our present great dependency on petroleum? Questions to be answered regarding these alternative sources include:

- How easily can they replace existing sources in terms of convenience of use?
- What is the initial cost, and the maintenance cost?
- Can they be obtained in significant quantity?
- How widespread around the world are they?
- Are they truly renewable, or are they, in the long run non-renewable alternatives?
- What non-renewable resources do they use in their production?
- What is their environmental impact?

The attempt here is to put these various alternative energy sources in realistic perspective. What is possible and practical in the foreseeable future? The conclusions expressed here may not suit everyone but it is hoped that the facts presented will be a basis for further exploration of this important subject. Expectations about what alternative energy sources might do tend sometimes to be overdone and lead to misconceptions which become popular. Some are noted here and in Chapter 27, *Myths and Realities of Mineral Resources.*

Extensive literature

The oil crises of 1973 and 1979 brought home to America the finite nature of petroleum together with the fact that the U.S. cannot now supply its own oil needs. This resulted in a plethora of literature dealing with alternative energy sources.

With such a large volume of literature, these two chapters dealing with non-renewable and renewable energy sources are relatively long. We justify this, however, on the basis that the subject is now becoming significant in both personal and national affairs, and will be even more so in the not too distant future.

Facts

It is important to know the realities of what energy prospects lie ahead. In the area of alternative energy sources, there exist in the general public high but not well-founded hopes, and considerable misinformation. Many people hold firm opinions and prejudices for or against various energy sources. (Solar energy and nuclear power seem to evoke some of the strongest feelings). As the Petroleum Interval, which we have the good fortune to be in, matures and then begins the slow inevitable decline, it is important in a democracy in particular that the citizens be well informed as to what can, and what probably cannot be accomplished in the realm of alternative energy sources. In the United States, in terms of domestic oil resources, the decline in oil production is here, having begun in 1970. The general public seems convinced that adequate substitutes will be found when they are needed. We here examine the possibilities.

Renewable and Non-renewable

Energy sources can be grouped in various ways. A common approach is to divide them into those which are renewable and those which are non-renewable. It is also noted that all energy sources ultimately come from the Sun except nuclear energy (of which geothermal energy is also an aspect), and tidal energy, the latter being a combination of the gravitational influences of the Moon and the Sun. Coal, petroleum, shale oil (from oil shale), wood and other biomass, wind, hydropower, ocean thermal gradients, and waves are all in various ways energy forms which have their source in the Sun. Alternative energy sources are listed in Table 12.

Table 12. Alternative Energy Sources

Non-renewable	Renewable
Oilsands, tarsands, heavy oil	Wood and other biomass
Gas hydrates	Hydropower[1]
Shale oil	Solar energy
Coal	Tidal power
Nuclear (fission, fusion[2])	Wave energy
Geothermal[3]	Wind power
	Ocean thermal energy conversion (OTEC)

[1]Renewable unless reservoir involved. Then renewable only to end of reservoir life. [2]Because fusion can use either of two hydrogen isotopes, one of which is in almost unlimited supply in the ocean, it might be regarded as renewable. [3]Proven to be renewable in direct use if reservoir managed carefully. May be renewable in electricity production if new reservoir re-charging technologies prove successful.

More sources?

One might ask if this list of energy sources is really complete, or is there some significant energy source or sources which we do not yet see or see only dimly? It is a fair question. Before the advent of the atomic bomb, the potential energy from the atom was only glimpsed by a few. Before Col. Drake drilled his well, and the Industrial Revolution made use of oil, the potential for this liquid energy was almost unknown and certainly not visualized in terms of the importance of oil today. Before electricity was generated commercially, the lightning in the sky was only an interesting phenomenon.

But we have come a very long way in a few decades in understanding what kinds of energy sources do exist. It may seem rash to state that all the major energy sources potentially available to us are known but we do seem to have a rather reliable and complete view of the energy spectrum which goes from wood and other biomass, to fusion which is the energy in the Sun. It is unlikely that there is anything beyond fusion which we could grasp and use, even if such existed. Between wood and fusion there seem to be no energy sources which we suspect, and have not evaluated in some way. Probably we know them all. In this regard, Dr. Albert Bartlett, nuclear physicist, writes:

> "The probability is very small that technological developments will produce new sources of energy in the next century, sources not already known in 1994, that will have the potential of supplying a significant fraction of the world's energy needs for any appreciable period of time."(11)

Making transitions

In going to alternative energy sources, the question is: now that we have known oil, how easily can we use other energy sources which may be more expensive, and importantly, may be less convenient than is oil? Each energy source is distinctive, and although one can draw Btu's, horsepower, or however one measures energy, from all energy supplies, the economic problem is the cost to get that energy and use it in a particular application. Substituting coal, wood, or solar energy for the gasoline energy in a car can be done, but it is not a simple matter to do. There are considerable technological problems. Oil is hard to replace as a versatile and convenient energy source. One of its main advantages is how much energy it can hold per unit weight. Being able to conveniently carry in a five gallon container the amount of energy which can propel a car more than 100 miles is quite remarkable.

Another exceedingly important advantage of oil (its derivatives, kerosene, diesel, and gasoline) is that it does not have to be used as it is produced, but can be used anytime later. This is in contrast to electricity which must be used immediately as it is produced (as in the case of a power plant connecting to a power grid). Or, if stored, the batteries which do this take a huge volume of space compared with the energy which can be stored in the same volume by oil. One does not grasp how large this difference is until one experiences the problem of storing electricity in quantity. It is a problem which so far has defied a satisfactory resolution.

The following paragraphs are brief summaries of alternative energy sources, suggesting how important they might be, and determining who has them around the world. (Not all of these are minerals — such as solar and wind — but as they all compete with one another, all energy sources, mineral and non-mineral, must be considered in order to judge how important energy minerals might be in the future).

There are variations of these energy sources, for example, fuel cells, but fuel cells are not an energy source in themselves but merely a different way of using energy materials, chiefly petroleum. The long word "magnetohydrodynamics" simply refers to a technology which can be applied to obtain electricity from coal, oil, or gas, rather than burning these under a boiler. In simple statement "magnetohydrodynamics, or MHD, is a process for the direct conversion of heat energy into electric energy by passing a hot ionized gas through a strong magnetic field."(34) The technique is still being researched.

In sequence are considered those alternative non-renewable energy sources which are the closest to conventional petroleum resources (petroleum obtained by drilling, from wells). Then energy sources which are more removed from petroleum — coal, oil shale, and geothermal — are examined.

Oilsands, Tarsands, Heavy Oil, Gas Hydrates

There are all gradations of oil from those which are a gas at room temperature to very heavy tar deposits. Oil which can be recovered by conventional means (flowing or pumping) is the oil supply which the world now uses for the most part. But here we discuss unconventional oil deposits as alternative energy sources, even though they are an evolutionary continuation of conventional oil deposits. These unconventional oil sources include oilsands, tarsands, and heavy oil. We make the distinction between the first two and the third on the basis that the former have to be mined, and the last can be produced with considerable effort from specially constructed wells with auxiliary equipment involving the injection of hot water or steam. The distinction between "oilsands" and "tarsand" is blurred, and the terms are frequently used interchangeably. If there is a distinction to be made it is that tarsand is simply a denser, less fluid hydrocarbon deposit than oilsands.

The term "bitumen" is a general term applied to solid and semi-solid hydrocarbon materials, and is sometimes used for oilsands and tarsands.

Oilsands and tarsands

These deposits are, in effect, ancient oil fields which have been uncovered by erosion or from which oil has migrated to the surface or near-surface, and has lost its lighter, more volatile elements. The largest of these deposits is in northern Alberta, the Athabasca oilsands a few miles north of Fort McMurray. These contain an estimated 870 billion to 1.3 trillion or more barrels of oil and some semi-solid hydrocarbons which could be called tarsands.(46) At the present time two plant complexes, the Syncrude plant (a consortium of companies, and including the Alberta government), and the Suncor operation (Sun Oil Company) are developing these deposits. There are three other smaller oilsand deposits in Alberta. Syncrude Ltd. states, "The production potential of all four oilsands deposits could be as high as 1.7 trillion barrels of bitumen (five times more than the conventional oil reserves in Saudi Arabia). The Athabasca deposit is twice the area of Lake Ontario."(46)

Other estimates of oil in the four oilsands deposits, covering about 48,000 square miles, are by the Alberta Energy and Utilities Board which states that there are 1.7 trillion barrels of bitumen in place, and the National Energy Board of Canada estimates there are 2.5 trillion barrels. Both organizations agree that about 300 billion barrels are ultimately recoverable, with about four billion barrels that can be produced under existing economic conditions.

The deposit consists of grains of sand each of which has a thin film of water and outside this water film there is a coating of oil. By a hot water flotation process the oil is stripped from the sand. The process Suncor now uses recovers about one barrel of oil from about two tons of sand.(45) Initially, on recovery, the hydrocarbon is a black viscous tar-like material. By several steps, chiefly involving the addition of naphtha, a petroleum distillate, this hydrocarbon is upgraded to a straw-colored synthetic crude oil which can be pumped.

At the present time, the production costs of some of this oil are competitive with the costs of conventional crude oil. How much of this oil can be recovered economically, however, is a question the answer to which really cannot be known precisely because economic circumstances change, as does technology. With the current technology and world oil prices, some 90 percent of these deposits are too deeply buried to be mined economically. However, a new method is being tried whereby two wells are drilled by one specially designed drilling rig from an underground chamber beneath the oilsands. The wells are drilled at a slant upward and then laterally through the oilsand layer. The upper well carries steam which heats the thick oil making it flow. The lower well bore collects the oil and takes it away.(16) Whether this rather expensive procedure can be made economical has not yet been determined, but the potential is large if it is successful.

There is certainly a large amount of oil in the oilsands which can ultimately be recovered, but there are problems. The presently employed laborious process of stripping off the over-burden, mining the oilsands, transporting it to the processing plant, treating it to recover the oil, and then disposing of the waste sand is costly in manpower and especially in energy used. Also in northern Alberta, winter comes early and stays long. Temperatures drop to 50 below zero Fahrenheit. Conducting mining operations in sub-zero temperatures, and keeping all the equipment working is difficult. The quartz sand in these deposits is harder than steel and it inevitably gets into the machinery and causes maintenance difficulties. It is a much different situation from drilling a well to 6,000 feet, and having oil freely flow out at the rate of 12,000 or more barrels a day in sunny, always warm Saudi Arabia. Also, the energy recovery efficiency of oilsands operations is very much lower than from conventional oil operations.

Three barrels yield one barrel

It is estimated that it takes the equivalent of two out of each three barrels of oil recovered to pay for all the energy and other costs involved in getting the oil from oilsands. Furthermore, the capital costs of the operation are high. Large plants have to be built and maintained; no such plants are required for an oil well, which either flows or can be pumped to get the oil out and into a tank. Just to put the oil in a tank from the oilsands requires a number of procedures and much equipment.

However, there is a mitigating economic factor in producing oil from oilsands rather than from wells. There are no exploration costs; the oil deposit has already been found. There is no need for costly and usually non-productive wildcat well drilling (less than one exploratory well in 11 is a discovery).

With considerable experience, oilsands production costs have been continually reduced and the program is expanding. Suncor is producing about 70,000 barrels a day and has been quite successful in reducing costs from $19 a barrel in 1992, to less than $14 today (Canadian dollars).(43,45) Present oilsands production is about 280,000 barrels of oil a day which is sent by pipeline 270 miles south to a refinery in Edmonton. But to put this in perspective, 280,000 barrels a day are about 1/64 of the oil now being consumed daily in the United States. A third plant is now in the planning stages which will increase this production figure by perhaps 100,000 barrels. Compared with the oil demand either in the United States or in Canada, production from these oilsands is modest. In 1995, production from the oilsands operation at Fort McMurray by Syncrude Canada Ltd. was a total of 73.9 million barrels for the year. This would be a little over four days supply for the United States.

Nevertheless, the Athabasca oilsands represent a large energy asset which Canada can draw upon for decades to come. This is particularly important as conventional crude oil reserves in Alberta, which is by far Canada's largest oil producing province, are now in decline.(4) As conventional oil becomes more expensive, more of the Athabasca oilsands will become commercial.(4) Currently about 21 percent of Canada's crude oil needs are supplied by this oil. Recommendations for expanding these operations, if implemented, would raise production so that about 50 percent of Canada's oil demand would be met by 2020.(6)

There are very few oilsands deposits in the United States. There are some tarsands in Utah and elsewhere. These are small, and the hydrocarbon is tar and not lighter oil, and therefore takes even more processing than do oilsands.

In general in the past, fuel production from oilsands and tar deposits has not done very well without large government subsidies, but the Canadian oilsands and heavy oil fuel industry is an exception. Perhaps it has succeeded while others have failed because it has rather high quality deposits. The future for these Alberta oilsands looks favorable.

Heavy oil in oil fields

In conventional oil fields, usually less than half the oil in place is now being recovered, and in general the heavier oil fractions are left behind. With higher prices, better technology, and applying new technologies, more may be produced than is now figured into "conventional proven reserves." This will help stretch out oil supplies for at least some time into the future, but the low-cost flush production which the United States and other mature oil producing countries have enjoyed is gone. There is, however, still a lot of oil available in various kinds of deposits both here and abroad, but at a price, and with a considerable time lag in development to put the needed equipment in place.

In California, which passed its peak of production a number of years ago, the heavy oil resources are the last to be developed because they are the most expensive. Northwest of Taft, site of one of the very early oil fields developed in that State, there is now a huge complex of steam generating stations which pipe steam into the ground to reduce the viscosity of the oil so that it can be pumped to the surface. This is far less efficient than drilling a well and having the oil flow to the surface, but is the final effort to get oil which was left behind by ordinary pumping methods of oil production.

Another huge oil field in North America, the Kuparak River Field, lies to the northwest of Prudhoe Bay. The oil reservoir is at a depth of about 7,000 feet. But above that is another potential oilfield, the West Sak. It is a sandy, shallow unit (about 3,500 feet deep), which contains an estimated 20 billion barrels of oil, almost twice as large as the Prudhoe Bay Field. But the oil is thick, and the reservoir rocks are a loose, sandy formation which would tend to clog up the wells. This is an example of an oil deposit which probably is technically "recoverable," but the cost would be high and the net energy which would be produced would be small after the energy inputs of the production processes are subtracted.

Other heavy oil

There are very large deposits of heavy oil in the world which never were developed as oil fields. This is oil which has lost its lighter fractions, or was initially composed of organic compounds which did not mature in the Earth as oil does, and never were very fluid. The two most notable of these deposits are in eastern Alberta and adjacent western Saskatchewan, and in eastern Venezuela.

In the Alberta/Saskatchewan area (particularly in the vicinity of Cold Lake, Alberta which is one of the four bitumen areas of Alberta), wells are directionally drilled in groups from specially constructed pads to a depth of about 1200 to 1500 feet. Hot water is then injected for several weeks and the wells are subsequently pumped for a time. The process may be repeated several times, but considerable oil still remains in the ground after all this is done. Some oil is recovered but it is more expensive than conventional oil. However, these oil deposits are now being developed and are able to marginally compete with conventional sources. There are at least 25 billion barrels, and perhaps several times that much in these deposits. How much can be recovered economically is not known, and the net energy recovery is low.

Venezuelan very heavy oil

The world's largest deposit of heavy oil is in Venezuela, estimated to be about two trillion barrels. This is exceedingly thick oil in an elongate deposit sometimes called the "cinturon de la brea" (belt of tar). There are two ways this oil is being produced for the market. One is to make a synthetic fuel out of it, by moving the heavy oil in its composition more toward conventional crude oil by addition of hydrogen or certain solvents. The other way is to make the heavy oil into an oil-water emulsion (end product called "orimulsion") which is sold to power plants as a fuel.(5,7) This seems to be a fairly successful product, for at present seven electric generating plants around the world use a total of five million metric tons a year. Three plants are in Japan, two in Britain, and one each in Canada and Denmark. Three other plants have agreed to use orimulsion, including one in Florida. The product leaves Venezuela by ship, and to compete with coal, only plants located on coasts are supplied with orimulsion as the cost of transporting it inland would not be economic. The deposits are said to be equal to 63.5 billion metric tons of coal equivalent, and enough to last 500 years at planned production rates.(5) An interesting aspect of this product is the presence of vanadium, nickel, and magnesium in the ash to the extent that a company, Orbit Metallurgical Ltd., has been formed to commercially recover these metals.

Because this oil cannot easily be recovered by conventional drilling, it would take considerable energy to produce a barrel, and it would have to be upgraded to be used as is conventional oil. It is not likely to enter the regular oil market, but it can displace oil now used to fuel electric generating plants and in that way stretch out conventional oil supplies.

Gas hydrates

Gas hydrates are also called gas clathrates. This is an interesting type of hydrocarbon deposit which contains very large volumes of natural gas.(8) It occurs as a solid substance composed of water molecules forming a rigid lattice of cages with most of the cages each containing a molecule of natural gas, chiefly methane. They exist as relatively thin zones interbedded with other sediments, and are known from two distinct areas —polar regions where it is cold enough for permafrost to be present, and in sediments on the edges of Arctic ocean islands, and at the edge of the continental shelf where there are cold bottom temperatures at depths greater than 300 to 500 meters.(27)

The amount of gas in these deposits is exceedingly large, with estimates ranging from 500 trillion to 1,200,000 trillion cubic feet.(18) There is a gas hydrate deposit off the coast of North Carolina which is the area of about 10,000 square miles and is estimated to contain 350 times the amount of energy consumed in the United States in 1989. The U.S. Geological Survey states:

"If current estimates are correct, gas hydrates contain more potential fossil fuel energy than occurs in conventional oil, gas, and coal deposits. Uncertain, however, is the proportion of this potential energy that can actually be recovered. Because of unsolved technological problems in producing methane from gas hydrates, wide-scale recovery of methane from these substances probably will not take place until sometime in the 21st century."(27)

Small amounts of methane have been produced from gas hydrates in the Messoyakha Field in the West Siberian basin of Russia, but the costs have been prohibitively expensive. From petroleum exploration drilling in the Prudhoe Bay-Kuparuk River area on the Alaska North slope, extensive gas hydrates have been discovered there.(18)

The most serious effort to develop gas hydrates is now being undertaken by very energy-short Japan, which is the second largest consumer of energy in the world. The Japan National Oil Company (which invests in oil wherever it can) is planning an $87 million project with the object of getting gas hydrates from the Sea of Japan.

Oil Shale — Shale Oil

The United States has the largest oil shale deposits in the world chiefly located in Wyoming, Colorado, and Utah. The richest are in the Piceance Basin of western Colorado and the Uinta Basin of eastern Utah. These are organic-rich rocks deposited in ancient lake basins beginning about 50 million years ago and continuing for several million years.

Impressive numbers

Enthusiastic reports on the potential of oil shale abound. In a splendid volume on U.S. energy resources, the statement appears: "Oil shale represents one of the largest untapped sources of hydrocarbon known to man. Deposits occur throughout the world on every continent, and constitute a greater potential energy source than any other natural material with the exception of coal. The Rocky Mountain area of the United States alone contains oil shale representing billions of barrels of oil. This resource, however, has remained virtually undeveloped because supplies of conventional crude have been available at lower development costs."(20)

The U.S. Geological Survey, in a pamphlet distributed for general public information, states:

> "The oil shale deposits of the United States can be considered collectively as an enormous low-grade source of oil, hydrocarbon gas, or solid fuel. Deposits with an estimated yield of 10 gallons or more oil per ton of rock contain more than 2 trillion barrels; their possible extensions may contain an additional 3 trillion barrels; and, speculatively, other unappraised deposits may contain several times as much oil.
>
> "At present, only the highest grade and more accessible oil shales are of commercial interest. Such potential resources are more than double the proven reserves of petroleum in the country. However, using demonstrated methods of extraction, recovery of about 80 billion barrels of oil from accessible high-grade deposits of the Green River Formation is possible at costs competitive with petroleum of comparable quality."(22)

A State of Utah report on the Piceance and Uinta basin deposits states that "the deposits are estimated to contain 562 billion barrels of recoverable oil."(48) In 1974, a report stated that now that the price of conventional oil had reached seven dollars a barrel, oil shale appeared to be economically viable. These optimistic appraisals indicate the great expectations which have frequently been pronounced upon oil shale, particularly during the oil crises of the 1970s. It has been widely reported in the press (as is the case of the State of

Utah, just cited) that there are about 500 billion barrels of "recoverable oil" in these deposits. So far, no significant amounts have been produced.

Oil Shale Reality Check

As explained in the following paragraphs, the term "recoverable oil" seems to be used rather loosely, and can be somewhat misleading in the sense that it may be ignoring environmental costs and basic economics. It might be possible to "recover" 500 billion or more barrels of an oil-like substance (not crude oil as we know it from wells). But would the recovery process use as much energy as is obtained from the product produced, and what would be the monetary costs and the environmental impacts? A look at the facts is in order.

No oil, not shale

The first fact is that there is no oil in oil shale. The organic material in oil shale is kerogen, a solid organic material which has not evolved to oil. The second fact is that the rock is not shale but what geologists call organic marlstone. But as one promoter put it, "New York bankers won't invest a dime in 'organic marlstone,' but 'oil shale' is another matter." So "oil shale" is a promotional term. This and other items about the history of the Colorado oil shale ventures are included in a fascinating volume by Harry Savage, titled *The Rock That Burns*.(39) Thus, except for the fact that oil shale neither contains oil nor is it shale, the term is very good. It can raise money at times for oil shale ventures, none of which so far, however, has proved successful.

Various recovery projects

For more than 80 years, numerous attempts have been made to develop a shale oil industry in the United States. These include Occidental Petroleum's project near De Beque, Colorado, which involved tunneling into the shale, excavating a room, and then blasting down shale from the ceiling. The room was than sealed off, and the fragmented shale set afire. The oil released from the shale by the fire was to be drained out through a trough previously cut in the floor. The project has been abandoned. Equity Oil and the U.S. Department of Energy formed a joint project wherein 1,000 degree Fahrenheit steam was injected into the shale through numerous wells under pressure of 1,500 pounds per square inch. Water, oil and gas were to be recovered from the injected zone through production wells. This project proved uneconomic. Unocal (formerly Union Oil Company of California) has been involved in oil shale technology since the 1920s, and one small experimental plant was built many years ago in Parachute Creek canyon, western Colorado, then abandoned. A much larger plant was built in the 1980s in the same area and also abandoned, as described later.

Recovery process

In the now generally accepted idea as to how to obtain what eventually by various means can be converted to substances something like conventional oil, the rock must first be mined (blasted out to begin with), and then loaded on trucks or by other means hauled to the plant where it is heated to a temperature of about 900 degrees Fahrenheit. From this, a tarry mass emerges to which hydrogen must be added in order to make it flow readily. Currently the chief source of hydrogen is natural gas which unfortunately brings us back to petroleum which we are trying to replace. Also, the rock, when heated, tends to pop like popcorn, so the resulting volume, even after the organic material (kerogen) is removed, is larger than the volume of rock initially mined. This makes for a large waste disposal problem. The waste material has to be hauled to somewhere. It has been suggested that the ideal situation would be to have a mountain of oil shale near something equivalent to the Grand Canyon, where the oil shale could be brought down the mountain largely by gravity, run through the processing plant, and then the waste material dumped into the adjacent canyon. Because there are various salt compounds associated with the oil shale, the waste pile of oil shale would have to be effectively sealed off from any stream drainages, or surface water contamination would result.

How much net energy?

Developing oil shale deposits involves huge materials handling and disposal problems. Also, when the energy costs of mining, transporting, refining, and waste disposal are all added up, the net amount of energy recovered from oil shale is relatively small. It does not begin to compare with the net energy reward now obtained through conventional oil well drilling and production operations. Some studies have suggested that the final figure for the net energy in oil recovered from oil shale is negative. At best it is not large, and surface mining for oil shale may disturb up to five times as much land as would be caused by coal mining for the same net amount of energy recovered. And, it would be much more destructive to the landscape than would be oil wells producing the same net amount of energy.

Water supply support

Another problem with regard to the Utah and Colorado oil shale deposits is that the processing and the auxiliary support facilities would use large amounts of water. The richest oil shale deposits are located in the headwaters of the Colorado River. This river now barely if at all reaches the Gulf of Lower California. Present demands are already more than the river can meet. Water supply would be a serious problem to any large development of oil shale for it would take at least two barrels of water for the shale processing and all the support facilities to produce one barrel of oil. Even before there has been any commercial development of shale oil, the states downstream from the oil shale deposits have already protested the withdrawal of Colorado River water for shale oil production. Any such development would immediately pit Colorado and Utah against California, Nevada, and especially Arizona.

These factors and the difficult technological problems of how to most efficiently mine and process the shale (organic marlstone) and dispose of the waste have combined to delay development of these deposits. So far very little oil, except on a pilot plant scale, has been produced. Chevron, Unocal, Exxon, Occidental Petroleum and others have made major efforts to develop a viable, economic, commercial operation but none has been successful.

The story of the most recent attempt to economically recover an oil-like substance from oil shale reached a rather astounding climax and conclusion in the 1980s and early 1990s. With the oil crises of 1973 and 1979 fresh in mind, both Exxon and Unocal launched huge projects in the area of Parachute Creek just north of the Colorado River.(24) Exxon, in 1980, made a major move into oil shale, and announced that production of 15 million barrels a day of synthetic fuels by 2010 would not be "beyond achievement."(2) To do this would require spending some $800 billion over 30 years and moving large numbers of workers and their families, and support industries to the West.(2)

Exxon, to get the project started, announced it would spend $5 billion on various preliminary projects, and would build a town for 25,000 workers.(44) To house this expected small city of employees which would be involved in their oil shale operation, Exxon built a model community across the Colorado River on a broad gently sloping upland called Battlement Mesa. It had everything including a recreation center. But about the time that the Battlement Mesa community was completed, Exxon concluded that the oil shale project was uneconomic. On May 2, 1982, called "Black Sunday" in the town of Parachute, Exxon announced it was abandoning the project.(44) The Battlement Mesa housing was never fully occupied by Exxon. Ultimately it was sold for a real estate development. The buyers have now made it into a rather nice upscale mostly retirement community.

Unocal, backed by a government production subsidy, persisted and built a plant just north of the town of Parachute (previously called Grand Valley). Construction began in December 1980 and was completed in August of 1983, at a cost of $654 million. In their 1987 annual report Unocal stated regarding this project: "The ultimate goal is to achieve steady production at design capacity — about 10,000 barrels a day." Scaling this up to where it would be a significant amount in terms of United States' oil consumption would be a tremendous task and at best many years in the future. Peak production of 7,000 barrels a day was achieved in October 1989.(3)

During its experimental testing program, the plant operated with the aid of a $400 million Federal subsidy. By 1991 Unocal had used $114 million of this subsidy, and received $42.23 a barrel for the oil produced at Parachute Creek, with the U.S. government paying $23.46 of that. Unocal's production costs were about $57 a barrel.(47)

On June 1, 1991, this $654 million plant was permanently closed, and the project abandoned.(3) Parts of the plant have been sold or moved to other Unocal operations. Much of the plant, remains, however, as a monument to the so far failed efforts to develop a viable shale oil operation.(47)

When oil was $5 a barrel, then $10 was the magic figure at which shale oil would be competitive, so said the industry. When oil reached $30 a barrel during one of its erratic oil crisis moves, the economically successful figure for shale oil was stated to be $40.

Adding up the water supply problem, the enormous scale of the mining which would be needed, the low, at best, net energy return, and the huge waste disposal problem, it is evident that oil shale is unlikely to yield any very significant amount of oil, as compared with the huge amounts of conventional oil now being used, in the near or even foreseeable future. Shale oil can, at most, supply only a small portion of what are now the world's oil demands. Also, shale oil, by its composition, is better adapted for use as a raw material for petrochemical plants than for the production of gasoline. As a petrochemical feedstock, which perhaps could justify a higher price than it would as a substitute for gasoline, shale oil may play a modest role in the future economy.

Oil shale exists in a number of other countries, notably Brazil, Scotland, and Estonia. China has some deposits. A modest shale oil production operation was conducted in Scotland for a number of years. In Estonia, the oil shale is simply shoveled into furnaces beneath boilers of power plants and burned without processing. This results in a huge amount of ash to be disposed of, but the operation seems to be economic enough to be useful.

It is doubtful that shale oil can ever play a significant role in replacing world oil supplies, if it can replace them at all. Shale oil cannot possibly make the United States energy self-sufficient in terms of liquid fuel. The extravagant statements which have been made to suggest that shale oil can make the U.S. oil independent are usually made by promoters who seize upon the fact that there are perhaps a half-trillion barrels of an oil-like substance which could be distilled from the kerogen in oil shale. But some of these beds are only a few feet thick and hundreds of feet deep. The financial economics and the energy economics are simply not viable. It is very clear that for much of these deposits the cost of recovery would exceed the economic return by a large amount, and, more critically, it would take more energy to produce the "oil" from oil shale than would be gotten out of it.

In 1946, along the U.S. highway near what is now the town of Parachute, a large sign read "OIL SHALE BOOM! Get in on the Ground Floor." It is still not too late to do that if you wish. Shale oil development is still no higher than the ground floor. The numerous unsuccessful attempts to produce an oil-like product from oil shale have elicited a variety of wry comments which include "economic shale oil is just around the corner, and will be for the next 40 years." Cuff and Young stated in 1980, "The future of oil shale development would seem to rest with four major oil shale projects. These are operated by the following companies: Occidental Petroleum, Union Oil of California, Atlantic Richfield and TOSCO Corporation, and Rio Blanco Oil Shale Company owned by Gulf Oil and Standard Oil of Indiana."(20) When Union Oil (Unocal) abandoned its operation in June of 1991, no other companies remained. All had departed, perhaps giving some credence to another wry comment: "Shale oil is the fuel of the future and always will be."

Coal

Coal is a form of biomass handed down to us from the geological past. It has been referred to as "buried sunshine" and in a sense it is. Coal currently provides 26 percent of the world's primary energy requirements, and until it was displaced by oil it was the major fuel of the industrial world in the recent past and it may become so again, for in terms of total energy content, coal, around the world, far exceeds the energy in the world's oil and gas. Russia appears to hold the world's largest coal deposits, but not all of them are recoverable at reasonable cost.(26) It may be that the United States has the most economically recoverable coal. China is apparently third in amount of coal resources, with western Europe fourth. Australia, Africa, India, and South America follow in that order, but the amounts in these latter countries are considerably less than those of the three world leaders, Russia, the United States, and China. In China, coal is now the principal fuel in use, with concurrent severe air pollution problems. And China is now having to expand its coal production to meet its growing industrial energy needs. Coal will be their principal source of energy for at least several decades to come, and the air pollution problem is likely to become even worse than at present.

Coal — large U.S. energy source

The United States is running out of oil but coal remains in great abundance. It has been estimated that in the total energy in coal, oil, and natural gas deposits of the United States, 90 percent of that amount is in the form of coal. Also, these coal deposits are already located; there is no expensive exploration work involved as in the case now for deeply hidden oil reservoirs. The coal deposits are a known quantity.

The use of coal, however, has some substantial environmental problems starting with the fact that underground coal mines are dangerous. Each year a number of miners are killed, and others have their health permanently impaired. Now, however, in the United States most western coal, and considerable eastern coal is mined by open-pit methods. Underground mines are becoming less common. Germany also mines much of its coal (which, in general, is rather poor quality) by the open pit method. But in many other parts of the world, much coal is still mined underground and safety precautions are frequently not adequate. The result is that world-wide, coal mining remains a hazardous occupation claiming lives every year.

Pollution

Environmental negatives of coal use also include air pollution and acid rain when coal is used for power production. Research done on these problems and very large expenditures of money have reduced them considerably but not entirely.

Coal as oil substitute

Coal in some uses, particularly as a fuel source for electric power plants, is an important alternative fuel to petroleum. However, coal cannot conveniently be put into the fuel tank of a car or truck. Coal can be liquefied but at a cost, and only by means of a huge physical plant complex if any great quantity of oil is to be produced from this source. A synthetic fuels plant using coal as the raw material was built in North Dakota by the U.S. government but it has since been shut down as uneconomical. South Africa has liquefied coal for years as a source of about half of the fuel for its vehicles, but the fuel is expensive and in relatively short supply. These coal-to-oil plants were built when South Africa, because of its apartheid political structure, was partially isolated economically and it feared a total cut-off of oil supplies. The coal is converted to oil by the Sasol Process, but it could not compete with conventional oil, so the government paid a subsidy for the synthetic fuel which from 1985-1994 amounted to $960 million. The future of this synthetic fuel industry has come under debate, because South Africa has eliminated apartheid and presumably now has no concern about being cut off from conventional oil imports. However, because South Africa produces no oil, some people want to preserve the synthetic fuel industry as an insurance policy for the future when oil supplies may be scarce. At present the coal-derived synthetic fuel accounts for about 46 percent of the South African market.(30)

If coal were to be used in the United States as a substantial substitute for oil by liquefying it, the cost of putting in place the physical plants which would be needed to supply the United States with oil as we use it now would be enormous. And to mine the coal which would have to go into these plants would involve the largest mining operation ever seen in the world.

Coal can be a substitute in part for oil but cannot completely replace oil as we use it today. However, if coal gains in importance again as a fuel in whatever form it may be used, then Russia, the United States, and China are the fortunate nations. Saudi Arabia and other Persian Gulf countries are not even faintly in the running, with the single exception of Iran which has some small coal deposits.

Fossil Fuel — a Brief Flash

The fuels just considered are fossil fuels, the accumulation of myriad animal and plant remains during a period, going back to the oldest, of more than 500 million years. It is sobering to realize that the most useful fossil fuels, coal and petroleum, which took geologic ages to be produced by nature, will be consumed in no more than a brief flash of Earth history, probably lasting less than 500 years. Even in terms of human history this will be recorded as a very brief and unique time. (Figure 6)

Oil has now given the individual freedom of movement unheard of in the past. To be able to do this 200 years from now will have to involve great technological advances which may or may not be accomplished, for the fuel sources will be much different than oil in such wide use today.

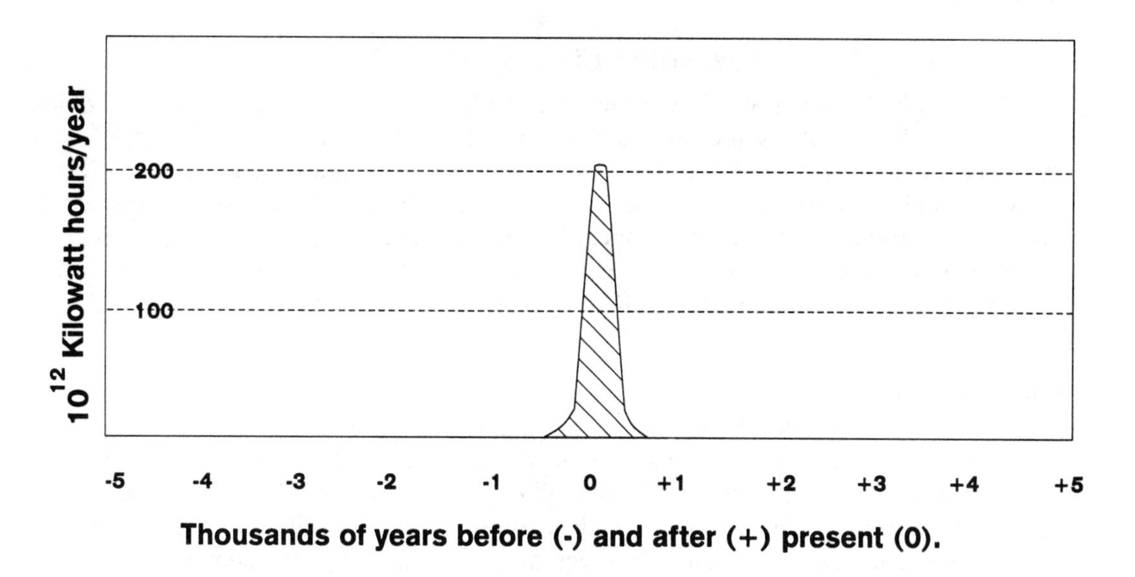

Figure 6. Time of fossil fuel use in perspective of human history, from 5,000 years past to 5,000 years in the future.
(Adapted from Hubbert, 1962)

Fossil Fuels and Food

The availability of fossil fuels for such a short time in human history is of special concern in relation to how they affect agricultural productivity. Pimentel and associates have made some interesting quantitative observations in this regard:

> "Energy use, particularly fossil energy, has played an important role in increasing food production in the world. In the United States, for example, crop yields have increased about 3.5 times during the last 60 years...The fossil energy inputs to achieve this food increase have grown 4-fold or to an input of about 1000 liters [6 barrels] of oil equivalents per hectare [per year]...In China, a similar technological change led to a 2-fold increase in food production the last 30 years, while fossil energy use increased nearly 100-fold or to an input of about 500 liters of oil equivalents per hectare [per year]...Energy and population growth are also interrelated. The major increases in human population growth rates during the last 300 to 400 years has coincided with the discovery and use of stored fossil energy resources such as coal, oil, and gas. Since about 1700, rapid population growth has closely paralleled the increased use of fossil fuel for agriculture and improvement in health."(36)

With the great importance of fossil fuels to the production of food, the gradual depletion of these resources and their ultimate disappearance poses some severe future problems. Can renewable resources really replace oil, gas, and coal? In the interim, as we try to make the transition to alternative energies, and these non-renewable resources continue to decline, the countries which have them will have an ever-larger role in the world economy. Conversely, those countries which do not have them or have the capability of importing them, will be increasingly at a disadvantage.

The Atom: Fission and Fusion

Although both fission and fusion are nuclear energy processes, the sources are quite different. In fission, the source is uranium, or it could be thorium, both of which are moderately abundant in a number of places in the world. The United States, Canada, South Africa, Namibia, Niger, and Australia have large uranium deposits. Lesser amounts are known from Brazil, Sweden, and France. Russia does not publish its uranium resource statistics, but is known to have some deposits which could be quite substantial given the large and varied geological terrains of Russia. Currently the largest uranium producing area in the world is in northern Saskatchewan, Canada.

Huge energy potential

An important fact is that the amount of energy potentially available in the known uranium resources of the United States, using the breeder reactor, far exceeds the combined energy potential in all the oil, gas, coal, and shale oil (such as it may be). But nuclear energy at present has a poor reputation in the United States, especially after the accident at the Chernobyl nuclear plant in Russia. However, at present nuclear power supplies more than 20 percent of U.S. electricity. Seven states and the city of Chicago get more than half their electricity from the atom, as does the province of Ontario, Canada.

Except for France, western Europe also has reservations on the matter of nuclear power. In 1980, Sweden voted to gradually phase out all nuclear plants with the last to be closed in 2010. However, what they will do after that no one seems to know. It may be that Sweden will have to reconsider that decision, for nuclear energy now provides 50 percent of Sweden's electricity.

In contrast, energy-short France has vigorously pursued nuclear power and now generates more than three-fourths of its electricity from the atom. Many nations outside of western Europe are pursuing nuclear power, including Brazil and Japan, and others which have no large coal or petroleum supplies.

Huge energy per weight

Uranium has a special attraction to energy-short countries because the amount of energy which can be shipped in the form of uranium is very large per unit weight compared with coal or oil. Weight for weight, uranium has more recoverable energy than any other source by far. When one short ton of uranium-235 is fissioned it produces the heat equivalent of 22 billion kilowatt hours, which is the heat in approximately three billion tons of coal. Shipping energy in the form of uranium is a very efficient operation.

Nuclear energy = electricity

Nuclear energy is transformed into electricity and therefore it can only be substituted for whatever fuel is used to generate electric power. (Although the waste heat from nuclear powered steam turbines is sometimes used in local space heating districts). In many places, oil or natural gas are now used for electric power generation. In that circumstance, nuclear energy can replace petroleum. But unless electric vehicles are invented which are efficient and used in quantity, uranium (used in power plants to produce electricity for vehicles) will not be a major substitute for gasoline or jet fuel. It is hard to visualize atomic powered or battery-run airplanes. The best fuel alternative for aircraft at present would seem to be hydrogen, but the extreme flammability of hydrogen causes that alternative to be viewed with some misgivings. Uranium can be used to produce electricity which in turn can, through the electrolysis of water, make hydrogen for fuel. However, each change in energy form from one to another involves a loss of energy and efficiency. Air transport beyond the age of oil presents a large technological challenge.

Transport in the form of railroads can use electric power, and as diesel fuel becomes more expensive, the electrification of the railroads of the world is likely to be the trend. Russia now makes extensive use of electricity for that purpose. Much of the Trans-Siberian Railroad has been electrified, thus saving diesel fuel for uses which electricity cannot replace, and for sale abroad to obtain foreign exchange. In the United States some railroads, once electrified as was the Milwaukee Road in the west, went to diesel engines, but as oil supplies become more costly, electric trains may make a comeback. The atom, used to produce electricity, could provide the power. Electric trains are now widely used in Europe.

As mass transit (which includes electrified buses and trolley cars) begins to take over more and more from the private transit system, especially personal automobiles, the atom can be increasingly used for transportation via the production of electricity. Mass transit, by whatever fuel, should be vigorously pursued.

Fusion

With fusion, the problem for practical purposes, is not fuel supply and radioactive contamination, but technology. It is generally believed that fusion involves creating and containing heat equal to the temperature of the core of the Sun by the fusion of either of two isotopes of hydrogen, deuterium, or tritium. Deuterium exists in great quantities in ordinary water, and tritium can be produced in an atomic reactor from lithium, an element that occurs in some granitic rocks and in some underground brines. If the energy in the deuterium of the Pacific Ocean was efficiently harnessed to generate power in fusion reactors, it could provide the energy to generate enough electricity to light and heat the whole world for literally millions of years.

Fusion and fission are both nuclear energy sources. Fission is based on limited supplies of uranium and thorium, and therefore considered as a non-renewable energy source. But, although there is also some limit to the deuterium in the oceans, it is, for practical purposes, limitless, and therefore the discussion of fusion might well have been included with renewable energy sources.

Uncontrolled fusion was accomplished in 1952 with the explosion of the thermonuclear bomb. So far, however, the process of using fusion on a controlled commercial scale has eluded some of the world's best scientists and engineers. In Europe, Japan, and the United States large teams of scientists have been pursuing the Holy Grail of fusion. With fission somewhat discredited because of fear of nuclear contamination and the problem of waste disposal, and, in the case of the use of fossil fuels the fact that they cause air pollution and are a finite resource, the lure of fusion as a cheap, clean, and virtually unlimited fuel source is hard to resist. But in spite of the best brains in the world combined with billions of dollars spent over many years, success still is not at hand.

At least one school of thought believes that the chances of ever doing this are no better than 50-50, and governments are beginning to wonder if the pursuit with the huge continued costs can really be justified. The United States, over a period of 40 years has spent more than $9 billion on fusion research, but expert opinion is that commercial fusion energy is still at least 50 years away, and may never be accomplished. However, the pursuit of fusion continues, and the United States has lost its leading research position to the Europeans who have combined to put twice as many scientists and engineers to work on the problem. The

Japanese, always very aware of their acute vulnerability in regard to energy supplies, have recently doubled their budget for fusion research. Ultimately because of the huge costs, it may be that fusion research will be conducted on an international cooperative basis.

Cold fusion?

Brief note should be taken of the claimed success of a so-called "cold fusion" process reported in March, 1989, by two chemists working at the University of Utah. Using an apparatus consisting of palladium and platinum metals immersed in water rich in deuterium, these scientists claimed that cold fusion had been accomplished and the amount of energy emitted from the experiment was four times the amount of electrical energy which had been put into the experiment.

This announcement created a great stir among physical scientists and energy experts. A large number of experiments were subsequently conducted to determine the validity of this claim, but so far the claim of "cold fusion" has not been substantiated.(37) The matter, however, seems not to have been entirely resolved.(33) The two chemists have now gone to France to continue their research which is funded largely by the Japanese. There is no recent report of what progress has been made.

Time to be commercial on large scale

Even if controlled fusion were to be accomplished, the time it would take to scale up this process to commercial size would be quite long. A rough estimate, perhaps even on the conservative side, is that it would take at least 30 to 40 years to scale it up, build plants, and make this energy source a significant factor in the energy supply of the United States. To bring it to lesser developed nations which currently use oil by which to generate electricity would probably take much longer.

Patterson Power Cell

Recently another somewhat enigmatic power source has been reported by James Patterson working in Sarasota, Florida. The power cell is reported to contain copper-nickel-palladium coated beads immersed in salt water. When electricity is put through the cell it is claimed that one watt put into the cell produces 200 watts of output. How or why this works seems not to be known. Whether this is actually a new and significant power source has not yet been established but several reputable industrial organizations are doing research on the matter.

Continued Use of Present Energy Sources

At present it appears that the discussion and evaluation of energy supplies must continue based on the now known and widely utilized energy sources. It might be added that even if fusion is accomplished, and can be scaled up to commercial size, the problem of the widespread application of the fusion-derived electrical energy to transportation — that is, to motor vehicles — is in itself a huge technological challenge, the solution to which is not immediately in sight. It would involve an entire revision of distribution of fuel supply stations. For longer range vehicles, it would require thousands of battery or hydrogen stations along highways, ultimately penetrating into the byways as well. The transition would proceed slowly at best because the hundreds of millions of motor vehicles now powered by gasoline or diesel fuel would have to be phased out. For more remote regions of the world, this might not be practical for many years into the future.

At best it is unlikely that fusion will be a commercial reality on any appreciable scale for at least several decades to come. Nevertheless, fusion probably will be pursued as an international long range goal for no technology on the horizon offers, theoretically at least, the energy potential of this process. Also, unlike fission, fusion does not have the risk of a runaway reaction and its by-products do not pose the disposal problems to the extent that fission does. And fusion uses an isotope of hydrogen which exists in the ocean in almost unlimited quantities. For all practical purposes, fusion may be regarded as a renewable energy source but, again, it may never be a useful reality. Furth, however, in 1995 expressed the most optimistic forecast for fusion thus far, stating "Fifty years from now engineers should be able to construct the first industrial plants for fusion energy...this schedule matches critical timescale of 50 to 100 years in which fossil-energy resources will need to be replaced.(25) The point should be made again that fusion produces electrical energy, and converting this energy form into all the uses to which petroleum and other fossil fuels are now used, especially in transportation, remains a very large technological challenge, not yet fully met. It cannot replace the petrochemical uses of petroleum.

Fission

If we continue to use the present fission technology for nuclear power, then a number of countries have uranium resources to obtain energy from this source in great quantities. The decision is currently a political/environmental one. For countries which do not have significant uranium supplies, the decision to go nuclear becomes largely one of economics, for uranium can be shipped in much more cheaply than can either oil or coal on a net energy basis. And, should an oil supply crisis occur again, large amounts of energy in the form of uranium can be more easily stockpiled and stored than can oil.

World trend toward nuclear

Like it or not, the world in general is going nuclear. In 1960, the atom provided less than one percent of the world's electricity. Today it provides more than 17 percent. The London-based Uranium Institute estimates that in the next decade the world's nuclear-derived electric generating capacity will rise from the present 261,000 megawatts to 347,000 megawatts. The International Atomic Energy Agency predicts that nuclear energy will provide at least 20 percent of the world's electricity within the next several years. Countries including Thailand, Turkey, Argentina, Brazil, Bulgaria, Japan, France, and Belgium now use nuclear energy and are developing more.

China's decision

In March 1995, Yao Quiming, the general manager of the Quinshan Nuclear Power Company stated that China had decided to "energetically develop nuclear power," changing its national power priorities from hydropower and coal. This was to help alleviate electricity shortages in coastal China. With the world's largest population, and the world's fastest growing economy, China's decision importantly adds to the long term international trend toward nuclear energy for central power stations.

Currently, outside of the United States, nearly 600 nuclear plants in 41 countries are in various stages of development or in operation. Even after the Russian Chernobyl nuclear plant accident, the Russians have continued construction on 15 atomic power reactors. Japan how has 49 nuclear plants and obtains 30 percent of its electricity from them. It plans to construct about 40 more and generate 42 percent of its electricity from the atom by 2010. South Korea has 11 nuclear power plants in operation with 19 more under construction or authorized. Taiwan operates six plants and is looking for sites for several more. Indonesia plans to build 12 plants.

The two big now-available power sources

As oil and gas supplies diminish, there are only two sources of energy large enough potentially to take up the slack on relatively short notice. These are coal and the atom (fission). However, due to increasing concern about the "greenhouse effect" and global warming caused by burning fossil fuels like coal (which now produces 37 percent of the world's electricity, and is a much more limited resource in total than is nuclear energy), the nuclear option is starting to again be looked upon more favorably by some. However, in the United States, nuclear power has such a poor reputation that coming back to that power source would be most difficult. At the present time in the U.S. no nuclear plants are under construction, and recently one partially completed was finished as a coal-fired plant.

A new type of nuclear power plant, the integral fast breeder reactor, now being developed produces much less toxic waste than present plants and reprocesses most of its own hazardous fission products. This technology may be a satisfactory route to further nuclear power development if fusion proves to be yet unattainable.

Nuclear energy and Japan as a test case

When oil is no longer available in large and relatively cheap quantities which can be conveniently shipped around the world as need be, countries like Japan will be severely impacted because it has no significant indigenous alternative energy sources on its islands. Japan's limited land area, together with its frequently overcast weather and its huge industrial energy demand, indicate that solar energy converted to electricity may not be available in sufficient quantity, even for just the energy uses of oil, not considering its other applications.

At the present time, Japan imports 90 percent of its total energy supplies. It is critically dependent on an oceanic highway of tankers, about one oil ship every 100 miles along the route from the Persian Gulf to the Japanese islands. This is a continuous procession that must go on every day of the year, year after year. Eventually, of course, it will stop.

It is likely that atomic power is the energy path Japan will have to pursue. Indeed, Japan already makes extensive use of the atom. Japan is in a difficult and precarious energy supply situation and may be the first country to illustrate what happens to an industrial nation which has no significant domestic energy base. In some fashion, after oil, Japan must continue to import large amounts of energy. Uranium would seem to be the best alternative as it is a highly concentrated energy source which can efficiently be shipped long distances.

Japan's position as the second largest industrial nation in the world is a very tenuous one based on fossil fuels. However, because Japan is already well along on the nuclear power road, it is possible that beyond the time of the Petroleum Interval Japan will do even better than they are doing now relative to other nations in the matter of energy. By being forced early to go to nuclear power which appears, along with coal, the only other source of energy immediately available which is large enough to partially replace petroleum, Japan will have pioneered the use of such power nationwide, and for the world.

An analogy already exists in that Japan, because of having no oil, developed smaller and more fuel efficient cars before the rest of the world did. As a result, Japan was ahead of other car makers, especially those in the U.S. when the price of oil went up dramatically and the demand for smaller fuel efficient cars arrived. Japan's successes and/or failures in ultimately replacing petroleum as the nation's chief fuel will be followed with keen interest by the rest of the world. But it should be noted that to go nuclear, Japan still has to import the basic energy source, uranium. Japan does not have it.

Geothermal Energy

Geothermal energy is heat from the Earth, apparently generated in most part by radioactivity — the decay of radioactive elements in rocks. This natural heat occurs everywhere in the Earth but only in certain places is it concentrated enough, and hot enough close to the Earth's surface to be commercially tapped. These areas tend to occur chiefly along the junctions of what we now know to be plate margins, as defined by the theory of plate tectonics, but there are certain so-called "hot spots," such as the Island of Hawaii and Yellowstone, which do not seem to be related to plate junctions. In general, most geothermal areas tend to be related to regions of active volcanoes.

Volcanoes and petroleum do not geologically co-exist very well. This generally means that areas which have volcanoes have little or no oil, but do have considerable geothermal potential. These areas include the northern California Coast Range, the Cascade Mountains of Oregon and perhaps the southern part of Washington (Mount St. Helens in 1980 provided a spectacular example of geothermal energy in action), the Coast Range of British Columbia, the Aleutian Range of Alaska, the Imperial Valley of California and extending into northern Mexico, Central America, Hawaii, Iceland, Indonesia, New Zealand, the Philippines, Japan, Italy, and eastern Africa. (One estimate is that there is enough geothermal power in the East African Rift Zone including Ethiopia, if properly harnessed, to meet all of Africa's present lighting needs.).

Types of geothermal energy occurrences

Geothermal energy exists in several geological circumstances in the Earth. There is a general heat flow from the Earth, and also a geothermal gradient which simply means that the Earth gets warmer with depth, generally about one degree Fahrenheit for each 60 to 100 feet. Geothermal energy exists in a few places in the form of dry steam. More often it occurs as hot water, in some instances so hot that when it is brought to the surface and the pressure reduced, it flashes into steam, and is called wet steam. This can drive turbines connected to electric generators, just as the dry steam form of geothermal energy does. The Geysers geothermal electric generating plants in California are dry steam operations. Lower temperature waters are more common and more widely distributed, appearing as hot springs.

Another way in which geothermal energy occurs is in the form of hot, dry rocks. There appears to be an appreciable amount of heat energy stored in the Earth in this form, but its recovery has been difficult. Near Los Alamos, New Mexico an experimental operation has obtained heat from hot dry rocks by drilling two wells and then interconnecting them at the bottom by a horizontal borehole through the hot rock. Water is pumped down one well and through the hot rock and is returned from the other well as hot water. So far this has not been done commercially to obtain heat energy.

Geothermal energy used early

Geothermal energy is certainly the first Earth-derived energy source to be used — much before oil, coal, or even wood came into use. In the form of hot springs, native peoples around the world have enjoyed relatively non-polluting geothermal energy. Native Americans were known to pitch their camps on warm ground in geothermal areas. The Maoris of New Zealand cooked their food in the very hot springs of the North Island.

Two ways to use geothermal energy

This energy can be used either directly in heating or in electric power generation in steam-powered plants much like conventional electric generating plants. Only the hotter waters (generally above 360 degrees Fahrenheit) can efficiently be used to produce electric power. A technology called a binary system, can use water of somewhat lower temperature but the efficiency is also lower. By far the most common use of geothermal energy is in the form of the lower temperature waters directly in space heating, or in various hot water facilities.

Geothermal waters: many uses

Currently the lower temperature geothermal waters are used locally in numerous ways. There are many places in the world where hot springs are used as therapeutic baths. In Japan people make more than 100 million visits annually to hotels located near the more than 1,500 hot springs in that volcanic country. In Italy more than 15 million people have been treated in one year at the more than 200 thermal clinics there. In Russia more than 10 million people are treated annually by thermal waters.(9) In the United States, Glenwood Springs, Colorado, Hot Springs, Arkansas, and Warm Springs, Georgia are famous thermal water localities used by people each year. Small hot springs are abundant in many parts of the world. The State of Oregon has more than 200 warm springs, and several guidebooks have been published showing the locations of the many hot springs in the western United States. These attract thousands of individuals and small groups each year, bringing income to local business establishments.

Other uses of thermal waters include heating buildings for the raising of pigs and chickens, growing mushrooms in sheds, and heating greenhouses producing a variety of crops in many places in the world. Iceland has an extensive greenhouse agriculture based on geothermal waters, and has reportedly grown bananas there. Crop drying also may use geothermal energy, as, for example, an onion processing facility in western Nevada. Geothermal waters have formed world famous scenic geological features. The geysers of New Zealand and of Yellowstone National Park in the USA attract multitudes of visitors each year.

In Japan, the combination at several places of alligator breeding, the raising of tropical fish, and greenhouses with tropical flora have become great tourist attractions. Direct use in space heating is a relatively efficient use of geothermal energy, and it is almost entirely non-polluting. At the present time 60 percent of the buildings in Iceland are heated in this manner, essentially pollution free. The capital, Reykjavik (which means "smoking bay" in Icelandic because of the many hot springs in the region) with a population of 145,000 is heated by a geothermal district heating system.

"Geothermal living"

Klamath Falls, Oregon, has long been known as a "city in hot water" because of its use of geothermal water for heating. In Klamath Falls it is possible to enjoy a cradle to grave geothermal life. One can be born in a geothermally heated hospital, be geothermally heated in schools from grade schools through college at Oregon Institute of Technology, get married in a geothermally heated church, and live in a geothermally heated house built from wood dried using geothermal energy. You can buy a car from a geothermally heated automobile agency and park it on your geothermally heated driveway. You can drive the car on geothermally heated pavement, and take the car to a car-wash using geothermal water. You can get medical attention in a geothermally heated doctor's office, and do your clothes in a geothermally heated laundromat. If you are of good character you can enjoy the facilities of the local geothermally heated YMCA; if not, you may be taken to a geothermally heated Klamath Falls jail. Winter snow and ice are repulsed on geothermally heated sidewalks. For later years there is a geothermally heated nursing home available, and finally one may be buried out of a geothermally heated mortuary.

The local Klamath Falls area low temperature geothermal waters are also used for greenhouse operations from which more than 7,000 flats of annuals and perennials are produced each year. Also, the effluent water from the greenhouses is used again in some 40 fish ponds from which some 1,000 tropical fish of several species are shipped by air each week to retail outlets in the San Francisco area. The outflow from the fish ponds is then used for livestock watering. This process of using the same water in decreasing temperature applications is called "cascading." It increases the efficiency of use of geothermal waters, and can be usefully employed in a variety of situations.

Many other places also employ geothermal water for space heating. Boise, Idaho heats government buildings and a mall geothermally. Greenhouses in Idaho are geothermally heated as are greenhouses near Susanville, California. Russia makes considerable use of geothermal water for space heating. In France, near Paris, 13,000 apartments are heated with hot water from wells in the vicinity.

Electric power generation

Geothermal energy was first employed to produce electricity when a light bulb was lighted in Italy in 1905 from a geothermal field some 70 miles north of Rome at Larderello. The field is still in operation. Italy now has developed enough geothermal electric power to provide the equivalent of what it takes to run all the electric trains in Italy, and almost all Italian railroads are now electrified. This is a great saving to Italy which has very little oil and would otherwise have to import large amounts of diesel fuel to run its railroads.

The world's largest geothermal electric generating development is at The Geysers, about 70 miles north of San Francisco. Electricity was first produced here in 1960. More than 2,000 megawatts were eventually being generated, which is about the equivalent of a very large coal-fired plant or two large dams, and is enough to supply the electric power needs of about one and one-half million people.

In a number of other nations, geothermal energy is being converted to electric power. New Zealand obtains 10 percent of its electricity from generating plants using the heat of the Earth as the power source. The Philippines is now using geothermally generated electricity and is planning a further large development of this energy source. With almost no oil, geothermal power is proving a great boon to that volcanic island country.

Mexico, at Cerro Prieto near the United States border, generates 150 megawatts from geothermal wells, half of which is used in northern Mexico. The other half is sent to the San Diego, California, area.

Locally important, site specific

Geothermal energy tends to be site specific. That is, unless it is hot enough to be used to generate electricity which can be transported to distant areas by power lines, it must be used fairly close to where it exists in nature. It cannot be shipped across oceans as can coal, oil, or uranium. Thus, locally, geothermal energy can be and is important, but as a help to the world's energy needs, it can have only minor impact. Still, there is considerable geothermal energy which can be developed.

Both Central America and the East African Rift have very large geothermal energy potential. Some of this has already been developed in Nicaragua, El Salvador, Costa Rica, and in Kenya. Chile and southern Peru in the high Andean volcanic areas hold good prospects for geothermal development. Chile has some capacity already installed. The Kamchatka Peninsula of far eastern Russia has large geothermal resources which have so far been developed only slightly. Neighboring Japan uses geothermal waters for both space heating and electric power production. There are very large prospects for geothermal developments in Indonesia

The Indonesian government has estimated the geothermal potential converted to electric power production (stated as megawatts electric — MWe) to amount to 8,100 MWe in Java and Bali, 4,885 MWe in Sumatra, 1,500 MWe in Sulawesi, and in other islands 1,550 MWe, totaling 16,035 MWe. Geothermal electric generating stations have been in operation in Indonesia since 1981, and an extensive geothermal electric power development is planned for several decades to come. Unocal's Geothermal Division has recently made some outstanding discoveries there. Other discoveries have been made by Amoseas, a company owned jointly by Texaco and Chevron, and together with Indonesian interests anticipate spending as much as $500 million the next decade in geothermal developments in Java.

In total, some 80 countries are now either exploring or developing and using geothermal energy for electric power generation. This use is bound to grow as demand for energy increases, and oil supplies eventually become more costly and less available.

Renewable?

As a technical matter, we have classified geothermal energy as a non-renewable resource. When geothermal energy is used for electric power generation, in the fairly short term it is non-renewable because ultimately the reservoirs of steam and/or hot water will be depleted to the point where they are no longer capable of sustaining electric power generation. The time to depletion is variously estimated to be from 40 to 100 years in most geothermal electric power fields. However, after being shut down, the field over a period of many hundreds or perhaps thousands of years will recover and could be produced again, for the heat will still be there; it is only the hydro system which gathers the heat from the fractured hot rocks and brings it to the well bore which becomes exhausted.

For this reason in a practical sense geothermal energy used for electric power generation does not now appear, under current operating practices, to be a renewable resource. There is, however, one technology under testing which may modify this conclusion. In some geothermal fields, waste water is being injected into the reservoir to see if the reservoir level and pressure can be maintained without adversely reducing the temperature. Results are yet inconclusive, but if successful, then geothermal energy for electric power may possibly become renewable, and with proper management it can be made into a sustainable energy source.

The Geysers Field

If it is to be a sustained energy source for a number of years, the geothermal field must be carefully engineered. The Geysers field, already cited as the world's largest geothermal electric power installation, unfortunately has come into difficult times. Apparently because this steam reservoir was not well managed, too many wells were drilled into it. The result is a decline in production to the point where in 1995, the capacity of the field had dropped 25 percent, and was expected to shrink even more. Half a dozen of the 24 plants were shut down and three were being dismantled and shipped to New Zealand, which has an expanding geothermal electric generating industry. Whether or not a geothermal electric generating facility can be maintained indefinitely as a power source is still basically unknown. Some places are doing better than others, but all geothermal fields now being used for electric power generation appear to be in decline at least to some extent.

Renewable for space heating?

When used in space heating, the use of the hot water may possibly be controlled so as to be kept in balance with the natural recharge of the hydro system which brings the heat from the permeable hot rocks to the well bore. In this case geothermal energy is a renewable resource. Thus, depending on its end use, geothermal energy can be thought of either as renewable or non-renewable. However, even in the use of geothermal resources for space heating, the reservoir can become depleted if it is over-used. Such seems to be the case in several of the district heating systems of southern Idaho, and at Klamath Falls, Oregon, where studies of this problem are underway. However, use of geothermal water for space heating is the most efficient use of such energy and should be pursued. The efficiency comes from the fact that there is no change in energy from one form to another, as is done when geothermal waters are used to generate electric power. Any change in energy form results in an energy loss. Using hot water directly for heating is more efficient than converting it into electricity which then might be used for heating. The lower temperature waters which can be used for space heating are also much more abundant and widespread than are the high temperature waters required for efficient electric power generation. Even where there is no especially warm water to use, the general heat flow of the Earth can be drawn upon by means of groundwater heat pumps which are efficient in most temperate areas, and better in some places than the usual air to air heat pumps.

Use of Earth's natural heat flow

It is possible, particularly in a hilly setting, to build a split level house with part of it below ground level. In colder countries in particular, the natural heat flow of the Earth will add to the warmth of the house in the winter. It also, because it is a steady temperature, will have a cooling effect in the summer relative to the high atmospheric temperatures which may prevail at times. Such an arrangement becomes a natural air conditioner. Some houses have been built almost entirely underground. Whereas this uses more power for lighting, there is a net reduction in power use because both heating and cooling take more energy than does lighting. It also is land efficient as some people grow gardens over the roofs of such dwellings. Others simply put in a lawn. In using the natural heat flow of the Earth in this fashion, geothermal energy is renewable.

A modest help

Locally, geothermal energy can be and is now in places very important, but in terms of the energy needs of the world as a whole, it is likely to be only a small amount. It has the advantage, particularly in direct use in space heating, of having a very low environmental impact. In electric power generation it has a far less environmental impact than do dams, or coal-fired plants.

Geothermal widely used

Although small in the total world energy picture, with the numerous hot springs, and in places very hot waters, together with the natural heat flow from the Earth, geothermal energy is widely used. It makes for easier and more pleasant living conditions for many people, and locally and even to a small country it can be exceedingly important, Iceland being a prime example. Even though the use of geothermal energy for electric power production may only be a small part of the world energy supply, the total contribution of geothermal energy particularly in its low temperature uses is of considerable local economic importance, and replaces other more polluting fuels such as wood and coal.

Non-Renewable — Renewable Balance

An interesting idea regarding replacement of non-renewable resources has been suggested by Daly.(54) He has proposed that non-renewable resources should be depleted only at a rate equal to the creation of renewable substitutes. The problems of implementing this program on even a modest, much less a global scale, would be very large. But simply considering the idea shows how far we now are from being in that position and how much we are dependent on non-renewable resources, with no replacement of the quantities and qualities needed in sight.

In the future it will indeed be necessary to try to find adequate quantities of renewable resource substitutes for non-renewable resources, particularly oil. But so far because we still have use of the non-renewables, we continue on the hope that when the time comes, something will be found. If a strict one-for-one replacement accounting would be required today, it would show us the magnitude of the problem. It is large, as discussed in the next chapter.

Non-renewable Alternative Energy Summary

There are a number of alternative non-renewable energy sources to the present chief source, petroleum. However, none is so versatile as is petroleum. Each has its advantages and limitations. The replacement of petroleum will require a mix of other energy sources and will involve a substantial reorganization of our economic structure. It would also cause a considerable change in personal lifestyles from what they are today in the industrialized countries. It is certain that changes will happen. The transition is inevitable and it is well to begin now to prepare for it before it is forced upon us by some crisis. Petroleum is a finite resource which will surely be mostly gone in less than 100 years from the present. Other non-renewables such as coal and uranium will last longer, but eventually they, too, will be depleted. Renewable energy sources are considered next.

"Culture advances as the amount of energy harnessed per capita per year increases, or as the efficiency or economy of the means of controlling energy is increased, or both."

L. A. White
The Evolution of Culture: The Development of Civilization to the Fall of Rome

BIBLIOGRAPHY

1 ABELSON, P. H., (ed.), 1974, Energy: Use, Conservation and Supply: American Association for the Advancement of Science, Washington, D. C., 154 p.

2 ANONYMOUS, 1980, Exxon Takes A Giant Step Into Oil Shale: Business Week, June 2, p. 27-28.

3 ANONYMOUS, 1991, Unocal to Close Sole U.S. Commercial Oil Shale Plant: Oil & Gas Journal, April 8, p. 38.

4 ANONYMOUS, 1993, Oilsands Play Larger Role in Canadian Oil Production: Oil & Gas Journal, August 2, p. 25-30.

5 ANONYMOUS, 1995, Venezuela Details Orimulsion Expansion Program: Oil & Gas Journal, April 3, p. 32-33.

6 ANONYMOUS, 1995, Production Increase Recommended for Alberta Oilsands: Oil & Gas Journal, May 29, p. 18.

7 ANONYMOUS, 1995, Conoco Poised to Embark on Two Heavy Oil Projects in Venezuela: Oil & Gas Journal, June 12, p. 33.

8 APPENZELLER, TIM, 1991, Fire and Ice Under Deep-sea Floor. Vast Undersea Deposits of Gas Hydrates May Play a Major Role in Climate Change and the Future Energy Economy: Science, v. 252, June 28, p. 1790-1792.

9 ARMSTEAD, H. C. H., 1978, Geothermal Energy: Its Past, Present, and Future Contributions to Energy Needs of Man: E. & F. N. Spon Ltd., London, 357 p.

10 AUBRECHT, GORDON, 1989, Energy: Merrill Publishing Company, Columbus, Ohio, 552 p.

11 BARTLETT, A. A., 1994, Reflections on Sustainability, Population Growth, and the Environment: Population and Environment, v. 16, n. 1, p. 5-35.

12 BISHOP, J. E., 1989, Physicists Outline Possible Errors That Led to Claims of Cold Fusion: The Wall Street Journal, May 3.

13 BROBST, D. A., and PRATT, W. P., (eds.), 1973, United States Mineral Resources: U.S. Geological Survey Prof. Paper 820, Washington, D. C., 722 p.

14 BROMBERG, J. L., 1982, Fusion: Science, Politics, and the Invention of a New Energy Source: MIT Press, Cambridge, Massachusetts, 376 p.

15 BROOKOUT, J. F., 1989, Two Centuries of Fossil Fuel Energy: Episodes, v. 12, p. 257-262.

16 CARLISLE, TAMSIN, 1995, Upside-down Wells Tap Vast Oil Sands: The Wall Street Journal, April 12.

17 CLOSE, FRANK, 1991, Too Hot to Handle. The Race for Cold Fusion: Princeton Univ. Press, Princeton, New Jersey, 376 p.

18 COLLETTE, T. S., 1992, Potential of Gas Hydrates Outlined: Oil & Gas Journal, June 22, p. 84-87.

19 CROWE, C. T., 1981, Our Energy Fix — No Quick Fix: Quest, Washington State Univ., Pullman, Spring issue, p. 14-17.

20 CUFF, D. J., and YOUNG, W. J., 1980, The United States Energy Atlas: The Free Press, Macmillan Publishing Co., Inc., New York, 416 p.

21 DICKINSON, K. A., 1991, Short Papers of the U.S. Geological Survey Uranium Workshop, 1990: U.S. Geological Survey Circular 1069, 56 p.

22 DUNCAN, D. C., 1981, Oil Shale: A Potential Source of Energy: (pamphlet issued for general public information), U.S. Geological Survey, Washington, D. C., 15 p.

23 FOREMAN, HARRY, (ed.), 1970, Nuclear Power and the Public; Univ. Minnesota Press, Minneapolis, 273 p.

24 GULLIFORD, ANDREW, 1989, Boomtown Blues: Colorado Oil Shale, 1885-1985: Univ. Colorado Press, Boulder, Colorado, 302 p.

25 FURTH, H. P., 1995, Fusion: Scientific American, September, p. 174-176.

26 HODGKINS, J. A., 1961, Soviet Power — Energy Resources, Production, and Potentials: Prentice-Hall, Inc., Englewood Cliffs, New Jersey, 190 p.

27 HOWELL, D. G., (ed.), 1993, The Future of Energy Gases: U.S. Geological Survey Prof. Paper 1570, Washington, D. C., 890 p.

28 HUBBERT, M. K., 1962, Energy Resources: National Academy of Sciences —National Research Council, Washington, D. C., 141 p.

29 KENWARD, MICHAEL, 1976, Potential Energy. An Analysis of World Energy Technology: Cambridge Univ. Press, Cambridge, 227 p.

30 KNOTT, DAVID, 1995, South Africa's Synfuel Issue: Oil & Gas Journal, June 5, p. 26.

31 LIPPMAN, T. W., 1993, Clinton Tax Plan Ignored Cheap Energy Tradition: The Washington Post, June 12.

32 MAKOGON, Y. F., 1981, Hydrates of Natural Gas: Pennwell Publishing Company, Tulsa, Oklahoma, 237 p.

33 MALLOVE, E. F., 1991, Fire From Ice. Search for the Truth Behind the Cold Fusion Furor: John Wiley & Sons, Inc., New York, 334 p.

34 MEYERS, R. A., (ed.),1983, Handbook of Energy Technology & Economics: John Wiley & Sons, New York, 1089 p.

35 MUFFLER, L. J. P., (ed.), 1979, Assessment of Geothermal Resources of the United States: U.S. Geological Survey Circular 790, Washington, D. C., 163 p.

36 PIMENTEL, DAVID, et al., 1988, Food Versus Biomass Fuel: Socioeconomic and Environmental Impacts in the United States, Brazil, India, and Kenya: Advances in Food Research, Academic Press, New York, v. 32, p. 185-237.

37 POOL, ROBERT, 1989, Skepticism Grows Over Cold Fusion: Science, v. 244, p. 284-285.

38 RUEDISILI, L. C., and FIREBAUGH, M. W., (eds.), 1978, Perspectives on Energy. Issues, Ideas, and Environmental Dilemmas (Second Edition): Oxford Univ. Press, New York, 591 p.

39 SAVAGE, H. K., 1967, The Rock that Burns: (Privately published, printed by Pruett Press, Boulder, Colorado), 111 p.

40 SEITZ, FREDERICK, 1990, Must We Have Nuclear Power? Reader's Digest, August, p. 113-118.

41 SLOAN, E. D., 1990, Clathrate Hydrates of Natural Gas: Marcel Dekker, New York, 641 p.

42 SMITH, J. W., 1984, We Need Oil from Oil Shale: Mineral & Energy Resources, Colorado School of Mines, v. 27, n. 6, p. 1-9.

43 SUNCOR, 1995, Annual Report for 1994, North York, Ontario.

44 SYMONDS, W. C., 1990, How Shale-Oil Fever Burned Colorado: Business Week, April 9, p. 12.

45 SYMONDS, W. C., 1995, Congratulations — You Struck Sand: Business Week, December 18, p. 90-91.

46 SYNCRUDE CANADA, LTD., 1994, Everything You Ever Wanted to Know about Syncrude: Public Affairs Department, Syncrude Canada, Ltd., Fort McMurray, Alberta, 24 p.

47 TURNER, RICHARD, 1991, Unocal's Loss-plagued Shale Oil Facility to Suspend Operations, Lay Off Workers: The Wall Street Journal, March 27.

48 UTAH (STATE OF) NATURAL RESOURCES DIVISION OF ENERGY, (no date given, approx. 1990), Oil Shale: Utah Department of Natural Resources, Salt Lake City, 14 p.

49 U.S. DEPARTMENT OF ENERGY, 1980, An Assessment Report on Uranium in the United States of America: U.S. Department of Energy, Report GJO-111-80, 150 p.

50 U.S. DEPARTMENT OF ENERGY, 1991, Geothermal Energy in the Western United States and Hawaii: Resources and Projected Electricity Generation Supplies: Energy Information Administration, U.S. Department of Energy, Washington, D. C., 70 p.

51 WILSON, C. L., (Project director), 1977, Energy: Global Prospects 1985-2000: McGraw-Hill Book Company, New York, 291 p.

52 WOLFSON, RICHARD, 1993, Nuclear Choices. A Citizen's Guide to Nuclear Technology: MIT Press, Cambridge, Massachusetts, 467 p.

53 WOODBURN, J. H., 1973, The Whole Earth Energy Crisis. Our Dwindling Sources of Energy: G. P. Putnam's Sons, New York, 189 p.

54 ZACHARY, G. P., 1996, A 'Green Economist' Warns Growth May be Overrated: [Interview with Daly] The Wall Street Journal, June 25.

CHAPTER 14

Alternative Energy Resources: Renewable

Our industrial and personal energy needs now are largely supplied from non-renewable sources, but for the very long term, we must depend on renewable energy sources. Although these resources are not regarded as minerals, they compete and may eventually replace energy minerals so their economics, the availability, and the dependability of these renewable energy sources must be considered along with mineral energy sources.

Wood and other biomass

Only a little more than a hundred years ago, about 1880, wood was the chief fuel used in the United States. It is still the principal fuel in many parts of the world. But the use of wood is deforesting many areas, causing huge and sometimes fatal landslides, devastating floods, as in Bangladesh in 1988, and widespread erosion and loss of valuable topsoil.(79)

Haiti was at one time nearly all forested. Now, only two percent of the land is wooded, and the demand for charcoal which is the chief fuel for the majority of the population exceeds the reforestation rate. Topsoil is being lost rapidly. As one flies into the Haitian capital of Port-au-Prince, the brown fringe of muddy water around that island country testifies to the disaster a lack of something else besides wood as an energy source has become for that impoverished land. Erosion of the denuded hills is already a catastrophe. Even stumps of trees now are being dug for fuel.

There is a severe and growing firewood supply crisis in many parts of the world. In the countryside around Bogota, Colombia, the highlands of Peru, and throughout much of India, Pakistan, Bangladesh, Nepal, and in many regions of Africa firewood is so scarce that local people are now reduced to having only one cooked meal a day. It is doubtful that the use of wood for fuel can be much expanded from its present worldwide volume. In many regions the forest must be allowed to re-grow to prevent further erosion and floods which already are severe in many deforested areas. Deforestation is a particular problem in the foothills of the Himalayas where forests once allowed a gradual run-off of water. Now that has been replaced by great floods in the lowlands to the south, especially in the densely populated lowlands of Bangladesh. In both India and Bangladesh, deforestation caused by firewood demand, has resulted in reservoirs being filled from increased soil erosion far faster than was projected for the dam sites.

Biomass other than wood is also used such as wastes from crops, for example, sugar cane stalks used to fire boilers at sugar refineries. Garbage, and animal excrement (cow dung in India and Africa) are locally important to the population but they are not high grade energy sources. Also, the use of animal wastes for fuel deprives the soil of much needed natural fiber and fertilizer. In India more than 200 million tons of cow dung are burned annually to cook food.

The conversion of wood to alcohol is possible, but the economics are marginal because the net energy thus obtained is small. The amount of fuel which could be obtained from this source is limited. Again there is the problem of deforestation, and the related effect of soil erosion if wood were used extensively for conversion to alcohol fuel. The volume of wood potentially available for this purpose is such that no significant contribution to liquid fuel supplies can come from that source.

Ethanol: Facts, Fallacies, and Politics

In regard to alcohol, special mention should be made of ethanol which is alcohol made from grain, commonly corn. Ethanol has been frequently and enthusiastically embraced as a happy, non-polluting, renewable alternative to gasoline.(30) Its use has been heavily promoted by companies which make ethanol. However, there has been considerable controversy over the use of ethanol as a fuel, and in particular in a 10 percent blend into gasoline, as a product commonly called gasohol. As a claimed improvement in air quality, the U.S. Environmental Protection Agency (EPA) has required that in certain areas during particular times of the year, this oxygenated (ethanol mix) fuel be sold to motorists.

Corn-producing states like Iowa and Nebraska are strong advocates of this fuel for the nation, as are politicians running for office in those areas. Even presidential candidates embrace gasohol while campaigning in the corn states, being either ignorant of, or simply ignoring the fact that the economics are not attractive, or even positive in terms of net energy recovered. The United States can no longer afford politics as usual where energy and minerals are concerned. That imperative began in 1970 when the United States could no longer supply its own oil needs. The surplus energy "cushion" allowing for indecision on implementing a strong national energy program is gone. A long term comprehensive energy policy for the United States should be drawn up and vigorously pursued, but it is sadly lacking to date.

Tax subsidy

At present what makes it economically possible to put ethanol in gasoline and make this mix, gasohol, competitive in price with ordinary gasoline with no ethanol is the subsidy by the federal government of 60 cents a gallon on ethanol. This subsidy is more than the basic cost of a gallon of gasoline, before taxes. There are also state subsidies for ethanol which bring the average total subsidy to 79 cents a gallon.(80) Again, this is a political matter because without these government subsidies, charged to the taxpayer, the price of oil would have to be at least $40 a barrel before ethanol would be competitive even with corn at the low price of $2 a bushel. For comparison, oil is currently priced between $16 and $24 a barrel, depending on the quality. Dr. T. Stauffer, a research associate at Harvard, in a lecture on "Economics of Energy" stated that "The bottom line is that using alcohol to stretch gasoline is like using filet mignon to stretch hamburger."

The U.S. government also waived the federal gasoline tax on gasohol. This tax pays for the maintenance of the federal highway system. By eliminating this tax, the federal government relieved the gasohol users of the cost of paying for federal roads, although they use the highways as do all other drivers. This gas tax exemption for ethanol currently costs the federal Highway Trust Fund about $475 million dollars a year, and is another form of subsidy to the ethanol producers.

Ethanol — net energy loss

Beyond its high cost, there is another basic problem in making ethanol from corn, the grain source most widely favored for ethanol production in the United States. When the energy cost of plowing, planting, fertilizing, and harvesting the corn is added to the transport of the corn to the processing plant, together with the energy involved in processing the corn to alcohol, the net energy result is negative. Less energy is recovered than it takes to produce the ethanol. Among sources to support this conclusion is a letter received from Iowa State University, in the great corn-producing State of Iowa, where agricultural engineers should know the economics of ethanol production from corn. Dr. Peter J. Reilly of the Department of Chemical Engineering kindly replied to my inquiry on the true net energy return from ethanol production. His letter is quoted here in its entirety as it is a definitive statement on ethanol energy economics, which he describes as a system:

> "The answer to your question depends on where the boundaries of the system are drawn. If the boundaries are drawn at the ethanol plant fence, then energy balance is approximately neutral...As soon as the boundaries are widened to take into account energy costs of transportation, fertilizer, plowing, making the tractors, and whatever, the balance becomes negative. The wider the boundaries, the more negative the balance becomes. The only valid claim that can be made is that the production of ethanol is desirable because inflexible solid fuels such as coal and corn are turned into a much more flexible liquid fuel. The only way to come out ahead on the energy balance is to burn the corn directly, which I understand was done around here during the Depression.
>
> "Of course, the greater problem with ethanol, at least when it is used for fuel, is that it is so much more expensive than gasoline and the various competing octane enhancers, and that a major use of ethanol to save petroleum would take all our corn supply and give us an unmanageable amount of corn protein or distillers' solids."(85)

A University of Nebraska study (Nebraska is also a major corn growing state) showed that it requires 131,000 to 135,000 British thermal units (Btu) of process energy to make one gallon of alcohol valued at 84,000 to 85,000 Btu.(7) Braun in the *Farm Journal* summarizes, "At the present time it takes more energy to produce a gallon of alcohol from crops than is contained in that alcohol."(20)

Pimentel has given ethanol economics careful study and concludes:

> "Proponents of producing ethanol from U.S. corn and other grains claim that it reduces oil imports and saves the nation money (ERAB, 1981) [Energy Research Advisory Board, U.S. Dept. of Energy]. Unfortunately, the opposite is true...a gallon of ethanol has only about two-thirds as much energy as a gallon of gasoline. To produce a gallon of ethanol in a large 60 million gal/year plant with all modern facilities requires an energy input of 35,046 kcal. A gallon of ethanol contains only 19,450 kcal. This means that it takes about 80% more energy to produce a gallon of ethanol than can be obtained in net fuel. Therefore, not only does the nation

> have to import oil from the Middle East to fuel this corn/alcohol system but ethanol production is costing the taxpayer huge sums of tax money in the form of subsidies and its production adds to environmental degradation of land, water, energy and biological resources."(81)

It should also be noted that ammonia fertilizer, widely used on cornfields, is now made from natural gas and therefore becomes a direct energy cost in growing this grain. Its use further depletes our petroleum resources.

Air quality

The stated reason that the Environmental Protection Agency has required the use of ethanol in gasoline, is that ethanol reduces air pollution, but the evidence does not seem to bear this out. Dr. Larry G. Anderson, University of Colorado Chemistry Professor, reported at the 1993 annual meeting of the American Chemical Society that burning ethanol puts several pollutants into the air including formaldehyde and nitrogen oxides. These pollutants contribute to the brown clouds which loom over Denver in the winter. Other studies also show that ethanol, in total atmospheric effect, is no better than neutral and may be a negative. Anderson stated:

> "Although ethanol has been advertized as reducing air pollution when mixed with gasoline or burned as the only fuel, there is no reduction when the entire production system is considered. Ethanol does release less carbon monoxide and sulfur oxides than gasoline and diesel fuels. However, nitrogen oxides, and alcohol — all serious air pollutants — are associated with the burning of ethanol as fuel mixture with or without gasoline. Also, the production and use of ethanol fuel contribute to the increase in atmospheric carbon dioxide and to global warming, because twice as much fossil energy is burned in ethanol production than is produced as ethanol."(82)

Environmental costs

A longer range cost, difficult to assess in precise terms, is that growing corn is hard on the land. Loss of fertility of the soil from years of corn cropping and erosion of the land are inevitable. Also, in terms of the total resources which production of corn requires, it is estimated to take about 3,600 gallons of water to produce a bushel of corn. Some of this corn is produced by irrigation which involves pumps which use energy. In many areas, the aquifers are being over-pumped, as is the case of the Ogallala Formation which underlies parts of eight midwest to western states. The total environmental costs of producing ethanol are high. It is a vastly different set of economics from drilling an oil well, and then simply opening the valve and letting the oil flow out.

Ethanol and politics

Ethanol production and its use as a blended gasoline motor fuel is clearly an economic, energy, and environmental negative. It is no help in the problem of petroleum supply, and because ethanol production uses more energy than it produces, it actually makes the petroleum import situation worse. So why have a large number of members of the Congress and several presidents supported the requirement that ethanol be used? The answer is that it is an example of politics overriding reason. The political block of the corn belt states holds votes crucial to elections, and companies which produce the ethanol in the United States have been some of the largest contributors to political campaign funds in recent years.(19,45,49,52,69,75,76)

Although ethanol is clearly uneconomic as a motor vehicle fuel and an energy negative, politics continued to replace reason in the 1996 U.S. Presidential campaign. Senators Robert Dole and Phil Gramm were both candidates for the Republican Party nomination. To enlist the farm vote, Senator Gramm told the corn growers, "I am an ethanol Senator." Senator Dole issued a press release calling himself "Senator Ethanol," and rented a truck outfitted with Dole signs to follow another presidential nomination contender, Steve Forbes, around Iowa to pledge loyalty to ethanol and carrying the statement "Phase out Forbe$, not ethanol."(*The Wall Street Journal*, February 9, 1996)

Pimentel summarizes the realties of ethanol as a fuel:

> "Ethanol production is wasteful of energy resources and does not increase energy security. Considerably more energy, much of it high-grade fossil fuels, is required to produce ethanol than is available in the energy output. About 72% more energy is used to produce a gallon of ethanol than the energy in a gallon of ethanol. Ethanol production from corn is not renewable energy. Its production uses more non-renewable energy resources in growing the corn and in the fermentation/distillation process than is produced as ethanol energy. Ethanol produced from corn and other food crop is also an unreliable and therefore a non-secure source of energy, because of the likelihood of uncontrollable climatic fluctuations, particularly droughts which reduce crop yields. The expected priority for corn and other food crops would be for food and feed. Increasing ethanol production would increase degradation of agricultural land and water and pollute the environment. In U.S. corn production, soil erodes some 18-times faster than soil is reformed, and where irrigated, corn production mines water faster than recharge of aquifers. Increasing the cost of food and diverting human food resources to the costly and inefficient production of ethanol fuel raise major ethical questions. These occur at a time when more food is needed to meet the basic needs of a rapidly growing world population."(80)

Using corn to produce ethanol as a fuel makes no economic, energy efficiency, environmental, nor ethical sense.

A move toward reality in the matter of ethanol as a motor fuel was made by a U.S. Federal Court in April, 1995, when it blocked an order from the Clinton Administration to require that roughly one-tenth of all gasoline sold in the U.S. contain ethanol.(74) However, in August, 1995, "The Clinton administration, in a move likely to benefit the beleaguered Archer-Daniels-Midland Co., [producer of 60 percent of U.S. ethanol] expanded the tax subsidies for ethanol, a corn-derived automobile fuel additive."(75) As of 1995, this subsidy to ethanol production cost the U.S. taxpayer an estimated $770 million annually. Later when the whole matter of farm subsidies came up for a vote in the U.S. Congress, the Congress voted to continue the ethanol subsidy program. The farm vote in national elections is very important. However, a dissenting review of the ethanol program was expressed in a column written by Borvard wherein he makes the concluding comment that, "Subsidizing ethanol to benefit corn farmers makes as little sense as a mandate forcing Americans to drink more grain alcohol. Washington should finally sober up and stop worshipping ethanol."(19)

Ethanol/methanol

Methanol (CH_3OH) and ethanol (C_2H_5OH) can both be made from biomass. Methanol is commonly now made from natural gas and from coal. It also has been made from wood and wood wastes. The U.S. Bureau of Mines was instructed by an act of Congress to construct a plant in Oregon to convert wood wastes to methanol. The Bureau did a creditable job of plant development, and proved that a liquid fuel could be made from wood wastes, but it was uneconomical (not the fault of the USBM), and the concept was abandoned. The problem also is that there are not enough wood wastes available to provide a significant amount of methanol even if the conversion process were energy efficient. Almost all so-called wood "wastes" left from production of dimension lumber are now used within the lumber industry as boiler fuel, or made into other wood products such as particle board as an effective substitute for plywood in several uses. Furthermore, the United States is now an *im*porter of timber and timber products.

As motor fuel

Arguments made for both methanol and ethanol as motor fuels are usually two. One, that they reduce air pollution. It is true that they can reduce some kinds of air pollution but they produce other kinds of air pollution so the effectiveness of this route to cleaner air is questionable, as just stated in the discussion on ethanol. The second argument is that when mixed with gasoline to form gasohol, this saves gasoline. This is not true. Both methanol and ethanol have a lower energy content (Btu) than gasoline. Methanol has 9,500 Btu/lb and ethanol has 12,780 Btu/lb. A gallon of ethanol has approximately 0.7 as much Btu content as does a gallon of gasoline.(26) Methanol has even less. Because these alcohols have substantially less energy content per unit volume than gasoline, there is a reduction in mileage when gasohol is used. More gasohol must be purchased than if pure gasoline were used. Some studies show that a 10 percent ethanol content in gasoline reduces mileage by as much as 17 percent. Use of ethanol or methanol in gasoline does not save gasoline and may actually increase the amount of gasoline which has to be used to achieve the same mileage.

Engine performance

Ethanol and methanol both raise the octane rating of gasoline, but there are some negative factors in the use of these alcohols for motor fuel. These include hard starting, reduced mileage, and phase separation (in the case of ethanol) because of water contamination. Ethanol is 100 percent soluble in water, and it will pull any water in a separate phase from gasoline in a car's fuel tanks. This can cause severe problems, one being that the motor will not idle, and will stop at low motor speeds. An experienced automotive technician and owner of a large automobile repair shop observed the results of ethanol/gasoline mixture, gasohol, used in vehicles he serviced:

> "Most vehicles don't perform nearly as well with alcohol blended into the fuel, and gas mileage usually suffers as well. It seems the American public has been deceived again by some of our leaders...The moral to the story is this: if your vehicle's performance or gas mileage is suffering, confer with your auto technician to see if the gasoline may be causing the problem. If it is, call your representatives in Congress and express your dissatisfaction with the use of alcohol in gasoline. Your vehicle will run better for it."(104)

The owner's manual of the 1991 Volkswagen Vanagon states: "We recommend that you use quality gasoline that is not blended with alcohol. The use of fuel containing alcohol can cause loss of fuel economy, and driveability and performance problems."

A recent test of methanol, made from coal, wood, or natural gas was run by United Parcel Service which has a large fleet of delivery vans. The experiment was done in Southern California. A blend of methanol and unleaded gasoline was used in 50 vans. The results were higher fuel costs, poor vehicle performance, more maintenance, frequent breakdowns, and lower mileage.

The impossibility of using methanol as a replacement for oil in the United States is summarized by Pimentel: "If methanol from biomass (33 quads) were used as a substitute for oil in the United States, from 250 to 430 million hectares of land would be needed to supply the raw material. This land area is greater than the 162 million hectares of U.S. cropland now in production."(83) [One hectare equals 2.47 acres.]

Biomass Not a Major Substitute for Liquid Fuel

Oil from seeds

One of the materials suggested from which to make a liquid fuel has been rapeseed, source of the popular cooking oil, canola. Using even an optimistic production figure, it takes from 13 to 18 units of energy to produce 10 units of energy from rapeseed.(83) And, these figures do not include the energy cost of processing the rapeseed.

There does not appear to be any crop which can be raised and processed efficiently enough into a liquid fuel to have an energy positive result. Even under optimistic theoretical conditions, the net energy gain would be small at best. Also, the degradation of the land by the cultivation of crops, and also the loss of soil nutrients (which cannot be entirely replaced by chemical fertilizers), are not considered when the cost of the fuel is calculated. They should be, for they are very real costs.

Considering all biomass sources, the total volumes of raw materials needed to produce enough liquid fuel to make a substantial substitute for oil really do not exist. The United States currently consumes more than 18 million barrels of oil a day. A barrel weighs about 310 pounds, which means that each day some 5.6 billion pounds (2,395,000 metric tons) of raw material would have to be grown and harvested each day, winter and summer, 365 days a year. With the energy costs of planting, plowing, fertilizing, harvesting, and processing, the net gain in energy would be very small, so the actual amount of net energy produced would probably be equal to less than 20% of the total oil produced. Replacing any very appreciable amount of conventional oil now used with liquid fuel from biomass seems quite unrealistic. Biomass cannot supply even a modest proportion of the current oil consumption in the United States nor anywhere else.

Crop-grown alcohol has locally been used to supply a minor need, but the true economics have been disregarded for the moment. The widely watched Brazilian experiment to produce ethanol from sugar cane has now almost been entirely abandoned. It cost the Brazilian government 33 cents a liter to produce ethanol which they were selling with a subsidy for 22 cents.(61,83)

At best, alcohol could supply only a small fraction of the world's mobile energy supplies, gasoline, kerosene, and diesel.

Food or fuel?

A very serious matter of concern if crops were used to produce alcohol is the impact of this use of land against the need for food. The competition between using crops for fuel rather than food was studied in detail by Brown.(23) He concludes that any large attempt to use crop-derived alcohol for vehicle fuel would have a very negative effect on the ability of nations to feed their populations, and is basically not feasible. Brown also raised the ethical question of whether the affluent car-owning minority in the world should be allowed to use grain for automotive fuel which would drive up world grain prices, noting that "for the several hundred million who are already spending most of their meager income on food, continually rising food prices will further narrow the thin margin of survival."

Brown adds, "It is either food or fuel, and the use of crops for liquid fuel would simply starve the already world's poor for the benefit of those who enjoy the automotive age."(23) And the energy economics are not there either for liquid fuels from currently grown biomass.

Pimentel's quantitative analysis of this situation is illuminating:

> "Assuming zero energy input for the fermentation and distillation process of ethanol production and charging only for the fossil energy expenditure to culture the corn (essential to have corn/alcohol produce net energy), the amount of cropland required to fuel just one U.S. automobile is enormous. Making these assumptions, more than 6 hectares of corn land would be necessary to fuel one automobile for one year. In contrast, one person is fed using only 0.6 hectares of cropland. This emphasizes the tremendous waste of agricultural resources when ethanol is produced from grains."(81)

Biomass as energy source: summary

The use of wood as an energy source goes back to primitive people. It is still a world-wide source of energy, chiefly for cooking and space heating. Biomass sources of energy, including wood and coal, have been converted to liquid fuels such as methanol and ethanol. In some cases, as with the concentrated biomass, coal, the conversion can be an energy positive, but only marginally so. In using current crops for conversion to alcohol, the energy balance appears to be negative. More energy goes in than is produced, especially when the energy costs of plowing, planting, fertilizing, harvesting, and transportation are included.

The use of certain plants for fuel, including sagebrush, has been proposed, and numerous articles appear from time to time which optimistically predict that biomass-derived liquid fuels can be a substantial replacement for oil.(1,2,4,18) The term "petroleum plantations" has been used to suggest that a substitute for oil can be grown in cultivated fields. However, despite these hopeful predictions, no appreciable amount of liquid fuel, relative to the world demand for petroleum, has thus far been derived from these sources.

Hydroelectric Power

The Sun initially heats ocean water and evaporates it. Wind, caused by the uneven heating of the Earth's surface by the Sun, carries the water vapor over the land where, as the atmosphere rises, and also due to seasonal temperature changes on the land, clouds are formed and the moisture falls chiefly as rain or snow. This fills streams, is stored by dams, and ultimately runs through turbines which turn generators to produce electric power. Thus hydroelectric power is derived from the Sun just as are virtually all other energy sources.

Hydroelectric power is an important and currently economic source of energy, widely used today. The history of the use of hydroelectric power in quantity goes back less than 150 years. At the time of the discovery of how to generate electricity, no rivers of any size had dams across them, although some small streams were dammed to run local grain mills. Big power dams are historically recent. Even Bonneville Dam, the first dam on the Columbia River, did not produce power until 1937.

The initial phase of Grand Coulee Dam on the Columbia was not completed until 1950, and final phases of construction on some of the dams on the Columbia, and even on the Bonneville Dam were not completed until just the past few years on this Pacific Northwest power and irrigation complex. Because big dams are a relatively new arrival on the scene, their ultimate environmental impact on a region has yet to be determined, but some concerns are beginning to appear. Dams, which have been commonly thought of as environmentally clean sources of power, actually have many negative consequences. Dams flood fertile lowlands. In more northerly regions and in mountainous areas, river valleys are important winter rangelands for wildlife, particularly big game animals. When these valleys are filled with water, the winter range is lost. Looking longer term, ultimately all reservoirs will fill with sediment.

Who has the hydroelectric power?

This power resource is widespread. The United States has a modest amount of hydroelectric power developments, but nearly all the large sites and many smaller dam sites have now been utilized. Canada, however, still has a large number of excellent unused hydroelectric sites, some of which, however, would involve destruction of important wildlife habitat if developed. South America also has considerable undeveloped hydroelectric power, but today much of it is along the eastern flank of the Andes Mountains in rather remote areas. Russia has done a great deal with large low-head dams, and still has enough undeveloped hydroelectric power to serve an estimated 100 million people. Asia has the largest potential resources of hydroelectric power with only about 5 percent developed. Africa is second with about 10 percent of its potential damsites used. Japan and Australia have very little hydroelectric power resources. Europe has developed about 50 percent of its hydroelectric power potential. For Switzerland, which has within its borders the higher gradient headwaters of some of western Europe's streams, electric power is one of their exports but the amount is small. China has a fair number of undeveloped dam sites, but not as many as one might expect from the size of the country. Also, large developments would effectively take out of production what are now some very important fertile agricultural lowland. But with 1.2 billion people now for whom to provide energy, China is planning some major hydroelectric projects including the world's largest dam across the Yangtze River.

Thus Canada, Brazil, Argentina, Russia, Asia, and certain countries in Africa are the great holders of major hydroelectric power sites. These resources will be important to those nations in the future. However, as the environmental costs of huge dams are becoming more apparent, the trend toward developing this energy source may be lessened.

Earlier enthusiasm for dam projects is diminishing. The World Bank used to generously support Third World dam construction, but, its financing of dams has dropped from an average of 18 a year between 1980 and 1985, to only about six a year between 1986 and 1992. The huge Yangtze River dam project has come under particular criticism because of its large environmental impact.

Hydroelectric power renewable?

Hydroelectric power is commonly classed as a renewable energy source, but the inevitable silting up of the reservoirs, although it may take hundreds of years, adds a reservation to the concept of hydroelectric power as continually renewable. Some small reservoirs have already been filled with sediment. But, for the present generations of people who have had large sites on which to build dams, the resource has been renewable and economic. Each year brings the seasonal precipitation to refill the reservoirs, but the water also brings in sediment which stays. The fact that, due to silting, the reservoir capacity is slightly less each year can be ignored for the time being. In the early life of big dams which we now enjoy, they do bring many benefits. But for the generations to come, with no new hydro sites to occupy, these existing sites, gradually filling with sediment, may prove not to represent a renewable resource, but a large muddy problem.

Solar Energy

Petroleum and coal are derived from ancient solar energy. But as generally defined and as used here, solar energy means energy collected by some method directly from the Sun. Solar energy is so obvious and so universal that it has very wide appeal to many people as an abundant and free energy source. It is also pollution free if the environmental impacts of the manufacturing processes to make the solar energy collecting equipment, and its ultimate disposal are ignored.

Solar energy produces no radioactive end products, no acid rain, and no ash to be disposed. It looks so good to many environmentalists in particular that many believe it is the great ultimate solution to our energy supply problems. Others are somewhat skeptical.(109)

Estimates of the potential for solar energy are impressive at first glance. Some 15,000 square miles of the United States, properly located, if covered by solar cells, would provide all the electricity the United States presently needs. Swan states "Photovoltaics in combination with a modicum of other renewable sources may replace nearly all electrical generation by fossil or nuclear fuels within 35 to 50 years. They may also be a major source of energy for transportation as well . . . There is little doubt that the photovoltaic industry will become a . . . multibillion dollar industry by 2000 . . ."(100) These statements generate much enthusiasm for solar energy, but a pragmatic view is more sobering.

The practicality of heating homes by solar energy has been researched over the years. This may appear to be a relatively simple problem. Locating a house to take advantage of the Sun by means of the "greenhouse effect" through the windows (whereby light enters the house in short waves but upon hitting a solid substance the rays are changed to long heat waves which cannot escape back through glass) is easy to do. But this can provide only a small amount of the total energy required to run a household, and is not effective on cloudy days or at night. Making a house entirely dependent on solar energy for heat during long periods of cloudy days, and in cold weather which also may have cloudy days and compounds the heating problem, has been much more difficult. In high northern and southern latitudes where days are seasonally short and nights are long, solar energy also has serious limitations.

The major problem is how to store solar energy. Some rather elaborate systems have been devised. Extensive solar heat collectors are placed on the roof and a heated fluid is piped to a large area in the basement where rocks are heated or the hot fluid is stored. But these systems have proved to be of limited value. Even in sunny Arizona, cloudy days may persist. One experimental Arizona home, very well fitted with solar energy storing facilities, after three days of cloudy, cool weather, was out of heat. At best, these home heating systems seem applicable in moderate climates rather than in colder regions.

Various other systems for storing solar heat have also been tried including using the principle of the latent heat of a phase change (for example, a solid to a liquid). McDaniels has discussed this stating: "Numerous difficulties with this approach remain to be overcome. The materials are usually expensive and corrosive. Some materials have short lifetimes due to decomposition when thermally cycled."(68)

It may be desirable to build as many houses as possible using solar energy as a source of heat and power. But to put this in perspective of total energy used in the United States, the installation of solar energy systems in one million homes to provide their entire heating and electric needs would displace less than 0.3 percent of total U.S. energy consumption. Recently, however, photovoltaic cells have been integrated into building materials, and these appear to offer prospects for a more widespread use of solar energy.(107)

When solar devices are used to produce electricity, storage is also a problem. For half or more of the 24-hour day the solar devices would not generate electricity. Large quantities of electricity would have to be stored, or alternative backup generating systems used which are costly. Feeding power smoothly from an electricity storage system into a major power grid would be difficult.

There is also the problem of maintaining the solar energy gathering equipment. Perhaps all these difficulties can be overcome in time, but the costs of producing electricity from the Sun are presently quite high, although some progress is being made to make solar cells more efficient. The cost of power from photovoltaic cells has dropped during the past 20 years from about $2 a kilowatt hour to about 18 cents, but this is still twice or three times more than the cost of electricity from the conventional electric power sources, and the true cost of long term maintenance and ultimate disposal of solar energy equipment is not yet completely known.

In remote areas, and in space satellites where cost is not a consideration, solar energy can be quite useful. In small local uses it can be economic such as for powering relay stations for message transmissions, and in the form of solar water heaters for domestic use. Israel has many solar water heaters now in use. Railroads and utilities use solar devices to aid in their communications systems, and many of us appreciate the use of hand held solar calculators.

Solar energy devices

To try to make solar energy for electricity competitive against other energy sources in developed areas such as the United States, a number of experimental projects have been undertaken. At Solar 1, Southern California Edison's experiment near Daggett, California, in the Mohave Desert, several thousand mirrors focus the Sun's rays on a large liquid filled tank raising the liquid to above the boiling point, and producing electric power by a steam turbine and generator. This system also has a superheated fluid storage arrangement which does allow some power to be generated for a few hours after the Sun goes down. Cost of the electric power thus produced, however, is higher than that generated by conventional means.

ARCO, the oil company, at one time was a major producer of photovoltaic (solar) cells but apparently decided payoff was too far in the future and sold their operations to Siemens in 1989. Another oil company, Amoco, however, is continuing to pursue solar energy development. Photovoltaic cells produce electricity directly from sunlight. Experimental banks of cells have operated in the Mohave Desert near Hesperia, California for some time. A variety of efficiencies have been achieved with experimental solar cells. Most range from 10 to 20 percent, but in 1988, scientists at Sandia National Laboratories built a cell which is 31 percent efficient. In 1989, the Boeing Company achieved 37 percent. These cells, however, require both costly technology and materials to produce. However, thin films of amorphous silicon can be deposited on metal or glass and produce a photovoltaic cell at relatively low cost. The problem, however, is that the efficiency of sunlight conversion to electric power by this technology is low, only about six percent, although some laboratory experiments have raised it to about 10 percent.(108)

Other experiments to produce solar electric power, in part already operational, include tubes of a working fluid which are centered in a sunlight reflecting trough, and the heated fluid is piped to a central generating station. One such operation is in service in the Mohave Desert of California.

The future

Solar energy will undoubtedly increase its share of the energy mix in the future. However, this is likely to come slowly. There is a trade-off in regard to solar cell cost and efficiency. The more efficient cells cost considerably more than the less efficient cells. Thus it may be just as cost-effective to use low-cost, low-efficiency cells as to use the high-cost, high-efficiency cells. It is predicted that the new generation of solar cells, called the "thin film" cells, will bring down the cost to less than 15 cents a kilowatt hour. This would be within striking range of a commercial situation.

Zweibel writes about thin-film photovoltaic cells: "A couple of decades ago, some people believed that solar technology would solve the world's energy problems, if only the oil companies would let it. But those of us involved in solar research have always known that making solar energy practical was less matter of corporate goodwill than of technological development...Progress in this has come in an area known as thin-film photovoltaics."(113) Zweibel goes on to note that conventional crystalline-silicon modules cost between $300 and $500 per square meter, and require from 1,000 to 2,000 kilowatt hours to produce per square meter. In contrast, thin-film photovoltaic cells require only about 100 to 300 kilowatt-hours, and cost between $50 and $150 per square meter. These advances will make solar-electric conversion much more competitive and allow a greater penetration of solar energy into the market, but it will not in itself solve the foreseeable energy problems. It will be a modest help as presently understood.

Who has the big solar resources?

All parts of the Earth are touched by the Sun's rays to a greater or lesser intensity at one time or another. But who has the quality solar energy where the Sun's rays are relatively direct most of the time instead of glancing in at an angle? And, where does the climate provide a large number of clear days? In this case, it is the rich getting richer, as the Middle East desert countries clearly are fortunate in this regard also. The Persian Gulf nations have abundant solar energy, as does Egypt and the other countries along the north coast of Africa all the way to Morocco. The now relatively economically useless Sahara and central Australian deserts have large potential solar energy resource. The use of deserts for this energy is practical because they are not used for growing foodstuffs. In other cultivated parts of the world, covering the land with solar collecting devices would render the land unsuitable for agricultural purposes.

Fortunately, for the future of solar energy, there are a lot of deserts. The Atacama Desert in northern Chile has an area which has never had a drop of rain recorded and should be a fine site for reception of sunlight. There are many other areas of the world with a high solar energy input. Most everywhere on the Earth there are times when at least a modest amount of solar energy is available. How much the presence of abundant solar energy or lack thereof might affect the future of a nation can hardly be visualized at this time. Much will depend on further developments in solar energy technology. However, solar energy will undoubtedly be increasingly important, and the countries which have good sites will no doubt benefit. Will an electric power grid from solar collectors on the Arabian Peninsula someday bring energy to Europe even as Saudi oil does today?

Solar energy enthusiasm

The concept of a solar energy economy has a large following among the general public. Publications which claim a near and great future for this energy source are very popular, even best sellers for the moment. *Energy Future* was one such volume which appeared in its first edition in 1979, the decade of the U.S. oil crises. The authors enlarged the scope of "solar energy" to include biomass, wind, hydroelectric power, and ocean thermal electric, and stated, "We believe it is possible that solar will end up little more than a mosquito bite —that is not provide any significant addition to America's energy mix. Yet we also believe that it is entirely possible for solar to meet as much as a fifth of the country's energy needs by the year 2000 — in absolute terms, the equivalent of the current amount of imported oil."(95) So far imported oil is still far ahead of solar energy. Solar energy may take over a substantial part of segments of the energy load of the future, but progress to date has been slow. It will need to be pursued more vigorously if a satisfactory bridge to the future from the present petroleum-dominated society is to be constructed before petroleum supplies become scarce. The size of world population at that time will also determine if solar energy can be harnessed in large enough quantities to meet the demand.

Solar energy is so evident that it seems to many people to be the obvious solution to our energy problems. "GO SOLAR" bumper stickers abound, as do others with the same enthusiastic message such as 'THE SOLAR SOLUTION TO NUCLEAR POLLUTION." One observer has suggested that solar energy is as much a religion as a science — one has to be a believer and have the faith. Solar energy is a very large and renewable resource, and if increasingly efficient ways can be found to harness it, solar energy will have a larger role in the future energy mix, but it is difficult with the known technology to visualize it as a complete replacement for petroleum.

Current solar energy income versus an inherited bank account

Living on only what solar energy comes in each day, and surviving the cloudy days, will be a vastly different situation from doing as we are today — drawing upon the stored sunlight of millions and millions of years deposited in an energy bank account in the Earth by geological processes. We now draw upon this very much concentrated high grade solar energy inheritance in the form of coal, oil, and gas, and are using it in a geological instant. The challenge of living on current income rather than our fortunate energy inheritance is very large, and not simple.

A major shift to a solar energy economy from today's oil-based economy would almost certainly involve some large changes in industry, and in the lifestyles of society. This cannot be done quickly. Solar energy if it does become a significant source of energy, will do so very gradually. It has been said, however, that solar energy does have one advantage over other energy sources — you can simply look up and see how much is left.

Tides, Waves, Wind, and OTEC

Tides, waves, wind, and simply the heat of the ocean are so widespread as to appear to be potential large sources of energy. Tides have been harnessed as an energy source in a few places, and wind more widely so. Wave energy and ocean temperature differentials have so far seen very limited development.

Tidal power

Watching the changes of tide in bays and estuaries, one cannot but be impressed with the amount of energy involved in moving all that water twice a day. Harnessing the tides, however, has been slow in developing. One reason is that tidal power sites are quite limited in number because such sites must meet certain requirements. They must have a fairly high tidal range, and there must be a favorable configuration of the land. There has to be an estuary into which the tide will funnel and where it can be trapped by a dam. If the tide is sufficiently high it can turn turbines which power the electric generators as the tide moves into and then out of the estuary.

High tides in the world are found at high latitudes, rather than near the equator. Most of the good sites which do exist tend to be in relatively remote areas. One example is Frobisher Bay in southern Baffin Island, west of Greenland, where I witnessed a 50-foot tide. It was impressive but a long way from where power in quantity would be needed. There are several potential tidal sites in southern Chile where the very rugged Andes Mountains reach the sea. These places are also remote from where the power could be used.

At present there are three places where tidal power has been developed. One is in northwestern Russia, at a site of an inlet of the Barents Sea with the nearly unpronounceable name of Kislayagobuga, near the city of Murmansk. Another is on the Rance River in France. The third is on Hog's Island near the mouth of the Annapolis River in Nova Scotia, Canada. This plant is somewhat different from the other two in that the tidal range is only from about 14.5 feet (neap tides) to 28.5 feet (spring tides), with the average being about 21 feet. Power is generated only during the out-going tide. This station is unique because of the relatively low head (vertical distance the water drops) by which it operates. It is claimed to be operative in a range of 4.6 feet to about 22 feet. To do this it uses a new type of low-head turbine called Straflo, developed by Escher Wyss and installed first in 14 low-head submerged stations in south Germany and Austria.

The project is still under evaluation. If the economics seem reasonable relative to what the cost of power may be in the future, this sort of plant may open up substantially more areas for tidal power development. For now, however, the cost of power at all tidal plants is fairly high, although the French Rance River operation appears to be successful and competitive with other forms of energy. Tidal power economics are definitely best where the tidal ranges are large. This fact combined with the need for a special shape of the coastline limits the number of potential sites. There are less than two dozen optimal tidal energy sites in the world, and, again, most of them are in fairly remote locations.

Furthermore, as tidal power is not a continual power source over a 24-hour period, but can be produced at only certain stages of the tide and these tidal stages change in time from day to day, there is a problem of integrating this varying power production into a large regular power grid. Tidal power, in the few special areas which have relatively high tides and a properly shaped coastline, can be useful. However, it is unlikely that tidal power can make a very significant contribution to world energy supplies.

The entire Bay of Fundy with its several arms in eastern Canada lying between New Brunswick and Nova Scotia, is being studied for future tidal power sites. However, the environmental impact of interfering with the free flow of the tides is not entirely known. The bay has abundant marine life, and is an important fishing area. It also is an important habitat for many waterfowl, both indigenous, and migratory. It seems likely that putting in major tidal power installations will appreciably affect the ecology. You cannot just do one thing in the complex interconnected systems of nature.

Waves

Along with the tides, the strength of waves is impressive, but, like the tides, harnessing this energy is difficult. At times waves can be 60 feet or more in height. Yet at the same location for weeks the ocean may be essentially calm. All sorts of mechanisms have been tried to usefully capture wave energy. Japan, surrounded by waves and very short of energy, has conducted research on using wave energy for electric power production since the 1970s. Several countries including Japan, Norway, Denmark, Britain, Belgium, and India have constructed a variety of devices to capture wave energy and convert it to electric power. These range from a tapered channel system (Norway), a barge wherein waves force air into chambers and then into turbines (Japan), to a float on the ocean surface attached to a piston on the ocean floor which moves water through a turbine (Denmark). Costs have been estimated to be as low as six cents a kilowatt-hour. However, the intermittent rather than continual wave movement is one of several problems with wave energy systems. Results have been mixed, but some studies continue. Present indications are that waves will not provide a significant part of the energy mix in the reasonably foreseeable future. Ross has written a comprehensive summary of wave research and prospects.(88)

Wind

The Dutch did not invent the windmill and in fact, were relative latecomers in its use. For a time they did construct windmills to make an asset of the miserably cold winds which blow in from the North Sea, but only a few are still in use. Denmark, also in a windy position by the North Sea, makes use of wind power for a small amount of energy.

The Persians, many centuries B.C., were the first to use windmills and employed them in pumping water for their arid land, now known as Iran. The windmill has kept the western United States rancher in business by pumping water for his cattle scattered across thousands of acres of dry range land. With the aid of a large number of batteries, windmills have provided the rancher and other remote people with some stored electric power which can be drawn upon in modest amounts.

In the U.S., windmills have recently been syndicated, and the concept of "wind farms" has given rise to limited partnerships which built windmills and sold the power to power companies, but these projects were subsidized by the federal government in the form of tax write-offs. Without them, windpower would not have been economical, at least at that time. Recently developed technology has made wind power more competitive with conventional power sources, and numerous windpower installations exist across the United States and around the world and are being integrated into regular power systems. Windpower installations are especially useful in more remote locations where electric lines have not reached.

However, wind farms are not universally admired, especially by the citizens who may live near them. Windmills are noisy. They also take considerable maintenance and can easily be damaged by the vagaries of the wind which at 50 or more miles per hour can change direction 90 degrees in less than five seconds. This is hard on equipment. And there are those who feel that lines of windmills on ridgetops and in mountain passes are not an attractive part of the scenery.

Wind energy can be locally important, and will no doubt grow in use. But, like the Sun which doesn't shine all the time, wind, which is caused by heating the Earth's surface by the Sun, doesn't blow all the time. It is undependable. One of the problems is how to efficiently use this energy source which is not available all the time in a power grid. Some advances have been made in storing the electric energy from wind, but the general problem of intermittent winds remains.

Britain's Department of Energy and Central Electricity Generating Board in 1989, launched a wind park program, estimating that wind could contribute up to 20 percent of Britain's electricity. However, it was calculated that with present technology it could take one hundred square miles of properly spaced windmills to produce only one percent of the country's electric power.

Large-scale commercial use of windpower, however, so far, in the United States at least, has produced mostly experience and some fairly costly electricity which, without tax credits, would not have been produced at all. There are also some environmental problems with windmills. Aside from noise and visual pollution, windmills also produce electromagnetic interference, much of it in the television bands. This has been in places a rather troublesome situation, but may possibly be reduced by use of fiberglass blades. But a more serious problem is in regard to killing birds. The very large windfarm at Altamont, California east of San Francisco has been monitored in this regard and "One study concluded that more than 500 raptors were killed in the wind farm over two years, including 78 golden eagles."(16) Because of these casualties, the U.S. Fish and Wildlife Service has considered filing suit against the manufacturers of wind-driven electric generators on the basis of possible violation of the federal Golden Eagle Preservation Act, and the Migratory Bird Treaty Act. Efforts to mitigate this problem have been only moderately successful.

In 1995, three demonstration wind power plants were proposed for the Columbia River Gorge area between Washington and Oregon. This drew immediate protest from the regional Audubon Society members who pointed out that the gorge was an important flyway for migratory birds and raised the question, "Can bird chopping turbines stacked along the Columbia be greener than their salmon chopping hydroturbine counterparts?" Wind power plants are commonly located where wind funnels through passes in higher terrains and these tend to be the flyways for birds both seasonally and throughout the year. Like all other sources of energy, windpower also has its environmental price. Windpower technology continues to advance, however, and ignoring environmental and aesthetic considerations, the basic economics are becoming more attractive. In 1995 in the United States, Texas Utilities company signed a contract for windpower to be produced by a private firm in the windy plains of West Texas at a price which it is claimed will be competitive with conventional power sources.

In theory, wind has considerably more potential for producing power than does hydro simply because of the greater number of sites available. Also, there is no eventual silting up of reservoirs, and wind farms do not flood the fertile lowlands. Windpower installations tend to be located in higher areas, frequently in rough terrain generally less suitable for agriculture. In land use, wind power also has an advantage over solar energy installations in that the collecting system does not completely cover the ground, and most of the land remains available for other purposes. Wind does not blow all the time, but useful solar energy also is only available for less than half the 24 hour day, and winds may blow both day and night. Windmills also can be directly hooked up to electric generators whereas solar energy involves somewhat more complicated conversion equipment. Wind energy will probably have a modest place in the future energy mix, and at the present time there is a moderate expansion of this energy-producing technology.

There are more than 5,000 clusters of wind power turbines now in California, Iowa, Minnesota, New York, and Texas. More are planned in Maine, Vermont, Washington, and Wyoming. Generating costs have been substantially reduced. Texas is planning a particularly aggressive wind power program, and Texas Land Commissioner Garry Mauro stated, "We're going to run out of oil and gas, but we're never going to run out of wind in west Texas."(90) Texas may well become an important wind power state, but a slight reservation may be made with regard to Mr. Mauro's statement — the implication that oil and gas can be replaced by wind power. That may be true for the electric power which is now generated by oil and gas, but there are many uses of oil and gas which wind power-generated electricity cannot replace, such as production of petrochemicals, and ability to replace gasoline as a motor fuel for vehicles which would operate as effectively as such vehicles do today.

In certain wind-favored areas, wind power use is increasing. The world wind electric generating capacity rose from 10 megawatts in 1960 to 1020 megawatts in 1985, to about 3,700 megawatts in 1994. However, the last figure still remains a very small amount in terms of world electricity use, and, to put it into perspective, is only equal to the output of about two to three conventional coal-fired generating plants.

Ocean thermal gradients

This is also called ocean thermal energy conversion or OTEC. The physics principle is that if there is a temperature difference between two masses there is a potential for power generation. In this case, the surface water of the ocean in the tropics is warm, but the depths are cool. The warm water is used to evaporate a low boiling point fluid such as ammonia and the vapor of the working fluid drives a turbine attached to a generator as in conventional power plants. Cool water brought up from the ocean depths is used to condense the vapor, just like the cooling towers of other power plants. But because the temperature differential in the ocean is not great, the efficiency of this operation is low, and huge volumes of water must be circulated to generate any significant amount of energy. A 250 megawatt plant (roughly a fifth or less of a conventional coal-fired plant) would have to use pipes about 100 feet in diameter, and circulate the amount of water equivalent to the discharge of the Mississippi River.(35)

The OTEC system has been tried experimentally on the Kona Coast of Hawaii, and on a small island off the coast of Brazil. It has not been tried on a power plant anchored in the open ocean as some have suggested. There are no OTEC sites adjacent to continental United States, as there is insufficient ocean thermal gradient. OTEC must operate in tropical or semi-tropical waters.

Although the physics principle of OTEC is simple, there are some major difficulties which make the practicality of the technology doubtful. Corrosion of pipes by salt water, is severe. Algae and barnacles grow on the heat exchangers which necessitates almost weekly cleaning. The tropical areas where the highest thermal gradients exist are subject to devastating hurricanes with hundred mile an hour winds, and thirty foot waves.(35) Anchoring what would be a large installation in the open ocean against the elements would be a problem. Also, carrying the power to the shore would involve very large cables laid at great depths which would have to withstand constant flexing of the platform from the wave motion. Using this technology on a large scale may not be practical.(22)

It has been suggested that OTEC might be useful in tropical island areas in the form of small power plants, but local environmental effects of bringing up large quantities of cold water to the surface are not known. These more nutrient-rich deeper waters would encourage more fish life, but might also stimulate algal growth.

In summary, the very large amount of equipment needed for significant OTEC plants, the maintenance problems, and the very low efficiency all suggest that OTEC, except in certain special local land-ocean situations, will not be a very useful technology.(35) However, the United States and Japan have each spent more than $100 million on OTEC research, and enthusiasm has occasionally run high among government officials. I have in my files, a 1987 letter from a Congressman in which he states "Forget about the oil companies, OTEC will be ready in five years." In spite of this assurance, the OTEC energy era has not yet arrived. Nor does it appear that this source of power will be developed to any appreciable extent in the foreseeable future.

A Hydrogen Economy

Hydrogen has been suggested as a great new fuel for the future in a visualized "hydrogen economy." Hydrogen is by far the most abundant element in the Universe, but on Earth it is relatively rare, being only 84/10,000ths of Earth's total composition. However, it is abundant on Earth in a very accessible source —water. Natural gas, methane (CH_4), is also a hydrogen compound in considerable quantity. However, hydrogen is not a primary source of energy because it has to be obtained by the use of some other energy source. Commonly this is done by electrolysis of water, or by processing natural gas. It can also be produced from coal. Agricultural and forest biomass is another source but the large scale use of biomass for this purpose would have a variety of environmental problems, and also would compete for agricultural land used for food production. Hydrogen, when burned, produces water, nitrogen oxides and hydrogen peroxide. The latter two are pollutants, but the amounts are about one-third lower than those produced from gasoline, so hydrogen is environmentally attractive as a fuel.(82) However, as hydrogen has to be produced by means of other energy sources, the environmental impact of these energies must be considered, and all have environmental problems. The most common concept of large scale production of hydrogen would be from water using nuclear-generated electricity, or solar-electric power.

Storing hydrogen

To be used as a fuel for transportation, hydrogen can be stored and carried in four ways: It can be compressed and contained in a high pressure tank. It can be cooled down to a liquid and kept refrigerated. It can be bound up with a metal in the form of metal hydrides. It can be in the form of a hydrogen-rich gas and used in a fuel cell.

If hydrogen is to be used in compressed form, it has to be compressed as high as 3,000 pounds per square inch, otherwise the container becomes excessively large. It can be kept in liquid form refrigerated at or below minus 253 degrees Celsius (approximately minus 423 degrees Fahrenheit). Above that temperature it will boil away. But it takes energy to keep it refrigerated. Liquid hydrogen gives three times as much energy as gasoline by weight, but it also takes up three times as much space. At the normal one atmosphere of pressure on gasoline and hydrogen, a car can go 3,000 times farther on gasoline, as the energy in the same volume of hydrogen at normal atmospheric pressure is very small compared to that in gasoline.(41) It is for this reason that hydrogen has to be compressed, concentrated by refrigeration, or be bound up in metal hydrides.

Metal hydrides appear to be the most practical approach for storing and transporting hydrogen. Hydrogen attaches itself to both iron and nickel to form a stable compound, from which hydrogen can be released upon heating. The German automotive firm, Daimler-Benz, has studied this system extensively.(36) Fuel cells can also use hydrogen supplied by a hydrogen-rich gas such as methane. The cells convert hydrogen directly into electric power.(22)

Hydrogen-powered vehicles

Recognizing the eventual exhaustion of oil resources, automobile and airplane manufacturers have studied hydrogen as a future fuel. Deutsch reports:

> "Actually in the early to mid-1970's, most car companies in the United States and abroad experimented heavily with hydrogen-powered cars. They all ran up against the same obstacles: it took more energy to get the hydrogen than the hydrogen itself provided. The processes for deriving that energy were polluting. And no one could figure out how to put enough gaseous or liquid hydrogen into a car to give it a decent cruising range."(36)

Some hydrogen powered vehicles have already been demonstrated, and tests funded by the West German Government proved that "the operation of vehicles with hydrogen is possible and controllable," according to the head of Daimler-Benz hydrogen research. The Russians have built an experimental plane that flies on liquefied hydrogen. Whether or not hydrogen is "tomorrow's limitless power source" remains to be seen.(32) It does have some advantages over electricity as a source of mobile power, because much more energy can be stored in less weight than with battery storage. The problem is to obtain hydrogen in an energy efficient manner.

Hydrogen to store solar energy

If solar energy is converted to electricity, if it is not immediately put into a power grid, there is a storage problem. Solar electricity is not generated at night so if that were the only local source of power, it would have to be stored in some fashion. Hydrogen offers a solution to this problem. If solar energy is used in the daytime to produce electricity which in turn is used to produce hydrogen, the hydrogen can be stored. Hydrogen is a fairly high energy per unit weight energy source.

William Hoagland of the National Renewable Laboratory endorses hydrogen's prospects stating that some decades from now "...we'll no longer be using fossil fuels. The use of nuclear energy is uncertain. And most technical people agree that eventually societies will be (primarily) using hydrogen and electricity as energy carriers."(72)

Mazda (Japan) announced in 1995, that it plans to market a hydrogen fueled automobile which will outperform electric vehicles. It would have a range of about 150 miles and a top speed of about 90 miles an hour. The timing of the marketing of the car would depend on a

Japanese government program to develop a hydrogen fuel distribution network. There was no statement as to how the quantities of hydrogen would be produced, or how hydrogen will be contained.

High cost at present

Aside from the problem of handling and storing hydrogen in compact useful amounts, the major obstacle at present to the use of hydrogen as a fuel is its cost. The California Energy Commission in 1990 estimated that hydrogen can be produced by electrolysis of water at a cost of about $18 per million Btu. This figure based on an assumed cost of electric power at five cents a kilowatt-hour, which is about the national average. But this cost of $18 per million Btu compares with the 1990 cost which utilities paid for a million Btu which was $1.46. for coal, $2.32 per million Btu for natural gas, and $3.38 per million Btu for oil and related liquids. However if an inexpensive method of producing hydrogen can be devised and a way to carry it in vehicles in a convenient and safe form, hydrogen could be a major transportation fuel, possibly even in aircraft.(8,40,41,71,72)

Fuel cell car

In 1996, the German automaker Daimler Benz announced what it claimed to be the first fuel cell powered car, which would be practical for everyday use. It was claimed to be a competitive, pollution free alternative to the internal combustion engine. The company said it could be in production before the year 2010. Named the Necar II, Daimler Benz said the prototype can carry six people at speeds of more than 100 km/hour with an operating range of more than 250 kilometers (155 miles). In the car's fuel cell, hydrogen reacts with oxygen to create electric energy for propulsion. Water vapor is the waste product. Daimler Benz indicated that an onboard liquid fuel such as methanol would be used which can be broken down by an onboard reformer into hydrogen and CO_2.

Though stated to be competitive with conventional combustion engine vehicles, the range of approximately 155 miles is less than half what a gasoline powered automobile can do. Also, methanol is a fuel which must be produced by the processing of coal or other organic materials which is relatively energy intensive, and if it is obtained from coal the mining of the vast amounts of coal necessary to be converted to methanol to power the huge number of present gasoline powered vehicles would be an enormous undertaking and disturb a lot of landscape. Mining the coal itself takes considerable energy as does reclaiming the disturbed land.

Gasoline, in its convenience of handling and storage, and high energy value per unit volume and weight, will be very difficult to replace. In all its total versatile aspects, gasoline may be irreplaceable.

The Electric Car

In discussing alternative fuels, the question commonly arises as to what might fuel motor vehicles, especially the personal automobile so dear to the heart of most people in today's industrial societies.

The most commonly considered car of the future is the electric car, about which numerous articles and several books have been written.(73,78,94) There is an Electric Vehicles Association of the Americas organized to promote such cars. Electricity would seem a logical energy for the car of the future because almost all renewable energy sources, solar, wind, hydro, and OTEC, produce electricity as the usable end energy. Given the importance of motor vehicles, the electric car has been under study for many years, and a variety of such vehicles have been produced on a limited scale or simply as prototypes. Electric vehicles actually predate gasoline powered ones.

Non-polluting?

Electric cars are touted as being non-polluting. But the U.S. Environmental Protection Agency says when the power plants required to produce the electricity for the electric cars are considered, the widespread use of the electric car would have considerable environmental impact.(98) Perhaps the State of California assumed that such power plants could be placed in more remote areas, and that it would be easier to control pollution from a few large power plants than from millions of roaming car tailpipes. Accordingly, California mandated that automakers convert five percent of their yearly sales in California to zero emission vehicles by the year 2001 (about 100,000 cars), and 10 percent by 2003.(10,11) New York and Massachusetts subsequently passed similar requirements, and several other states considered them.

The battery problem

Electric cars to date suffer from the drawbacks of the large expense of good storage batteries and their limited driving range requiring frequent recharge. The sets of batteries required for each car are expensive, costing several thousand dollars. They also have a limited lifespan and must be replaced every few years which compounds the cost. The most serious and advanced effort on this problem has been pursued by the U.S. Advanced Battery Consortium, formed by the Big Three automakers, General Motors, Ford, and Chrysler. Its aim is to develop batteries that will last five years and cost $6,000. Ford has calculated that their operating cost would be equivalent to gasoline at $3.72 a gallon figuring a penny a mile for recharging. So far such batteries have not been developed.

Recharging batteries also takes time, usually several hours. Although there are some technological advances which suggest this might be done in 15 to 20 minutes, this is still not so quick as a stop at the service station. To eliminate this, it has been suggested that "battery stations" be built and located just as gasoline stations are now, where a quick change from an exhausted battery set to a fully charged set of batteries could be made. To make this practical, because cars are distributed and driven from one end of the country to the other, it would be necessary to design a standard car with a standard battery set. It would not be feasible to have many different types of batteries for a variety of cars.

Research to develop other batteries than the conventional lead-acid battery has included cadmium and lithium batteries.(47) But neither of these elements is abundant in easily recoverable ores in the Earth, and they are considerably more expensive than lead.(106) Also, these batteries which are more powerful than the conventional lead-acid batteries have proved to be dangerous and unstable running at temperatures three times the boiling point of water and liable to explode. Other types of batteries under development are nickel-metal hydride, nickel-iron, sulfur-aluminum, zinc-air, zinc bromine, lithium polymer, and sodium sulfur.(25) The sodium-sulfur battery has been in development for more than 20 years, but it operates at temperatures between 600 and 700 degrees Fahrenheit, requiring expensive insulation. Also, sodium, as well as lithium are explosive when in contact with moisture.(73) Regardless of the type of battery used, the disposal problem of so many batteries would be enormous, although the nickel-metal hydride battery is reported to be recyclable.(77)

Carey, et al. summarized the battery problem:

> "Researchers have yet to come up with a battery without materials that are either toxic, nonrecyclable, or potentially flammable. The main roadblock to wide use of electric vehicles, however, has been that each kilogram of a conventional battery has only one-hundredth the stored energy of a kilogram of gasoline — which means that keeping the battery to a practical size limits the range an electric vehicle can go without recharging."(25)

With the present battery technology, weight for weight, gasoline stores 80 to 100 times more energy than does a battery. It takes a one-ton lead-acid battery to give an electric car the same energy as a gallon of gasoline.(36) It should also be noted that batteries cannot be run down completely to the last bit of energy in them for two reasons. This tends to greatly shorten the life of a battery, and also when the last part of the electric charge is drawn upon, it is weak, and the car will not move very fast if at all. This is in contrast to gasoline as the power fuel which, to the last drop has as much energy as does the first drop. What also must be considered is the extra energy required just to move the one ton of storage batteries in the car, as compared with the minute amount of energy required to move the one gallon of gasoline in the tank along with the car. Improved batteries may have a better performance, but the difference between gasoline and electric storage batteries is likely to remain large. A liter of gasoline provides at least 100 times the energy of a battery taking up the same amount of space.

Both cold and hot weather present particular problems for electric vehicles. The batteries do not hold up well in cold weather, and there is the problem of heating the car's passenger compartment in the occasional colder weather. Even in California's generally moderate climate, in what was called a "bow to reality," automakers won a waiver from the California Air Resources Board which allows them to sell electric vehicles with fuel-fired heaters — tiny engines which burn gasoline or diesel fuel for the purpose of heating the vehicle's interior.(97) Driving in the colder parts of the country, with short winter days, requiring that the cars be driven in the dark with lights on and heater running would cause an energy drain which no batteries at present can meet.

In hot weather, the use of a car's air-conditioner made one automotive engineer comment: "If you have the air-conditioner on, you aren't going to move." This does not say that eventually these various electric car problems cannot be solved, but progress so far has been rather slow. Europeans also have been working on the electric car, but the results there have not been very satisfactory either.(27)

Batteries and the environment

Although electric cars themselves are touted as being environmentally benign compared with gasoline powered vehicles, studies related to the batteries of electric cars indicate there are large environmental problems with the production of the necessary batteries. Lave, et al. state:

> "Electric cars have been criticized for their cost and poor performance as compared with current cars. The more fundamental problem is that these vehicles do not deliver the promised environmental benefits. A 1998 model electric car is estimated to release 60 times more lead per kilometer of use relatively to a comparable car burning leaded gasoline...Electric vehicles will not be in the public interest until they pose no greater threat to public health and the environment than do alternate technologies using low-emissions gasoline. Nickel-cadmium and

nickel metal hydride batteries are much more expensive and highly toxic; they do not appear to offer environmental advantages. Sodium-sulfur and lithium-polymer technologies may eventually be attractive."(64)

Setting aside comments about the lack of true environmental advantage of electric cars, but simply considering the practicalities of these cars, then light-weight electric vehicles can be useful for short runs such as city service, where recharging stations or battery exchange stations are nearby. The heavier electric car with a driving range comparable to gasoline powered vehicles is not yet at hand. There is considerable doubt among the major car manufacturers about the electric car ever replacing today's conventional automobile. In a *Forbes* magazine interview, the chief executive officer of the Ford Motor company states: "'We've spent hundreds of millions on things that didn't work,' says Alex Trotman, Chief executive of the Ford Motor Co., 'But when we spent the money, we didn't know they wouldn't work.' Now he complains, 'Ford is spending hundreds of millions of dollars on something its people know in advance won't work — the electric car.'"(44)

Castaing, an automotive engineer with Chrysler Corporation, has stated that the electric equation for cars just does not work out. "When it is fully charged [the electric car] gets no more energy than when you get into a normal car and the gasoline gauge is on red. It may have some applications for old folks in small retirement communities or as golf cars. Beyond that, for me the EV [electric vehicle] is incomprehensible."

In June 1995, General Motors, Chrysler, and Ford urged the State of California to roll back its electric vehicle mandate stating that the high price and low driving range of any vehicles which might be developed in the near future at least, made them unattractive. In July 1995, these automakers made the same request to New York, Massachusetts, and adjacent states.(97) (It is interesting to note that in 1993, General Motors and Boston Edison announced a project to make electric vehicles use widespread in the northeast.(94) But two years later GM had apparently lost enthusiasm).

The reluctance of car manufacturers to produce a certain percentage of their cars in the form of electric vehicles is because of the high cost and poor performance. People are unwilling to buy them, and the auto makers foresee increasingly large numbers of cars for which there is no market. Reason for opposition to the California mandate on electric cars has been summed up: "The irony of the mandate is this: Auto makers will be required to make electric cars, and auto dealers will be required to stock them, but nobody will be required to buy them."(13) The cost of retooling and setting up assembly lines for a new car is high, and this cost probably could not be recovered if there were few sales. One solution suggested was that the cars be sold at less than cost and the difference to the automakers be made up by having them raise the price of conventional cars. In this way the regular car owners would subsidize the buyers of the electric cars. This idea did not seem to have much popular support.

Recognizing to some extent the problems of electric cars, John Dunlap, chairman of the California Air Resources Board, the rule-setting agency for the law, said that the board planned to relax enforcement of the mandate which had been set for a quota of electric cars. In March, 1996, The California Air Resources Board by unanimous vote, relaxed the previous rules in regard to electric car quotas. The original mandate would have put 22,000 cars into California automobile dealers' lots starting in 1998. The new agreement drops that number to 3,750. the requirement was further modified in the future to "market-oriented" goals under which automobile manufacturers are supposed to "strive to sell thousands of electric vehicles." The Board stated that the agreement will still put 800,000 electric-powered vehicles on California roads by 2010, but there does not seem to be strong evidence that the public will buy the cars.

However, there was still a requirement that a percentage of the cars offered for sale would have to be electric vehicles. In compliance with this in late 1995, Chrysler and Ford agreed on what electric cars they would produce. Chrysler's minivan will carry 27 batteries and limit drivers to only 60 miles between long rechargings. In 1995, Ford unveiled an electric-powered Ranger pickup which costs about $30,000 compared with the $12,000 cost of a gasoline-powered vehicle. It carries 2,000 pounds of batteries which last only 58 miles between rechargings which take six hours. Every three years the batteries have to be replaced at a cost of between $2,000 and $3,000. (Associated Press Dispatch December 12, 1995). At the present this is the state of the art of electric cars. And, although the companies have to make these cars there is no law which requires the public to buy them. The laws mandating building a percentage of the cars as electric, provide that a fine for each such vehicle not produced will have to be paid. "...if the laws stay on the books, automakers say privately they would rather pay the $5,000-per-vehicle annual fine than build the white elephants no one will buy."(102)

General Motors and EV1 (Electric Vehicle 1)

Despite the problems with electric cars, General Motors, in December, 1996, perhaps to ensure its position in the lucrative California car market, launched its EV1, with some restrictions as to the buyers, and some physical limitations.

The car will not be sold, but leased for three years at a cost of between $520 and $620 per month, depending on credits from smog agencies. The car will only be leased in Southern California and Arizona, because the batteries do not function well in colder climates. The person wishing to lease the car (termed "candidate" by the salesman), must be interviewed to see if the car and candidate fit. The candidate profile: Environmentally minded consumer with a minimum $120,000 annual income and two other gasoline-powered cars.

The physical limitations, beyond being unable to function in a cold climate, include a lightweight aluminum cage frame which replaces the conventional body of welded steel. This frame would not sustain a collision as would a steel body, and the car is not structurally strong enough to be towed with a tow bar. If it has to be towed for any reason, it is so low that it cannot be lifted up by a conventional tow truck, because one end would drag. It must be carefully lifted onto a flatbed truck and hauled away. It will go between 70 and 90 miles on a single battery charge, and then needs three to twelve hours to recharge. A recharging device costs about $2,000, plus installation charges, which typically run from $1,000 to $3,000.

As a newspaper headline describing the vehicle stated, "EV1 Not For Everyone." The contrast between this vehicle and all other proposed electric cars, and the conventional automobile remains large.

In January 1997, Chrysler Corporation announced plans for a "hybrid electric car." It is said to be powered by a fuel cell which produces electricity from hydrogen extracted from gasoline while the car is being driven. It is "expected to be at least 50 percent more fuel efficient and 90 percent cleaner than a modern gasoline-powered internal combustion engine." However, it does not solve the dependency on oil, as it would still rely on gasoline, a non-renewable fossil fuel.

The convenience and relatively low cost of fifteen gallons of gasoline weighing less than 100 pounds which will allow a car of today to travel 400 or more miles at 60 miles an hour without stopping will be hard to replace. Transition to alternative fuels will be more difficult than most people realize. You cannot put solar energy into your gasoline tank and have it be a simple substitute for gasoline, even if the solar energy has been converted to electricity. Oil, at present, supplies 97 percent of the energy needed to move transportation in all its various forms. If there was a viable equivalent substitute for it, it probably would have been on the market by this time. None is apparent.

Other vehicles

Even with its problems, the car might, theoretically at least, be the easiest vehicle to become electric. Much larger difficulties are foreseen in producing electric vehicles such as the great variety of trucks which have to travel many miles in remote areas on steep grades with very heavy loads, or the semi-tractor trailers with two or three trailers on interstate highways, and on transcontinental hauls. Producing electric heavy farm equipment or large bulldozers which could be used in remote areas far from any battery exchange or recharging stations also present problems. The convenience of hauling large quantities of diesel fuel or gasoline with a gasoline or diesel powered truck for use in remote locations is unmatched by any other known energy source. Cargo ships and ocean liners can be powered by electricity using nuclear reactors to generate steam to power turbine-generators, but powering an airplane in that fashion seems unlikely. Although it has a number of problems chiefly related to fuel storage, which the use of gasoline does not have, hydrogen at present appears to be a better alternative fuel than electricity for vehicles.

In Broad View

The most important conclusion which comes from this survey of alternative energy sources is that there is no single adequate energy substitute on the horizon for the conventionally produced quantities of oil and gas which we use in the world today. Also, replacing petroleum by other energy sources, to any significant degree, will involve very large investments in new technology and equipment. It will also cause considerable changes in lifestyles. The ease with which oil and its important transport fuel derivatives, gasoline, diesel, and jet fuel (kerosene), can be handled and used is unmatched by any other fuel source. There is nothing available which can produce 62 million barrels a day of a liquid fuel to replace that amount of crude oil which the world now uses in myriad ways. This includes the making of petrochemicals. Electricity will not do it.

Energy and the U.S. economic engine

U.S. economic history shows that the cheap energy which the country had the good fortune to inherit has been a major force in growth of industry and all that goes with it. Lippman states "...the hard fact [is] that low-cost energy is the great engine that empowers American industry and makes possible the air-conditioned, multiple-car American way of life. In the world of manufacturing and transportation, cheap energy is not some environmentally incorrect frill, it is an essential ingredient that gives U.S. industry an edge in global competition."(25) Now increasingly dependent on foreign oil, can the U.S. maintain that economic edge?

As the world's largest user of oil, both on a per capita basis, and in total national consumption, the development of alternative energy sources is critical to the future of the United States. Little progress has been made. Senator Mark O. Hatfield, member of the U.S. Senate Committee on Energy and Natural Resources, stated in 1990: "Current debates over where and how to drill for oil in this country soon may be rendered irrelevant by a nation desperate to maintain its quality of life and economic productivity. War over access to the diminishing supply of oil may be inevitable unless the United States and other countries act now to develop alternatives to their dependence on oil."

Future energy mix

Because no single alternative energy source appears able to replace oil, the future energy mix will be considerably more diversified than it is at present with petroleum (oil and gas) now carrying more than half the load. Also, it is unlikely that other energy sources will be as convenient to use as petroleum is today. And they may be more expensive. The energy mix has changed markedly the past two hundred years for the world as a whole. Some areas, however, still use energy sources (biomass largely) which they have used for thousands of years. In the kinds of energy sources now being used, parts of the world are centuries apart from other regions. This disparity will probably continue for decades.

Nations and the changes in energy mix

The survival of nations with regard to energy will have a great influence on how well they survive in general. It will involve an increasing amount of technology combined with every energy resource they can obtain and a willingness of the nation's people to gracefully and orderly adjust to the change in lifestyle which alternative energy sources will require.

It seems clear that the future will see a less concentrated and more diffuse mix of energy sources than at present. Petroleum now carries a very large load in the industrial world. But as this energy form is gradually depleted that portion of the energy demand which petroleum now supplies must be gradually assumed by several rather than one other energy source. This may actually result in healthier national economies because no one energy supply would be so critical as petroleum is now. Nations may then not be so vulnerable to cut-offs of energy supplies as has been the case in the recent past.

If one takes the very long view of the energy situation, renewable energy resources must ultimately be the energy sources. But in the interim, which probably means several centuries to come, other energy minerals will still be the most important alternatives to the major role which petroleum now plays. Coal, shale oil (maybe), oilsands, tarsands, heavy oil, geothermal resources, atomic minerals (chiefly uranium and thorium), and hydroelectric power will be utilized. Renewables will gradually increase in use. How changes to alternative energy sources can be accomplished by the different nations will have much to do with their futures. The energy mix for any given nation and for the world will continue to change, offering many challenges to both technology and world economics in the process. International competition for what energy resources do exist will continue, and probably intensify. The Gulf War is one indication of what may lie ahead.

Population

In looking at all alternative energy sources, the factor of population should also be considered. Will the alternatives be of significant use considering the size of the population demanding energy? World population is expected to double by 2050. Can double the amount of energy used today be available then? Even if this were possible it would not mean any increase in the amount of energy available to the low-energy users in the Third World. In the excellent, comprehensive contemporary review of alternative energy sources, the word "population" does not appear.(22) Perhaps in a purely technical book on energy sources, such a consideration should not enter in, but somewhere the connection has to be made. In considering the move to alternative fuels, it is doubtful the reasonably foreseeable technology applied to all alternative energy sources would maintain the present world population in its standard of living, much less raise that of the half of the world which now has only a substandard existence.

Environment and energy

One of the presumed attractions some alternative energies have over conventional energy sources is their lower level of environmental impact. This is an important issue and requires honest evaluation. Hubbard makes a point: "One important fact highlighted by the [Department of Energy] studies is that no energy technology is completely environmentally benign. For example, although photovoltaic cells emit no pollutants during operation, their manufacture requires large quantities of hazardous materials, and their ultimate disposal could release toxic elements, such as arsenic, and cadmium to the environment."(55) Every energy source has some environmental impact. The largest environmental impact on the Earth is by population, and its continued growth, which demands more and more energy in whatever forms it can be obtained.

With Some Reservations

As the reader may have noticed, this discussion of alternative energy sources has shown only moderate enthusiasm for some of them, based on present information and known technology. When the huge storehouse of fossil energy accumulated over millions of years is now having to be drawn upon in a geological instant to sustain the present population, the prospects of transferring from that energy source to sunshine and wind on merely a daily input basis appear daunting. Ocean thermal electric conversion (OTEC), also a renewable resource, likewise is not visualized as becoming a major energy supply. Nuclear fusion, if accomplished, would change this outlook, but without that technological breakthrough, to have only solar, wind, biomass, hydro and OTEC carry the world economy with its present and rather rapidly expanding population seems difficult indeed. Biomass is regarded by some as a renewable energy source. But the soil degradation together with its competition with food supplies do not suggest biomass is a renewable resource for the longer term to sustain the energy demand now being supplied by 71 million barrels of oil (crude oil and natural gas liquids) each day or the more than 76 trillion cubic feet of gas used annually in the world.(6)

The two giant steps

The transition from petroleum to alternative fuels involves two very large steps. The first is going from drilling for a liquid or a gas, as is done now, to producing energy from sources other than from wells. This is not just a large step, it is a huge step. Eliminating the prospect that biomass will ever supply very much liquid fuel, the second giant step is to realize that the other energy sources ultimately produce electricity, not a convenient liquid fuel. Using electricity as an alternative to oil as it is used in transport fuels involves not a short step, but a very long one. The public does not generally realize the magnitude of these necessary two steps, tending to regard them as relatively easy, simple transitions. They are not.

Energy source transition

In what Starr et al. call a realistic outlook on energy sources, they observe that historically it usually takes about 50 years to significantly shift fuel patterns. They state:

> "The potential role of the nonfossil and renewables in the future global energy mix depends on their developing economic competitiveness. This category included biomass, solar, wind, geothermal, and the two commercial electricity sources, hydro and nuclear. Only hydro and nuclear are significant contributors today, with hydro about 20% of global electricity and nuclear about 17%. There are practical upper bounds for the potential contribution of the nonfossil and renewable sources summarized here.
>
> "Both the energy input to manufacture the renewables and their initial capital cost are the issues. A basic consideration is net energy output, or the output minus the energy input from other resources required for their manufacture. This is particularly relevant to biomass, where the energy input for their growth (for example, fertilizer and irrigation) and processing are substantial...As yet, renewables such as solar, wind, and biomass have been able to penetrate only limited niche markets, with much uncertainty about their net energy contribution."(93)

Net energy, and the energy profit ratio

An essential consideration in evaluating the usefulness of alternative energy sources, and in fact, any energy source, is the question of net energy recovery. All energy sources take energy to produce. Even solar energy requires equipment which uses energy to manufacture the solar energy equipment. Gever, et al., have made a special attempt to determine what they term the "energy profit ratio" which is the ratio of the amount of energy which goes into the production of a particular energy source as compared with how much comes out.(46) As Starr, et al., noted, the net energy recovery from biomass is relatively low. In the case of ethanol it is an energy negative. The enthusiasm for biomass, in particular, as a major energy source as advocated by Johansson et al.,(59) and Driscoll,(37) is not generally shared by Starr, et al., Pimentel, and others.

Alternative energy sources will increasingly have to be used. It is imperative to determine realistically how useful they can and will be in terms of the size of the population which they could support, and in what standard of living and lifestyle.

Energy and the future

There are a number of alternative energy sources available. But developing them will take time. They will not in many cases be either as convenient, safe, nor inexpensive as currently used energy sources. And, they may not be able to completely fill the gap left by depletion of the non-renewable sources drawn from great accumulations of the past.

Whatever energy sources are used in the future, they will be, as they are now, the key to unlocking all other physical resources. How much energy each person can command directly or indirectly determines the individual's material standard of living. Minerals, including soil and water are the basics to human survival. Energy determines how efficiently these material resources can be obtained and used. Energy sources and quantities potentially available must be examined realistically and projected into the future with respect to population. Failure to do so will be disastrous.

The U.S. Federal Energy Information Administration (a Division of the Department of Energy) has estimated that if present trends continue in the U.S., within 10 years the United States will be importing at least 60 percent of its oil, and have an annual oil import bill of $100 billion which will undermine any action to reduce the trade deficit. At that time the Persian Gulf nations will control two thirds of the world's oil available for export, and will have an annual income of $200 billion. From this it is clear that alternative energy sources should be immediately and vigorously pursued.

"...the concepts of 'civilization' and 'controlled use of energy' are inseparable."

Harrison Brown
The Human Future Revisited

BIBLIOGRAPHY

1 ABLESON, P. H., 1976, Energy From Biomass: Science, v. 191, n. 4233, March 26.

2 ABLESON, P. H., 1979, Bio-Energy: Science, v. 204, n. 4398, June 15.

3 ABLESON, P. H., 1980, Energy from Biomass: Science, v. 208, n. 4450, June 20.

4 ABLESON, P. H., 1991, Improved Yields of Biomass: Science, v. 252, n. 5012, June 14.

5 ABLESON, P. H., 1995, Renewable Liquid Fuels: Science, v. 268, n. 5213, May 19.

6 AMERICAN PETROLEUM INSTITUTE, 1995, Basic Petroleum Data Book: American Petroleum Institute, Washington, D. C., v. 15, n. 1, (no pagination, large volume).

7 ANONYMOUS, 1979, Gasohol: A Mirage in Energy Desert: Oregonian, Portland, Oregon, August 23.

8 ANONYMOUS, 1991, Fill 'Er Up — With Hydrogen, Please: Business Week, March 4, p. 59.

9 ANONYMOUS, 1993, Charge! Electric Cars Have Been Tomorrow's Technology for a Long Time. Is Tomorrow About to Arrive?: The Economist, October 23, p. 105-106.

10 ANONYMOUS, 1994, California Forcing Electric Car Sales: Oil & Gas Journal, May 9, p. 17.

11 ANONYMOUS, 1994, California Electric Vehicle Mandate Reaffirmed Amid Controversy: Oil & Gas Journal, May 30, p. 25.

12 ANONYMOUS, 1995, Alternative Motor Fuels: A Slow Start Toward Wider Use: Oil & Gas Journal, February 20, p. 25-28.

13 ANONYMOUS, 1995, California's Electric Car Mandate Draws Fire: Oil & Gas Journal, April 24, p. 44.

14 ANONYMOUS, 1995, Court Tosses Out EPA's Renewable Fuel Rule: Oil & Gas Journal, May 8, p. 32.

15 AVERY, W. H., and WU, C., 1994, Renewable Energy From the Ocean. A Guide to OTEC: Oxford Univ. Press, New York, 480 p.

16 BAILEY, JEFF, 1995, 'Perpa Power': The Wall Street Journal, May 17.

17 BARTLETT, A. A., 1978, Forgotten Fundamentals of the Energy Crisis: American Journal of Physics, v. 46, n. 9, p 876-888. Note: A 65 minute video titled "Arithmetic, Population, and Energy," based on the above article, and narrated by Dr. Bartlett, can be purchased from Academic Media Services, Campus Box 379, University of Colorado, Boulder, Colorado 80309-0379.

18 BERNTON, HAL, et al., 1982, The Forbidden Fuel. Power Alcohol in the Twentieth Century: Boyd Griffin, New York, 269 p.

19 BOVARD, JAMES, 1995, Dole, Gingrich and the Big Ethanol Boondoggle: The Wall Street Journal, November 2.

20 BRAUN, DICK, 1979, Gasohol Goes Commercial in the U.S.: Farm Journal (Western Edition), p. 33A-33B.

21 BRAUNSTEIN, H. M., et al., 1981, Biomass Energy Systems and the Environment: Pergamon Press, New York, 182 p.

22 BROWER, MICHAEL, 1994, Cool Energy. Renewable Solutions to Environmental Problems: MIT Press, Cambridge, Massachusetts, 215 p.

23 BROWN, L. R., 1980, Food or Fuel: New Competition for the World's Cropland: Worldwatch Paper 35, Worldwatch Institute, Washington, D. C., 43 p.

24 BROWN, L. R., et al., 1994, State of the World 1994: W. W. Norton & Company, New York, 265 p.

25 CAREY, JOHN, et al., 1994, Where's the Juice?: Business Week, May 30, p. 110.

26 CHEREMISINOFF, N. P., 1979, Gasohol For Energy Production: Ann Arbor Science Publishers, Ann Arbor, Michigan, 140 p.

27 CHOI, AUDREY, and STERN, GABRIELLA, 1995, The Lessons of Reugen: Electric Cars are Slow, Temperamental and Exasperating: The Wall Street Journal, March 30.

28 CLARK, WILSON, 1974, Energy for Survival: The Alternative to Extinction: Anchor Press, Garden City, New York, 652 p.

29 CONANT, MELVIN, 1979, Access to Energy — 2000 and Beyond: Univ. Kentucky Press, Lexington, 134 p.

30 CORSON, W. H., (ed.), 1990, The Global Ecology Handbook. What You Can Do About the Environmental Crisis: Beacon Press, Boston, 414 p.

31 CRABBE, DAVID, and McBRIDE, RICHARD, (eds.), 1978, World Energy Book. An A-Z Atlas and Statistical Source Book: Nichols Publishing Co., New York, 259 p.

32 CRUVER, P. C,, 1989, Hydrogen. Tomorrow's Limitless Power Source: The Futurist, November/December, p. 24-26.

33 CUFF, D. J., and YOUNG, W. J., 1980, The United States Energy Atlas: The Free Press, Macmillan Publishing Co., Inc., New York, 416 p.

34 DANIELS, FARRINGTON, 1964, Direct Use of the Sun's Energy: Yale Univ. Press, New Haven, Connecticut, 374 p.

35 DUEDNEY, DANIEL, and FLAVIN, CHRISTOPHER, 1983, Renewable Energy. The Power to Choose: W. W. Norton & Company, New York, 431 p.

36 DEUTSCH, C. H., 1990, As Oil Prices Rise, the Hydrogen Car is Looking Better: New York Times, August 26.

37 DRISCOLL, W. L., 1993, Fill 'Er Up with Biomass Derivatives: Technology Review, August/September, p. 74-76.

38 DUNKERLEY, JOY, et al., 1981, Energy Strategies for Developing Nations: Resources for the Future, Washington, D. C., 265 p.

39 DUNN, P. D., 1978, Appropriate Technology. Technology With a Human Face: Schocken Books, New York, 220 p.

40 DUNNE, JIM, 1994, The Alternate Fuel Report: Hydrogen-Powered Miata: Popular Mechanics, October, p. 112.

41 FLAMSTEED, SAM, 1992, H_2OH!: Discover, February, p. 78-80.

42 FLAVIN, CHRISTOPHER, and LENSSEN, NICHOLAS, 1990, Beyond the Petroleum Age: Designing a Solar Economy: Worldwatch Paper 100, Worldwatch Institute, Washington, D. C., 65 p.

43 FLAVIN, CHRISTOPHER, and LENSSEN, NICHOLAS, 1994, Power Surge. Guide to the Coming Energy Revolution: W. W. Norton & Company, New York, 382 p.

44 FLINT, JERRY, 1994, Don Quixote, Meet Alex Trotman: Forbes, January 3.

45 FUMENTO, MICHAEL, 1987, Robber Barons: National Review, March 13, p. 32-38.

46 GEVER, JOHN, et al., 1986, Beyond Oil. The Threat to Food and Fuel in the Coming Decades: Ballinger Publishing Company, Cambridge, Massachusetts, 304 p.

47 GLANZ, JAMES, 1994, Lithium: Battery Takes to Water — And Maybe the Road: Science, v. 242, May 20, p. 1084.

48 GORDON, DEBORAH, 1991, Steering A New Course. Transportation, Energy, and the Environment: Union of Concerned Scientists, Cambridge, Massachusetts, 228 p.

49 GREISING, DAVID, and HONG, PETER, 1992, Big Stink on the Farm. The Issue: Ethanol Subsidies. The Players: Agribusiness and Big Oil: Business Week, July 20, p.31.

50 GRIGGS, G. B., and GILCHRIST, J. A., 1977, The Earth and Land Use Planning: Duxury Press, Belmont, California, 492 p.

51 HAGEL, JOHN, 1976, Alternative Energy Strategies: Constraints and Opportunities: Praeger, New York, 185 p.

52 HASS, NANCY, 1992, Faust Triumphant: Financial World, December 8, p. 22-23.

53 HAYES, DENIS, 1977, Rays of Hope. The Transition to a Post-Petroleum World: W. W. Norton & Company, New York, 240 p.

54 HILLER, E. A. and STOUT, B. A., 1985, Biomass Energy. A Monograph: Texas A&M University Press, College Station, Texas, 313 p.

55 HUBBARD, H. H., 1991, The Real Cost of Energy: Scientific American, April, p. 36-42.

56 HUBBERT, M. K., 1962, Energy Resources: National Academy of Sciences, National Research Council Publication, 1000-D, Washington, D. C., 141 p.

57 HUBBERT, M. K., 1971, The Energy Resources of the Earth: Scientific American, September, p. 60-70.

58 HUBBERT, M. K., 1976, Energy Resources: A Scientific and Cultural Dilemma: Bull. of the Association of Engineering Geologists, v. 13, n. 2, Spring, p. 81-124.

59 JOHANSSON, T. B., et al., (eds.), 1992, Renewable Energy: Sources for Fuels and Electricity: Island Press, Washington, D. C., 1160 p.

60 JOHNSON, ROBERT, 1993, Electric Utilities Study an Old, New Source of Fuel: Firewood: The Wall Street Journal, December 2.

61 KANDELL, JONATHAN, 1989, Brazil's costly mix: Autos and Alcohol: The Wall Street Journal, September 28.

62 KNOTT, DAVID, 1994, Hydrogen: The Fuel of the Future?: Oil & Gas Journal, May, p. 26.

63 KRANZBERG, MELVIN, et al., 1980, Energy and The Way We Live: Boyd & Fraser Publishing Company, San Francisco, 496 p.

64 LAVE, L. B., et al., 1995, Environmental Implications of Electric Cars: Science, v. 268, May 19, p. 993-995.

65 LENSSEN, NICHOLAS, 1992, Empowering Development: The New Energy Equation: Worldwatch Paper 111, Worldwatch Institute, Washington, D. C., 57 p.

66 LEVINE, J. S., 1991, Global Biomass Burning. Atmospheric, Climatic, and Biospheric Implications: MIT Press, Cambridge, Massachusetts, 569 p.

67 LINCOLN, J. W., 1980, Driving Without Gas: Garden Way Publishing, Charlotte, Vermont, 150 p.

68 McDANIELS, D. K., 1994, The Sun: Our Future Energy Source, Second Edition: Krieger Publishing Company, Malabar, Florida, 346 p.

69 McMENAMIN, BRIGID, 1991, Political Greenmail: Forbes, May 27, p. 72.

70 MEYERS, R. A., (ed.), 1983, Handbook of Energy Technology & Economics: John Wiley & Sons, New York, 1089 p.

71 MILLER, MARSHALL, 1994, A Fuel for All Reasons: Nucleus, Fall issue, p. 1-3.

72 NAIK, GAUTUM, 1993, Scientist Envisions a 'Hydrogen Economy": The Wall Street Journal, February 11.

73 NAJ, A. K., 1993, Battery May Make Electric Vehicles Much More Viable: Wall Street Journal, August 20.

74 NOAH, TIMOTHY, and KILMAN, SCOTT, 1995, Archer, Ethanol Industry Dealt a Blow as Court Blocks EPA Gasoline Order: The Wall Street Journal, May 1.

75 NOAH, TIMOTHY, 1995, Clinton Expands Ethanol Tax Break in Boon for Archer: The Wall Street Journal, August 7.

76 NOVAK, R. D., 1995, The Ethanol Outrage: Washington Post, October 5.

77 OVSHINSKY, S. R., et al., 1993, A Nickel Metal Hydride Battery for Electric Vehicles: Science, v. 260, April 9, p. 176-181.

78 PERRIN, NOEL, 1992, Life With an Electric Car: W. W. Norton & Company, New York, 191 p.

79 PIMENTEL, DAVID, and KRUMMEL, J., 1987, Biomass Energy and Soil Erosion: Assessment of Resource Costs: Biomass, v. 14, p. 15-38.

80 PIMENTEL, DAVID, 1991, Ethanol Fuels: Energy Security, Economics, and the Environment: Journal of Agricultural and Environmental Ethics, p. 1-13.

81 PIMENTEL, DAVID, 1993, Environmental and Economic Benefits of Sustainable Agriculture: Socio-economic and Policy Issues for Sustainable Farming Systems: Cooperativa Amicizia S.r.1., Padova, Italy, p. 5-20.

82 PIMENTEL, DAVID, 1994, Renewable Energy: Economic and Environmental Issues: BioScience, v. 44, n. 8, p. 536-547.

83 PIMENTEL, DAVID, 1995, Letter dated June 6.

84 RAPP, DONALD, 1981, Solar Energy: Prentice-Hall, Englewood Cliffs, New Jersey, 516 p.

85 REILLY, P. J., 1988, Letter dated March 10.

86 RENNER, MICHAEL, 1988, Rethinking the Role of the Automobile: Worldwatch Paper 84, Worldwatch Institute, Washington, D. C., 70 p.

87 RIDGEWAY, JAMES, 1982, Powering Society: The Complete Energy Reader: Pantheon Books, New York, 365 p.

88 ROSS, DAVID, 1995, Power From the Waves: Oxford University Press, New York, 224 p.

89 RUEDISILI, L. C., and FIREBAUGH, M. W., (eds.), 1978, Perspectives on Energy. Issues, Ideas, and Environmental Dilemmas: Oxford Univ. Press, New York, 591 p.

90 SANCHEZ, SANDRA, 1995, Movement Is In the Air As Texas Taps the Wind: USA Today, November 15.

91 SLESSER, MALCOLM, and LEWIS, CHRIS, 1979, Biological Energy Sources: E & F. N. Spon Ltd., London, 192 p.

92 SPERLING, DANIEL, 1988, New Transportation Fuels. A Strategic Approach to Technological Change: Univ. California Press, Berkeley, 532 p.

93 STARR, CHAUNCEY, et al., 1992, Energy Sources: A Realistic Look: Science, v. 256, May 15, p. 981-987.

94 STIPP, DAVID, 1993, GM and Utility Mount Charge on Electric Cars: The Wall Street Journal, May 19.

95 STOBAUGH, ROBERT, and YERGIN, DANIEL, (eds.), 1979, Energy Future: Ballentine Books, New York, 493 p.

96 STOBAUGH, ROBERT, and YERGIN, DANIEL, (eds.), 1983, Energy Future (Third Edition): Vantage Books, Random House, New York, 459 p.

97 SURIS, OSCAR, 1994, Cold Weather Still a Problem in Electric Cars: The Wall Street Journal, March 8.

98 SURIS, OSCAR, 1994, Electric Cars Also Pollute Air, EPA Study Says: The Wall Street Journal, April 5.

99 SURIS, OSCAR, 1995, Big Three Fight Sales Mandates for Electric Cars: The Wall Street Journal, July 3.

100 SWAN, C. C., 1986, Suncell. Energy, Economy & Photovoltaics: Sierra Club Books, San Francisco, 237 p.

101 TAYLOR, ALEX, III, 1993, Why Electric Cars Make No Sense: Fortune, July 26, p. 126-127.

102 TAYLOR, ALEX, III, 1995, Why the Electric Car is Running Out of Energy: Fortune, November 13, p. 56.

103 TELLER, EDWARD, 1979, Energy From Heaven and Earth: W. H. Freeman and Company, San Francisco, 322 p.

104 TURNER, DAVID, 1993, Letter to the Editor, Register-Guard, Eugene, Oregon, March 12.

105 VEZIROGLU, T. N., and BARBIR, F., 1992, Hydrogen: The Wonder Fuel: International Journal Hydrogen Energy, v. 17, p. 391-404.

106 VINE, J. D., (ed.), 1970, Lithium Resources and Requirements By the Year 2000: U.S. Geological Survey Professional Paper 1005: Washington, D. C., 162 p.

107 WICKELGREN, INGRID, 1995, Sunup at Last for Solar? Business Week, July 24, p. 84, 86.

108 WILSON, C. L. (project director), 1977, Energy: Global Prospects 1985-2000: McGraw-Hill Book Company, New York, 291 p.

109 WINSTON, D. C., 1979, There Goes the Sun: Newsweek, December 3, p. 35.

110 WOODRUFF, DAVID, and REGAN, M. B., 1994, Is It Too Soon to Jump-Start Electric Cars?: Business Week, March 21, p. 36.

111 WOODRUFF, DAVID, et al., 1994, Electric Cars. Will They Work? And Who Will Buy Them?: Business Week, May 30, p. 104-107, 110-111, 114.

112 YERGIN, DANIEL, and HILLENBRAND, MARTIN, 1982, Global Insecurity. A Strategy for Energy & Economic Renewal: Houghton Mifflin Company, Boston, 427 p.

113 ZWEIBEL, KEN, 1993, Thin-film Photovoltaic Cells: American Scientist, v. 81, p. 362-369.

An excellent comprehensive review of renewable energy sources is by Richard Golob and Eric Brus, 1990, *The Almanac of Renewable Energy. The Complete Guide to Emerging Energy Technologies:* Henry Holt and Company, New York, 348 p.

CHAPTER 15

Water — Life's Essential Connection to the Earth

There are two Earth resources which are absolutely essential to human existence. One is water. The other is soil. Water is the connection between humans, and indeed all life, with the Earth. No organism can live without water. Water is the medium by which minerals in the soil are put into solution to be taken up by plants and in turn by animal life all of which is dependent on plants. Water, therefore, connects all life to the minerals in the Earth on which life depends. The average adult human body is approximately 60 percent water. It is a common expression that "blood is thicker than water," but not by much. Blood which carries nourishment to all cells in the body is more than 90 percent water.

People existed many thousands of years before they used metals. They existed many years before they used oil, and presumably the human race will exist many thousands of years after oil is gone. There are some substitutes for some of the various uses of metals and oil. Where there are no substitutes, this fact would be very inconvenient, but it would not be fatal. But every living thing depends on water, for which there is no substitute.

Water enters into our daily life in myriad ways. It takes about 40 gallons of water to put one egg on the table, 3,600 gallons of water to grow a bushel of corn, 150 gallons to produce a loaf of bread, 375 gallons for five pounds of flour, and about 5,000 gallons to make one pound of beef. For a summer barbecue, two pounds of beef represent 10,000 gallons of water, four ears of corn are 240 gallons, French fries are 24 gallons, and four servings of watermelon used 400 gallons. The family of four can then sit down to a 10,664 gallon of water meal.(4)

The Sunday paper has taken about 280 gallons of water to produce, and one automobile, in all the processing of materials which went into its manufacture, used about 100,000 gallons of water. Then there are all the water usages around the home. In the United States and Canada, and also to a considerable extent in Europe, people have generally had the luxury of ample water supplies. But this is not true of many other parts of the world such as north and east Africa, the Middle East, and central Australia.

Chronic water shortages are a fact of life for some 40 percent of the world's population living in eighty countries.(8) Africa has the most water-short countries, and as the fixed supply of water must support a rapidly increasing population, water shortages will only become more severe.(18) In the semi-desert areas of the Middle East, water has been in critical supply for centuries and has controlled the course of civilization there. It promises to be even more critical in the future and the probable cause of conflict.(11)

Wide differences now in domestic water use

Personal water use differs greatly among countries. According to the World Resources Institute, the four nations which use the most water daily per capita are Australia 476 gallons, United States 159 gallons, Canada 125 gallons, and Sweden 120 gallons. At the other end of the range of water usage are many African countries including Rwanda three gallons, Ethiopia three gallons, Uganda two gallons, Somalia two gallons, and The Gambia one gallon. More than half the world's population has to make use of less than 25 gallons a day, and there is no assurance that the water is safe to drink.

World water distribution

Fresh water is what is useful to humans and this makes up only a small fraction of the total world water supply. The oceans hold 97.5 percent. The other 2.5 percent is fresh water.(20) Of this 2.5 percent fresh water, more than 99.5 percent is locked up in the form of ice in the Antarctic and Greenland ice caps, and in glaciers elsewhere, and deep within the Earth, or in the atmosphere. How much is ice is illustrated by the U.S. Geological Survey's calculations that if the Antarctic ice cap were gradually melted it could keep all the rivers of the world flowing for 750 years. It is estimated that only about .003 percent of all the water on Earth is usable water, and of the fresh water, less than one percent is readily available for human use.

Fresh water resources are divided into two categories: surface waters and waters under ground — groundwater. Each has certain special uses, and problems of development and usage.

Groundwater

Groundwater is water which exists in the Earth. About 22.4 percent of the freshwater of the world is groundwater.(21) Some of it is at relatively shallow depths, but about two-thirds of it is believed to occur below 2,500 feet. Although we are discussing fresh water, this is not pure water. Pure water does not occur in nature. Rain as it falls picks up dust particles and carbon dioxide from the air. Water in streams and lakes naturally contains dissolved mineral material.

Groundwater mineralized

Groundwater always has dissolved mineral material. In general, the deeper the groundwater is, the more dissolved minerals it contains to the point where it may not be usable without special treatment. This increased mineral content with depth is due to two factors. The Earth is hotter deeper. This increase in temperature with depth is called the geothermal gradient, and heats the groundwater which gives it a greater ability to dissolve minerals. Also, groundwater at greater depth is more likely to have been in the ground longer than water at shallower depth. This allows more time for the deeper, and hotter waters to dissolve more minerals. Deeper waters tend to be more mineralized than those at shallower depths, sometimes to the extent that they cannot be used for human consumption or growing crops.

The solution and subsequent depositional action of groundwater in some places has provided spectacular sights. One such is the great travertine terraces near the north entrance to Yellowstone National Park, still forming from the millions of gallons of the mineral-rich hot water which pour out every day. Dissolving the mineral calcium carbonate, limestone, and then redeposition of some of it, has formed the world's largest cavern system near Carlsbad, New Mexico, with huge stalactites and stalagmites, and other forms of dripstone.

How groundwater occurs

Contrary to some common ideas, groundwater does not exist as smaller streams, rivers or lakes in the ground, but exists in pores and cracks in rocks and sediments. Groundwater is commonly recovered for use by humans by means of wells. Some is obtained from springs. The United States uses more than 100 billion gallons of water a day, and of this about 20 percent is groundwater. In some areas when the shallow groundwater supplies are depleted, deeper and deeper wells are drilled. Pumping water from deeper levels takes considerable energy. This energy cost may increase with depth eventually to the point where the costs of pumping may exceed the value of the water for whatever purpose it may be used. So there are practical limits as to how deep usable and economic water can be obtained.

Groundwater recharge

The source of all fresh water, whether surface water, groundwater, or ice, is precipitation in the form of rain or snow. Groundwater resources in some circumstances are replenished quite rapidly. However, in other geologic and climatic settings it may take decades or even centuries to replenish. The amount of precipitation which becomes groundwater in an area is a relatively small part of the annual rainfall. If a region gets 30 inches of rain a year, it is estimated that less than an inch is stored as groundwater. In hot desert regions, the little rain which reaches the ground will almost all evaporate before it can percolate into the ground.

Mining groundwater

If recharge of water to an aquifer does not equal the amount which is being withdrawn, the water is said to be mined. This is an important reality for the arid and semi-arid regions which now depend on wells for irrigation water. If an area does not have sufficient annual rainfall to grow an annual crop, then withdrawing enough water to grow a crop means withdrawing for one yearly crop the water which accumulated over a number of years. In some places the water being used has been in the ground for thousands of years. Using many years of groundwater recharge in just one year cannot continue indefinitely. The water supply will eventually be exhausted, or the level will drop so low that it cannot be economically brought to the surface. There is an exception to this rule. In some cases the recharge area of an aquifer beneath an arid area may be in a more humid region, perhaps in mountain foothills some distance away. The water migrates to beneath the more arid region where it is used. However, migration of groundwater is slow at best, and it is easy to pump out more water than can migrate in. Accordingly, new areas brought into irrigation from groundwater may not be sustainable over the longer term. The groundwater then is like oil, it is a one-time crop. This is already evident in a number of regions. Ironically, the country which is pumping the most oil, Saudi Arabia, is also a leader in pumping out water which accumulated in past more moist times. This water lies deep and it is non-renewable in the human scale of time. There is a high energy cost of raising water from considerable depths but Saudi Arabia has the energy. However, much of the water is used to grow crops which Saudi Arabia could more cheaply purchase than pumping the water to grow them. Wheat is one. Saudi Arabia is now the world's seventh largest exporter of wheat. It must be sold at world market price which is about one-quarter of the Saudi cost. The government absorbs the loss and in this way is actually subsidizing the mining of its groundwater.(28) The use of a non-renewable resource to produce a surplus of a product which has to be sold at less than cost is rather short-sighted, especially when the non-renewable resource is water in a desert.

Importance of Groundwater

Groundwater is a very important resource in many parts of the world. In most underdeveloped countries, wells are the chief source of domestic water supply. More than half the world's population depends on groundwater for drinking water. In desert areas, with no permanent surface waters, groundwater is the only source of water available year around. For thousands of years, desert oases with a spring have sustained populations which otherwise could not have existed there. In semi-arid regions groundwater is an important supplement to the less than adequate rainfall, and allows crops to be grown where they otherwise could not survive. The High Plains of western interior United States are an example.

Groundwater in the United States

Both the United States and Canada have been exceedingly well endowed with groundwater resources. The rapid and widespread settlement of eastern and central United States was greatly facilitated by the availability of groundwater. Over one-third of the irrigated land in the United States is watered by groundwater, and more than 90 percent of people in rural America use groundwater for their domestic purposes. The windmill combined with wells drilled in more remote areas of western United States have kept ranchers in business, as cattle have been watered which otherwise, without the wells, could not have survived. About 20 of the larger cities of the United States get most if not all of their water from wells, and in twelve states groundwater provides more than half of the total public water supplies. In the United States, nearly half the population still obtains its drinking water from groundwater, and farms and smaller communities almost all obtain their drinking water from the ground.

The United States has been blessed with some exceptional natural aquifers. The glacial drift of the northern states is an easily-tapped shallow groundwater aquifer of excellent quality, which tends to be renewed each year with the rains and melting snows. And there are some deeper aquifers. The famous Dakota Sandstone of South Dakota at one time sustained flowing wells all the way east to the Minnesota border, but as more wells were drilled, this aquifer has had its pressure reduced substantially and many once-flowing wells flow no more.

The Ogallala Formation of the High Plains

The extensive Ogallala Formation, a huge area of water-saturated strata of sand and gravel, lies at shallow depth beneath parts of northern Texas, Oklahoma, New Mexico, Kansas, Colorado, Nebraska, Wyoming, and South Dakota. It is the principal geological unit in the High Plains aquifer which underlies 174,000 square miles. This aquifer, the largest and most heavily pumped groundwater system in North America, has a maximum saturated thickness of about 1,000 feet, and an average thickness of 200 feet.

According to the U.S. Geological Survey, the Ogallala aquifer contains about 3.25 billion acre-feet of drainable water (one acre-foot equals one foot of water over one acre, which is 325,851 gallons). Recent count shows that more than 170,000 wells have been drilled into this aquifer to irrigate about 13 million acres of land. This acreage produces about 15 percent of the nation's total value of corn, wheat, cotton, and sorghum, and about 38 percent of the livestock. Using the Ogallala aquifer, Nebraska produces 700 million more bushels of corn annually and Texas grows two million more bales of cotton than they would without the aquifer.(34) The added value of products based on the Ogallala water is about $20 billion dollars annually. A major aid in the U.S. foreign exchange problem is agricultural exports, and the Ogallala based produce is a significant factor in these exports. However, estimates are that about 24 million acre-feet are now being withdrawn each year, and the recharge is only about three million acre-feet.

The U.S. Geological Survey reports that since the beginning of the use of this aquifer, the volume of water in storage has decreased about 166 million acre-feet to 1980. From the 1940s to 1980, the water level in the Ogallala aquifer dropped about 10 feet on average, but in places it dropped more than 100 feet. During the period 1960-1980, the water pumped from the southern part of the aquifer in west Texas and eastern New Mexico exceeded natural recharge by more than 30 times.(9) In the 1980s, the average decline was only a foot because of favorable precipitation and a more regulated water program. Although efforts continue to better manage this valuable resource, computer models project a continuing net water loss. In most regions, pumping continues to withdraw more water than is replaced by rain.(34) The aquifer is being mined, and the saturated water thickness has decreased by more than 25 percent over an area of 14,000 square miles. Because the drop in water level increases pumping costs, irrigation has been discontinued on about 367,000 acres(18), an ominous trend.

Clearly, water mining in the Great Plains cannot be allowed to continue. To alleviate the problem to some extent there is a long standing proposal to bring water from as far away as the Mississippi River, but this idea has run into opposition from Arkansas and Louisiana residents who do not want the Mississippi River flow reduced, especially after the severe navigational problems on the Mississippi during the drought of the summer of 1988. Texas taxpayers have voted down various water financing proposals in the belief that the cost of nearly $4 billion was not worth it. In the meantime, the underground water, like the oil in Texas, is being depleted. In the Texas High Plains overlying the Ogallala aquifer, irrigated acreage between 1974 and 1989 decreased 34 percent because the cost of pumping the water from the declining aquifer was greater than the value of the food produced.(8)

Other U.S. groundwater problems

A somewhat similar situation exists in the upper Great Plains area where South Dakota now takes water from the Missouri River, partly to try to preserve the Dakota Sandstone aquifer. But, now South Dakota is faced with demands from Nebraska, Iowa, and Missouri who want to take more of the Missouri River water, as their groundwater supplies drop. The Missouri River in South Dakota is not a very large stream so even this supply is limited, and South Dakota does not wish to relinquish its current water withdrawal from the Missouri to help out the downstream states.

These problems arise when an agricultural system is based on a resource which is depletable, in this case, groundwater which is not recharged as fast as it is withdrawn. The water is being mined. It may be that the situation will ultimately be very much like that of metal mining areas where the ore has been taken out and now only ghost towns or greatly reduced populations remain.

For a number of years, the groundwater situation in Arizona has grown increasingly critical. Each year, Arizona uses an estimated 2½ million acre feet more groundwater than is replenished by natural means. Arizona has been mining its groundwater, and the demands on water grow as the state's population continues to increase.

Land subsidence

In addition to lowering the water table when groundwater is used faster than it is recharged, in many places there is another problem. When fluids, oil or water, are taken out of the ground, the ground may subside. This is an environmental problem which is discussed further in Chapter 22. Here, the important point is that a collapsed aquifer cannot be effectively restored.

The Groundwater Heritage

From time to time, especially in the Middle East, large groundwater sourced irrigation projects are proposed and a number already exist. It is doubtful, however, that these will be viable very long into the future because most of them are drawing upon water inherited from the geologic past, which is not being replaced by the present sparse rainfall.

Statements are sometimes made to the effect that beneath the desert sands there is a huge underground lake. The lake, of course, is simply a sand or gravel deposit or a porous rock which is saturated with water. Almost always in the deserts, the groundwater took decades if not thousands of years to accumulate. Withdrawing this water at a rate to sustain substantial agriculture over any long period of time is probably not possible. It is an interim solution at best.

In the United States, the early settlers occupied a land where there had been no withdrawal of water except by the surficial springs, and seepage into streams. Withdrawal and recharge of groundwater were in natural balance. But the groundwater heritage from ages past is now being used in many places in quantities which clearly cannot be sustained. In other parts of the world also, growing populations are drawing more and more heavily on groundwater supplies. The efficient management of these resources as a continually renewable resource is vital. Unfortunately, this is not being done in many and perhaps in most areas. In many places irreparable damage has already been done to some aquifers.

Groundwater problems now are usually not of large enough scale to greatly affect a nation as a whole, although in smaller countries and those in more arid regions, as, for example, in Israel and in parts of Africa, the effect of reduced or destroyed groundwater supplies can be significant. In Israel, groundwater supplies are now very much a national issue, and increased pressure on limited supplies which must be shared by several nations further increases tensions in that unstable region. Nevertheless, even in larger countries such as the United States where the Great Plains, and the highly productive San Joaquin Valley of California are becoming affected, the long term adverse results of mining groundwater are likely to be substantial and become important on a national scale.

Groundwater in Summary

Like oil or minerals, groundwater in places can be a one-crop resource. If the rate of recharge to replace the withdrawn water takes years or even centuries, that groundwater for all practical purposes is a crop which can be harvested only once. If excessive groundwater withdrawal results in collapse of the aquifer, then a greater or lesser amount of the groundwater of that aquifer will become a permanently mined out situation. If a region is developed on the basis of overdrawing the groundwater, this can continue for a time, but ultimately the development must stop and then contract unless other water supplies are brought in. Some areas are now using groundwater on a sustained yield basis, but, unfortunately, other places are not, creating seeds for future social and economic problems.

Communities can be and have been abandoned because of misuse of groundwater. The problem should not be allowed to spread.

Surface Water

Throughout history, the distribution of surface water, streams, rivers, and lakes, has largely determined where people lived. The Nile Valley, sometimes called the cradle of civilization, is a classic example. That thin ribbon of blue which traverses the desert has been and continues to be the lifeline for all Egypt. The Tigris and Euphrates rivers of the Middle East have similarly nourished civilizations. The Columbia River system of the Pacific Northwest United States, with its great salmon runs, provided the chief food supply for many Native Americans.

Rivers also have provided the initial access to many areas, and continue to provide transport and travel routes. Lewis and Clark traveled up the Missouri River, and then down the Columbia. Until recently when a few roads have been cut through the jungle, the Amazon River was the only pathway of consequence through the heart of Brazil. It remains a main artery by which ocean-going vessels can travel 2,000 miles through the interior of Brazil all the way to Iquitos in northeastern Peru.

Surface water in many places is a renewable resource, but it has some limitations. One is that it is not an infinite resource. At any given time it is in fixed supply, and this supply differs from year to year. Also, as it is a limited quantity at any given time, it can be used faster than it can be renewed. Surface waters can be overdrawn, and they are today in many areas.

Some details of surface water distribution and use

Groundwater can be used for small to medium sized populations, but the answer to any major water supply need is surface water. This is illustrated by the fact that nearly all large cities are located on or near large surface water supplies. Examples are London and the Thames; Paris and the Seine; Vienna and the Danube; the Rhine River with its many cities including Bonn, Germany; New York and the Hudson; Minneapolis/St. Paul, St. Louis, New Orleans all on the Mississippi; and Montreal and Quebec City on the St. Lawrence.

Some cities draw from a network of smaller streams. Denver, Salt Lake City, and Seattle all obtain water from a number of streams in adjacent mountain areas. Some cities are located next to major freshwater lakes like Chicago on Lake Michigan, Detroit on Lake Huron, Irkutsk in Siberia on Lake Baikal.

A few cities are not on major bodies of water. In the United States these include Phoenix and Tucson, Arizona, and most notably, Los Angeles and all its satellite communities. These essentially desert cities first located on small streams that provided a water supply, but the cities quickly outgrew these limited supplies and now must draw water from distance sources. Obtaining sufficient water is now a continuing problem, both to these cities, and to the areas from which the water is taken. The water demands of the Los Angeles area have reached hundreds of miles into rural regions, and numerous local conflicts of interest have occurred.

The growth in water demand in cities has been phenomenal, caused by the huge migration populations from rural to urban areas. In the Los Angeles area, the residents in 1900 used 26 million gallons of water a day. Now they use much more than that in one hour.(27)

Diversion of water

Areas which are water deficient may attempt to remedy the situation by diverting water from other drainage basins. On the east side of the Sierra Nevada in Owens Valley numerous signs posted on small streams say that the water belongs to the Los Angeles water system. These streams do not now reach the valley floor but are fully diverted. The same situation has existed in the Mono Lake basin, which has been gradually drying up because the streams which would replenish it have been tapped and taken 350 miles away to Los Angeles. Los Angeles has changed the natural environment in several regions east of the Sierra Nevada, particularly in both the Mono Lake and Owens Valley areas. Recently, an agreement has been reached to reduce the Los Angeles take of the Mono Lake drainage water, so that the lake can be stabilized at the present level.

Aral Sea

This is perhaps the most spectacular example of the results of water diversion. Almost 100 percent of the waters of the Syr Dar'ya and Amu Dar'ya rivers in the former Soviet Union have been diverted to grow crops, chiefly cotton which requires a great deal of water. As a result, the Aral Sea, formerly fed by these rivers, lost 40 percent of its surface area between 1926 and 1990. The volume was reduced by 65 percent, the salinity more than tripled and all 24 species of native fish died. The Aral Sea once had an important fishing industry which produced 44,000 tons of fish annually, supporting 60,000 jobs. Now all that is gone.(19)

Irrigation water

Land which does not have sufficient rainfall must import water, either from the ground beneath or from a distant source to irrigate the fields. Irrigation projects use very large amounts of water, and have an environmental impact on the place from which the water is drawn. The water may come from an adjacent stream or lake, another drainage basin, or it may be from the ground beneath the irrigated fields. That water beneath a field may also be water drawn through the aquifer from fairly distant areas. Moving water for irrigation from one place to another, or drawing it out of a river upstream from another area or country which might also need the water can cause conflict.

With these factors in mind, how important is irrigation water? It is very important. Seventeen percent of the world's cropland is irrigated, but, because it is managed more carefully than other land, it produces one-third of the world's crops. Half of China's cropland is irrigated, and nearly all of Egypt's crops are irrigated. Accordingly, being able to move water from one site to another for irrigation is now and will continue to be an important activity in the future of many nations.

Egypt's future is entirely controlled by water, almost all of which originates in other nations. Egypt has made a water agreement with Sudan regarding the Nile, but the seven other upstream nations of the Nile basin have not signed.(28)

Irrigation water: the long term problem

Water reaching farmland as rain contains no dissolved minerals, but both surface water and groundwater do. Over the long term, the use of surface water and groundwater will result in the accumulation of salts. If these salts cannot be flushed out, the soil will eventually become too salty to be productive. Currently, just in the State of California, $500 million is lost each year in the form of fields which must be abandoned due to excess salt, and because of stunted crops grown on fields which are becoming salty from evaporation of irrigation waters. This problem is pursued further in Chapter 17, *Topsoil—the Most Valuable Mineral Complex.*

Water for Power and Industry

Running water dropped through turbines is an important source of energy, particularly in countries which have limited fossil fuel resources. These include Brazil and much of Africa, the latter region which has a very large hydropower potential. However, development of this energy source must also be weighed against environmental impacts on the region, and the fact that dams have a limited lifespan.

Water for energy production

Water is not only directly involved in producing hydroelectricity, but it is also important in producing almost all other sources of energy, except passive space heating like sunlight through the window. Water is used in mining of coal and uranium. It is used in water and steam flooding of oil fields to obtain the maximum amount of oil from the strata. Water is used in the cooling towers of electric generating plants with coal, natural gas, oil, or the atom as fuel.

Synthetic fuels: shale oil, oilsands, and water

The State of Colorado has the world's largest and richest oil shale deposits. It also has a unique distinction in that it is the only State in the Union which has no streams flowing into it except for a short 30 mile loop which the Green River makes into the extreme northwestern corner. Water supplies, therefore, are exceedingly critical. The very slender Colorado River carries only 5 cubic miles of water a year. The Amazon River, by comparison, flows 4 cubic miles in a day. Development of the large oil shale deposits of the western slope of Colorado, if economics ever make them profitable to recover, would require large amounts of water. The Colorado River would have to be beheaded. There are already more water drafts against the river than there is water to supply them. It is estimated that from four to nine times more water would have to be used than the amount of oil produced from the oil shale. In part, this water could probably be reclaimed, but some would inevitably be lost, and would be water which does not go through the hydroelectric turbines, irrigation systems, or municipal water supply pipes downstream on the Colorado. Water supplies of Arizona and Southern California would be adversely affected. Immediate conflict would arise. It is probably lack of a sufficient water supply which will keep the oil shale of western Colorado from ever becoming a major fossil fuel source, even if the present very marginal energy profit ratio of the recovery process could be improved.

At the Athabasca oilsands of northern Alberta, about eight tons of water are required for every ton of final oil product. This does not include the water needed to supply the demands of the residential and commercial community which is the support facility to the oilsands operation. At present, there is ample water to support the operations, but to greatly expand this industry, water might become the limiting factor.

Water in all industry

Throughout industry, water is a vital ingredient. Paper-making, textiles, oil refining, petrochemical manufacturing, food processing, concrete buildings and roads, and electronic industries all require vast amounts of water. A modest computer chip factory will use two to three million gallons of water a day to control the environment of this operation. Water is the true life fluid of industrial civilization.

Water and Conflict

Conflict over water is recorded as far back as 4500 B.C. when the two cities, Lagash and Umma of Mesopotamia (now largely Iraq) were in dispute over river water supplies. Water has the potential for an increasing number of small disputes and some major international conflicts. Some conflicts are now occurring. Gleick states, "Today, the realization that renewable fresh water availability is finite, while population growth will continuously increase water demand, is starting to cause serious concern among developing nations."(8) Even within the relatively well watered United States, problems have arisen and are growing. At one time, in a dispute with California over water, Arizona actually called out the National Guard.

El Paso vs. New Mexico

In 1991, the city of El Paso, Texas, and the bordering State of New Mexico finally settled an eleven year old water rights suit which cost $20 million in legal fees. Ed Archuleta, general manager of El Paso Water Utilities said, "There's only so much water out here. There's been a long history of people doing their own thing. That's now got to be in the past."(15) That comment now applies to the world at large.

Water crossing borders

Both surface water and groundwater migrate. Because of this, a single water source may involve many people, many communities, and, in some cases, several countries. Ninety-seven percent of flowing water in Egypt comes from outside that country. For Hungary, it's 95 percent; for Syria, 79 percent; for Iraq, 66 percent; and for Germany, 51 percent. Around the world, 47 percent of the land is involved in international water systems. There are 214 multinational drainage basins. This means 40 percent of world population lives and depends on water from international drainages. As populations grow, the potential for international conflict over water supplies increases. And technology cannot make more water. The Indiana Center on Global Change and World Peace predicts that over the next half century, water will replace oil as the prime trigger for international conflict.

Rural versus urban

Within the next five years, more than half the world's population will be located in cities. In only the 30 year period from 1950 to 1980, cities such as Bogota, San Paulo, and Managua have tripled in size. In this same time period, the move to cities in Africa has been even more dramatic. Nairobi, Dar es Salaam, and Lagos have increased in population more than seven times.(8) When the demand for water in urban areas increases, agricultural areas face competition for water supplies, yet, without the rural agricultural areas, the cities could not survive. In California this problem is growing, and because the cities have the votes to determine where water will be used, the rural areas are losing. Cities are destroying their own life base by this process, but cities keep growing, which is sure to cause increasing conflict over water.

Some Current International Water Problems

Israeli-Palestinian groundwater problem

Along with the political problems of the West Bank of northern Israel, is the fact that the hills there are the water recharge area for the major source of water for Israel's underground aquifers. The rainwater which soaks into the hills of the West Bank area moves down beneath the lands which are used by most of the people in the central portion of Israel. This amounts to about one-third of the total water supply for that country. Increased Arab settlements in the West Bank area with their corresponding larger demand on both the surface and underground waters of the area are causing water supply problems to the south in Israel, and will continue to do so.

Seizure of the West Bank by Israel in the war of 1967 goes back to 1964 when the Arab states, especially Syria where the Jordan River begins, announced plans to divert the water to their own use. Israel made several military strikes against the project. Animosities intensified and contributed to the 1967 war at which time Israel seized the West Bank and the Golan Heights. At that time, an Israeli official stated that Israel was so dependent on this water that it could never return its control to the Palestinians. One of the first actions Israel took was to declare the water resources of the West Bank under military control.(8) However, in 1994, Israel did return the West Bank to the Palestinians. While Israel held the West Bank, the wells were strictly controlled. The Palestinians complained that Israel drilled deep wells which dried up the shallower Palestinian wells. At that time on the West Bank, all of the Israeli settlements had running water, but more than half the Palestinian settlements did not. Now, the Palestinians have control of the West Bank, and both surface and groundwaters of this region remain major issues. It is estimated that about half of Israel's water supplies come from this aquifer. With control of the West Bank, the Palestinians now plan to substantially increase their irrigation projects using underground water.

Water issues will grow as both Palestinian and Israeli populations continue to increase against a finite water supply which cannot be expanded. Israel also competes with adjacent Jordan for water. Jordan's chief negotiator on water rights, Munther Haddadin, has stated, "...it's a zero-sum game. What is taken by Israel is taken away from other people. And what is taken by other people is taken from Israel."

Again, the problem is that water is being withdrawn faster than it is being renewed, and, in this case, across political boundaries. This is an almost certain circumstance for more trouble. Recent statements by Israeli Water commissioner Tzemach Vishai that, "we want...Israel to control the water supply," and by Israeli Defense Force Col. Ra'anan Gissin that, "we are always ready to do battle over water" suggest that difficulties lie ahead.

Euphrates River: Turkey, Syria, and Iraq

This river originates in the mountains of southern Turkey and flows through the very water-short countries of Syria and Iraq, to the Persian Gulf. The river is important for hydroelectric power, drinking water, irrigation, and industrial use. In 1974, Iraq massed troops and threatened to bomb the Al Thawra Dam in Syria claiming the dam had reduced the water flow to Iraq.

Currently, Turkey is building a huge water project on the Euphrates which concerns Syria and Iraq. It will ultimately consist of 22 dams used to irrigate 1.5 million hectares, and will also produce electric power. The huge Ataturk Dam is already in place, and water from its reservoir will go through the world's two largest irrigation tunnels 25 feet in diameter to irrigate the Harran Plain 40 miles distant.(19) But, this diversion will reduce the flow of water into Syria by an estimated 40 percent, and to Iraq by as much as 90 percent.(8)

Desalinization Plants

In some parts of the world, notably the oil-rich but water-poor countries of the Persian Gulf region, fresh water is made by de-salting sea water. There are several ways this can be done, but all of them involve considerable amounts of energy. In these nations, which are sometimes said to have more oil than water, cheap energy is available.

Saudi Arabia has the world's largest number of desalination plants, 22, which produce 30 percent of all desalted water in the world.(28) However, economic usefulness of desalination in any region is directly related to the cost of energy there. At present, desalinized water costs between $1 and $8 per cubic meter, compared to between $0.01 and $0.05 per cubic meter paid for conventional water supplies by western U.S. farms, and $0.30 charged to users in U.S. cities.(8)

There are two processes by which salt water can be converted to fresh water. They are Multi-Stage Flash distillation (MSF) and Reverse Osmosis (RO). The MSF process introduces hot salt water into a chamber at such a reduced pressure that part of it flashes into vapor which is then condensed as fresh water. The remainder of the salt water, now slightly cooler, then passes to a second chamber at a lower pressure so more is evaporated. This process is repeated in a number of stages, usually 20 to 30, until the salt water is cool. This is an energy intensive procedure widely used in the arid Middle East where the chief fuel used, natural gas, is abundant.

In other areas, the RO process is more common. This process uses no heat, but depends on the ability of semi-permeable membranes under certain conditions of pressure to let water molecules through but to restrict the passage of salt molecules. The energy cost in the RO process is due to creating and maintaining the pressures required to make the process work.

The world over, there are more than 3,500 desalination plants, but in total they supply just one-thousandth of water use. This process can produce enough water for households, but it can hardly become a major factor in world water supply. In terms of total fresh water supplies, the amount obtained by desalination will remain negligible.(20) As petroleum resources diminish in the Middle East, water supply by the MSF desalination process eventually may not be feasible.

Water and Health

The one specific material required by the human digestive system is water. Water is also the medium which can and does most easily transmit a variety of serious, even fatal, diseases such as typhoid, amoebic dysentery, and cholera. In many parts of the world safe drinking water is still not consistently available. It may be contaminated by disease-causing organisms, and it may also be contaminated by inorganic materials. In the State of West Bengal, some 600,000 people obtain their water from wells whose water contains arsenic at three times the safe level for human consumption. In the years, 1991-1992, 700,000 cases of cholera transmitted by contaminated water were recorded in the Caribbean region, causing 6,400 deaths. In India, an estimated 1 1/2 million children die each year of diarrhea caused by polluted water.

The future of any nation depends on the health of its people and the most fundamental necessity is safe drinking water. In this regard, the immediate future of the industrial nations which now supply consistently healthful water to their citizens is much brighter than the futures of countries which do not have safe water supplies.

Water Wars

It has been said that the ultimate battle in the Middle East will not be fought over oil because the oil will long since be gone. It will be fought over water. Postel identifies nine out of the 14 countries in the Middle East which now have water-scarce conditions.(18) It is likely in the interim, before the "ultimate battle," that local wars may also be fought over this vital resource. The former General Secretary of the United Nations, Boutros Ghali, an Egyptian national, predicted, "the next war in our region will be over the waters of the Nile, not politics." Jordan's King Hussein has stated that water is such a vital issue that it "could drive nations of the region to war."(25)

With many other areas beginning to exceed their resources, water wars may not be confined to the Middle East. It is becoming a worldwide problem.

Water and the Future

Beyond all other Earth resources, water and soil are the most important. After oil and other mineral resources have been depleted, water and soil, essential to all life, will remain to be managed and used. Water, for the most part, is a renewable resource but its rate of renewal differs greatly from place to place. In the case of groundwater, some is so deep and is replenished so slowly as to be for all practical purposes non-renewable. In some regions, groundwater supplies are very frankly recognized as non-renewable with the recommendation that it be mined just like other non-renewable resources such as coal and metals.(12) But, to build an economy in part on that resource must be fully recognized as temporary. Unfortunately, use of such non-renewable water for food production, as is now done in some places, gives a false sense of security, and populations increasing on that base ultimately must make severe adjustments for which they may not be prepared. The same applies to populations supported by overdrawals of renewable groundwater. Using groundwater faster than it is replenished is now occurring in parts of China, Russia, India, Mexico, in all countries of the Middle East, in many countries in Africa, and in the United States. China faces an especially severe problem. With its people numbering 22 percent of the world's population, China has only eight percent of the world's renewable fresh water.(18)

The preservation of water supplies, and maintaining their quality, together with the comparable stewardship of soil will have more to do with the destinies of nations than will all other activities related to the Earth. Nations which have good water supplies will have a much more stable future than those who do not. Among such nations are Canada, Russia, most European countries, and the United States except for its southwestern portion. Major nations with severe problems include India and China. The Mideast nations and many in Africa clearly have long term water problems which will rapidly increase in severity. Syria is already short of water. Syria's Director of International Waters, Majed Daoud, in 1993, stated, "we will have 25 million more people by 2010, and these people will need food."(27) But, assuming water can be found to grow food for the additional 25 million people by the year 2010, what then? Another 25 million people in 17 years or less for whom to find more water? But water is already in short supply.

Since 1955, use of fresh water around the world has tripled. Water, for which there is no substitute, may be the ultimate limiting factor in the growth of world population.

BIBLIOGRAPHY

1 BABBITT, BRUCE, 1991, Age-old Challenge: Water and the West. The Wasteful Use of Water Resources May be How the West is Lost: National Geographic, v. 173, n. 6, p. 2-34.

2 BISWAS, A. K., et al., (eds.), 1994, Water For Sustainable Development in the Twenty-First Century: Oxford Univ. Press, New York, 290 p.

3 BLOOMQUIST, WILLIAM, 1992, Dividing the Waters: Governing Groundwater in Southern California: Institute for Contemporary Studies, San Francisco, 413 p.

4 BRANK, GLEN, 1991, While Reservoirs Run Dry, Plates Runneth Over: Water Digest, Klamath Falls, Oregon, May, p. 4-5.

5 CLARKE, ROBIN, 1993, Water. The International Crisis: MIT Press, Cambridge, Massachusetts, 208 p.

6 CLIFFORD, FRANK, 1994, Water Wars Could Drown State's Recovery: Los Angeles times, October 4.

7 GARDNER, GARY, 1995, From Oasis to Mirage: The Aquifers That Won't Replenish: World Watch, May/June, p. 30-36.

8 GLEICK, P. H., 1993, Water in Crisis. A Guide to the World's Fresh Water Resources: Oxford Univ. Press, New York, 473 p.

9 HAMMOND, A. L., (ed.), 1992, World Resources 1992-1993. Toward Sustainable Development: World Resources Institute, New York, 385 p.

10 HARDIN, GARRETT, 1991, From Shortage to Longage: Forty Years in the Population Vineyards: Population and Environment, v. 12, p. 339-349.

11 HILLEL, DANIEL, 1994, Rivers of Eden. The Struggle for Water and the Quest for Peace in the Middle East: Oxford University Press, New York, 368 p.

12 ISSAR, A. S., and NATIV, R., 1988, Water Beneath Deserts: Keys to the Past, A Resource for the Present: Episodes, v. 11, n. 4;, p. 256-262.

13 KLIOT, NURIT, 1994, Water, Resource and Conflict in the Middle East: Routledge Inc., New York, 368 p.

14 MAURITS la RIVIERE, J. W., 1989, Threats to the World's Water: Scientific American, September, p. 80-94.

15 NEGRON, SITO, 1992, El Paso, NM Finally Agree on Water Plan: El Paso Times, November 13.

16 PICKFORD, JOHN, (ed.), 1987, Developing World Water: Grosvenor Press International, London, 586 p.

17 POSTEL, SANDRA, 1989, Water for Agriculture: Facing the Limit: Worldwatch Paper 93, Worldwatch Institute, Washington, D. C., 54 p.

18 POSTEL, SANDRA, 1992, Last Oasis. Facing Water Scarcity: W. W. Norton & Company, New York, 239 p.

19 POSTEL, SANDRA, 1995, Where Have All the Rivers Gone? World Watch, May/June, p. 9-19.

20 POSTEL, S. L., DAILY, G. C. and EHRLICH, P. R., 1996, Human Appropriation of Renewable Fresh Water: Science, v. 271, February 9, p. 785-788.

21 POWLEDGE, FRED, 1992, Water. The Nature, Uses, and Future of Our Most Precious and Abused Resource: Farrar Straus Giroux, New York, 423 p.

22 PRINGLE, LAURENCE, 1982, Water. The Next Great Resource Battle: Macmillan Publishing co., Inc., New York, 144 p.

23 REISNER, MARC, 1993, Cadillac Desert. The American West and Its Disappearing Water: Viking Penguin, Inc., New York, 582 p.

24 ROGERS, PETER, 1993, America's Water: MIT Press, Cambridge, Massachusetts, 216 p.

25 SHAH, TUSHAAR, 1993, Groundwater Markets and Irrigation Development. Political Economy and Practical Policy: Oxford Univ. Press, Oxford, 538 p.

26 SUN, R. J., and JOHNSTON, R. H., 1994, Regional Aquifer-system Analysis Program of the U.S. Geological Survey, 1978-1992: U.S. Geological Survey Circular 1090, Washington, D. C., 126 p.

27 TOBIAS, MICHAEL, 1994, World War III. Population and the Biosphere at the End of the Millenium: Bear & Company Publishing, Santa Fe, New Mexico, 608 p.

28 VESILIND, P. J., 1993, Water. The Middle East's Critical Resource: National Geographic, v. 183, n. 5, p. 38-71.

29 WALKER, T., 1991, Another Middle East Issue of Life and Death: Financial Times, London, May 8.

30 WEEKS, J. B., et al., 1988, Summary of the High Plains Regional Aquifer-System Analysis in Parts of Colorado, Kansas, Nebraska, New Mexico, Oklahoma, South Dakota, Texas, and Wyoming: U.S. Geological Survey Prof. Paper 1400-D, Washington, D. C., 30 p.

31 WINPENNY, JAMES, 1994, Managing Water as an Economic Resource: Routledge Inc., New York, 176 p.

32 WORSTER, DONALD, 1985, Rivers of Empire. Aridity & Growth of the American West: Pantheon Books, New York, 402 p.

33 YOUNG, G. J., et al., 1994, Global Water Resource Issues: Cambridge Univ. Press, New York, 213 p.

34 ZWINGLE, ERLA, 1993, Ogallala Aquifer. Wellspring of the High Plains: National Geographic, v. 183, n. 3, p. 80-109.

CHAPTER 16

Minerals from the Ocean

Minerals have been taken from the sea for thousands of years. The Chinese obtained salt from the sea as far back as 2200 B.C., and India's rulers appointed a "superintendent of ocean mines" as early as the fourth century B.C. From the shore areas and shallow ocean shelves, diamonds and other precious stones, corals, conch shells, pearls, and salt were obtained.

Nations and coastlines.

There are two categories of nations regarding minerals from the ocean. First, there are nations which have a coastline and thereby extend their territorial rights out into the sea. Earlier, these rights commonly extended for three miles, which was the range of cannons then. Now these territorial rights have usually been extended to 200 miles.

Second, there are nations which have little or no coastline. These countries would have to obtain minerals from the sea by having the technology to go to the deeper ocean areas beyond the usual 200 mile limit and in some fashion mine the ocean floor. They would presumably rent marine harbor facilities from some accommodating coastal country. Or these landlocked nations could by treaty get a share of what would be mined by other nations.

Landlocked countries

For many years, there has been a proposal in the United Nations by landlocked countries to have just such a treaty enacted, but so far it has not passed. Many of these landlocked nations are also the smaller and less developed countries. Their lack of technology and in many cases the lack of finances makes them unable to go out and mine the ocean floor. Larger nations with an extensive coastline are the potential developers of ocean minerals. These nations also, as might be expected, oppose any treaty stating they must share the wealth with landlocked countries. The United States is one which has voted "no".

U.S. 200 mile limit

The United States has plenty of territory to explore. In 1983, President Reagan proclaimed the ocean area from a line three miles offshore of the coast of the United States and its island territories to a distance 200 nautical miles out to be the U.S. federal government Exclusive Economic Zone. Ocean area within the three miles of the shore belong to the states. Presumably this 200 mile offshore zone is U.S. territory for mineral exploration and development. Especially with its islands in the Pacific including the Hawaiian chain, Midway Island, Wake Island, Guam and the Northern Mariana islands, American Samoa and Howland and Baker islands, the United States has a huge amount of ocean floor on which to prospect. The total area is 3.9 billion acres. This compares with the total onshore United States territory of 2.3 billion acres.

How valuable this may be is uncertain because the sea floor is known in detail only in very few places. The side of the moon facing the Earth is much better mapped than are the ocean floors. The economics of recovering minerals from this 3.9 billion acres is also not well known. It is a vast frontier for the future.

Petroleum search going seaward

In recent years, increasing amounts of petroleum are being recovered from beneath the sea, as prospective land areas are becoming drilled up. The first major offshore oil production came from Lake Maracaibo, Venezuela, beginning in the 1930s. In the late 1930s, U.S. offshore oil production began in the Gulf of Mexico at the High Island Field near Texas. In 1946, the first oil was produced in the Gulf of Mexico out of sight of land and since then wells have been drilled more than seventy miles offshore.(4)

In Norway, there is no on-shore petroleum production nor will there ever be, and in Britain there is very little. The North Sea is the oil province for both countries, and for Denmark and The Netherlands as well which each have a small share.

Even the countries around the oil-rich Persian Gulf including Saudi Arabia, Kuwait, Iran, Qatar, and the United Arab Emirates are increasingly moving oil exploration offshore. The Gulf of Thailand and the South China Sea have recently come into production, and Brunei gets most of its oil from offshore. Much Indonesian oil is produced from beneath the sea. Australia, Nigeria, Trinidad, and Angola produce most of their oil offshore. Canada has moved offshore Newfoundland to develop the Hibernia Field as there are few onshore prospects in eastern Canada. Canada is also moving exploration north of the Arctic coast. Mexico's big oil discoveries in recent years are offshore, and the United States has been drilling in the Gulf of Mexico for many years, as well as offshore California, although the latter area is now ruled off limits to further oil exploration.

In Alaska, the Cook Inlet area, adjacent to Anchorage, has been a petroleum producer for years, and such oil exploration frontiers that exist in Alaska are along the coast and offshore. Alaska has a land area of about 586,000 square miles, but given the geology of Alaska, it is exceedingly unlikely that any significant amount of oil will ever be found more than 50 miles inland. Some of the Alaska North Slope production is offshore.

Gas hydrates

There is a type of gas deposit called gas hydrates. These are solids composed of water molecules which form a rigid lattice of cages, most of which contain a molecule of natural gas. Gas hydrates exist worldwide in polar regions and certain colder ocean areas. Volumes are very large, but no way is now known to recover this energy source economically.(2,5,10)

Placer deposits

For many years, the shallow near-shore saltwater areas have produced tin in Malaysia — the Malaysian placer tin deposits. Elsewhere, along the coast of Africa, similar types of deposits are mined for diamonds and for gold. Shorelines which were formed at lower stands of sea level during glacial times, and are now in shallow near-shore waters, were the basis for a brief gold rush to Nome, Alaska about the turn of the century. Japan has, from time to time, mined off-shore iron-bearing sands. However, in total value the tin sands and diamonds have been the most important items recovered from shallow ocean placer deposits but the amount has not been large.

Sand and gravel

These ordinary materials in annual total value are the most valuable resources now being recovered from the shallow ocean shelves. In areas where development of such resources on-shore for urban areas is difficult because of environmental regulations, and the resources on land which can be developed are much more distant, simply dredging sand and gravel from the adjacent ocean floor is convenient and does not make a visible impact on the environment. Because the materials are bulky, movement by barge on ocean waters makes their transport economic. The United States, Britain, and Japan are the chief exploiters of marine sand and gravel, using them in such large metropolitan areas as the east coast regions of New York and Boston, and in London and vicinity. Shallow, near-shore deposits of sand and gravel supply 16 percent of the construction needs of Britain, and about one-third of those of Japan.(12)

Manganese nodules

There are various other materials on the sea floor of potential value. Notable are the very large deposits of manganese nodules which lie in the mid-Pacific, between the two great industrial nations, Japan and the United States. These nodules, about the size of a golf ball, are abundant in some places to the extent of a pound and a quarter per square foot, and may cover many square miles.(7) Although called manganese nodules because this is the predominant metal in them, the nodules contain other valuable metals also. Generally, the content is about 23 percent manganese, 6 percent iron, as much as 1.6 percent nickel, and lesser amounts of cobalt.

If these deposits were on land they would definitely be an ore deposit. However, “ore” is not a geological term but an economic one, and defines a deposit from which a metal or metals can be extracted at a profit. Whether or not the manganese nodules can be recovered for their metal content at a profit is still uncertain. However, the quantity of metal in this mid-Pacific deposit is exceedingly large, running into the billions of pounds. Also, an intriguing fact is that these nodules are continually forming today. As a result, even with extensive mining operations, it is doubtful that as many nodules could be mined as are formed in a given year. It is an interesting situation where a mineral deposit gets larger even as it is mined. There aren’t many mines like that!

Some experimental equipment has been built to recover these nodules. Lockheed Aircraft designed and used a vessel for that purpose. Results have been mixed. Also, the question has arisen as to who owns these nodules, for they are beyond the claimed territorial rights of any nation. The Law of the Sea Treaty has not been endorsed by all countries. Non-signers include the United States, West Germany, Great Britain, and Japan. At present, it seems that the “law of capture” prevails — that is, whoever can get the nodules can have them.

Phosphorite nodules

There are other mineral deposits on the sea floor including phosphorite nodules. These exist in substantial amounts off both coasts of the United States, the west coast of Central and South America, in waters adjacent to Argentina, Japan, and South Africa, and off the central and northern coasts of New South Wales (Australia). These Australian deposits are fairly thin and so far poorly mapped, but some are locally known to have a quality almost equal to those of onshore deposits, with a P_2O_2 content reaching 29 percent.

Mineral-rich brines

Considerable interest has recently been shown in the brines now forming around rift zones in the ocean floor. At these places, hot waters coming from the Earth's interior reach the ocean floor and precipitate a variety of metals. In the bottom of the Red Sea there are rift zones in three distinct basins. Here metalliferous muds are relatively rich in various metals, principally manganese, lead, zinc, copper, silver, and gold. Drill cores show that in some areas the muds are as thick as 300 feet. Metal concentrations here are from 1,000 to 50,000 times greater than in ordinary sea water. Analysis of some of the better deposits show a content (by weight) of 0.12 percent manganese, 0.16 percent lead, 0.70 percent copper, and 2.06 percent zinc. However, they are still unable to compete economically with land-based deposits, but they may become economic some time in the future.

Mineral crusts are also formed by rising hot mineral-laden waters and these have been observed in a number of places where "chimneys" of sorts are built up by what oceanographers call "black smokers" which actually are mineral-charged vents of hot water. Such occurrences are known along the Gorda Ridge off northern California and southern Oregon coasts, as well as in a number of other places on the ocean floors. Some of these areas, including the California-Oregon localities, have warranted preliminary mapping. From what is now known about these occurrences, they probably are not large enough or economically accessible to justify pursuing at present. However, it appears that this is the geologic origin of some of the mineral deposits which are now being mined on land which was once part of the ocean floor.

Ocean water

Ocean water itself is a tremendous storehouse of elements. In fact, most of the known elements occur in sea water in greater or lesser amounts, mostly lesser. Gold, for example, exists in sea water in the amount of about 37.5 pounds per cubic mile, but this concentration is only 0.0005 percent. The energy cost to process that much sea water to produce that little gold is inordinate. The same is true of nearly all other elements in sea water. Concentrations are so low that it does not pay to obtain them from this source.

There are, however, some exceptions. Both bromine and magnesium have been obtained commercially from sea water. About two-thirds of the magnesium used in metal alloys comes from sea water, as does most of the bromide which is used in pharmaceutical products in certain types of photography, and in gasoline.(11,12) And, of course, common table salt is obtained from the ocean in many areas by allowing the water to flood into shallow salt pans at high tide. Then the pans are diked off from the sea and the water is allowed to evaporate. Mexico, Thailand, and a number of other countries obtain salt in this fashion. One of the world's largest salt mines is at Guerrero Negro in Baja California, Mexico, covering 49,000 acres, and is a vast salt water evaporation system. From this, Mexico exports six million tons of salt annually to Japan.(1) In total, about one-third of the world's supply of common salt is evaporated from sea water.(12)

The Future

What role all these marine deposits — petroleum, phosphorite and manganese nodules, hot brine muds, mineral crusts, minerals in solution including salt, bromine, and magnesium, and placers —will play in the future for various nations is difficult to assess. Petroleum is a special situation and even though the costs in energy and money are high for offshore development, rewards can be large for successful drilling. But because of costs, it takes a larger field offshore than onshore to justify development. Many offshore fields too small to be commercial, would be economic onshore.

Marine mineral deposits, except for the richer placers now being dredged in relatively shallow waters and the sand and gravel deposits, at present require too much energy to be economically recovered. In the case of sea water, as already noted, the amount of energy required to obtain various elements by processing the huge volumes of sea water which would be involved is such that except for a very few, the economic recovery of these elements from that source is quite unlikely for the foreseeable future. Until the less costly resources on land are exhausted, these marine sources will not be seriously considered. Even then they cannot be developed if energy is not cheap enough to make it worthwhile.

For the immediate and reasonably near future, except for petroleum, the mineral resources of the sea will not be an appreciable factor in the futures of nations. Because of this, there have not been any great conflicts over sea-floor minerals like the conflicts which are now arising over the exploitation of fish. However, in areas where petroleum resources exist or are believed to exist, and there have been no clear demarcation of national boundaries, conflicts are appearing as in the South China Sea. Ownership of two areas in the Gulf of Mexico are now under discussion between the United States and Mexico.

BIBLIOGRAPHY

1 BORGESE, E. M., 1985, The Mines of Neptune: Minerals and Metals From the Sea: Harry N. Abrams, Inc., New York, 158 p.

2 COLLETT, T. S., 1992, Potential of Gas Hydrates Outlined: Oil and Gas Journal, June 22, p. 84-87.

3 EARNEY, F.C.F., 1980, Petroleum and Hard Minerals from the Sea: John Wiley and Sons, New York, 244 p.

4 ENGLISH, T. S., (ed.), 1973, Ocean Resources and Public Policy: Univ. Washington Press, Seattle, 184 p.

5 HOWELL, D. G., (ed.), 1993, The Future of Energy Gases: U.S. Geological Survey Professional Paper 1570, Washington, D. C., 892 p.

6 KASH, D. E., et al., 1973, Energy Under the Oceans: Univ. Oklahoma Press, Norman, Oklahoma, 378 p.

7 MAJORAM, TONY, et al., 1981, Manganese Nodules and Marine Technology: Resources Policy, March, p. 45-47.

8 MCKELVEY, V. E., and WANG, F. H., 1970, World Subsea Mineral Resources: U.S. Geological Survey Map I-672, Washington, D. C.

9 MERO, J. L., 1964, Mineral Resources of the Sea: Elsevier Publishing Company, Amsterdam, 312 p.

10 PIRIE, R. G., 1996, Oceanography: Oxford University Press, New York, 448 p.

11 SKINNER, BRIAN, and TUREKIAN, K. K., 1973, Man and the Ocean: Prentice-Hall, Inc., Englewood Cliffs, New Jersey, 149 p.

12 WEBER, M. L., and GRADWOHL, J. A., 1995, The Wealth of Oceans: W. W. Norton & company, New York, 256 p.

13 WEYLE, P. K., 1970, Oceanography. An Introduction to the Marine Environment: John Wiley & Sons, Inc., New York, 535 p.

CHAPTER 17

Topsoil — The Most Valuable Mineral Complex

To walk across a field or through the woods is to walk on a complex of minerals to which we owe our very existence. Without the mineral complex called soil, no useful crops could exist. It has been accurately said that, "civilization depends on the top six inches of the Earth's surface." The world over, topsoil averages less than one foot thick.

Soil and civilization

"Betraying ecological illiteracy, most people are unaware of their dependence on a thin sheet of topsoil," writes Eckholm. He notes, "Soil destruction has contributed to the fall of past civilizations, yet this lesson of history is seldom acknowledged and usually unheeded. Today's cropland losses impair the well-being of the living as well as the generations to come. Yet in this matter as in others societies seem incapable of acts of foresight."(16)

History is replete with evidences of how soils, their use, and abuse, have affected the destinies of regions and nations. Bennett, in his 1939 classic volume on soil, *Soil Conservation*, has detailed these events from Asia, Europe and to the Americas.(4) Hyams describes the effect of soils on the history of peoples of the Middle East.(20) Eckholm has reviewed the history of soils and their effect on cultures.(15) These authors cite Mesopotamia (now chiefly Iraq) as a land which was once incredibly productive. This was probably the first region in which irrigation was practiced. But, today, the legacy of that irrigation is salt deposits. "Vast areas of southern Iraq today glisten like fields of freshly fallen snow; from 20 to 30 percent of the country's potentially irrigated land is unusable."(15) Silt loads also overburdened canals and reduced the flows to mere trickles, and correspondingly "reduced a leading world civilization to a few dusty villages."(15)

Hillel describes a number of striking examples of the destruction wrought in ancient times by the poor management of soils. One example is Utica, a city some 30 kilometers northwest of Carthage in North Africa. It was founded by the Phoenicians at the mouth of the Bagradas River. But because of excessive sedimentation the remains of the city are now seven kilometers from the coast and under about 10 meters of silt.(19)

Six thousand years ago much of the Middle East was covered by a mixed oak forest. This was cut down to make way for growing wheat. When the soil was too depleted to grow wheat, sheep were turned onto the land, and when it could no longer support any vegetation, the loose cover was washed away down to bedrock and is now barren. This is the history of many of the rocky plains and slopes which are visible there today.

The dust storms of the Great Plains and the southwest United States in the 1930s degraded huge agricultural areas. This caused the abandonment of farms and mass exodus of people. The social impacts are recorded in literature such as the classic *Grapes of Wrath.* Fortunately, with careful soil management since that time, much of this region has been reclaimed, demonstrating that proper soil conservation practices can make a significant difference. But, overall, the U.S. is losing the race to prevent soil loss. Pimentel, et al. state:

> "About 90% of U.S. cropland is losing soil above the sustainable rate. About 54% of U.S. pasture land (including federal lands) is overgrazed and subject to high rates of erosion...One-half of the fertile topsoil of Iowa has been lost during the last 150 years of farming...Similarly, about 40% of the rich Palouse soils of the northwest United States has been lost in the past century."(29)

In other parts of the world soil degradation continues at an alarming pace. The World Resources Institute, in collaboration with the United Nations Environment Programme reports, "It is estimated that since World War II, 1.2 billion hectares, or about 10.5 percent of the world's vegetated land, has suffered at least moderate soil degradation as a result of human activity. This is a vast area, roughly the size of China and India combined." The report goes on to state that the greatest degradation is in Africa.(39)

Loss of grain production

The ominous result of worldwide soil erosion is that the capacity to grow food crops has been impaired to the point where it cannot keep up with population growth. Pimentel, et al. say, "Because of erosion-associated loss of productivity and population growth, the per capita food supply has been reduced over the past 10 years and continues to fall. The Food and Agriculture Organization reports that the per capita production of grains which make up 80% of the world food supply, has been declining since 1984."(29) Pimentel, et al. further state:

> "Not only is the availability of cropland per capita decreasing as the world population grows, but arable land is being lost due to excessive pressure on the environment. For instance, during the past 40 years nearly one-third of the world's cropland (1.5 billion hectares) has been abandoned because of soil erosion and degradation. Most of the replacement land has come from marginal land made available by removing forests. Agriculture accounts for 80% of the annual world deforestation.(29)

Soil formation

Soil is the result of slow weathering processes which start with solid rock and eventually reduce it to a material in which plants can grow. Combining the Sun's energy with Earth materials including water through the process of photosynthesis, produces vegetation on which all animal life ultimately depends. The food chain always goes back to plants, and land plants, with the exception of a few low forms such as fungi and lichens, depend on soil. More than 97 percent of the world's food comes from land rather than the ocean or other aquatic systems.(26)

Lands which are continually leached by heavy rainfall eventually become relatively infertile, the classic example being the Amazon Basin. Immediately along the rivers where sediment has washed in from other areas, the land can be fairly productive, but on the higher areas away from the river flood plains the land has been continually leached literally for millions of years and almost no usable mineral nutrients remain. This has given rise in the past to the "slash and burn" type of agriculture which has been commonly employed in the tropics. The land is cleared, the native vegetation burned, and crops planted in the slightly enriched ashy soil. This process gives one or two years of meager crops. Then the few nutrients in the soil and ash are exhausted, and the process is repeated somewhere else. More recently, attempts have been made to clear the jungle and plant conventional crops, but generally this has not been successful. The leached residual red clays, when cleared of their normal jungle cover, tend to bake like a brick under the tropical sun and be essentially nonproductive.

The renewing of minerals in the soil for an agricultural area is important. Today, this is usually done by humans through the application of fertilizers such as ammonium nitrate and sulfate, and various compounds of potassium and phosphorus. But, in many parts of the world these are not available or cost more than the local farmers can afford.

Volcanic fertilizer

In some regions nature does renew these leached soils. On the islands of Indonesia, particularly Java, which has one of the highest population densities in the world, volcanic activity frequently dusts the countryside with a fresh layer of ash that quickly weathers in the warm moist climate to release new minerals, producing a fertile soil. Without these volcanic eruptions it is doubtful that Java could sustain its present population, and there would surely be some trace element deficiencies without the mineral renewing effects of the ash falls. The flanks of Mount Vesuvius have seen several destructive volcanic eruptions, but immediately after each eruption the people there quickly move back up the slopes and replant the vineyards and other crops. The newly deposited mineral-rich ash soon breaks down to produce bumper harvests.

The explosion of Mount St. Helens in the State of Washington in 1980, spread large amounts of ash over central and eastern Washington's great wheat growing area, and on into northern Idaho and beyond. The immediate effect of this was negative, but farmers in this region now look forward (given normal moisture conditions) to excellent crops for many years to come as a result of this extensive and free natural fertilizer. In that sense, the volcanoes are a valuable mineral "mine," and a unique mine in that its minerals, unlike other mines, are renewed from time to time with each volcanic eruption.

Mineral-rich glacial soils

In a different geologic process nature has recently (in geologic time) renewed the fertility of a substantial portion of northern Europe, and the Upper Midwest area of the United States, including that very fertile State called Iowa. Iowa and adjacent parts of America's agricultural heartland produce wonderfully rich crops full of the nutrients needed for good health. The roots of these crops reach into glacial drift — a deposit of varied minerals transported from the north by glaciers. These fresh rock and soil materials were derived in large part from granitic areas which have an abundance of the minerals needed for healthy plant and human growth. After depositing this rich load, the glaciers retreated and weathering began to make these elements available to plant life. Also, in these climates, the cold winters preserve the rich black humus from the decayed vegetation of the previous year. It was this rich mineral complex inheritance from nature which the first settlers found, and from which people in the Upper Midwest still benefit, and will continue to do so for generations to come. It is a mineral heritage brought down duty-free from Canada by the ice sheets.

One can readily see, however, that it would have been a vastly different story without the glaciers. The road cuts and quarries of this area show that beneath the relatively thin cover of glacial drift lie hundreds to thousands of feet of leached marine clays and nutrient-poor limestone laid down in ancient seas which covered this part of the United States some 300 to 500 million years ago. Fossil corals, cephalopods, brachiopods, bryozoans, trilobites, and other forms of marine life lie beneath the glacial debris in Iowa in relatively infertile clays and limestones which would be difficult material in which to grow 150 or more bushels of corn to the acre. Also, corn that might be grown would be deficient in a number of vital trace minerals.

However, a warning should be sounded. Already in places, due to careless plowing and lack of soil conservation practices, the red and gray ancient marine clays are beginning to show through the tops of hills in Iowa and elsewhere, indicating that this rich glacial heritage may already have begun to be lost. Unlike volcanoes of other lands, the glaciers are not likely to return very soon to again refurbish the mineral content of the soils.

Degradation of Soils

Soils are formed by the weathering of rock materials. Weathering is the first stage in the process of erosion, which, geologically, is defined as the loosening and removal of material at the Earth's surface. In nature this is, for the most part, a beneficial process. By weathering, rocks are broken down into loose particles which become soil, and with further weathering and the accumulation of organic material, it becomes valuable topsoil. When soil is slowly removed by natural processes it is carried into lowland areas to form marshes and deltas which are home to myriad life forms. Some of the sediment is carried into basins where vegetation is buried to eventually form coal, or, in other geological settings, to become part of the process of forming oil.

The slow natural removal of soil commonly allows time for replacement soil to form, and a balance is kept so there is no net loss of topsoil. The key word here is "slow." It is when these processes are accelerated that the problems begin. For the most part, this acceleration across the world's agricultural areas has been caused by human action.

Erosion

Soil scientists generally use the term erosion to mean only the active transport of soil by various processes, and not including the weathering and breaking down of rocks. It is here used in that context. The weathering of rocks which breaks them down into small particles is a beneficial process, but too rapid removal of this material leaves unweathered rock which is agriculturally non-productive. Soil can be removed by running water, wind, or simply gravity in the form of soil creep, slump, or landslides. This factor of soil movement is greatly reduced if the ground is vegetated, but if the soil is bare, after plowing, for example, erosion can be quite rapid. At present, statistics are clear that the removal of topsoil exceeds its formation worldwide.(6)

A subtle activity

Soil erosion in general is an insidious process. It is for the most part almost imperceptible. Each raindrop which hits the Earth has the potential of moving a tiny bit of soil, and given the presence of gravity, that movement will be slightly downhill. Wilken has calculated that six metric tons of soil removed from one hectare will reduce the level of topsoil by only one millimeter.(38) The U.S. National Academy of Sciences has stated that "Soil erosion is slowly nibbling us to death."(23) In more rugged areas, such as Haiti, the Andean region of

South America, and in parts of the foothills of the Himalayas, the process is more obvious, and sometimes dramatic, where sudden mass soil movement is marked by great scars and gullies. In areas of low relief, soil erosion is a slower but significant process. Regardless of how the soil is removed, it is the unfortunate fact that the top thin zone of soil which contains the most nutrients is what goes first.

The effects of soil erosion can be devastating. All loose material may suddenly be removed above the bedrock making the area immediately unsuitable for agriculture for hundreds if not thousands of year. Or erosion may be a very slow process which reduces the productivity of the soil just slightly each year. Between these two extremes there is a process of sheet washing by heavy rains which on moderate slopes can remove a great deal of soil over a period of just a few years. An Alabama field which once produced a half-bale of cotton to the acre, suffered so severely from sheet washing that it would have taken more than 200 acres to produce a bale.(36)

Soil degradation is much like a cancer on the Earth. It is not easy to detect in its early stages, but if untreated it ultimately will be fatal. Over many parts of the world, soil degradation continues unchecked, frequently in areas already over-populated relative to the carrying capacity of the land. The problem is particularly severe in many parts of Africa, and the Middle East, and India.

Related effects of erosion

Aside from the actual loss of soil by increased erosion, that process has other negative effects. Reservoirs behind dams are rapidly filled with sediment by excessive erosion of the soils of the drainage area. For example, the Archicaya Dam in Colombia was three-quarters filled after only 10 years.(35) Waterway navigation channels must be dredged because of the deposition of sediments, and fish spawning areas may be destroyed by sediment. Water storage facilities are lost. Total costs of such damage in the United States has been estimated to be about $4.1 billion a year.(26)

A dollar spent on soil and water conservation returns its investment many times over in saving reservoirs from siltation, saving rivers from dredging, saving fish spawning grounds, preventing flood damage to property, and reducing the need for building levees.(26)

Besides the erosion of soil, other degradation processes are salinization, waterlogging, acidification, compaction, and leaching of nutrients.

Salinization

Some soils are destroyed in place, without being eroded away. Soils can be damaged by becoming salty, not simply by common salt but also from a large variety of other compounds, and from a variety of processes. A rising water table can bring salts in solution to the surface where, from evaporation, they remain as residue. Contamination of the soil by gypsum is a common result of this process.

Continued irrigation seems to be the chief cause of salinization, and as more and more lands are being irrigated the problem is increasing. Worldwide, irrigated land area expanded from eight million hectares in 1800 to 48 million in 1900, and irrigated land now has increased to more than 100 million hectares (about 247 million acres).

All water except precipitation contains minerals in solution. River water carries varying amounts of salts, and because of agricultural irrigation and use of fertilizers upstream, the rivers generally become more salty downstream. This is true of the Colorado which in its lower portions has had the salinity increased by 30 percent in the past 20 years.(22) It became so salty before it entered Mexico that complaints from Mexico prompted the U.S. government to build a desalinization plant near the border.

One of the richest, year-around agricultural regions in the world is the Imperial Valley of Southern California. But, already in this relatively recently developed agricultural area, some farms have had to be abandoned because of salinization.(30) And the salinization problem is increasing. Miller has reported that "The All-American canal, which brings water to the Imperial Valley from the Colorado River also carries about three million tons of salt per year to California's southern coastal plain — equivalent to importing a 210 car trainload of salts every day. Improperly managed fields produce only one useless crop — 'Imperial Valley snow' — a white crust of sodium sulfate."(22) If the various salts cannot be removed in some fashion, their accumulation may either reduce the productivity of the soil or completely destroy it for agricultural purposes. This has happened and is now occurring in many areas of the world, especially in the Middle East. In the Euphrates River Valley in Syria, for example, some 110,000 hectares (272,000 ares) of irrigated land have been degraded by salinization, of which about 20,000 hectares (50,000 acres) had to be totally abandoned. Salt degraded soils are found on all inhabited continents, and now cover about 10 percent of the total surface of dry land.(33) Combating this problem is expensive, and in the example of California, already more than 1,800 miles of drainage canals have been built for this purpose.(24)

Water lost in transit

In some cases, irrigation water is brought hundreds of miles from its source to the fields which use it. In California much water is brought from the northern part of the state to the southern regions. During transit, considerable water is lost through evaporation and by seepage loss in canals.(24)

Groundwater and salinization

Irrigation projects using groundwater are especially susceptible to salinization because groundwater usually carries much more dissolved solids than does surface water, particularly as deeper and deeper well waters must be used. Borgstrom is of the view that "ground water reserves *never* should be used for regular crop production but be *held in abeyance* for drought relief in critical times."(5)

Waterlogging

Soils may hold an excessive amount of water to the point where they cannot be worked agriculturally, and regular crops cannot be grown. Waterlogging may be the result of poor surface drainage, poor subsurface drainage, or a rising water table.

Acidification

This can result from a variety of factors. It usually seems to be the result of continual leaching of soils from heavy rainfall. Soils in the moist tropics are the most obvious examples of acid soils.(21) Acid soils may develop from the reaction of some bedrock minerals with the atmosphere and moisture. Also, it may result from acid rains, just as lakes have become acidified by sulfuric acid resulting from the mixture of moisture in the air combined with industrial sulfur fumes. Rain comes down slightly acidic by picking up carbon dioxide which naturally occurs in the air. Organic acids may be released from vegetation.(30) Exchangeable hydrogen and/or aluminum associated with colloidal material in the soil may produce an acid condition.(21) Whatever the cause may be, strong acidification of soils usually results in a marked decrease in plant productivity.

Compaction

The use of heavy machinery on agricultural land may cause compaction, but the worldwide major cause of compaction is the heavy grazing by livestock on the land. This can greatly reduce the ability of the soil to absorb and hold moisture. Also, as the animals crop the vegetation, the soil is subject to increased erosion.

Nutrient removal

Continued rainfall on a soil which is derived from the ancient underlying bedrock, instead of being recently carried in glaciers or wind, will gradually leach away the trace elements until only a laterite, a mixture of hydroxides of iron and aluminum, remains. Laterites are a very poor base for rooted plantlife. Most of the important nutrients exist in the organic material which is mostly at or near the surface of laterites and is subject to early removal by water or wind.

Present net result

Unfortunately, in nearly all agricultural areas today, the total net result of all processes affecting soils is that soils are being lost faster than they are being formed. Pimentel states "Worldwide degradation of agricultural land by erosion, salinization, and waterlogging is causing the irretrievable loss of an estimated 6 million hectares each year."(26)

Population growth and loss of topsoil

The sheer growth of population in many ways is a major destroyer of the soil, and this action is suicidal. The ultimate loss of use of all soil in an area can be accomplished by simply paving it over. This is being done everywhere all over the world. "Growth" is a favorite word of chambers of commerce, and developers of various sorts. Presumably growth is the evidence of economic success and progress. This includes the building of houses, apartment complexes, factories, parking lots, and shopping malls. Every day in California more than 100 acres of farmland are lost to various kinds of developments. In the United States as a whole, one million acres of cropland have been lost annually since 1945.(25) Loss of soil by "erosion" through population occupancy across the land is beginning to rival other causes of soil erosion.

Reclaiming Soil

Natural production of new soil is so slow that in human terms it cannot be a factor in restoring soils to an area. It is estimated that under tropical and temperate conditions, it takes from 200 to 1,000 years for the renewal of 2.5 centimeters (about an inch) of soil.(26) Soils which are carried by streams are lost to the region from which they are eroded, but to some extent some of these soils are reclaimed — used again. This may occur in the flood plains along the rivers, and in delta areas. These areas are the recipients of the topsoil from areas upstream which were eroded to furnish them, and are relatively fertile. Sediment builds deltas and wetlands which are the nurseries for many forms of wildlife.(14) People, accordingly, are lured to these areas because of their productivity. However, this may have a negative side effect in loss of human life and property because of the floods which bring the soil. Also, storms in these low-lying delta areas, which occur frequently in such places as the delta of the Ganges in Bangladesh, can be catastrophic.

Floods also may bring in sediment which is not mature soil and not very fertile.(26) This may cover more fertile lowland areas, as happened in the lower Mississippi River valley in the great floods of 1993.

But can soils which are degraded but not yet totally lost be reclaimed by human effort? They can to a degree, but studies indicate that even with the widespread use of a variety of fertilizers, these are not a complete substitute for the nutrients originally in the soil.(26) It has been demonstrated that "...raising organic matter [in the soil] by annual application of manure or by use of leguminous green manures increased yields of wheat, sugar beet, and potatoes beyond any that were achieved with equivalent inorganic fertilizer — no matter how much additional nitrogen was added."(21) Also, synthetic fertilizers are made in part from fossil fuels and by processes that use fossil fuels. These fuels and fertilizers have to be purchased. The inhabitants of many poor lands which might benefit from fertilizer, cannot afford them.

The use of fertilizers on once highly productive lands to restore some of their productivity or to raise the productivity of land which is still in good condition also gives a temporary illusion of productivity which really does not exist. The acres of productive "land" which fertilizer, in effect, creates are sometimes referred to as "ghost acres." The sad fact is that in many farming areas of the world, agriculture has developed a chemical dependency. As long ago as 1977-1979, over 96 percent of U.S. corn acreage received nitrogen fertilizers, 88 percent received phosphorus fertilizers, and 82 percent received potassium fertilizers. For wheat the comparable numbers were 63 percent, 42 percent, and 18 percent.

Brown and LeMay state:

> "Of all the chemical reactions that humans have learned to carry out and control for their own purposes, the synthesis of ammonia from hydrogen and atmospheric nitrogen is the most important. Plant growth requires a substantial store of nitrogen in the soil, in a form usable by plants. The quantity of food required to feed the ever-increasing human population far exceeds that which could be produced if we relied solely on naturally available nitrogen in the soil. Ever-larger quantities of fertilizer rich in nitrogen will be needed in the decades ahead."(8)

Hillel observes, "The two major innovations [of the industrial revolution] were the introduction of motorized machines and the advent of chemical fertilizers and pesticides. Machinery and chemicals became the hallmarks of the new agriculture as it entered the modern era."(19) Bartlett states: "...modern agriculture is based on petroleum-powered machinery and on petroleum-based fertilizers. This is reflected in a definition of modern agriculture: 'Modern agriculture is the use of land to convert petroleum into food.'"(2)

Domesticated plants, the basis for civilization today, are plowed, planted, cultivated, fertilized, and harvested by using the energy of oil. Ammonia, the basis for the various ammonia fertilizers including the important hydrous ammonia and urea, is now produced almost entirely using natural gas as the source of the hydrogen in the ammonia compound, NH_3. It can also be obtained from coal, another fossil fuel. Beyond that, hydrogen would come chiefly from hydrolysis of water, requiring large amounts of electricity which must be generated from some other energy source, such as dams, or oil or coal-fired plants, or by natural gas turbine-generators. The pesticides and, especially, herbicides (selective weed-killers) are almost entirely produced from petroleum. Thus, the high productivity of modern agriculture is another unanticipated result of the recently arrived, but due to be short-lived Petroleum Interval. That world population and its vital agricultural base has become so dependent on this finite source of raw material is the cause for much concern.

Beyond ammonia, the next most important fertilizer used in agriculture are compounds of phosphorus. Phosphorus is found in commercial quantities in only a few places, with Morocco, the United States, and Russia having the great majority. The third fertilizer widely used is potash which is the term used for various compounds of potassium, an element needed in every living cell. Potassium deposits are fairly widely distributed, the largest of which are in Saskatchewan, Canada. The ocean would be an immense source if energy were cheap enough to process the ocean water for potassium.

Ammonia compounds, phosphate, and potash are now basic to worldwide agriculture. The use of artificial fertilizers is a habit, like other chemical dependencies, which may be hard to break, and coming off of it for the human race may be difficult. Also, it has been clearly demonstrated that beyond a certain amount, additional fertilizer will not increase soil productivity, but may actually be detrimental to the soil as well as increase the salinity of the runoff water to the streams. Increasing the amount of fertilizer used may be an effort with diminishing returns or no additional returns at all. We have completed that agricultural fertilizer-leap forward in many areas and more fertilizer will not help.

However, much can be done to preserve the soil by processes other than the use of fertilizers. The acidity of a soil may be reduced by the addition of lime, which is not a fertilizer but simply a natural rock material. Commonly, it is just finely crushed limestone. And, beyond that, there are many ways by which soil can be preserved and improved including contour planting, building terraces, strip cropping, crop rotations, no-till planting, tree-shrub hedges, use of mulches, and other management procedures.(26) These are useful in all agricultural areas but are particularly valuable in the less developed countries which do not have access to affordable fertilizers. Training students from these lands in relatively simple and low cost agricultural practices is probably some of the best money spent for the benefit of future generations worldwide.

Situation of Some Nations

Soil management and conservation is of utmost importance to all nations, because soil combined with water is the basis for life. A comprehensive survey of worldwide soil conditions is presented in Pimentel.(26) Europe is the largest best area in terms of productive soils. Canada and the United States are also very fortunate, geologically endowed with some of today's richest soils in the world. But studies show that these regions continue to lose soil far more rapidly than would allow for sustainable future agriculture. Soil degradation is winning over conservation.

In the Middle East, Asia including India and adjacent countries, and in Africa, the problems of soil conservation and fertility are the most severe. In part, this is because of the long history of agriculture in these regions, and partly because of the increasing population which tries to cultivate more and more marginal lands. Marginal agricultural lands in general are those on steeper slopes, with less soil initially, and are subject to much more rapid erosion than are the more arable lands of low relief.

South America, particularly with the loss of forest cover in the Amazon Basin, is another area of major soil degradation, although the Amazon Basin soil initially is not very fertile. The steep terrain of the Andes is another region of rapid soil erosion. However, using some of the techniques which were developed during the time of the Inca empire, some localities are managing their soil relatively well. But, worldwide, land degradation is proceeding much above the rate of soil replacement.

Even the relatively recently settled Australian continent is already facing massive soil degradation. The Prime Minister has stated that soil degradation now occurs on more than half of Australia's agricultural land.

Population and Soil

The basis of human life is the soil, and as population grows, the amount of soil available to support each individual declines. Soil formation now is not keeping up with either erosion or population growth. From 1951 to the end of this century, the amount of arable land to support each person in the world will have been reduced by more than 50 percent.(24) This is primarily due to population growth. When the factor of soil degradation is added, the problem becomes even more severe.

The effect of population growth on the land has a number of harmful aspects. As marginal lands are invaded, forested steep hillsides are cut for firewood, or to clear more land for cultivation. In either case, the loss of forest cover results in rapid erosion of whatever soil exists. As fuel becomes more scarce, all vegetation that can be used for fuel is picked up, resulting in a great loss of nutrients to the soil which previously were added from this organic material. Increasing numbers of livestock raised by more and more people further degrades the soil by compaction, and by destroying the vegetative cover necessary to prevent erosion. Also, animal excrement by which the vegetative nutrients can be recycled to the soil is increasingly used as fuel when forest resources are no longer available. These problems compound upon each other. Brown and Wolf have summarized the situation: "Throughout the Third World increasing population pressure and the accelerating loss of topsoil seem to go hand in hand."(6)

It is significant and ominous that the regions with the poorest soils, Asia and Africa, are experiencing the highest growth rates of population. From 1961 to 1984, the population increased 70 percent but only 15 percent more arable land was able to be developed. During this period the world over, population increased 55 percent but only nine percent additional arable land was made available.(14)

Increased industrialization and the great desire for countries to "get on wheels" is converting large areas of productive farm land to non-farm uses. As in the United States, the land is being paved over. China, where population is now outstripping domestic food production, is nevertheless heading toward a motor vehicle economy which will mean loss of cropland. In 1994, the Chinese government indicated that automobiles were scheduled to be one of four growth industries in the next two decades. Cars in China which numbered 1.15 million in 1990, are projected to reach 22 million in 2010.(7) China is now embarked on a massive highway construction program. Brown summarizes the situation stating: "The bottom line is that there is now an army of bulldozers cutting their way through the Chinese countryside building highways across rice paddies and wheat fields, leaving the nation with ever less cropland to satisfying its rising demand for food." He adds, "In addition to factories, housing, and roads, farmland is also being claimed by shopping centers, tennis courts, golf courses and private villas."(7) Population continues to grow, but farmland, the vital base for population, is diminishing. Asphalt is assaulting agriculture. It cannot continue without dire results, but when and how does it stop?

Soils and the Future

The world, in total, is vitally dependent on soil. It is more valuable than oil or any metal, and its value for human existence is only equaled by that of water. If productive soil does not exist in rocky areas, or is lost in some fashion from other regions, the population of these places must still be supported by soil which exists somewhere else where food can be produced and shipped to non-productive areas. M. Rupert Culter, formerly U.S. Assistant Secretary of Agriculture has stated: "Asphalt is the land's last crop." People living in the paved-over land of cities have to depend on soil elsewhere. Urban dwellers should be concerned about the soil basis for their existence wherever that "elsewhere" is. But city denizens seldom regard soil erosion as their problem. It very definitely is. Because the world is becoming a more and more integrated economy, loss of productive soil in one place has a rippling effect in that it causes a need for other areas of soil to produce more food to make up for the loss.

Export of soil

Soil cultivation results in a loss of soil greater than what would occur if the areas were not farmed. This means that with every bushel of wheat or corn, or any other agricultural product, some soil nutrients are shipped along with it. One of the most important exports of the United States, to mitigate to some extent its current very negative balance of payments, is agricultural products. Being a young land in terms of exploitation, the United States still has large regions of rich soils. But, as soil loss currently exceeds soil formation, the United States is clearly also, in effect, exporting some of its soil along with its agricultural products.

Soils and nations' futures

The existence of large tracts of fertile soil available for farming was a major factor in luring people from the Old World to the New. These New World lands have been and remain a very great asset. With a growing population, the possession of large areas of productive soil will have a very positive effect on the future of the United States. This is also true of Canada, Australia, Argentina, and other relatively young countries in terms of agricultural usage. Preserving these productive soils is one of the most important things these nations can do. Humans can live without oil. Water (except for mined groundwater) is replaced by precipitation. But, soils, absolutely vital to human existence, form so slowly as to be non-renewable for all practical purposes.

Other countries where agriculture has been practiced for much longer periods of time, particularly India, and China, home to nearly 40 percent of the world's population, now have much less fertile soils on average than do the more recently settled regions. Their future is more precarious. In some places, loss of soil is already a disaster. Many areas now have to increasingly meet the dual problems of a decreasing soil productivity and a rising population. Wilken makes the general calculation that "if the amount of land in production remains constant over the next 40 years, farmers will nearly have to double their yields to feed the growing population."(38) It is doubtful that this can be done. The generally flat lands are already being farmed. The unattractive prospect is to put more marginal land into production, and these terrains tend to have relatively short useful agricultural lives. The matter of soil conservation must be given increasing concern in all countries, but it is unfortunately not getting enough. A researcher on worldwide soil trends and conditions, Pimentel states, "The public has clamored for action to plug the ozone hole over Antarctica, stop acid rain, halt the greenhouse effect and save the forests — yet little has been said about the soil loss that could destroy the hopes of a stable and fruitful earth a century from now."(26)

There is no more important concern in regard to mineral resources and the destinies of all nations than the matter of soil preservation. That world now is in many conflicting divisions, resulting in huge expenditures for armaments rather than using what would be lesser amounts of money for the vitally important long range objective of soil conservation. This is deplorable evidence of human folly. Unfortunately, with present worldwide trends, we are quite literally continuing to lose ground.(15)

BIBLIOGRAPHY

1 BARROW, C. J., 1991, Land Degradation. Development and Breakdown of Terrestrial Environments: Cambridge Univ. Press, Cambridge, 295 p.

2 BARTLETT, A. A., 1986, Forgotten Fundamentals of the Energy Crisis: American Journal of Physics, v. 46, n. 9, September, p. 876-888.

3 BATIE, S. S., 1983, Soil Erosion: Crisis in America's Cropland: Conservation Foundation, Washington, D. C., 136 p.

4 BENNETT, H. H., 1939, Soil Conservation: McGraw-Hill, New York, 993 p.

5 BORGSTROM, GEORG, 1973, World Food Resources: Intes, New York, 237 p.

6 BROWN, L. R., and WOLF, E. C., 1985, Soil Erosion: Quiet Crisis in the World Economy: in, SOUTHWICK, C. H., (ed.), Global Ecology: Sinauer Associates, Inc., Publishers, Sunderland, Massachusetts, 323 p.

7 BROWN, L. R., 1995, Who Will Feed China?: W. W. Norton & Company, New York, 163 p.

8 BROWN, T. L., and LeMAY, H. E., 1988, Chemistry. The Central Science: Prentice Hall, Inc., Englewood Cliffs, New Jersey, 1028 p.

9 CARTER, V. G., and DALE, T., 1974, Topsoil and Civilization: Univ. of Oklahoma Press, Norman, 292 p.

10 CLARK, E.H., II, 1985, The Off-site Costs of Soil Erosion: Journal of Soil and Water Conservation, v. 40, p. 19-22.

11 COOKE, G. W., 1977, The Roles of Organic Manures and Organic Matter in Managing Soils for Higher Crop Yields — A Review of the Experimental Evidence: Proc. International Seminar on Soil Environment and Fertility Management in Intensive Agriculture: Society of the Science of Soil and Manure, Tokyo, p. 53-64.

12 CRAIG, G. M.,(ed.), 1993, The Agriculture of Egypt: Oxford Univ. Press, Oxford, 516 p.

13 DONAHUE, R. L., MILLER, R. W., and SHICKLUM, J., 1983, Soils. An Introduction to Soils and Plant Growth: Prentice-Hall, Inc., Englewood Cliffs, New Jersey, 667 p.

14 DUDAL, R., 1982, Land Degradation in a World Perspective: Jour. of Soil and Water Conservation, v. 82, p. 245-249.

15 ECKHOLM, E. P., 1976, Losing Ground. Environmental Stress and World Food Prospects: W. W. Norton & Company, Inc., New York, 223 p.

16 ECKHOLM, E. P., 1982, Down to Earth. Environment and Human Needs: W. W. Norton & Company, New York, 238 p.

17 ELLIS, STEVE, and MELLOR, TONY, 1995, Soils and the Environment: Routledge Inc., New York, 336 p.

18 FANNING, D. S., and FANNING, M. C. B., 1989, Soil: Morphology, Genesis and Classification: John Wiley & Sons, New York, 395 p.

19 HILLEL, D. J., 1991, Out of the Earth. Civilization and the Life of the Soil: The Free Press, Macmillan, Inc., New York, 321 p.

20 HYAMS, EDWARD, 1952, Soil and Civilization: Thames and Hudson, New York, 312 p.

21 MILLAR, C. E., et al., 1951, Fundamentals of Soil Science: John Wiley & Sons, Inc., New York, 491 p.

22 MILLER, G.T., Jr., 1975, Living in the Environment. Concepts Problems and Alternatives: Wadsworth, Belmont, California, 382 p.

23 NATIONAL ACADEMY OF SCIENCES, 1972, The Earth and Human Affairs: Committee on Geological Sciences, Division of Earth Sciences, National Research Council, Washington, D. C., printed by Canfield Press, San Francisco, 138 p.

24 OWEN, O. S., 1980, Natural Resource Conservation. An Ecological Approach: Macmillan Publishing Co., Inc., New York, 883 p.

25 PIMENTEL, DAVID, and KRUMMEL, JOHN, 1977, America's Agricultural Future: Ecologist, v. 7, n. 7, p. 254-261.

26 PIMENTEL, DAVID, (ed.), 1993, World Soil Erosion and Conservation: Cambridge Univ. Press, Cambridge, 349 p.

27 PIMENTEL, DAVID, 1993, Environmental and Economic Benefits of Sustainable Agriculture: Socio-Economic and Policy Issues for Sustainable Farming Systems: Cooperative Amicizia S.r.l., Padova, Italy, p. 5-20.

28 PIMENTEL, DAVID, and GIAMPIETRO, MARIO, 1994, Food, Land, Population and the U.S. Economy: Carrying Capacity Network, Washington, D. C., 81 p.

29 PIMENTEL, DAVID, et al., 1995, Environmental and Economic Costs of Soil Erosion and Conservation Benefits: Science, v. 267, February 24, p. 1117-1123.

30 ROWELL, D. L., 1994, Soil Science: Methods and Applications: John Wiley & sons, Inc., New York, 1994, 350 p.

31 SPOSITO, GARRISON, 1989, The Chemistry of Soils: Oxford Univ. Press, New York, 277 p.

32 STEINER, F. R., 1990, Soil Conservation in the United States: The Johns Hopkins Univ. Press, Baltimore, 249 p.

33 SZABOLECS, ISTVAN, 1989, Salt Affected Soils: CRC Press, Inc., Boca Raton, Florida, 274 p.

34 TOBIAS, MICHAEL, 1994, World War III. Population and the Biosphere at the End of the Millennium: Bear & Company Publishing, Santa Fe, New Mexico, 608 p.

35 UNEP [United Nations Environmental Programme], 1982, Development and Environment in the Wider Caribbean Region: A Synthesis: UNEP Regional Reports and Studies, No. 14, New York.

36 U.S. DEPARTMENT OF AGRICULTURE, 1938, Soils and Men: Yearbook of Agriculture: U.S. Government Printing Office, Washington, D. C., 1232 p.

37 WILD, ALAN, 1993, Soils and the Environment: Cambridge Univ. Press, New York, 306 p.

38 WILKEN, ELENA, 1995, Assault of the Earth: World Watch, March/April, p. 20-27.

39 WORLD RESOURCES INSTITUTE, 1994, World Resources 1994-1995. A Guide to the Global Environmental: World Resources Institute, Oxford Univ. Press, New York, 400 p.

CHAPTER 18

Minerals and Health

Most of this book concerns the economic and military health of nations with respect to the distribution and availability of mineral resources. But there is another important aspect of mineral distribution that affects the physical health of individuals and of domestic and wild animals which humans utilize.

The weathering of bedrock, or in some areas the presence of glacial drift or other transported Earth materials ultimately determines what elements are locally in soil for plant nutrients, and which will then be eaten by animals and humans.(14)

The bedrock or overlying transported rock materials can be quite different from place to place. In the past, and even at the present time, many populations derive their sustenance from relatively small and local areas. Therefore, there is a possibility that vital trace elements needed for proper nutrition can be missing from the diet of a population in a particular locality. This can apply to regions as large as some countries.

Diet, Disease, and Trace Elements

The human body is structured on a mineral compound — the skeleton — which is chiefly calcium phosphate with minor amounts of other minerals. Without iron in our systems we would not survive more than a few minutes as iron in the blood is what carries life-giving oxygen to all the cells. Minerals are part of us. There are at least seventeen minerals which are essential to human nutrition. The study of the importance of trace elements in human as well as animal and plant nutrition has revealed some interesting facts about distribution of trace elements and their relationship to diseases in various parts of the world.

Human Health and Minerals

There is evidence from the geographic distribution of thyroid disease, hypertension, arteriosclerosis, cancer, tooth decay, and from several diseases of animals that a definite relationship exists between the minerals of the Earth in those places, and these medical conditions. Trace elements in human diets are very important. Trace elements are related to regulating the dynamic processes of enzymes, and minute amounts are needed to modify the kinetics of enzyme reactions.

However, excessive amounts of certain minerals can have a negative effect on health. The vegetables grown in New York and Maryland soils are relatively high in iron, manganese, titanium, arsenic, copper, lead, and zinc compared with most other soils. Helen

Cannon of the U.S. Geological Survey has correlated this fact with the occurrence of certain diseases.(7) Another study in an area known for abnormal concentrations of metals suggested that the high mineralization was a possible factor in an unusual cancer-mortality pattern in that area.(24)

Salt

It is interesting that two elements, sodium which is highly explosive if placed in water, and chlorine which is a deadly gas, combine to form common salt which is needed in human nutrition. This mineral is probably the first mineral which humans specifically sought for use in their diet. In Medieval times humans used 10 to 20 times as much salt as today's average American. Salt, at that time, was sometimes highly prized and kept in elaborate precious metal containers as much as two feet high. Salt in solution is literally part of the lifeblood of humans. As salt travels through the various human organ systems, it has a variety of effects. The extensive research and literature of salt in human metabolism is beyond the scope of this discussion. Two brief mentions are made here.

In heavy physical work where sweating is extreme, the replacement of salt by salt tablets has been commonly done. Loss of salt can cause weakness, nausea, and even fainting. However, excessive salt may also have undesirable effects. Salt in drinking water has been correlated with exacerbating high blood pressure. A number of studies have shown that people who live in areas where the drinking water has a high sodium content, have a markedly greater risk of high blood pressure.(18) One of the current recommendations for the control of blood pressure is the reduction of common salt in the diet.

Iodine

Iodine deficiency is one of the most widespread mineral medical problems in the world. Lack of a very minute amount of iodine in the diet can stunt both physical growth and mental ability. Iodine is essential to life. Iodine enables the thyroid gland to produce the hormones necessary to develop and maintain the brain and nervous system. When the levels of thyroid hormones fall, the heart, liver, kidneys, muscles, and endocrine system are all affected adversely. Lack of iodine in the diet of pregnant women can adversely affect the baby. Seafood and food grown in iodine-sufficient soils provide adequate iodine in human diets. But, it is estimated that about 1.5 billion people in at least 110 countries are threatened by iodine deficiency. The chief regions where deficiency occurs is in mountainous regions and areas prone to frequent flooding which washes out the iodine in the soil.

The solution to the problem, however, is very simple. Iodine capsules can be eaten, or in many countries iodine is simply added to common table salt (iodized salt). Making iodized salt widely and cheaply available to iodine-deficient areas is being done by both government and private organizations. The Kiwanis International has such a program. The amount of iodine needed is exceedingly small — one teaspoonful of iodine consumed over a lifetime in tiny amounts every day is all that is needed to prevent iodine deficiency caused disorders such as goiters.

Selenium

This is an element which seems to both cause and cure a variety of human ailments. A study of 45,000 Chinese reviewed the occurrence of Keshan disease. This is a form of heart disease, mostly affecting children up to the age of eight or nine years. Its symptoms are enlargement of the heart, low blood pressure, and a fast pulse. A high death rate was found to be clearly related geographically to the amount of selenium in the soil. The disease occurs in a wide band of land running from the northeast coast of China towards the southwestern border of the country. In this area, the soil and crops grown on it are deficient in selenium. Within this region, children given selenium showed a lower incidence of the disease, but it did not diminish in other affected areas where the children were not treated.(25) It was found that, "...the dramatic responses to Se [selenium] supplementation by individuals suffering from Keshan disease suggest that selenium may yet help mankind overcome two of its most damaging disease conditions."(20) The other disease referred to is a form of cancer to which selenium appears to be a useful trace element in treatment.

In the United States, an area of the coastal plain of Georgia and the Carolinas has come to be termed "the stroke belt." It also has a higher than normal incidence of heart disease. This area, too, is low in selenium. Although studies are not yet complete, it appears that death rates from a variety of cancers are lower in areas of the United States where local crops take in larger amounts of selenium from the soil. A report from Finland concluded that men with lower levels of selenium in the blood were more likely to develop cancers of the lung, stomach, and pancreas. Women also had a marginally higher risk of these ailments, and the report noted that the Finns do not get much selenium in their natural diets.

Selenium poisoning

Too high a concentration of some elements, however, can become a negative health factor. Selenium poisoning can occur from an excess of this element. In late 1988, a general selenium poisoning warning was published by the *Sacramento Bee* (California) reporting investigations which found selenium contamination in the marshes, lakes, and streams. Fish and game in Wyoming, Colorado, Utah, Montana, and Nevada, as well as in California contained excessive amounts of selenium. Eighty-one percent of the trout, carp, perch, catfish, and goose eggs collected throughout the West exceeded the 200 microgram safety limit and 67 percent were over the 500 level of toxic effect. The samples averaged 974 micrograms, or nearly double the level at which poisoning symptoms begin to appear in healthy human adults. The most notable case of excessive selenium was in Sweitzer Lake at the state recreational area of the same name near Delta, Colorado. The water tested 51 parts per billion of selenium which is 10 times higher than the Environmental Protection Agency's limit for protecting freshwater fish species.

Products for human consumption were studied and half the foods tested such as steak, liver, poultry, eggs, and vegetables from areas in Oregon, Montana, South Dakota, Nebraska, Wyoming, and Colorado were found to exceed the safe level of 200 micrograms of selenium. The true magnitude of this situation in the western United States has yet to be established but the clues already indicate that the problem could be large. However, it should be stated that in spite of all the studies which have been conducted, the precise role of selenium in human health, particularly with relation to heart disease, has still not been conclusively determined. The research continues.

Chromium

Chromium is a relatively rare element in the United States which imports nearly 100 percent of its supplies. There are small deposits in Montana and Oregon and in a few other places, but for the most part, both North and South America are markedly deficient in chromium. Is this chromium deficiency identifiable in humans, and is it significant in terms of health? The answer appears to be "yes" to both questions.(10) People in the Middle East have about 4.4 times as much chromium in their bodies as Americans, and Asians have five times as much. In studying death rates and causes, it was found that chromium in the aorta was too low to be detected in every person dying of coronary artery disease, one form of arteriosclerosis. Chromium's presence has been found to be important in reducing cholesterol levels, which apparently relates to its absence in people dying of coronary artery problems.

Chromium also appears to be helpful to athletes where high muscular strength and endurance are important. Studies have shown that strenuous physical exercise can deplete the body of chromium. The fuels which muscles use are carbohydrates which need chromium and insulin for proper metabolism.(10) However, some studies suggest that chromium picolinate, the most popular of the chromium supplements sold widely in health stores, may cause chromosomal damage that ranged from threefold to eighteenfold the amount that occurred in other cells exposed to other chromium compounds. Such damage is considered an indication of a substance's cancer-causing potential.(4) So, just as other minerals can be helpful or harmful in various amounts and forms, chromium also has a variety of impacts on human health.

Iron

This element is absolutely vital to human life. It is the vehicle in the blood which combines with oxygen to send life-supporting oxygen to the entire body system. Although iron is a common element in the Earth's crust, some geological regions have very little. Children living on certain iron-deficient soils in Florida (soil derived almost wholly from limestone) were found to be anemic. Treatment of the children with iron greatly improved their condition, usually as soon as four to six weeks.

Phosphorus and potassium

Every living cell must have two elements, potassium and phosphorus. Phosphorus tends to get locked away in mineral compound form quite easily instead of freely circulating. Bones and teeth are calcium phosphate. Simply to build the bones of 1.2 billion Chinese takes a lot of phosphorus. It has been suggested that lack of phosphorous accounts for the typically small-bone structure of many Asians. Keeping phosphorus in circulation so that all people can get their needed share is somewhat of a problem already. It may be more so in the future.

Luther Tweeten, Professor of Agricultural Marketing at Ohio State University, at the 1995 meeting of the American Association for the Advancement of Science stated that the world "currently uses about 150 million metric tons of phosphate rock a year." This use is now increasing at the rate of four percent a year. There is an estimated 34 billion metric tons of phosphate rock in the world reserve. Phosphate is used as fertilizer for almost all food crops and thus into the human system. But if the current usage trend continues, the world's supply of phosphate may be depleted by 2050. If world population growth can be slowed to just one percent annual increase (compared with present 1.7 percent), the phosphate supply would last 82 years. Tweeten further noted that phosphate is a basic building block of plants and for which there is absolutely no substitute.

Morocco has almost half the world's known phosphate deposits.(2,21) Geology may have destined Morocco to become an especially critical nation to the world economy even as are the oil-rich nations now.

Potassium is as vital as phosphorus but it is much more widespread in the Earth. Most diets, even in fairly poor soil areas, have sufficient potassium. Canada has more than half of the world's known deposits of potash, the principal commercial source of potassium used in fertilizers.(19)

Arsenic

Arsenic in drinking water is a fairly common problem in several parts of the world. In an area of the southern Willamette Valley in Oregon near the town of Creswell, there are excessive amounts of arsenic in groundwater used for local water supplies. The source of the arsenic is not precisely known, but it possibly comes from grains of the mineral arsenopyrite in the aquifer. Arsenopyrite is a fairly common arsenic-bearing mineral in regions where basalt or other basic igneous rocks are abundant which is the circumstance near Creswell. In Taiwan, drinking water from wells caused arsenic poisoning in hundreds of persons. The arsenic exceeded 70 parts per million and many of the affected Chinese developed skin cancers as well as arsenic poisoning.

Lead

Lead is a metal in use for more than 2,000 years. The Romans made their water pipes from lead, some of which are still very well preserved in the ruins of the City of Pompeii. It has been suggested that the Romans suffered as a population from lead poisoning because of this use of lead in their water supply systems. Pliny, the Roman naturalist and writer, described a disease among the slaves that was clearly lead poisoning. The U.S. Geological Survey, in analyzing municipal water supplies of one hundred largest cities of the United States looked for 23 trace elements, and of these, 16 were found in sizable quantities. Of these, five are essential and 10 are biologically inert. Only one, lead, is toxic over a lifetime. Maximum concentration of lead found was 62 parts per billion which exceeds the Federal lead safety standard which is 50 parts per billion. The lead that was found, occurred in water supply systems which included some lead pipes where the water was soft (low in minerals) and acid. The lead was not an initial constituent of groundwater, all of which suggested that the Romans might have indeed picked up lead from their water systems.

Radon gas

Recently, especially in the United States, considerable concern has arisen regarding radon gas as a possible health hazard. This gas is derived from a series of radioactive decays which start with uranium. Actually, radon gas itself has a very short life. It is the subsequent decay products which cause cancer. These can accumulate in houses and other closed structures.

The occurrence of radon gas and its derivatives is definitely related to the geology of the area. Uranium in particular is found in certain black shales like those which underlie parts of Kansas and some other Midwest regions, and in granitic rocks. The granitic rocks are the original source of the uranium and, upon weathering, the uranium is transported in the oxidized form and then is immobilized in the reducing environments of organic-rich muds which when compacted are black shales. Both black shale and granitic bedrock areas usually show a higher than average radon gas presence. Certain other rocks may also produce radon, and the geological associations of radon are becoming well documented. Geological mapping is proceeding to identify the rocks which are potential sources of radon.

"Hard" water

The term "hard" water is loosely applied to water which contains more than normally large amounts of minerals, particularly calcium in soluble form. Calcium exists in large quantities in the form of calcium carbonate, the mineral which forms limestone. Limestone is widely distributed in the eastern, midwest, and southern portions of the United States. The State of Texas has a thick blanket of limestone over much of its 265,000 square miles, and water in Texas is notably hard. The teeth of the natives of Texas are also notably better than the teeth of people living in soft water areas such as the United States Pacific Northwest.

Teeth are composed of calcium phosphate, and children growing up in calcium-deficient water areas tend to have poor teeth. In contrast, in Deaf Smith County, Texas, with its hard water from wells in limestone, the dental decay rate (caries) is the lowest in the nation.

Heart disease has been studied perhaps more thoroughly than any other human ailment, and there seems to be at least one conclusive relationship to minerals. In England, Wales, the United States, Sweden, and Japan numerous studies have shown that areas with relatively soft water, water with few dissolved minerals, have a higher than average death rate from heart disease. Hard water areas, where water is drawn from wells in limestone, appear to have populations which have significantly less heart disease, including both cardiovascular disease and coronary heart disease, than areas with fewer minerals in the water.

In Japan, a relatively small country in area, this relationship of minerals to cardiovascular disease has been defined quite well. Northeastern Japan has an abundance of sulfur-rich volcanic rocks. Rivers there carry relatively soft water because the high sulfate produces an acid water which is soft water. These areas proved to be places where the death rate from apoplexy (death due to rupture of a blood vessel) was substantially higher than in areas to the south where the river waters were relatively hard. Studies in various regions of the United States show this same negative correlation between heart disease and the hardness of the local waters. Soft water may be fine for your car battery, and for washing your hair, but it is bad for your teeth and heart. It should be noted, however, that the precise factors involved in this correlation between heart disease and hardness (or softness) of water are still undetermined. Just exactly what hard water does to inhibit heart disease is unknown; the correlation simply exists.

Miscellaneous minerals and heart disease

Other relationships between heart disease and the geology of an area have been noted. For example, in Georgia, nine northern countries have a notably low rate of heart disease, whereas in south-central Georgia the death rate from heart disease is appreciably higher. A detailed geochemical analysis of the soils and the plants in the two areas disclosed that in the northern low death rate area, manganese, vanadium, copper, chromium, and iron are more abundant than in the high death rate area. These elements are among those trace elements which are known to have beneficial effect on heart disease.

Minerals and cancer

The scourge of mankind, cancer, also appears to have some definite geological associations, but, as in the case of heart disease, exactly what causes many cancers is unknown. In West Devon, England, it seems clear that a high incidence of cancer is related to groundwater obtained from a distinctive rock in the area from which exceptionally highly mineralized water is obtained.

The geography of esophageal cancer near the Caspian Sea in Iran also shows a distinctive pattern related to soil types. The soils of the eastern portion of this area are saline, in a relatively dry area. Westward, the amount of rainfall increases and the salinity of the soils markedly decreases because the rainfall washes away the salts. Along with this trend is a marked decrease in esophageal cancer. A follow-up to this study found that other areas of the world with a high rate of esophageal cancer have much the same nutrition including mineral content of the soils as does the population of the high-risk areas in Iran. One such locality was found in Puerto Rico. Another area of excessive esophageal cancer, with a climatic environment which in turn affected the mineral environment (high salinity) similar to the high cancer incidence area in Iran, is located in the Turkoman semi-desert region of the former USSR.

Silicon, uranium, and lungs

Very locally, the effect of certain minerals in an environment can be quite striking. Women living in a particular silver mining area in Bohemia would outlive three or more husbands who worked in the mines. These silver mines are now a major source for uranium. The miners did not know it at the time, but they were being subject to excessive radiation while working underground, and also were subject to lung problems from silica dust resulting in silicosis. The fine silica dust also appears to have contained uranium contamination.

Fluoride

It appears that fluoride, like many other minerals can be beneficial or toxic depending on the amount. Found naturally in some drinking waters, in moderate amounts (about one part per million) fluoride appears to be a substantial deterrent to tooth decay. On the other hand, excessive quantities of fluoride in drinking water causes undesirable bone changes, as well as mottling of the teeth. The function of fluoride in both teeth and bones is to help retain calcium. Where water is deficient in fluoride such as in certain areas of North Dakota, there is an increase in osteoporosis. This is a decreased strength of bones due to leaching of calcium, an ailment which tends to affect older persons, especially women. There also is evidence that taking calcium fluoride pills can decrease the incidence of hardening of the aorta. Deficiencies of magnesium, calcium, and lithium are correlated with an increase in cardiovascular troubles in general. Hard water appears to reduce these problems.

A study in Oregon by a number of dentists concluded that fluoridation of water was a special benefit to the poorer people who do not have as much money to spend on regularly scheduled dental care as do more affluent individuals. Getting free dental care from the fluoridated water was to their advantage. However, it should be noted that there is considerable controversy about the total effects of fluoride on the human body. Some studies indicate it may cause some genetic damage, and there are possibly other negative effects, and therefore critics of a fluoridated water supply say that the use of fluoride should be an individual decision. Use of a fluoridated toothpaste is an option.

Local Differences in Geology Important

Differences in mineral content of soils in areas not far apart can result in appreciable health differences. In New Zealand there have been substantial contrasts in amounts of tooth decay between children in the City of Napier and the nearby City of Hastings. Soil analyses from the areas adjacent to the cities showed considerable variations in trace elements. There was also corresponding variation in the trace elements in vegetables grown in the two areas. Tooth decay was lower in the Napier area where the soil was higher in molybdenum, aluminum, and titanium, and lower in manganese, copper, barium, and strontium compared with the soils of the Hastings area where the vegetables grown there reflected these differences. The teeth of the Napier children were higher in molybdenum than those of the Hastings children. Also, the Medical Research Council of New Zealand, the U.S. Public Health Service, and the U.S. Navy found that areas with a low fluoride content in the soil and groundwater contained less than average calcium and boron, and where there was a low calcium/magnesium ratio, the susceptibility to dental decay was relatively high.(7)

Minerals in Medical Procedures and Treatments

A variety of minerals are used in medical procedures, as for example, the employment of radium in treatment of certain cancers. Radioactive tracers are used to obtain information about circulatory problems, and in a number of other medical procedures. In studying the human digestive tract with a fluoroscope, a barium compound is used. It may be noted that barium is also used as a material to weight drilling muds in oil well operations. It is a versatile mineral!

Minerals in Agriculture — Health of livestock

Selenium

The effect of selenium deficiency is one of the earliest cases of mineral deficiency observed. Marco Polo described this during his travels in China, noting that hoofs of cattle went bad and were dropping off. He did not know, of course, that it was due to lack of selenium but he did relate the disease to a particular region. Subsequent studies of that region of China have identified it as selenium-deficient.

In British Columbia, cattle are injected with selenium to cure muscular dystrophy. In New Zealand, where there is no natural selenium in the soils, selenium must be included in the fertilizer to prevent decimation of the flocks of sheep on which New Zealand is dependent for substantial income. Selenium, it was found, is a catalyst to the enzymes which, combined with vitamin E, operated to insure the production of viable offspring.

Too much selenium

It has been observed for a long time that in some parts of South Dakota cattle may be victims of selenium poisoning due to excess of the metal in the local soil. This has resulted in what has been called "alkali disease" or "blind staggers." It also has become well established over the years that Wyoming ranchers cannot successfully graze sheep on a particular geological stratum, specifically the Niobrara Chalk, or on certain areas of the Mesaverde Sandstone in the spring, especially in wet years, due to selenium poisoning. The sheep would lose muscular control and frequently died. Selenium poisoning is quite widespread around the world, being known in Canada, Mexico, England, Wales, Colombia, Argentina, Israel, Zaire, Nigeria, Kenya, India, South Africa, Australia, and Japan.

An extreme problem of selenium poisoning developed in the northwestern part of the San Joaquin Valley of California, when the San Luis Drain was not completed into San Francisco Bay, but instead was emptied into the Kesterson Wildlife Refuge wetlands area. These waste irrigation waters were high in selenium, and as they accumulated and evaporated, the selenium content in Kesterson increased to highly toxic levels. Deformed birds of many species resulted. Men were hired to go through the refuge and discharge firearms to prevent waterfowl from staying there and nesting. So far, the toxic effects are confined to the wildlife of the refuge, but there is concern that the selenium will ultimately get into the groundwater of the area, which is used for human consumption.

Molybdenum

Molybdenum poisoning has been observed in several areas in eastern California, notably in Mono and Inyo counties. The soils which contain excesses of molybdenum are usually alkali, salty, and occur over high water tables. Toxic molybdenum levels are most likely to occur in plants growing on valley floors, near rivers, streams, lakes, or sinks. Excess molybdenum inhibits the storage of copper in the bodies of livestock (mostly cattle, but also in sheep and horses), and creates a copper deficiency. Symptoms are brittle bones and joint abnormalities (lameness), weakness and listlessness, a dry, rough coat, and hair color changes. There is also a low level of conception and more frequent abortions. The cure is relatively easy — simply add copper sulfate to the feed supplements.

Iron, copper, and cobalt

A number of years ago, the term "salt sick" was applied to animals which exhibited weakness, failed to fatten, and in general were anemic. It was found that cattle developed this trouble when they foraged only on grass grown on certain light-colored sandy soils and some peat soils. Examination of the soils showed they were deficient in both iron and copper compared with other soils. When iron, copper, and cobalt were added to the soils, the problem disappeared.

A deficiency of copper may cause a cardiovascular disorder in cattle called "falling disease." It ultimately may result in heart failure, and the animal may suddenly die, frequently after excitement or strenuous exercise.(13)

Magnesium

Another example of the influence of a particular element on livestock is known as "grass tetany," which has been reported from Ohio. This is the result of the lack of magnesium in the diet to maintain normal magnesium levels in the serum of the blood. Cattle do not have a readily usable magnesium reserve, and the symptoms of grass tetany can develop quite rapidly, within 24 to 48 hours. The cattle may die within a few hours. Female cattle are particularly subject to this problem shortly after giving birth to their calves. It was found that this affliction was chiefly in a 26 county area where the soils are primarily derived from sandstones and shales, both of which in that area are deficient in magnesium. The apparent solution to this problem was to increase the magnesium content of the soil by applying dolomite, a high magnesium content limestone. Fortunately, this is also a long term solution because dolomite weathers slowly and gradually releases magnesium to the soil over a period of as much as 10 years. The immediate solution to the problem was to dust pastures with magnesium oxide.

Phosphate

Bones and teeth are made of calcium phosphate. Calcium is widespread but phosphate occurs much less widely. Phosphate deficient areas include parts of Argentina, South Africa, Australia, and the United States. In the 1920s, it was noted that cattle in South Africa tended to chew on old bones from dead animals. Giraffes in East Africa did the same thing. It was shown that this action was caused by a phosphate deficiency in the soil. Gnawing on bone is a well-know problem called osteophagia, and occurs in other phosphate deficient regions.(15)

Sodium

There is a large moose population living in Isle Royal National Park in Lake Superior. The island vegetation is so low in sodium, that it would supply only about a tenth of the needs of the moose. However, the moose have met this problem by eating aquatic vegetation which has up to 500 times the amount of sodium which the island's land plants have.(15)

Minerals in Agriculture — the Plants

Many minerals are critical in soils for plant growth, or in some cases may inhibit plant growth. A few are cited here.

Boron

This element can be quite toxic to some plants, and is carried in irrigation waters derived from boron-rich geological terrains. Because boron is quite soluble, water in these boron-rich terrains will pick up that element quite easily, The problem with boron is that the range between nutritionally deficit and toxic levels is very narrow. As boron cannot be precipitated or easily removed from water by other methods, the solution is to dilute it with water with very little boron in it. Insufficient boron can also be a problem. It has been found that peanuts with a boron deficiency tend to be hollow, a symptom called "hollow-heart." Boron deficiency also negatively affects the growth of white clover.

Copper, zinc

These are important trace elements for many plants. In some areas, for example, the addition of 10 pounds of water soluble copper sulfate per acre more than tripled the production of oats. Correcting a zinc deficiency markedly increased corn production.(7)

Salt

There are many kinds of salts but the one which causes most agricultural problems is sodium chloride. Many plants are highly sensitive to high chloride concentrations. Salty waters, either from surface or groundwater sources, can be very destructive of plant life. When the ancient Romans finally destroyed Carthage, the area of the city and its surroundings were plowed and then covered with salt. The build-up of various salts in irrigated areas can eventually destroy the soil's ability to grow plants. Rainfall is devoid of salt, but all irrigation water which must be either surface water or groundwater has some dissolved mineral materials in it. Salts not flushed out will destroy the soil's plant growing capability. Some well water is too salty to be used for irrigation.

Fertilizers in general

This is a very large subject in itself. Many detailed studies have been made on plant nutrition and how various fertilizers can affect plant growth. Just as animals including humans need certain minerals to survive and grow properly, the same applies to plants. Soils are analyzed and the deficiencies of minerals relative to the crops to be grown on them are provided for by the proper added fertilizers. Chief among the elements which have to be added are copper, iron, zinc, magnesium, nitrogen, and phosphorus.

As more and more studies are made of the details of soils, their mineral constituents, and the requirements of plants and animals which depend on these soils, the missing minerals are now being replaced by specifically designed fertilizers. Thus the problem of mineral deficiencies is gradually being reduced.

Fertilizers not a complete substitute

Farming to some extent, however, is more concerned with the crop yield —the quantity of the product — rather than the quality. High crop yields may not be the same as a high nutrient content. Fertilizers may stimulate crop yield, but may not completely replace nutrients taken from the soil by successive croppings. Some crops taken from more poorly yielding unfertilized soils may have more mineral nutrients in them than the soils which, by artificial fertilizers, are high yielding.(15) This may be a serious and growing problem, for the use of commercial fertilizers in the United States increased 16 times between 1955 and 1980, and continues to increase. The answer to this problem appears to be in the use of organic fertilizers such as animal manure, and "green" manures —plant debris — and to plant legumes which fix nitrogen in the soil. But the manure approach to fertilizing is more labor intensive than simply the convenience of artificial fertilizers, and therefore may not come into widespread practice, given the economic factor. Some smaller farms are using the organic method, but the products are often higher priced than those from larger operations.

Mineral-Rich Glacial Soils

Geological processes through glacial deposits have renewed the fertility of a substantial portion of the Upper Midwest area of the United States including that very productive State of Iowa. Iowa and adjacent parts of America's agricultural heartland grow wonderfully rich crops full of the nutrients needed for good health. The roots of these crops are in glacial drift, a deposit of varied materials transported from the north by glaciers in the quite recent geological past. These fresh rock and soil materials were derived in large part from granitic areas which have an abundance of the essential minerals. After deposition, the glaciers were gone and weathering took over to make these elements available to plant life. Also, the cold winters preserve the rich black humus from the decayed vegetation of the previous year.

The apparent importance of these mineral-rich glacial soils to longevity was reported by a panel headed by Dr. Howard Hopps, Professor of Pathology at the University of Missouri. The study compared death rates of men ages 35-74 in two 100,000 square mile areas. One area was in the glaciated Upper Midwest mineral-rich soil and groundwater area, and the locality was in the southeastern coastal area of parts of Virginia the Carolinas, Georgia, and central Alabama. This latter area has a meager supply of minerals in its drinking water and soil. The report stated that "for every 100 men in this age range who died in a given year in the Upper Midwest region, 200 died in the coastal area."

The panel found that cardiovascular diseases, primarily heart attacks and strokes, accounted for most of the differences in deaths between the two areas. Hopps noted that the Upper Midwest was left rich in minerals and trace elements by the glaciers that "ground up the rocks and made minerals in them available." These minerals include iron, copper, manganese, fluoride, chromium, selenium, molybdenum, magnesium, zinc, iodine, cobalt, silicon, and vanadium. In the Southeast, Hopps stated that "the minerals have been leached out of the soil for millennia." He also observed that the differences were consistent, stating that "no county in the Minnesota part of the region, for example, was above average in deaths. It seemed to be an inescapable conclusion that a lot of people in the Upper Midwest must be living a lot longer." The study focused on white men to rule out the possibility of different racial makeups of the regions, affecting the results. The study concluded that apparently trace minerals in the soil and water do contribute to relative longevity for persons living in this area of glacially transported materials, as compared with other areas without these fresh new rocks from which to weather out vital elements into the soil.

Soil Types and Bacterial Diseases

The presence of certain bacterial diseases can also be related to soil types. For example, the distribution of anthrax has been shown to be geologically controlled to a considerable extent. Anthrax is an infectious disease of warm blooded animals such as cattle and sheep, and can be transmitted to humans. It is frequently fatal. It is a disease which develops in areas where the soil pH is more than 6, and the minimum temperature is 60 degrees Fahrenheit. The soil conditions favorable for anthrax are associated with limestone terrains, alkaline alluvial soils, and clay soils where there is a hardpan layer. The occurrence of anthrax is rare on well-drained sandy soils or shale soils.

Minerals and National Health

Can minerals in the soil, or lack thereof, affect the health of a population as large as a nation? Perhaps. Some of the recently established smaller island nations of the Pacific are simply atolls — that is, coral reefs. Corals are very selective in what they use to make their reefs. The material is nearly pure calcium carbonate. If not for the fish which the inhabitants of these islands have incorporated into their diets, their daily meals would have been markedly deficient in vital minerals for soil formed from pure limestones is very poor in essential trace elements.

More recently, with the development of international trade, the more isolated nations throughout the world can now import a variety of foodstuffs which augment local diets and provide a greater spectrum of minerals. However, some nations still do not have the financial ability to import substantial quantities of food, and remain dependent on the local plant life. As more and more intensive agriculture further depletes the minerals of the soil in these communities, malnutrition becomes evident in populations. This is especially true in Africa.

Minerals and Healthy Nations

In general, mineral deficiencies would more likely be a problem in a small nation where there would be no great diversity of geology and therefore of minerals as compared with a nation of larger area. But even on a larger scale, lands which have been leached by rainfall over thousands and in some cases millions of years such as many parts of the Amazon Basin of Brazil clearly are not nearly so fertile as the relatively fresh glacially derived soils in the Upper Midwest of the United States and of northern Europe. Correspondingly, the populations of the these respective regions in all likelihood would be expected to have different levels of health. The advantage is clearly with the population deriving its sustenance from the fresh, mineral-rich glacial soils. Over the centuries, the vitality or lack of vitality of the citizens of a nation due to the mineral content of the bedrock and its derived soils, or its transported soils, may well be a contributing factor to its destiny. Nations with the better minerals in their soils will be nations with the healthier populations. People living off the glacial drift-derived soils of Illinois, Minnesota, and Iowa or in northern Europe will definitely be better nourished than those on the leached laterites of Brazil and parts of Africa. The health of its citizens can greatly affect the economic strength of a country. Examples are frequently shown across the screens of world-wide television, as under-nourished peoples migrate back and forth to just seek bare sustenance.

Recently the relationship between minerals in the soil of an area or country and the health of the people living there has received increasing international attention. In 1993, the Association of Geoscientists for International Development and the Society for Environmental Geochemistry and Health met in London. This was a forum of geologists and scientists in the medical professions. The focus was on developing countries where health problems related to the geology of the areas were particularly widespread. A summary of some of the addresses was made by McCall.(15) In one address, Jane Plant of the British Geological Survey (BGS) described BGS developments in making high-quality data available for studies of the relationship between the health of humans and animals and the distribution of geochemical elements and species in an environment. J. D. Appleton of the BGS described the importance of drainage geochemical maps being prepared by the survey. These will aid in studying mineral deficiencies and excesses affecting the health of grazing ruminants in tropical countries.

Wildlife

Most of this chapter has concerned the importance of minerals in human nutrition, but minerals also are important in the diets of wildlife.

Serengeti Plains

It had long been observed without knowing the cause, that certain regions of the Serengeti Plains of East Africa attracted a particularly large number and variety of animals. Botany Professor Samuel McNaughton of Syracuse University measured the levels of 19 minerals in some of the Serengeti grasses and found that the grasses in regions which the animals favored had much higher levels of minerals, particularly phosphorus, magnesium, and sodium than did the grasses in adjacent areas.

Salt licks

Around the world, animals have shown the importance of minerals to wildlife by visiting natural mineral licks. These are commonly called "salt licks" although other minerals besides common salt may be involved. There are a variety of minerals which animals need and take up in their diet. Jones and Hanson made an extensive study of the effect of minerals on the health of the large game animals in North America based on salt licks.(14)

A large salt lick in Kentucky attracted many kinds of recently extinct animals, such as the mammoth. Thomas Jefferson, Second President of the United States, heard about this area and collected large quantities of bones from the salt lick in what is now called the Big Bone Lick, in Boone County, Kentucky.

Salt licks were also important to native populations because these natural features lured animals to them and therefore were excellent places in which the humans could successfully hunt. In that way these salt licks aided in the livelihood of aboriginal societies. The licks also were used by the native populations for their own supplies of salt.(23) After the country was settled, licks continued to be important to the local residents as a place to harvest game, and at one lick in Northern California, more than 10,000 deer had been killed there.(6)

Summary

It is said, "you are what you eat." We are made from what we take from the Earth to exist. The complex chemistry of the human body is still, in many ways, a great unknown. But among the things which are known is that minerals, in varying amounts down to very minute traces, are critical to good health. Obtaining these minerals in food is vital to survival. The geology of the Earth controls the natural distribution of these minerals and in turn affects the futures of people.

BIBLIOGRAPHY

1 BALLENTINE, RUDOLPH, 1978, Diet and Nutrition. A Holistic Approach: The Himalayan International Institute, Honesdale, Pennsylvania, 634 p.

2 BEALL, J. V., and MERRITT, P. C., 1966, Phosphate & Potash. Minerals to Feed the World: Mining Engineering, October, p. 75-114.

3 BEESON, K. C., 1941, The Mineral Composition of Crops with Particular Reference to the Soils in Which They Were Grown: U.S. Dept. of Agriculture Misc. Pub. 369, Washington, D. C., 164 p.

4 BRODY, J. E., 1995, Chromium Diet Supplement Fails Lab Tests: New York Times News Service, November 25.

5 BRYAN, O. C., and BECKER, R. B., 1935, The Mineral Content of Soil Types as Related to "Salt Sick" of Cattle: Jour. American Society of Agronomy, v. 27, p. 120-127.

6 BRYANT, H. C., 1918, Deer Licks in the Trinity National Forest Game Refuge: California Fish and Game, v. 4, n. 1, p. 21-25.

7 CANNON, H. L., and DAVIDSON, D. F., (eds.), 1967, Relation of Geology and Trace Elements to Nutrition: Geological Society of America Special Paper 67, New York, 68 p.

8 DICKEY, R. D., and BLACKMAN, G. H., 1940, A Preliminary Report on Little-leaf of the Peach in Florida — A Zinc Deficiency: Florida Agricultural Experimental Station Bull. 344, p. 1-19.

9 DONAHUE, R. L., MILLER, R. W., and SHICKLUM, J., 1983, Soils. An Introduction to Soils and Plant Growth: Prentice-Hall, Inc., Englewood Cliffs, New Jersey, 667 p.

10 FISHER, J. A., 1990, The Chromium Program: Harper and Row Publishers, New York, 311 p.

11 FLEISCHER, MICHAEL, 1962, Fluoride Content of Ground Water in the Conterminous United States: U.S. Geological Survey Misc. Geol. Inv. Map I-387.

12 GILBERT, F. A., 1957, Mineral Nutrition and the Balance of Life: Univ. Oklahoma Press, Norman, Oklahoma, 350 p.

13 HOWARD, J. L., (ed.), 1981, Current Veterinary Therapy: W. B. Saunders Company, Philadelphia, 1233 p.

14 JONES, R. L., and HANSON, H. C., 1985, Mineral Licks, Geophagy, and Biogeochemistry of North American Ungulates: The Iowa State Univ. Press, Ames, Iowa, 299 p.

15 McCALL, JOE, 1994, Health and Environmental Geochemistry: Geotimes, January, p. 5.

16 OWEN, O. S., 1980, Natural Resource Conservation. An Ecological Approach: Macmillan Publishing Co., Inc., New York, 883 p.

17 SCHMIDT-NIELSEN, KNUT, 1994, How Are Control Systems Controlled?: American Scientist, v. 82, p. 38-44.

18 SCHROEDER, H. A., 1993, The Trace Elements and Man: The Devin-Adair Company, Old Greenwich, connecticut, 171 p.

19 SEARLS, J. P., 1985, Potash: Preprint from Mineral Facts and Problems 1985 Edition, U.S. Bureau of Mines, Washington, D. C., 17 p.

20 SPALLHOLZ, J. E., MARTIN, J. L., and GANTHER, H. E., (eds.), 1981, Selenium in Biology and Medicine: AVI Publishing Company, Inc., Westport, Conn., 573 p.

21 STOWASSER, W. F., 1983, Phosphate Rock: U.S. Bureau of Mines, Mineral Commodity Profile, Washington, D. C., 18 p.

22 TISDALE, SALLIE, 1988, Lot's Wife. Salt and the Human Condition: Henry Holt and Company, New York, 214 p.

23 UNDERWOOD, E. J., 1962, Trace Elements in Human and Animal Nutrition, (Second Edition): Academic Press, Inc., New York, 429 p.

24 WARREN, H. V., 1965 Medical Geology and Geography: Science, v. 148, April 23, p. 534-537, 539.

25 YUDKIN, JOHN, 1985, The Penguin Encyclopedia of Nutrition: Viking, New York, 431 p.

CHAPTER 19

Strategic Minerals — How Strategic are They?

From time to time the press reports there is national concern about supplies of "strategic minerals." There is unrest in South Africa, location of almost all the world's platinum. There is political instability in Zaire, source of much of the world's cobalt. How important are these minerals to the rest of the world, particularly the highly industrialized nations? And, in turn, how much of a power struggle might be caused over these minerals?

Considerable difference of opinion exists about the importance of strategic minerals. It has been suggested that Germany's dependence on Swedish iron ore might have been Germany's Achilles Heel which could have prevented or stopped World War II.(14) Some continue to say that strategic minerals are of great and vital concern, and their free flow through international trade must be maintained at virtually any cost. Others now contend there are adequate substitutes, and these, combined with government and private stockpiles of important minerals make the prospect of the cut-off of strategic minerals at any given time of little consequence, particularly as the world has now become more and more an integrated interdependent economic unit.

"Strategic Mineral" Defined

To understand the issue, the term "strategic mineral" should be defined. The U.S. Bureau of Mines advises by letter dated February 1, 1988, regarding "strategic" and "critical" that:

> "There have been many attempts to define each term by a large number of individuals, however, the Bureau does not make a distinction between the two terms. The Defense Production Act of 1950 regards 'strategic' and 'critical' essentially as the same. Section 12 of the 1950 Defense Production Act states: For purposes of this Act: the term 'strategic and critical materials' means materials that (A) would be needed to supply the military, industrial and essential civilian needs of the United States during a national emergency, and (B) are not found or produced in the United States in sufficient quantities to meet such a need. The act goes on to designate energy as a 'strategic and critical material'."

Under this rather broad definition, nearly all minerals are critical which are not produced in sufficient quantity in the United States, or in any industrial nation, to make the nation independent of foreign supplies. Nearly all mineral materials and energy minerals are needed in some fashion in industrial and military establishments. However, in a more narrow definition and a common usage of the term "strategic" or "critical" they would apply to certain metals which are irreplaceable in vital industrial and military operations and which exist in only a few deposits in the world, and therefore are not readily available to most countries. That definition is used here. Oil is a strategic non-metal consideration, and is taken up separately.

Strategic metals in a jet engine

A common example of minerals which are in this category has frequently been given as a list of metals which go into a jet airplane engine, a piece of machinery essential to both the military and civilian economies of the world. For these metals there are no adequate substitutes: aluminum, chromium, columbium, cobalt, nickel, tantalum, and titanium.

Of the metals listed, cobalt is perhaps the one which could be regarded as the most strategic because it is produced in so few places, and it is absolutely vital in jet engines.(5,10) There is no substitute for cobalt in this use. In 1978, during a civil war in Zaire, rebels invaded the Shaba Province where 70 percent of the world's supply of cobalt is produced. In May that year the price of cobalt rose from $6 a pound to more than $40.

Computers

Computers are essential to all modern industry and commerce, as well as to military operations. The list of critical metals needed in computers for which there are apparently no substitutes includes platinum, gold, silicon, rhenium, selenium, strontium, tantalum, gallium, germanium, beryllium, and yttrium.

Critical minerals by nations

What minerals are critical and those that are not depends to some extent on the country. A metal may be critical in a particular use but if the country has plenty of that metal, its situation in that country is not critical. For example, the United States has no operating nickel mines, but Canada has ample sources. Nickel is critical in the United States but not in Canada. The United States has lots of molybdenum, a key ingredient in steel, but Japan, which makes lots of steel, has no molybdenum. Molybdenum, therefore is a critical metal to Japan. It is not a critical metal in the United States.

Each nation has its own resources and its own needs, and thus has a particular list of what might be called critical minerals. The vulnerability of various countries and regions (European Economic Community) to being cut off from particular minerals differs widely with the commodity. In minerals usually listed as strategic, Russia is by far the most self-sufficient, and the United States is intermediate.

Relative Importance of Critical Metals

There are also wide differences in the importance of each mineral, and the degree to which it can be substituted. Antimony, for example, has no major critical uses and its absence would hardly paralyze a country. On the other hand, manganese is a key metal in the making of steel, and there appears to be no substitute. Tungsten is a critical material for which there are several specific uses and for which there appears to be no alternative. But, bauxite, the chief ore of aluminum, is critical only in the sense that it is the ore of aluminum from which that metal can most easily be recovered. Aluminum constitutes eight percent of the Earth's crust. There is no shortage of aluminum in the ground virtually anywhere; ordinary clay is an aluminum compound. In the case of bauxite, nature has partially done the work of breaking the tight aluminum bond with other elements, thus reducing the amount of energy needed to recover the metal. If energy were cheap enough, obtaining aluminum from common clay would be no problem.

Generally Regarded Strategic Metals

Table 13 is a selected list of what are generally regarded as strategic metals and which are held by South Africa and Russia. This tabulation is essentially correct, but it should be noted in the case of chromium that there are small deposits in a number of countries around the world which have not been included in the figures. Also, recent discoveries of uranium in both Canada and Namibia, where large deposits are already known, may change these figures somewhat. However, they are accurate in showing that South Africa and Russia are fortunate to possess some of the more critical metals.(20)

Table 13. Percentage of Selected Strategic Metal Reserves Held by South Africa and Russia

Metal	South Africa	Russia	Combined
Platinum group	86	13	99
Manganese	53	45	98
Vanadium	64	33	97
Chromium	95	1	96
Uranium	27	13	40
Titanium	5	16	21

(Source: U.S. Bureau of Mines 1994)

Iron, aluminum, zinc, lead, and copper are also important, but their distribution is so widespread that they are unlikely to be in critical supply at any given time.

Supplies of some minor metals are particularly important. "Without manganese, chromium, platinum, and cobalt, there can be no automobiles, no airplanes, no jet engines, no satellites, and no sophisticated weapons — not even home appliances."(11)

Oil Becomes Strategic in the USA

As noted in the early part of this chapter, oil is excluded from "strategic minerals" list because it is a special resource, and is given separate treatment in Chapter 12, *The Petroleum Interval*. But oil is certainly a strategic mineral for the countries which have none. It is becoming so for other countries which have some oil, but as industrialization increases and domestic oil supplies decrease, more and more oil must be imported. This is the situation of the United States. Until 1970, the United States was self-sufficient in oil, and it was not a strategic mineral but it has surely become such since that time as the U.S. now imports nearly eight million barrels a day, which is more oil than it produces.

Oil, in 1920, was not a U.S. strategic material for in that year the United States produced nearly 80 percent of the entire world's oil, and no one had ever heard of Saudi Arabia for the country of that name did not exist until 1932.

Now the United States accounts for less than 12 percent of world oil production, and its reserves are dwindling rapidly. So what was a very abundant and non-strategic material in 1920, is now in domestic short supply and will only become more so as time goes on.

Today, oil is the most strategic mineral for the United States, and a strategic supply has been stockpiled in some abandoned salt mines in salt domes in the Louisiana coastal area. Thus the concept of what are strategic minerals, using the criterion of adequate supply within domestic borders, changes as exploitation and depletion of these materials proceeds, and demands increase with a growing population.

On this basis, in the world's major industrial countries the number of strategic minerals can only increase with time. Thus in terms of location of demands and sources of these materials, the balance of power will shift toward those countries and areas which have the critical raw materials. In considerable number, these are today's lesser developed nations.

Uneven Geographic Distribution of Minerals

The distribution of strategic materials is the result of myriad geologic events of the past which had, of course, no relationship to present national boundaries. Mineral resources and energy mineral resources are very unevenly distributed around the world. The great silver deposits of the globe are largely in the Western Hemisphere near the west edge of both North and South America (Peru, Mexico, and the United States). Platinum, on the other hand, is chiefly found in just two places, South Africa and a relatively small area in Russia. (Minor deposit in Montana). These diverse locations of minerals combined with present political boundaries cause the problems of access to minerals by the world's economies.

Technology

Two factors, however, might possibly modify the problem of uneven mineral distribution. One is that perhaps an advance in technology can devise substitutes in particular situations for what are now regarded essential strategic minerals. In December, 1988, Ford Motor Company announced a substitute for platinum in automobile emission systems. However, no substitute is yet known for platinum in its many other vital uses.

Changing Societal Demands

The second possibility is that changing lifestyles, or different social and economic systems will reduce or eliminate the need for a particular material. However, given the desire of people to maintain or even increase their standard of living, this suggested solution seems unlikely. Demand will continue to grow for minerals of all kinds..

The strategic minerals each country needs today will probably continue to be strategic. With the increasing worldwide industrialization, the number of nations for which certain minerals become strategic will increase. As demand increases against finite supplies, higher grade ores will be depleted first, leaving the more expensive lower grade deposits for exploitation. The supply problem of strategic minerals will grow and their cost will increase.

Strategic Mineral Stockpiles

From time to time, nations set up strategic mineral "stockpiles." This concept stems from experiences in World War I and World War II, and is commonly thought of as a military measure. However, because of the horrors for all sides of atomic warfare, major conflict may be avoided in the future. In the event of atomic warfare, the end would probably come so quickly that a strategic minerals stockpile would be of no value. In either case, like building or re-commissioning battleships, we would probably be fighting the last war with the mineral stockpile concept, rather than the next war. As suggested elsewhere in this volume, the next war is now at hand in the form of economic warfare. Today's warfare is the competition among economies to see which ones are the more productive and efficient, and can do the most for its citizens. Strategic mineral stockpiles, if they are to be built at all, should be oriented primarily toward civilian economic demands.

A reasonably sized stock of strategic minerals can be maintained for use against the occasional disruption of such supplies by political upheavals in the source nations. And simply for the day to day maintenance of an industrial economy, it is necessary to maintain modest supplies of strategic metals in order to keep production lines flowing smoothly.

By law, the United States government is directed to maintain a strategic mineral stockpile.(15) This is currently being done more or less half-heartedly. But in today's realities, the concept of strategic mineral stockpiles for war may well be an anachronism.

The answer to the concern about strategic minerals today is simply free trade. Free trade is likely to prevail as most of the nations that produce strategic minerals which have a limited geologic distribution, such as cobalt, are developing nations in need of foreign exchange. Most metallic minerals are rather widely distributed, and are available to any nation through various international trade channels if they are not produced domestically.

BIBLIOGRAPHY

I AGNEW, A. F., (ed.), 1983, International Minerals. A National Perspective: American Association for the Advancement of Science, Washington, D. C., 163 p.

2 BAKER, ARTHUR, et al., 1972, Forecasts for the Future — Minerals: Nevada Bureau of Mines and Geology Bull. 82, Reno, Nevada, 222 p.

3 BATES, R. L., and JACKSON, J. A., 1982, Our Modern Stone Age: William Kaufman, Los Altos, California, 132 p.

4 BROBST, D. A., and PRATT, W. P., (eds.), 1973, United States Mineral Resources: U.S. Geological Survey Prof. Paper 820, Washington, D. C., 722 p.

5 BROWN, D. S., 1989, Exploration Versus Exploitation: Strategic Mineral Battles: Address (for U.S. Bureau of Mines) May 17, 1989, at 27th International Affairs Symposium, Lewis and Clark College, Portland, Oregon.

6 CAMERON, E. N., (ed.), 1972, The Mineral Position of the United States 1975-2000: Pub. for Soc. of Econ. Geologists Foundation, Inc., by Univ. Wisconsin Press, Madison, Wisconsin, 159 p.

7 CAMERON, E. N., 1986, At the Crossroads. The Mineral Problems of the United States: John Wiley and Sons, New York, 320 p.

8 CASTLE, E. N., and PRICE, K. A., (eds.), 1983, U.S. Interests & Global Natural Resources: Resources for the Future, Inc., Washington, D. C., 147 p.

9 CHALIAND, GERARD, and RAGEAU, JEAN-PIERRE, 1990, A Strategic Atlas. Comparative Geopolitics of the World's Powers: Harper & Row, Publishers, New York, 224 p.

10 CROCKETT, R. N., et al., 1987, International Strategic Minerals Inventory Summary Report: Cobalt: U.S. Geological Survey Circular 0930-F, Washington, D. C., 54 p.

11 DORR, ANN, 1987, Minerals — Foundations of Society (2nd ed): American Geological Institute, Alexandria, Virginia, 96 p.

12 HOLMES, H. N., 1942, Strategic Materials and National Strength: The Macmillan Company, New York, 106 p.

13 JAMES, DANIEL (ed.), 1981, Strategic Minerals: A Resource Crisis: Council on Economics and National Security, Washington, D. C., 105 p.

14 MILWARD, A. S., Jr., 1967, Could Sweden Have Stopped the Second World War? Scandinavian Economic History Review, v. 15, p. 127-138.

15 MORGAN, J. D., 1960, U.S. Strategic Materials Stockpile and National Strategy: Mining Engineering, v. 12, p. 925-928.

16 ROUSH, G. A., 1939, Strategic Mineral Supplies: McGraw-Hill, New York, 485 p.

17 SEELY, ARTHUR, 1981, How You Can Profit From Strategic Metals in the 1980s: Real Equity Diversification Publishing Company, Denver, Colorado, 279 p.

18 SINCLAIR, J. E., and PARKER, ROBERT, 1983, The Strategic Metals War: Arlington House/Publishers, New York, 185 p.

19 SNYDER, G. H., 1966, Stockpiling Strategic Materials. Politics and National Defense: Chandler Publishing Company, San Francisco, 314 p.

20 STRASS, S. D., 1979, Mineral Self-sufficiency — The Contrast Between the Soviet Union and the United States: Mining Congress Journal, p. 49-54, 59.

21 TAYLOR-RADFOD, R. S., and PUGSLEY, J. A., 1983, The Metals Investment Handbook: The Common Sense Press, Inc., Costa Mesa, California, 171 p.

22 UNITED STATES BUREAU OF MINES, 1983, The Domestic Supply of Critical Minerals: Washington, D. C., 49 p.

23 UNITED STATES BUREAU OF MINES, 1985, Mineral Facts and Problems: U.S. Bureau of Mines Bull. 675, Washington, D. C., 956 p.

24 UNITED STATES BUREAU OF MINES, 1989, Mineral Commodity Summaries 1989: U.S. Bureau of Mines, Washington, D. C., 191 p.

25 VOSKUIL, W. H., 1955, Minerals in World Industry: McGraw-Hill, New York, 324 p.

26 WALLACE, H. A., 1941, The Silent War. Battle for Strategic Materials: Colliers, Nov. 22, p. 14-15.

27 WARD, J. M., et al., 1975, In Short Supply. A Critical Analysis of World Resources: National Textbook Company, Skokie, Illinois, 342 p.

28 YODER, H. S., Jr., 1982, Strategic Minerals: A Critical Research Need and Opportunity: Proc. American Philosophical Society, v. 126, n. 3, p.229-241.

29 YOUNGQUIST, WALTER, 1980, Investing in Natural Resources (2nd Edition): Dow Jones-Irwin, Homewood, Illinois, 281 p.

CHAPTER 20

Nations and Mineral Self-sufficiency

Previous chapters have examined the relationships of minerals to nations in a variety of ways, including how minerals affect economic development, minerals in warfare, the concept of strategic minerals, minerals and health, and minerals and the money of nations. But, basic to all of these considerations is the extent to which a nation is self-sufficient in minerals. A nation that does not have resources it needs, must import them and find a way to pay for them. Otherwise, the country is destined to have a low standard of living.

Mineral self-sufficiency involves several variables. A largely agricultural nation, which has little manufacturing, needs very few minerals directly. So mineral self-sufficiency is a function of the nation's needs, at its particular level of economic and industrial development. Does one view the degree of self-sufficiency in terms of what it might be if the country were to embark on an industrialization program? To some extent it is a circular situation. A country does not need many minerals because it is not industrialized, and it may not be industrialized because it does not have many minerals. A nation may be virtually self-sufficient in minerals at a given standard of living, but if it chooses to increase its living standards or its population it could be very short of minerals. A highly diversified, highly developed industrial economy is likely to be short of a great variety of minerals. No industrialized nation is mineral self-sufficient. Thus there is the variable of how much and how fast a nation is becoming industrialized. Also, increased population may bring about shortages in minerals which were in sufficient supply to take care of the earlier demand from a smaller population. For example, the United States through the middle of the Twentieth Century had enough oil to supply its needs, but against a rising demand, it subsequently became a permanent oil importer. Just a steady demand for a given mineral will eventually deplete a nation's domestic supplies to the point where the mineral has to be imported. Or high grade ores may become depleted and a nation's economy may not be able to pay the higher costs of producing metals from lower grade ores. Mineral resources from other countries with higher grade and less costly minerals must then be imported. In some circumstances at a price beyond world prices, a nation can still remain, or regain self-sufficiency in a mineral.

These variables mean precise statements as to the degree of national mineral self-sufficiencies are conditional. It is a continually changing situation. An analysis of mineral self-sufficiency nation-by-nation becomes a large task. Once completed it would be almost immediately out of date. However, the degree of mineral self-sufficiency can be framed in general terms for several of the larger nations.

It is clear that smaller nations in terms of population, cannot very well be considered in this discussion. With a few exceptions such as Switzerland, if it is small, a nation will not have an industrial base of any consequence, and therefore mineral self-sufficiency is not a major consideration.

Diverse Geological Terrains Required

From a geological point of view, it is evident that to have a variety of domestic mineral resources, a nation must have a substantial land area. The geological occurrences of various minerals including energy minerals, are varied. Copper and oil occur in very different geological environments. Therefore, only countries with fairly large areas which include varied geological terrains can hope to have a broad spectrum of mineral resources.

Russia

The clear leader in the possession of large and diverse energy mineral and mineral resources is Russia. Because it is the world's largest country in land area, it has many different geological terrains wherein a large variety of minerals occur.

Russia has the world's largest known gas reserves, the world's largest coal deposits (although it may be that the United States has the largest deposits which can be economically recovered), large oil reserves (about two and one half times those of the U.S.), and enough undeveloped hydropower to supply the needs of 100 million more people.(2)

However, Russia is beginning to face one problem. As the once-prolific oil fields of southern Russia are now in marked decline, Russian oil exploration is having to move northward into more hostile territory. The better oil deposits have already been developed, and Russia is apparently no longer able to significantly expand its oil production. Discoveries that are made merely maintain production at current levels or raise it only slightly.

Russian oil field management has not been of high quality and the Russian oil industry has been declining. Production now within Russia proper (not including the total former USSR) is markedly lower than it was ten years ago. To prevent further decline in production, Russia has negotiated joint operating agreements with western oil and oil service companies, to take advantage of the West's advanced technology to better manage the fields more efficiently. It is possible that this Russian dependency on the West will have the benefit of making military conflict less likely. Thus, minerals may be a cause for peace as well as war.

It is unlikely, however, even with western aid, that Russia can regain its position as world top oil producer. Saudi Arabia now has that distinction and is likely to keep it. Russia is now passing from being self-sufficient in oil to becoming a net importer. It has been in oil production for a long time relative to the Middle East, and many of its fields are into old age. Russia's interest in the Persian Gulf region will surely increase as its own oil reserves decrease. Russia is conveniently located not far from the Gulf.

Russia's very large coal deposits, perversely, may have been a factor in the unsettling of the Soviet political system in the late 1980s. Soviet coal miners went on strike for a variety of improved wage and working conditions and benefits. The official news agency, Tass, reported that the strike threatened "catastrophe" in the steel and power industries, very heavily dependent on coal. To satisfy the workers, the Soviet leader, Gorbachev, promised to double the workers' pensions, increase the number of holidays, shorten the work week, and provide a variety of consumer goods previously difficult to find or in some cases almost unobtainable. These included meat, soap, cloth, shoes, cars, refrigerators, and furniture. But to do this meant taking these already scarce items from other parts of the country. The previous year, in 1988, Russian shoe production had fallen four percent, and cloth by six percent. Meat had been rationed in many parts of the USSR for several years, and for the general population there was a several year wait to buy an automobile.

The importance of Russian dependence on its huge coal reserves and the resultant dependence of the Soviet Union on them, enabled the miners to pressure the government into actions which it would otherwise not have taken. Settlement of the strike, which threatened to paralyze the country, set a very costly precedent and further upset the Soviet economy, but moved the country further towards democracy, where the citizens had a greater influence than before. Thus, the Soviet coal deposits directly affected the course of the Soviet economy and its political direction, and coal was used as the weapon by which the miners attacked the government.

No other nation equals Russia in metal self-sufficiency. Russia is in short supply in only a few minor metals. Its iron, copper, zinc, and lead deposits are large and not nearly so fully developed as those of the United States. Lack of transportation to remote areas of the Siberian Shield, a large area of potential mineral production similar to the Canadian Shield, has delayed exploitation of the deposits. Russian metal needs can probably be met from domestic sources much longer than by any other industrial nation.

China

China already has a fairly large industrial base and plans to expand it further. It must expand to provide for its still growing population. China does have a wide variety of minerals, including the world's largest deposits of tungsten, a most useful industrial metal in which almost all other countries are in short supply. China has large iron deposits and the world's third largest coal deposits. It was these latter two resources which impelled the Japanese to invade northern China (Manchuria) in 1931. China lacks oil, currently producing only about 3 million barrels a day compared with Russian production of about 7 million barrels, and the 6.4 million barrels of the United States. At best, China is expected to be able to produce only about 4 million barrels a day in the next decade. Although China now exports some oil to obtain foreign exchange, the demand for oil in China will soon exceed supply. Indeed, if China were to use oil to replace human labor and in vehicles equivalent to per capita oil use in the western industrial world, China would now face a huge oil shortage. There is now only about one car for every 2,000 people in China. This compares with one car for every two people in the United States. The chances of China equaling either current U.S. or Russian oil production capabilities appear remote.

While it may be said that China is self-sufficient in oil, it is only in the sense that there has been no major attempt to shift the burden of work from the human back to the machine. China, in terms of its population, and against the criterion of western world mineral use, is very short of energy and mineral resources, and will remain so.

Coal is China's current principal fuel, accounting for about 75 percent of the energy supply. This situation is likely to continue well into the next century. China has recently embarked on a program to have nuclear power meet some of its additional energy needs. It also is starting the huge Yangtze River dam project. But, coal is likely to continue for some time to be the source of at least half of China's energy supplies. China has announced plans to increase electrical generating capacity by 50 percent within the next few years. Under this plan, coal-fired plants will remain the main generating facilities and account for two-thirds of the capacity.(1)

However, increasing coal usage may be difficult. Even though possessing abundant coal, China somehow seems unable to develop supplies fast enough to meet demands. Lack of sufficient energy has become almost a way of life in China as power shortages have from time to time idled nearly 20 percent of its industrial capacity.

China's greatest problem relative to minerals and all other resources is simply that population growth has been outpacing the resources which can be found and developed fast enough to meet demand. Whatever energy mineral and mineral resources that China may have will be hard pressed just to keep even with the daily needs of its 1.2 billion people. Meeting the demands of a further increase in population will be a Herculean task, which may not be possible.

United States

At the beginning of this century, the United States enjoyed an abundance of nearly all energy mineral and mineral resources needed by a highly industrialized society. This abundance continued until the 1960s when growing demand and lower and lower grade remaining deposits made the U.S. increasingly a net importer of nearly all metals. Prior to that time, it already imported cobalt, nickel, chromium, platinum, tungsten, vanadium, aluminum ore (bauxite), and manganese. In 1982, Chile replaced the United States as the world's largest copper producer. In the early 1970s, the United States became hostage to other nations in terms of oil supplies being unable to supply its needs from domestic sources. Clearly, the United States will henceforth be a net importer of more and more mineral and energy resources, and the problem of balance of payments and the value of the dollar will continue to be serious.

Brazil

This industrial giant of South America has tremendous deposits of iron ore, and some good aluminum ore, but both of these take large amounts of energy to process. Brazil is short of energy resources, particularly coal and oil. This has caused severe problems in Brazil's international balance of payments and has put the country in a very large debtor position. Brazil must import almost all of its oil. During the great rise in oil prices, in the 1970s, Brazil's oil bill grew very large and the legacy remains in the form of a huge foreign debt. More recently, Brazil has had some minor successes in oil exploration but it still does not have nearly enough oil for its own needs, and almost certainly never will.

Western Europe

This region, including Great Britain, enjoyed a fair degree of mineral self-sufficiency in the early stages of the Industrial Revolution. But it soon became apparent to Germany, for example, that it did not have the resources to enter the industrial age for the long term. This spurred its aggressive territorial expansion plans which led to two World Wars. Great Britain, although a relatively small country, had a rather remarkable variety of mineral resources for its use including coal. To have this in a fairly limited geographic area worked out well, because the various resources did not have to be transported far. The geology of Great Britain is remarkably diversified for its less than 90,000 square miles. The many different geological terrains in a relatively small country was also probably the main reason why the science of geology was first developed in Great Britain. It is a splendid compact natural geological laboratory. However, although varied, Britain's mineral deposits were not large, and the demands of the Industrial Revolution soon depleted most of them. Britain turned to her colonies for supplies.

Now, without colonies, Britain is dependent on free trade and mineral imports. The relatively late discovery of oil in the British sector of the North Sea has, for the moment, made that country not only self-sufficient in oil but a net exporter. However, this surplus is almost gone and Britain will be a permanent oil importer within the next few years.

Canada

This second largest nation in the world in area, now has only a modest industrial base, perhaps because it has a relatively small population and still has many aspects of a frontier area, but the industrialization of Canada is growing. To support this, Canada has a fairly good variety of minerals including iron ore in great quantities. It also has large undeveloped hydropower resources, but not all of these may be utilized because of environmental reasons. One example is the huge James Bay project which has been at least temporarily delayed.

Today, Canada is self-sufficient in oil, but will lose that sufficiency within the next decade. Although Canada has some of the world's largest oilsands deposits, these can only be developed rather slowly. The capital costs are large, and the ability to produce oil in quantity from this source (as well as the heavy oil deposits there) takes time to put the additional facilities in place. The unconventional oil deposits can hardly equal the production which is now obtained from Canada's wells. Nonetheless, Canada has energy and mineral resources which should support it well for many years. Canada's uranium deposits can only be described as huge. Large uranium mines are already developed in Ontario and Saskatchewan. Even larger deposits have been discovered and not yet put into production, and are there for the future. Currently and historically, Canada has been a supplier of raw materials especially to the United States and to many other countries. But Canada will gradually retain and process more and more of its own resources to sustain its growing industrial base.

India

The world's second most populous nation, with now nearly a billion people, India lacks energy resources, especially oil. It is unlikely to find major oil deposits in its territories. India has a fairly wide variety of metals, but lacks, as do most other countries, such things as chromium, tungsten, platinum, and cobalt. India and China are very similar in having very large populations which now use virtually all available mineral resources simply to sustain a relatively low standard of living. India, therefore, is "self-sufficient" only in the sense that it does not import a great many minerals because it cannot afford to. It must live with what it has available. India does have iron and coal, and this supports a fair-sized steel industry. It is very short of oil, with proved reserves of 5.9 billion barrels, and total probable additional discoveries of 3 billion barrels.(12) Even with the limited amount of oil that is used in India, substantial amounts must be imported. India is one of the 20 top consumers of commercial energy in the world, but this consumption is spread over a great many people making the per capita energy consumption only about one-eighth of the world average.(9) Per capita energy consumption is a good index of standard of living. India illustrates how a large population can depress living standards.

Indonesia

Indonesia has the fourth largest population in the world. It is spread over several thousand islands. Indonesia is much like China and India in having modest mineral resources which are more or less adequate for the demand. But, the demand is probably conditioned by the availability of the minerals. That is, if more were available, more would be utilized. Indonesia does have some nickel, copper, aluminum ore, silver, gold, a little iron ore, and tin. Of these, tin and copper are the only significant exports. The chief mineral resource of Indonesia is oil, which provides more than half its export income. Indonesia is currently the 10th largest oil producer in OPEC and is the largest oil producer in the Far East.(6) However, it does not really have very large oil reserves, with only an estimated 5.8 billion barrels.(12) It is simply that it has more than most of its Asian neighbors. In terms of the total industrial mineral spectrum, Indonesia is not very well endowed, especially when its large population is considered.

South Africa

Although the Republic of South Africa occupies only a small portion of the African continent (some 471,000 square miles, about one-eighth the size of the United States), it accounts for more than half of the gross industrial production of the continent. One of the reasons it can do this is because it has a remarkably wide variety of minerals, including such rare items as platinum of which it holds about 80 percent of the world reserves. Also, South Africa is one of the few countries in the southern hemisphere with sizable coal deposits which it uses in its steel industry. It also has used the coal to supply about half of its oil needs, by converting coal to oil. This is a fairly expensive process, but South Africa has pursued it for that government wished to remain self-sufficient in liquid fuel as much as possible, fearing oil supplies might be cut off because of its racial policies.

Oil is the principal mineral resource which South Africa lacks, and this is not likely to change, except for the possibility of some modest offshore discoveries. However, with the new multi-racial government in place and no longer in fear of international economic boycotts, South Africa, with foreign exchange obtained from its minerals, is free to import the oil it needs, and it is reconsidering its use of expensive coal-derived oil. Other than for oil, South Africa is markedly self-sufficient in minerals.

Japan

Japan has an area smaller than California, and a substantial part of it is mountainous and volcanic. In terms of mineral and energy minerals, among industrial nations, Japan is the most deficient. Yet Japan is the second largest industrial nation in the world. It has achieved this despite the fact it has virtually no mineral or energy mineral resources. Japan is nearly 100 percent dependent on imported oil, it has very limited coal deposits, and it has no metals in any substantial quantity (some minor copper).

How then does Japan survive and do so well? Japan survives by importing resources, upgrading them to finished products, and then exporting them. Without large and continuous supplies of raw materials, Japan would be paralyzed industrially. To pay for these imports, Japan must have a large value of exports. Having lost in the effort to obtain its needed resources by military action in World War II, Japan had to embark on a program of rapid and high quality industrialization. The success of this program is represented in nearly every home and office in the industrialized world, and in the many garages where Japanese automobiles are parked.

Japan knows it must import — and export — to survive. It is simply a factory on a few islands. That is the reason why Japan is so concerned about maintaining its strong position as an exporter. This causes continual friction with many other nations of the world, particularly the United States with which Japan has been enjoying a huge balance of trade surplus.

Degrees of Mineral Dependence

Table 14 states the degree of dependence of the United States, Japan, the European Economic Community (E. E. C.), and Russia on the principal industrial metals.

Table 14. Dependence of Major Industrial Nations on Imports of Selected Metals

Metal	U.S.A.	Japan	E.E.C.*	Russia
Niobium	100%	100%	100%	0%
Manganese	99	97	99	0
Bauxite (aluminium)	97	100	86	30
Tantalum	90	100	100	0
Chromium	88	99	100	0
Platinum	85	98	100	0
Nickel	75	100	100	0
Tin	72	96	92	24
Silver	58	93	93	18
Zinc	53	53	81	0
Tungsten	48	68	100	14
Gold	43	96	99	14
Iron ore	36	99	90	0
Vanadium	14	78	100	0
Copper	7	99	99	0
Lead	0	73	74	0

(Source: U.S. Bureau of Mines 1994)

*European Economic Community

From Table 14 it is evident that no nation is entirely self-sufficient in terms of metals. Russia comes the closest in this regard, and has the further advantage of still having a large amount of territory yet to be geologically explored in detail. In 1928 Russia was 60 percent dependent on imports for its mineral needs. But since that time it has embarked on an extensive geological exploration program and as a result of numerous significant mineral discoveries it is now almost entirely independent of outside supplies.

World Minerals and Outlook

Viewing the world as a whole, no mineral supply shortages are likely to occur in the near future, if international free trade prevails. And if large quantities of cheap energy can be made available (perhaps from fusion) the mineral supply problem would be greatly enhanced. With cheap energy, large low grade mineral deposits which cannot now be mined economically, could be exploited. But there is no assurance of cheap energy in the future.

The industrial nations, with the exception of Russia, are now, and will be increasingly dependent on foreign sources of raw materials. Where minerals exist because of the geological processes of the past, and where these minerals are now in demand, are, in many cases, quite different places. Can free trade compensate for this? Also, although there will be no immediate worldwide shortage of energy minerals and minerals, as nations increase their resource demands individual nations will have problems of paying for these materials. As higher grade deposits are depleted, lower grade, higher cost mineral sources must be used. This raises the cost of minerals. A further transfer of wealth from the mineral-consuming countries to the mineral-producing nations will occur. Some nations will not be able to pay the bill and both their standard of living and their industrial development therefore may be adversely affected. It will be an ever-changing world economic scene, in which minerals are destined to continue to play a *controlling* role.

BIBLIOGRAPHY

1 ANONYMOUS, 1995, China to Raise Electric Capacity: The Wall Street Journal, July 31.

2 CENTRAL INTELLIGENCE AGENCY, 1985, USSR Energy Atlas: Central Intelligence Agency, Washington,D. C., 79 p.

3 CRESSEY, G. B., 1962, Soviet Potentials. A Geographic Appraisal: Syracuse Univ. Press, Syracuse, New York, 232 p.

4 DERRY, D. R., 1980, A Concise World Atlas of Geology and Mineral Deposits: John Wiley and Sons, New York, 110 p.

5 GERASIMOV, I. P., et al., 1971, Natural Resources of the Soviet Union: Their Use and Renewal: W. H. Freeman and Company, San Francisco, 349 p.

6 HALBOUTY, M. T., 1981, Energy Resources of the Pacific Rim: American Association of Petroleum Geologists Studies in Geology Number 12, 578 p.

7 HAMMOND, A. L., 1990, World Resources 1990-1991: Oxford Univ. Press, New York, 383 p.

8 HAMMOND, A. L., 1992, World Resources 1992-1993: Oxford Univ. Press, New York, 385 p.

9 HAMMOND, A. L., 1994, World Resources 1994-1995: Oxford Univ. Press, New York, 400 p.

10 INTERNATIONAL ENERGY AGENCY, 1993, Energy Balances of OECD Countries 1990-1991: International Energy Agency, Paris, 218 p.

11 KELLY, W. J., et al., 1986, Energy Research and Development in the USSR: Duke Univ. Press, Durham, North Carolina, 417 p.

12 RIVA, J. P., Jr., 1994, Domestic Oil: Past, Present, and Future: Congressional Research Service, The Library of Congress, Washington, D. C., 16 p.

13 SKINNER, B. J., 1986, Earth Resources: Prentice-Hall Inc., Englewood Cliffs, New Jersey, 182 p.

CHAPTER 21

International Access to Minerals — Free Trade Versus the Map of Geology

No nation is completely self-sufficient in energy and mineral resources. These resources are geographically distributed quite unevenly by myriad geological processes and events which have determined where oil, gas, coal, metals, and other Earth resources are now located. When this reality encounters the boundaries of nations, the result frequently is a substantial difference between the places where the resource is used and the source site of the resource. This causes problems. In the past, the possession of some minerals at times has meant the difference between life and death — for example, copper and tin to make bronze weapons which the Romans used so effectively against their less well-armed adversaries. That is why the Romans fought for control of the tin mines of Cornwall.

Access by war

Today, possession of or access to minerals, especially energy minerals, largely determines our material standard of living. Access to minerals in the past was frequently settled by warfare, and this has continued to the present. Oil supply was the immediate cause of Japan going to war in 1941. Japan's expansionist program in the 1930s was primarily to secure access to mineral resources, especially the coal and iron ore of Manchuria.

The Gulf War of 1990-1991, was a conflict over the threat to oil supplies for the industrial nations of the West and Japan. For the moment at least, the unequal distribution of minerals is accommodated by peaceable trade. Whether this will be true in the future, or whether the history of conflicts over mineral resources will be repeated is uncertain. There is a continued uneasiness in the Persian Gulf where at least half of the world's present most important energy source is located. Even as this is being written, ships of the U.S. Navy are cruising the Persian Gulf to ensure that the industrial nations will continue to have access to oil. The United States has 20,000 troops stationed more or less permanently in the area. Although the United States is paying the bill for the military patrol, most of the oil from the Mid-east still does not come to the U.S., but goes to other nations. The U.S. now imports only a modest amount from this area, but the amount will increase as U.S. domestic production continues to decline, because the Gulf nations are those with the longest lasting reserves.

When oil supplies were shut off to the United States in the 1970s, the cry was heard "send in the Marines!" Violence occurred in gasoline lines the shortage created in the supposedly civilized United States. Civilization is a rather thin veneer. When survival, or even the loss of the convenience of the automobile apparently is at stake, the veneer can disappear very fast. Makes one thoughtful about the future.

Cartels

From time to time there have been attempts to disrupt free trade by means of cartels. A cartel is defined by Webster as "a combination of independent commercial enterprises designed to limit competition." It was an Arab-Iranian cartel which precipitated the oil crises of the 1970s. An examination of the history of some resource cartels is worthwhile to see if they can, for a protracted time, limit trade.(7)

There are two kinds of mineral cartels. One is natural and one is artificial. A natural cartel is the result of geological processes which have created a relatively limited distribution of a particular resource. An example is platinum with which South Africa and Russia are particularly well endowed, and the rest of the world has very little. A single nation or association of perhaps two or three which have such a natural cartel has the possibility of severely controlling prices. This power can only be challenged by the ability of the consuming nations to discover or develop effective substitutes, or take the resource by force.

An artificial cartel is one in which the resource is reasonably widespread but where the countries holding a substantial portion of the resource agree to try to limit production to bolster the price, or to simply collectively raise the price. The best known modern example of this is OPEC — the Organization of Petroleum Exporting Countries. Another example, not so well known, was the tin producers cartel, which, however came apart in 1985.

Natural cartels may succeed, but artificial cartels almost always find themselves in disarray at some time or another. The pattern of a cartel rise and fall seems to be a fairly standard one. First, the cartel will push prices up beyond what the normal price would be. This encourages more production by marginal producers and smaller suppliers outside the cartel. It also encourages substitution and conservation by the end users. The cartel members will then generally try to hold down supply by agreeing among themselves on individual quotas. Almost always, this does not work. Because of domestic political or economic demands for money from the sale of the particular commodity, individual members of the cartel will tend to cheat. Ultimately, the whole cartel is in disarray. This is what happened to OPEC in the drop of oil prices in the mid-1980s.

First, the steep rise in oil prices stimulated development of new, and previously economically marginal prospects outside of OPEC. These were particularly in the United Kingdom and Norway (North Sea area), Mexico, and the Soviet Union. These countries had never joined OPEC but they soon became significant exporters of crude oil that competed with the oil cartel. Second, additional natural gas production became economic with the increased price of oil, and gas began to displace the markets previously held by oil. The use of coal and nuclear energy to generate electricity further reduced the demand for oil in many countries, including the United States. Third, rising prices created a surge in oil exploration in the United States in particular. The number of wells drilled each year rose annually for eight consecutive years starting in 1974 with 33,000 wells, and peaking in 1982 at 88,000. Fourth, rising prices encouraged conservation efforts throughout the oil importing world. Large sums of money were invested in factories, homes, and other buildings to install various devices, including more insulation, to conserve fuel. A new generation of much more fuel-efficient cars began to replace the old gas-guzzlers. An economic recession in the early

1980s also reduced demand for oil. Demand for OPEC oil dropped 12 million barrels a day between 1979 and 1982.

Finally, as demand declined, OPEC was forced to reduce production in order to maintain oil prices, but some of the poorer countries with large populations, such as Nigeria, decided to use their idle oil production capacity and increase their oil income. They began to cheat on the cartel's official price, and offered discounts and produced oil beyond their quota.

For a while, Saudi Arabia, with the largest potential oil production, single-handedly tried to support oil prices by cutting back on its own production, which, in 1980, was about 10 million barrels a day. Saudi Arabia had cut it back to about 2.3 million barrels a day by the summer of 1985. But the Saudis warned the other OPEC members that they would not indefinitely be the "swing producers" in order to maintain prices. In the fall of 1985, Saudi Arabia began to run out of patience and also money. Ultimately, the cartel began to come apart as the Saudis increased production and OPEC members abandoned efforts to hold up prices. Each went out for a "fair share" of the oil market, regardless of price. The price of oil came down rapidly from about $28 a barrel to less than $10, briefly, in early 1986. The price subsequently seesawed back and forth, gradually working higher to a more stable price of between $13 and $18 a barrel by 1988. But there was a world oil glut for the moment.

Although the short term effect of the price cutting by Saudi Arabia was devastating to the income of all the oil-producing nations, the architect of this program, Sheik Yamani, the Saudi oil minister, did strike a fairly effective blow toward ultimately putting OPEC back in the driver's seat some time in the future on oil prices. By cutting the price so low and by keeping up production in the face of shrinking demand, the marginal wells in many parts of the world, particularly the United States, could not be economically produced. About two million barrels a day, at that time, of the approximately nine million barrels of daily U.S. oil production were from "stripper" wells, defined as wells which produce less than 10 barrels a day. The reservoirs, mostly sands, from which these wells produce are relatively tight and have low pressure. When pumping of these wells is stopped because they are no longer economical due to the low oil price, they generally cannot again be put back into production. They are lost permanently.

During the interval of low oil prices in the mid-to late 1980s, a significant amount of stripper well production was permanently lost, putting the United States in a less competitive position once the demand for oil increased. In this way, Sheik Yamani and OPEC made a long-term negative impact on U.S. oil production, and put OPEC in a stronger position relative to the oil producing ability of the United States. This was part of Sheik Yamani's strategy, and it was a good plan. But perhaps because it was a longer term strategy, and the immediate effect of lower prices greatly reduced OPEC's income, Yamani and his policy came into disfavor and ultimately he was fired from his job in 1986. However, OPEC will benefit over the long term from the oil production that was lost in other countries due to the low oil prices of Yamani's administration.

OPEC

OPEC is the classic cartel attempt of the Twentieth and on into the Twenty-first Century. The final word on its success or failure is not yet in. Sometime in the future, it is quite likely to work better than it does at present, because more and more world oil production will ultimately be concentrated in the Persian Gulf area. It is so destined by the geology of world oil deposits.

OPEC was established in 1960. The event which caused its formation was the great flow of cheap oil in the late 1950s. In 1959, the major oil companies twice cut the prices of crude oil in the Middle East and Venezuela without conferring with the governments of those countries. This so irritated the governments of Iraq, Iran, Kuwait, Saudi Arabia, and Venezuela that they formed OPEC to restore the price structure and to force the international oil companies in the future to consult with OPEC members about any oil price changes.(4) OPEC was a formidable association at that time because its members controlled 90 percent of world exports outside of the communist countries.

OPEC membership gradually expanded from its beginnings in 1960 in Venezuela. Qatar joined in 1961, followed by Indonesia, Libya, the United Arab Emirates, Algeria, Nigeria, and Ecuador. Gabon joined in 1975. Ecuador has since dropped out of OPEC, and Gabon did so in 1996. Ecuador has only small oil production, but because Ecuador is a small country, and oil is Ecuador's main export income it did not want to be bound by OPEC export quotas. Loss of Ecuador was of not great concern to OPEC as Ecuador's production is insignificant on world markets, as is also the case of Gabon.

Non-OPEC countries which export more than 100,000 barrels of oil per day, and therefore theoretically can compete with OPEC, are Canada, Mexico, the United Kingdom, Norway, Angola, Egypt, Oman, Russia, China, Malaysia, and Brunei. However, considering the potential for increasing oil production, the non-OPEC countries are in a much weaker position than are the OPEC nations. At present, non-OPEC countries have substantial production but it cannot be maintained at present levels much longer. OPEC's influence will rise.

Of the non-OPEC countries, Mexico, Angola, Oman, China, Malaysia, and Brunei might be able to increase production from current levels but only slightly. But China is going to need all its possible increased production internally to meet its program of growing industrialization. And China's population continues to increase. The two biggest non-OPEC producers are the U.S. and Russia. The United States peaked in production in 1970, and has declined considerably since then. Russian production has dropped markedly. It is very unlikely that the United States can ever significantly increase oil production over the long term. Russia, with the aid of western technology, may be able to increase production to some extent for a time. But Russia's production will probably be almost entirely used internally. The world's significant oil export potential is now, and will increasingly continue to be, in the control of the Persian Gulf OPEC nations. Saudi Arabia is producing about eight million barrels a day, but has a potential to produce 10 million barrels a day or more by just opening the valves. It also has more areas to explore including deeper undrilled zones in existing fields. Saudi Arabia by itself in the OPEC group has the ability to strongly influence if not control oil prices in the years to come, and provide the member nations of OPEC with an optimistic future.(3)

Kuwait to a lesser extent, also has the potential for increasing oil production by which to influence prices, but Kuwait has usually been a compliant OPEC member. Iraq may not be a very reliable member of OPEC, and may ignore production quotas at times. Iran, similarly, because of pressing internal needs, may also violate OPEC quotas. Iran and Saudi Arabia are historic enemies and do not get along well.

OPEC and the future

OPEC will be somewhat restrained until the more marginal non-OPEC producers have depleted their resources. Then the cartel will evolve from being an "artificial" cartel which is chronically insupportable, as it has been at times, to a "natural" cartel which has a much better basis for existence. With more than 60 percent of the world's oil in the countries bordering the Persian Gulf, all of whom belong to OPEC, this natural cartel has a good basis for surviving.(3) When it does become a natural cartel it will be potentially much more powerful than today.

Other Cartels

Can OPEC-like cartels be successfully formed with other minerals? There has been considerable discussion about this among Third World countries which in many cases are so vitally dependent on just one mineral resource. Almost all of those which have been tried so far have not met with any great success.(7) This is apparently because the resources are amenable to substitution more readily than is oil, and the resources are not so concentrated geographically.

International tin cartel

The International Tin Cartel, in the late 1960s and early 1970s raised the price of tin substantially. As a result, non-members such as Brazil and China stepped up production. Even Great Britain was able to economically open some of the long-closed ancient tin mines of Cornwall, at one time worked by the Romans. Tin users found they could quite easily turn to substitutes including plastics, glass, cardboard, and aluminum. In 1972, about 80 percent of the beverage cans in the United States were made of tin plate. However, by 1985, almost all American beer cans and 87 percent of the soft drink cans were made of aluminum. The cartel came apart in 1985. Most "tin cans" no longer have any tin.

Copper cartel

At one time, there was an attempt to set up a copper cartel. Four major copper producing countries, Chile, Peru, Zaire, and Zambia, in 1967, formed the Intergovernmental Council of Copper Exporting Countries (CIPEC). In 1974, seven producers of bauxite, the ore of aluminum, set up the International Bauxite Association (IBA). These countries were Australia, Guinea, Guyana, Jamaica, Sierra Leone, Suriname, and Yugoslavia. Similarly, in 1975, seven iron-exporting countries, Algeria, Australia, Chile, India, Mauritania, Peru, and Venezuela organized the Association of Iron Exporting countries (AIEC). None of these cartels was successful.(7)

Cartel — an old idea

The cartel idea goes back at least 3,000 years, but in almost every case free trade prevailed. The wheat, uranium, and sugar cartels are all history. The natural latex rubber cartel bounces along but is not doing very well at present. Coffee cartels have been tried but seem less effective in controlling prices than is the temperature in Brazil. A hard freeze does more than can price agreements to bolster the market. Conversely, good weather usually creates a crop surplus which cannot be stored indefinitely, and when sold depresses the price.

Diamonds and De Beers

There is, however, one mineral cartel which has been effective. Even though a zircon or the synthetic gemstone, cubic zirconium, both of which closely simulate the appearance of a diamond, can be bought for a fraction of the cost of a genuine diamond of the same size, the concept that "diamonds are a girl's best friend" seems to have prevailed. The diamond cartel is essentially one company — De Beers — because they control and have controlled about 80 percent of the world diamond production for more than 50 years.(7)

Two other minor sources of diamonds, Zaire, and Russia, have made half-hearted attempts to break the cartel, but they are such small producers that the effect of their efforts were negligible. Realizing that, for the most part, these other two sources have joined De Beers in controlling production and prices. What makes the De Beers-run cartel successful is the fact that it is one of the best natural cartels. Gem-quality diamonds are found in only a very few places in quantity and De Beers controls the majority of these sources. Also, De Beers is virtually the only company in the business. If there were two or three major diamond producers, this might make a difference, as total agreement all the time would be unlikely. But, as there is only De Beers, it has only to agree with itself. It is not subject to differences of opinions and diverse economic circumstances that disrupt OPEC from time to time.

Other critical minerals

There are some other minor strategic minerals such as vanadium and cobalt which are now in natural cartels, because they are found in quantity in only a very few places. But, none of these is a major item of commerce, and, to a considerable extent, the nations owning these minerals are in rather urgent need of foreign exchange, and therefore not likely to cut off shipments for very long.

Free Trade in Minerals Likely

It seems probable that free trade in minerals will prevail for the foreseeable future. Only one major resource cartel is likely to be viable, if the members wish it to be. It is OPEC, and only some of the present members of OPEC will be in the final OPEC cartel — those countries in the Persian Gulf region. The other members have substantially smaller oil reserves and will not be able to sustain oil production nearly so long as will the Persian Gulf nations. The Persian Gulf countries of OPEC are projected to become a strong natural cartel beginning within the next decade as non-OPEC members' production dwindles. OPEC will have the potential of being very effective then if its members stay together.

At present, however, OPEC members are almost all cheating on their quotas.(5) The lure of, or the need for, current income is stronger than the concept of saving oil for the future. In March 1996, the OPEC total oil quota for all members was 24,520,000 barrels daily. The actual production was 26,010,000 barrels.(5) It would be in the long run benefit of an oil exporting country to keep its production down as it seems certain that oil several decades ahead will command a higher price than oil today, even adjusted for inflation. Oil will bring a relatively better price then, but the social and political realities of the intervening time may preclude saving oil to provide income for later generations. What the internal income demands of each OPEC member will do to the structure of OPEC remains to be seen.

Economic Interdependence

OPEC may not use its strength to economically damage the major oil consuming nations. The world is becoming so economically interdependent that it is doubtful that the OPEC countries around the Persian Gulf would enforce their cartel to an economically injurious extent because they have large investments in the countries where they must sell their oil. Substantial harm to the economies of these countries would not be in OPEC's interest. Furthermore, looking very long term, when the last several nations of OPEC run out of oil — which means the Persian Gulf area nations — they may have to depend quite substantially on investments made in other countries with more diverse economies. The largely desert regions of Kuwait, Saudi Arabia, the UAE, Qatar, and Oman have very few resources to fall back upon. As their oil production declines, they will be increasingly dependent for income on their foreign investments.

Free Trade Can Compensate

In summary, free trade can erase or at least ease the effects where geology has placed the mineral and energy mineral resources. Free trade will also adjust the balance in standards of living to some extent among countries. The resource producers' standards of living will probably rise relative to the standards of living in the consuming countries. This will be the result of the continuing transfer of wealth from the industrialized nations to the major oil-producing countries. Although at times going to war over access to mineral resources has been successfully used, this cannot be regarded as the ultimate solution to the problem. In some fashion free trade in vital minerals and energy supplies must prevail in the future. If not, however, war could result, but this would never put things into a stable relationship. The Middle East has a history of unrest. There are political and religious extremist factions there who might create crisis situations, but peaceful exchange of resources and goods is in the best interest of all countries. Economic interdependence of the oil producers and consumers will probably be the overriding factor in maintaining free trade in mineral resources without resort to warfare, or by economic warfare through the use of cartel power.

BIBLIOGRAPHY

1 ECKES, A. E., Jr., 1979, The United States and the global Struggle for Minerals: Univ. Texas Press, Austin, Texas, 353 p.

2 CHOUCRI, NAZLI, 1982, Power and Politics in World Oil: Technology Review, MIT Press, Cambridge, Massachusetts, p. 24-36.

3 FARRELL, CHRISTOPHER, 1989, It Looks as if OPEC May Have the Last Laugh: Business Week, September 25, p. 34.

4 MANSFIELD, PETER, 1991, A History of the Middle East: Viking, New York, 373 p.

5 REIFENBERG, ANNE, 1996, OPEC Members Boldly Violate Quotas: The Wall Street Journal, April 25.

6 ROSTOW, W. W., 1980, The World Economy. History and Prospects: Univ. Texas Press, Austin, 833 p.

7 SPAR, D. L., 1994, The Cooperative Edge: The Internal Politics of International Cartels: Cornell University, Ithaca, New York, 273 p.

8 TANZER, MICHAEL, 1980, The Race for Resources. Continuing Struggles over Minerals and Fuels: Monthly Review Press, London, 285 p.

9 WOLFE, J. A., 1984, Mineral Resources. A World Review: Chapman and Hall, New York, 293 p.

CHAPTER 22

Mineral Development and the Environment

Humans, like all other organisms are made of materials which come from the Earth, and use resources to exist. Everything we use each day to survive, except for sunlight and the air we breath, comes from the Earth, and must be obtained someplace by someone. This applies to all human beings whether they be lawyers, small business men, corporation presidents, factory workers, teachers, homemakers, conservationists, farmers, or environmentalists. We all use materials from the Earth — each of us, and every day. Every human has a continuing impact on the Earth from birth to death.(42) This is becoming an increasing worldwide concern as population continues to grow at the astounding rate of about a quarter of a million people a day.

Public involvement

Increasingly, people are expressing their views on resource development and management. It is therefore important in a democracy that the various aspects of resource discovery, extraction, and utilization be widely understood.(47) The great growth of the environmental movement in the past few decades has markedly influenced the regulations under which natural resource enterprises must operate in the developed countries, and these constraints are spreading to the lesser developed nations.

Resources for living

At the same time that environmental rules are enacted, it is important to remember that society has been brought to its present state of affluence by the use of Earth resources. A higher standard of living, in material terms, means the use of more energy and mineral resources. Environmental impacts of obtaining these resources can be mitigated to some extent, but in order to drive an automobile, somewhere holes in the Earth have to be dug to obtain the iron, aluminum, copper, and glass to build the car. Energy has to be obtained to process these materials into the final form of the car (all the materials listed have to be smelted which is an energy intensive process). Obtaining this energy as well as drilling to get the oil to run the car involves environmental impacts. To lead the good life, or any life, Earth resources must be used.

Resources are Site-Specific

The first fact to recognize about mineral resources is that they are site-specific. They are located where geologic processes put them millions and in some cases billions of years ago. They are not located with any regard to where human habitation is today. Petroleum, coal, gravel, and metals have to be produced where they occur — not where one might like them to conveniently be. As population increases, humans are increasingly living in or near places where useful mineral deposits are already known or later discovered. As people occupy more and more land, or set it aside for wilderness areas, there is growing conflict between resource development and the population which it supports. Many areas are now designated as permanently off-limits to any resource production. The Earth is finite and there is only so much area. Decisions have to be made as to what purpose each area will be put. It is important to recognize what are the logical priorities. Presumably, providing the basics of human existence comes first.

The development of mineral resources uses only a very small part of the Earth's surface. Metal mines in the United States occupy less than one percent of the land surface, but they are vital to our economy. The locations of these resources cannot be moved for the convenience of humans, and civilization as we know it today cannot exist without these materials. The accommodation of these two facts has caused some problems in the past, but is now causing many more problems as human population increases and spreads over much of the Earth's limited surface.

Human Habitation and Mineral Development

A hundred years or so ago in the United States, there seemed to be ample space for mineral resource development and human habitation. The west was thinly populated. Mining, for example, did not greatly impact inhabited areas. Frequently it was quite the other way. Mineral deposits were discovered in remote areas, and communities grew around the mining or energy mineral producing sites. Now, however, oil development and mining projects meet with increased opposition. One hundred years ago there were no such things as designated wilderness areas although there were a few national parks. Now, there are many national parks and monuments, many wilderness areas have been established, as well as numerous state and local parks and preserves.

Furthermore, many offshore ocean areas are now off limits to mineral resources exploitation, which chiefly affects petroleum operations. This is notably true off the California coast, a state which is the largest single consumer of gasoline. There are several stated reasons for this. There are occasional oil spills connected with ocean drilling. Also, ocean view property is extremely desirable and expensive. Tourism in California is important to the economy. Therefore, the value of a pristine view, unobstructed by drilling rigs or petroleum production platforms in the ocean, is thought to be more valuable than the resource which might be developed.(38) California is highly dependent on imported oil, so it is a clear example of "dirty someone else's backyard, not ours, for the resources we use."

Locking up Resources

The net result of numerous environmental rules and actions is that in the United States, and increasingly in many other parts of the world, more and more land is being set aside for various purposes, and cannot be used for mineral production despite the inescapable fact that materials obtained from the Earth are the basis for human existence. More and more people are becoming dependent on less and less land. Again, it should be clearly recognized that everything we have including ourselves comes from the Earth. The current trend of more and more people trying to survive on less and less land as more areas are locked up from resource development cannot continue indefinitely. The human population increases but land supply and mineral and mineral energy resources do not grow.

Continuing this trend will result, in a material sense, in our standard of living inevitably being eroded. There is serious question as to whether the material standard of living will increase for coming generations even in the industrialized countries. And there now are the several billion people who live marginally or simply just at the bare subsistence level. To bring these humans up to what could be regarded as reasonable living standards would involve greatly increased demands on energy supplies and mineral resources, and correspondingly increased environmental impacts.

Exporting Environmental Problems

Unfortunately in our society the environmental impact each of us makes is now frequently either not recognized or conveniently ignored. Sometimes the environmental effects of producing these vital materials for our existence are simply moved to other areas out of local or national sight — "out of sight, out of mind." Minerals and energy mineral resources which are needed for our daily use are closed to development in some areas. But because they are needed nevertheless by one and all including the most ardent environmentalists, these resources are simply imported from other areas which may have less restrictive laws for mineral resource production. The citizens of New York City draw resources for their support from many distant places. But very few if any New Yorkers are aware of the environmental impacts on those distant places caused by their daily use of these resources. This is the situation in all industrialized countries. In the United States, for example, oil is produced offshore Alaska and offshore Louisiana to be shipped to and used in California where offshore development has essentially been stopped.

This lack of awareness or concern for the impact on a distant environment which an individual may have is now global. It is probably more evident among urban residents than among rural dwellers, as the latter live closer to the land, and they have a better appreciation of the relationship of the Earth to their existence. Wackernagel and Rees state:

> "With access to global resources, urban populations everywhere are seemingly immune to the consequences of locally unsustainable land and resource management practices — at least for a few decades. In effect, modernization alienates us spatially and psychologically from the land. The citizens of the industrial world suffer from a collective ecological blindness that reduces their collective sense of 'connectedness' to the ecosystems that sustain them."(89)

Moving the environmental impact of resource extraction to regions beyond where the resources are used is in part economic.

There are mineral deposits in the United States which, if they were developed under the rules — or lack of rules — by which some foreign deposits are produced, could be economically mined here. But because of environmental restrictions, it is cheaper to obtain these materials abroad. For example, there is a huge base metal (lead, zinc) deposit discovered a number of years ago in northern Wisconsin which because of environmental lawsuit after lawsuit may never be developed. In the meantime, the U.S. imports base metals from Peru, Chile, Bolivia, and several African nations which have fewer environmental restrictions.

Somewhere the Earth Must be Disturbed

To produce the needed resources, somewhere holes have to be drilled, mines have to be developed, and the waste rock disposed of. Land, or submerged portions of the continent called the continental shelves, must be occupied by drilling rigs to explore for oil and gas, and sulfur — all vital materials for our industrial society. Again, the simple fact must be recognized that these materials which support us in many ways every day have to be obtained somewhere from the Earth.

Gasoline for our cars does not magically appear at the service station. The iron of the nails which hold our houses together, and the copper which is in the plumbing by which we get water into our houses have to be dug from the Earth somewhere. The pipe which drains away the wastes is made of clay, or plastic (from oil). All had to come from somewhere in the Earth. Copper also is what brings electricity to homes, factories, hospitals, office buildings, and stores. Copper is the wire by which automobiles, refrigerators, radios, telephones, and television sets operate.

These things have become the very material fabric of our civilization and affect each of us many times every day. Most people tend to take the Earth materials which do these things for granted, but if they were to follow these metals and fuels to their source, they would find an underground mine, or an open pit hole in the ground, or a well for which they were, in effect, partly responsible by their personal use of these resources.

Importance of gravel pits

An example of a basic resource we use which comes from nearby localities is gravel. Gravel pits are commonplace and generally not highly regarded. Yet we are all highly dependent on them. In all our homes, and all the buildings of towns and cities, and in all the highways and byways all across the country there is a very important group of materials called aggregates — sand and gravel. They are used in very large quantities and they are heavy. Hauling them long distances is expensive because of the energy costs, so it is important that nearby sources be used. The development of gravel pits is a frequent subject of contention, but it is necessary that they be provided for. Gravel pits can sometimes become a continuing asset to the community when they are no longer needed or the supply of aggregates is exhausted. They can and have been graded and landscaped into parks, or made into ponds for local recreation.

Again, to provide all these everyday materials, the Earth has to be disturbed somewhere. If wells are not drilled or mines are not dug in your backyard, this will have to be done in someone else's backyard. This may occur where the local population urgently needs the money, both for jobs and for public revenues. This concept applies not only to such things as sand and gravel but to all Earth-derived resources. On a global scale, some smaller nations without diversified economies will export anything of value and ignore environmental problems in order to obtain money for badly needed foreign exchange.

Environmental honesty

If the environmental movement is to be morally honest about these matters, it should recognize that by locking up domestic resources from development, the problem does not disappear. It does "go away" — away to some other place where the hole has to be dug to produce the resource. It then becomes simply a case of "dig the hole in somebody else's backyard, not mine." This has become such a common attitude as to generate the acronym NIMBY: "not in my backyard." One might suggest that if the environmental movement is to be absolutely "pure" in the sense of not disturbing the Earth at all, houses, hospitals, automobiles, and factories should not be allowed, and we should all go back to living in caves. Unfortunately, like other mineral resources, the supply of caves is also limited. As the world becomes more and more populated, and as the native populations of what have been regarded as undeveloped nations have begun to also become environmentally conscious, the matter of environmental impact of mineral resource development is becoming a worldwide concern.

Demands of increased population

Substantially adding to the problem of the impact of resource development on the environment, is that population continues to increase. Currently a population equal to more than twice the population of Spain is added to the world each year. The additional resources to support these people must come from somewhere. Also, many relatively undeveloped countries are now striving to achieve a higher material standard of living. So there is not only the problem of providing additional material resources for the nearly 100 million people added each year, but to provide increasing amounts of Earth resources for the many people already here who aspire to a better existence. The so-called Third World or lesser developed countries represent almost half the world's population. The resources necessary to appreciably raise their living standards would be enormous, and in fact may not be available. The problem is immense and has the potential for serious conflict.

Environmental Concerns

Environmental concerns must be intellectually honest and be accommodated to the fact that our very existence depends on obtaining resources from the Earth. The solution is bound up with the concept of a "sustainable society" which is taken up in Chapter 28. Presented here are examples of both negative and positive environmental impacts of mineral development.

Environmental Impacts of Mineral Production

Every day we need metals and energy supplies, and we use them in myriad ways, but obtaining them from the Earth can and does have a variety of environmental impacts. Many of them have come to be regarded as negative. But it may come as a surprise that some environmental impacts of mineral developments in the past, and more recently an increasing number may have a useful impact.

Earlier impacts mostly negative

At the beginning of the Industrial Revolution which saw such a great increase in demand for mineral and energy mineral resources, the emphasis was simply on obtaining these resources. This was due in part to the fact that these resources were seen to immediately provide a much increased standard of living for the general population. Increased use and industrialization offered not only more and better paying jobs than the earlier largely agriculturally based economy provided, but also supplied newly invented wonders such as cars, trains, household appliances, central heating, and many other items for the average citizen. The rush to develop resources to produce these things did not usually consider the environment. In the initial stages of industrial development in the United States, the population was small, the land was uncrowded, and there were large wide open spaces.

The early days of mineral exploitation in the United States proceeded with little regard to environmental consequences, and the results of this are still visible in places. This had merit in the sense that it allowed the rapid building of the United States and at minimum cost at the time, and produced a large rise in the physical standard of living of the general population, to where it became the envy of the world. So great were the volumes and quality of virgin resources available at the time that this could have been accomplished anyway, even if environmental costs were included, but the cost would have been greater, and somewhat fewer people would have benefitted. Now there is a bigger demand on lesser grade resources, and we increasingly are adding in the cost of environmental impacts. This trend is now becoming world-wide. To maintain our standard of living will become more expensive.

This early impact on the environment was locally severe, and some effects remain. Mine dumps in Colorado in places still leak toxic metals into streams. The smelters which were set up to process the ores spewed destructive fumes over the landscape, killing the vegetation and giving it a desolate moon-like appearance in such places as the Bunker Hill zinc smelter near Kellogg, Idaho, and the Ducktown, Tennessee region. The waste material dug from underground mines remains as huge piles on hillsides or in valleys, and the holes which were the open pit mines will be visible for many centuries to come.

The relatively high sulfur content which is characteristic of eastern U.S. coals and associated sediments has caused havoc with many streams in West Virginia, Kentucky, and Tennessee. The underground coal mines of Pennsylvania still have fires burning in them in some places. The result of this and other mine workings not on fire but which become unstable over the years, is that cave-ins occur in unexpected areas, sometimes beneath occupied houses. Coal exists in 37 states, and is mined underground in 22 states.(45) Underground mining produced 40 percent of the coal mined in 1980. These underground mines represent a long-term hazard to the landscape which overlies them. It is the price paid for the benefits which the mined coal produced. It is estimated that eventually underground coal mining in the United States will involve 40 million acres, eight million of which have already seen underground mining. Ground subsidence over coal mines is already noted on more than two million acres. The U.S. Bureau of Mines states that nearly 400,000 acres of land in urban areas in 18 states may be subject to subsidence, and the total costs to stabilize these lands would be about $12 billion.(49) Already, land subsidence over abandoned coal mines has caused severe damage to houses and other structures. Pennsylvania and West Virginia are particularly affected.

The classic novel *How Green Was My Valley* described the changes wrought in a Welsh coal-mining area of Britain at a time when environmental matters were not considered. Effects still exist. Landslides from unstable mine dumps have caused destruction of buildings and loss of lives.

Gold placer mining in California, Colorado, Idaho, Montana, and to a lesser extent in Oregon, has left hillside erosional scars, and unsightly piles of rock in stream valleys. Nature will eventually heal or at least modify these scars to some extent if left alone, but it will take thousands of years to do so.

The Present

The present world environmental scene with regard to mineral resource development is mixed. In some areas the situation is not good; in other places strict laws are minimizing the impacts. On the negative side one might cite the recent central Amazon basin gold rush.(56) Tens of thousands of people have invaded the area and set up crude mining facilities. The panning and sluicing operations have put tons of sediment into the local streams much to the detriment of the fish. But possibly even more destructive is that in most operations mercury is used as an agent to recover the fine gold. This mercury has gotten into parts of the Amazon drainage and can become a deadly contaminant to the aquatic life, and ultimately becomes part of the food chain which leads to humans.

Geothermal energy is relatively clean, and for the most part has a good environmental reputation as an energy source. However, to illustrate the fact that every resource development does have some environmental impact, it may be noted that in New Zealand's Wairakei geothermal field, begun in 1950, ground subsidence up to 4.5 meters has occurred due to fluid withdrawal. Also, there have been horizontal ground displacements up to half a meter.(45) In The Geysers geothermal field in northern California, subsidence up to 0.13 meters has occurred. However, at Larderello, Italy, a geothermal field which has been producing since 1905, there have been no ground displacements nor have any occurred at the Cerro Prieto Field in northern Mexico.

Studies on numerous oil and gas fields, show that ground subsidence over these resource developments is generally quite slight, if at all.(45) The oil fields in the Long Beach area of California are one exception, where in some places subsidence has been as much as 28 feet, dropping the oil field surface below sea level so that dikes have had to be built to keep out the ocean. Subsidence of the land from excessive groundwater withdrawal has been considered in Chapter 15.

At the present time in the United States, and also in Canada, Australia, and in Europe, and beginning to some extent in the lesser developed countries, there are increasingly stringent regulations designed to protect the environment. Many companies on their own, beyond what the laws may require, are now protecting the environment as a matter of being responsible good neighbors. The cost of compliance with these regulations can be very high and time-consuming, but it is being done. An example is the coal mine near Centralia, Washington, where re-grading the mined area and the planting of trees is making the land as fully timber productive and attractive as it was before it was mined. Native vegetation is taking over very fast.

Some beneficial environmental impacts

There have been some unexpected benefits to the environment of mining and petroleum development. In Wyoming where strip-mining for coal and uranium has been conducted, the reclaimed areas where the pits have been filled now have unleached rock and soil materials at the surface which have been re-seeded with the natural vegetation. The unleached new soil produces much better vegetation than the old leached surfaces. A somewhat unexpected result is that deer and antelope much prefer to graze on the reclaimed areas.

All across the semi-arid plains and valleys of Wyoming where oil drilling has been extensive, pits dug in connection with drilling operations have been left. These catch moisture and provide water for birds, small animals, deer, and especially antelope whose numbers have benefitted substantially from these features.

Open pit coal mining areas have been landscaped in Alabama to create lakes providing good fishing and habitat for waterfowl. In rather featureless southeastern Kansas, open pit coal mine areas have been modified in places to form slightly hilly areas which are now wooded with intermingled small ponds affording a variety of natural habitats for wildlife.

Offshore drill rigs

Offshore drilling has been banned in many coastal areas, especially California. Where it has been allowed it has seen considerable opposition. But more recently there are people who endorse it. These are fishermen. The legs of the drilling platforms in the ocean offer a place for marine growth to accumulate, and along with this come fish. The fishermen off the Gulf Coast of the United States are particularly enthusiastic about the effect of the drilling rigs on the fishing, and frequently fish as close to these structures as possible because of the enhanced fish habitats. Studies by the Louisiana Artificial Reef Program show that 20 to 50 times more fish live around a drilling platform's underwater legs than on the soft bottoms nearby. Some tropical fish never previously seen in Louisiana waters now are abundant in the reef-like habitats beneath some of the state's 3,600 offshore drilling rigs and production platforms.

When an oil or gas field is depleted and the drilling and production platforms are no longer useful, the cost for oil companies to remove them has been about $2 million each. But legislation now exists whereby Louisiana can take title to the equipment and relieve the oil companies of further obligations. It is a happy solution whereby everyone benefits, including the fish. Generally, the structures are dynamited and dropped into the ocean where they attract and afford a home for a variety of marine organisms. With the greatly increased marine life around these artificial reefs, recreational scuba diving has become popular. Other coastal states, Texas, Mississippi, Alabama, and California are considering this concept.(2)

The U.S. Minerals Management Service predicts that about 2,500 offshore drilling rigs and production platforms will be abandoned by 2014, and state that these structures will be "a great opportunity to construct artificial reefs that we will never have again." Thus the oil industry will leave in this way a positive legacy which will last for many hundreds if not thousands of years.

Dams: Mixed Gains and Losses

In the case of dams, environmental impacts have been very mixed. Originally dams were thought to be environmentally rather benign, producing no noxious fumes such as do fossil fuel power plants. Used for irrigation, dams would make arid lands produce crops. In the United States more than 100,000 dams block normal stream flow. About 5,500 of them are more than 50 feet high.(24) But dams have proved to have diverse and in some instances quite unexpected effects on the environment.

The positives

Dams have provided cheap electric power and flood control, although nature can overwhelm even dams, as happened on the Mississippi during the great Midwest United States floods of 1993. Dams have brought water to previously unproductive arid areas which now produce a variety of crops. Dams with their reservoirs also provide water recreation of various sorts, and often afford good fishing and hunting. The elongated arms of the reservoirs reaching back into the hills supply water for wildlife which would otherwise not be there, or at least not in such abundance. Dams, particularly in their early life, have meant big economic gains for their adjacent areas.

The Negatives of Dams

As communities and regions live with dams over the longer term, some strong negative effects may become apparent. Some examples are described here.

Columbia River/Snake River

No river system has been more dammed than the Columbia/Snake with 30 major dams. The renowned Columbia River with its Bonneville Power System has been the world's largest hydroelectric power producer, and at one time supplied about 40 percent of all the hydroelectricity in the United States. It has been the pride of the Pacific Northwest giving that area the nation's cheapest electric power, and luring electric-intensive industries, particularly aluminum refining, to the region. The Columbia River system also had a world famous and highly productive salmon and steelhead (a sea-run rainbow trout) fishery, but power and fish have come into sharp conflict.(48,93)

Salmon versus dams

The Columbia drainage salmon fishery, long important to the Pacific Northwest, has nearly disappeared, and many individual salmon runs up particular streams tributary to the Columbia are now extinct. In 1884, 42.2 million pounds of salmon and steelhead were caught in the Columbia River by commercial fishermen. In 1994, 1.2 million pounds were caught. There are other causes besides the dams for the decline of the salmon, but the evidence is strong that the dams are the major problem.(64,93) Redfish Lake and the very small local streams which drain into it in central Idaho are the ultimate destination of some of the salmon of the Columbia drainage. The lake was so named because of the myriad sockeye salmon in their red spawning colors as they returned from the sea to the lake and its tributary streams to complete their life cycle. But the dams have taken their toll. To get to Redfish Lake the salmon now have to pass four dams on the Columbia River and four dams on the Snake River. Keith Johnson at the Eagle Hatchery at the head of Redfish Lake has supplied the following figures on salmon returns:

Year	Male	Female	Total
1990	0	0	0
1991	3	1	4
1992	1	0	1
1993	6	2	8
1994	0	1	1

Declining salmon runs have been apparent for some years and various attempts have been made to mitigate the problem. The dams are problems for the salmon in both their upstream migration, and also in the downstream migration to the sea of the smolts — the young salmon which have been hatched in the far upstream spawning areas and grow there to fingerling size. They must go back to the ocean as part of their life cycle, but to do this they have had to pass through the turbines which turn the electric generators in the dams. This tends to result in homogenized salmon smolts.

To eliminate this situation, turbines have been shut down at times, and water is simply spilled over the dam when the smolts are migrating seaward. This results in a loss of electric power generation. A relevant cartoon appeared in the regions' newspapers. It showed salmon in a fish market with a sign which read "Salmon, 700 kilowatts a pound."

This procedure of spilling water is only partially effective, however, as the turbulence of the water spilled over the dams creates a high nitrogen content that kills the young salmon. An alternative has been to truck salmon around the dams, and indeed a whole fleet of trucks has been doing so. The fish have also been barged around the dams. At Lower Granite Dam on the Snake River, tributary to the Columbia, each year the U.S. Army Corps of Engineers funnels from six to nine million salmon smolts into barges and then hauls them past seven other dams, eventually dumping them into the lower Columbia River beyond the last dam. Many scientists think that barging kills more smolts than it saves.(24)

The idea of continuing to truck and barge salmon around the dams seems almost an absurdity. The situation has been described as a "critically ill patient on life support."(8) Is this procedure supposed to continue indefinitely? Clearly it is not the answer to the impact of dams on the migrating salmon.

In the 1990s a concerted effort was begun to somehow restore at least part of the great Pacific Northwest salmon runs of the past. The plan has many facets, and the cost in terms of lost electrical power will be substantial. It will be years before the success or failure of this effort will be known. In 1997 there was serious discussion of the possibility of taking out some of the dams which lie across the historic salmon spawning routes. But the cost of doing this, together with the lost power make this alternative unlikely. However, the issue is not dead and discussion continues. Dams are clearly one of the main culprits in the decline of the salmon runs in the Columbia River system.

Salmon and California

To a considerable extent the salmon which have inhabited the streams of central California have suffered a fate similar to those of the Columbia River. The Friant Dam on the San Joaquin River eliminated a salmon spawning run which had been 150,000 fish. In the Sacramento River basin, the huge Shasta Dam along with those on the Sacramento River tributaries reduced the 130,000 winter-run salmon to about 2,000 by 1987.

Power and population

The problem, of course, is to supply power to the West. And more, not less, power will be needed, for the population in the area of the Bonneville Power System is expected to grow from its present size of 9.7 million to as much as 15.7 million by 2015. Until 1993, the Bonneville Power System had an electricity surplus, but in that year the surplus was gone, and all the generators had been put in place which were needed to harness the available water. The population may nearly double by 2015 from its 1995 level, but there is no more hydroelectric power available. The carrying capacity of the region's river system in terms of hydroelectricity has been reached. In the process the salmon fishery has almost been lost.

Loss of fertile lowlands

In both mountainous areas and in plains regions, dams may flood what were the most fertile parts of the country, the lowlands adjacent to the rivers. In parts of Kansas the uplands have little topsoil. The bedrock is close to or at the surface, whereas the adjacent broad river floodplains have rich alluvial soil. These areas are flooded and lost to agriculture when dams are built. In some cases, to put in the dams the government had to expropriate private farmland, and feelings may run high in this regard. Along one highway which borders a flooded river valley in Kansas, the local citizens erected a series of billboards which read: "Stop This Big Dam Foolishness."

Flooded winter range

Another negative effect of dams may be the loss of wildlife habitat. In some places, the various extended arms of a reservoir may promote growth of trees and other vegetation and in turn encourage wildlife. This is true primarily in relatively flat and fairly arid regions such as the Great Plains. However, in more mountainous regions of high uplands and deep canyons, the flooding of the canyon areas destroys the vital wintering grounds of big game animals, and fish spawning areas.

At the upper end of the Columbia River System, the rising waters behind Libby and Hungry Horse dams have flooded more than 90 miles of tributary streams, destroying more than 50,000 acres of wildlife habitat. Relocation of railroad tracks also destroyed 2,100 acres of wetland and riverbank habitat.

Dams and the Colorado River

The life-giving artery of the southwestern United States is the Colorado River. Once flowing strongly to the Gulf of Lower California, it reaches the Gulf now only as an occasional trickle and sometimes not at all. Today there are more demands on the Colorado River than there is water to meet them. To control the Colorado and to allow for irrigation and city water supplies, a series of dams has been built. The two most famous are Hoover Dam and its reservoir, Lake Mead, the first one below the Grand Canyon, and Glen Canyon Dam and its reservoir, Lake Powell, just above the Grand Canyon. Both dams with their reservoirs are exceedingly important to the Southwest as sources of power, irrigation water, and municipal water supply. They also provide a variety of recreational activities. But these benefits are not without some cost.

Raising and lowering reservoir levels at various times of the year has created unstable water conditions for both animal and plant life. There have been marked changes in shoreline vegetation. In particular, the growth of an exotic bush, the tamarisk has been encouraged. This plant forms dense thickets, accumulates litter, attracts obnoxious flies, and produces large amounts of hay-fever-causing pollen.(72) Also, the sediment which the Colorado River once carried through to the sea is now accumulating behind the dams. In the lower several miles of the Grand Canyon large banks and terraces of mud are forming due to low flows of water. Major flood control releases of water from the dams scour out river bottom vegetation vital to fish life, and also remove the natural sandy beaches, but may bury other areas in mud. Water levels in the Colorado River fluctuate as much as 13 feet in a day, and as a result about half the beach areas of the original pre-dam river have been destroyed. The dams have changed the entire regimen of the river and the process has altered the environment. Nevertheless, at the present time it seems clear that the beneficial effects of the dams outweigh the negative environmental impacts. The Southwest United States could not exist as it does today without having harnessed the Colorado River.

However, to provide whatever benefits dams in the U.S. produce, they have flooded an area equal to the size of New Hampshire and Vermont. And the only major river (defined as 600 miles or longer) still entirely free flowing in the 48-adjacent states is the Yellowstone.(24)

Aswan High Dam — Saad el Aali

One of the most famous dam projects in recent times is the great Aswan High Dam across the upper Nile in Egypt. There is a narrow zone of greenery bordering both sides of the Nile. Beyond this there is almost total desert. Egypt is the Nile and the Nile is Egypt. Planned for three purposes, a power source, a way to irrigate a million acres of desert, and to control the flooding of the lower Nile, the Aswan High Dam was built with the aid of the then Soviet Union which wanted to gain more influence in Egypt. The dam did these things, but the environmental effects have been profound, and largely negative.

With the completion of the dam in 1970, the annual flood of the lower Nile Valley was entirely eliminated.(81) These flood waters had been very useful because they left behind layers of new, fertile soil each year in the most important agricultural area of Egypt. Now the land must be fertilized artificially, and the approximately one million tons of synthetic fertilizer applied annually cannot equal the 100 million tons of silt previously deposited by the floods each year.(81) And no synthetic fertilizer can quite equal the nutrient balance of materials, particularly the very important organic constituents, in the sediment brought by the Nile. Now, the generally poor farmers who previously got fertilizer free each year from the Nile must buy it. Many do not have the money. Also, along the Nile the annual floods used to wash away salts from the soil which would otherwise injure plantlife. Now the salts are accumulating.(81)

Because the silt load of the Nile is now trapped behind the dam, the Nile delta in the Mediterranean, which formerly was stable and in places growing outward, is now being eroded. The Nile, before the Aswan High Dam, annually carried 42 billion cubic yards of water, with its sediments, to the delta area. Now the flow has been reduced to 7.9 billion cubic yards.(71) The sediments carried by this much smaller amount of water are insufficient to maintain the delta against erosion by the sea. This is an exceedingly productive agricultural area, but about one-third of the delta edge is now being cut back at the rate of as much as two meters a year. Before the dam, there was a balance between the building out of the delta and its erosion by wind and water which was favorable to the sustenance and ecology of the region. That balance has now been lost.(79,81)

The Nile floods also brought nutrients into the eastern Mediterranean, which fed a rich plankton population. But, by the effect of the dam cutting off much of the nutrient supply the plankton, which are the start of the marine food chain and which were fed in part by the Nile nutrients, have greatly diminished. The result is that crustacean, mackerel, and sardine populations have also shrunk. The sardine catch in the eastern Mediterranean has dropped by 83 percent.(71) Some 30,000 Egyptian fishermen have lost their livelihood, and the Egyptian food supply has correspondingly decreased. The dam changed the river system with marked effect on the freshwater fishery. "Out of 47 commercial fish species thriving in the Nile prior to the dam's construction, only 17 are being harvested a decade after the dam's completion."(71)

The Aswan High Dam does provide irrigation by canal to some 2,800 square kilometers of formerly desert land. But there has been an enormous build-up of the infamous water hyacinth which is choking the canals, causing a greatly increased water loss through evaporation by plant transpiration.(81) The stagnation of the waters by this weed, and the lack of the scouring effect of the floods has resulted in this area, as well as below the dam, in a great rise of the Bilharzia larvae which cause schistosomiasis, a very debilitating and often fatal blood fluke infection in humans. This disease is increasing rapidly. It had been brought under control before the dam, largely by the efforts of the Rockefeller Institute after World War II, but the disease now has acquired new virulence and is spreading.

Losses of water by evaporation have been far higher than expected due to the very high desert winds. The silt which was to seal cracks under the dam apparently went into cavernous areas in the bottom of the lake, and did not prevent seepage as expected. The result was that Lake Nasser behind the dam did not fill until 1985, and only about a third of the million acres expected have been brought under cultivation. Only three of the twelve turbines are presently operational. The life of the dam determined by the rate of silt accumulation had been predicted to be 500 years, but now is thought to be 100 years or less.

Analyzing all the various factors involved, some observers have concluded that the dam and reservoir, in total, are an ecological disaster.(41) Perhaps the observation by the Greek historian Herodotus, about 450 B. C., who traveled through the Nile region may be the last word. He wrote, "...especially in the part called the Delta, it seems to me that if...the Nile no longer floods it, then, for all time to come, the Egyptians will suffer."

Population growth absorbs dam's benefits

It is significant to note that Egypt's population is now increasing at the rate of about one million a year. The additional food supply provided by land irrigated by the Aswan High Dam was totally absorbed by the population increase which occurred just during the period of time during which the dam was being built.

The Aswan High Dam and Egyptian population growth is an excellent example of the problem of the race between resource development and a continually growing population. Egypt, already the Arab world's most populous country, continues to increase at a far faster rate than most other countries. The birth rate in Egypt is 40 per thousand people whereas in the United States it is 15 and in Britain it is 13 per thousand. In Cairo, the population in the most densely inhabited part of the city is 240,000 per square mile. At the current rate of population increase, within the next five years or less Egypt will have to import at least one seventh of all the surplus wheat in the world to feed its people. In 1992, Egypt, just behind Israel ($4.74 billion), was the second largest recipient of foreign aid from the United States, in the amount of $2.54 billion. The late Egyptian President Nasser stated, "I am not a believer in calling on people to exercise birth control by decree or persuasion. Instead of teaching people how to exercise birth control, we would do better to teach them how to increase their land production and raise their standard...If we direct our efforts to expanding the area in which we live instead of concentrating how to reduce the population we will soon find the solution." This simply ignores the fact that there are no more Nile's to dam and only so much water in the Nile with which to irrigate desert lands. The resource base is the ultimate control, and Egypt's destiny is tied to the waters of the Nile. Obtaining money from other governments by which to import food supplies simply allows Egypt to ignore this basic fact, but it is not the long term solution. The carrying capacity of Egypt cannot support the population.

Brazil and China

Brazil, the largest nation in South America, has very little fossil fuel. To meet the demand for electricity, large dams have been constructed. But in the process a number of native cultures have been displaced, and large areas of rainforest have been flooded. As the population continues to grow, more dams are contemplated.

China, with 1.2 billion people, is in need of more electric power. To meet this demand for electricity, and also for irrigation water, China has implemented a huge dam construction program over the past 50 years. In 1950, there were just two large dams higher than 50 feet in China. By 1985, there were 18,820. Now China has begun what may become the world's largest water impoundment by the construction of a dam across the Yangtze River, the Three Gorges Dam project. It will displace more than a million people, it will inundate 13 cities and 140 towns, and it will flood large areas of fertile river lowlands. It will also flood the spectacular scenery of the Yangtze River gorges, and numerous archeological sites. Scheduled for completion in 2009, the dam will be 600 feet high and one and a half miles wide with a 400 mile long reservoir. It will be the world's largest single electric generating dam.

However, numerous questions about the environmental impact of the dam caused the United States to stop giving technical assistance on the project in 1993. There also was a question as to whether or not it would be cost-effective, producing more benefits than were lost, according to Daniel Beard, Commissioner of the U.S. Bureau of Reclamation. In 1996, Beard stated that the U.S. had gone out of the business of promoting dam building in other countries, and urged countries to "avoid repeating our mistakes," and consider alternatives to large dams. In 1996, Beard, commented, "The legacy of the dam-building era is that we have cheap water and cheap power and expensive environmental impacts." That same year, the U.S. Export-Import bank said it would not help finance companies bidding on China's Three Gorges Dam project. The president of the bank stated that the information they had received, "...fails to establish the project's consistency with the bank's environmental guidelines." The dam is now under construction.

Dams have a lifespan

When one looks at the recently built huge dams with their reservoirs, it is not readily apparent that they are not permanent features just as they stand. We do not see this, for the life of recently constructed huge dams with their reservoirs is longer than the lifespan of one or perhaps several human generations.

However, both here and abroad many smaller impoundments are now simply concrete waterfalls, filled to the brim with mud. At least 2,000 irrigation dams in the United States are now useless having been filled with sediment.(48) A dam built in India with the high hopes of supplying both electricity and irrigation water long term, had its reservoir half filled with sediment in 30 years, and the power producing facilities were rendered almost useless. A study of 132 dams built 30 to 50 years ago in Zimbabwe indicated that now over half are more than 50 percent filled with sediment.(34)

Eventually all dams and their reservoirs will suffer the same fate. In the epilogue to Potter and Drake's book on Lake Powell and the Glen Canyon Dam, Professor Orson Anderson has considered the future and writes:

> "Will Lake Powell still be a sparkling blue lake winding among the spectacular red Triassic rock canyons, or will it be a meandering segmented river coursing through sediment-filled marshes of tamarisk? Will there still be a functioning dam, controlling and distributing the flow of the Colorado River, or will there be a cascade of water over a crumbling concrete ledge, a broken dam whose penstocks have long since been filled with mud?"(72)

Glen Canyon Dam, which lies upstream from the Grand Canyon, has had a marked effect on the Colorado River through the canyon. Beaches have been gradually lost as the sediment which would replenish them was deposited behind Glen Canyon Dam. In an effort to restore the river banks in the Grand Canyon, in March, 1996, a huge artificial flood was released from Glen Canyon Dam, over 117 billion gallons of water during one week. This is the first scientifically arranged artificial flood, designed to mimic seasonal flows restricted by the 33 year-old dam. Bruce Babbitt, Secretary of Interior, stated that it is, "a new era for ecosystems, a new era for dam management, not only for the Colorado but for every river system and every watershed in the United States." A member of the Hualapai Indian tribe indigenous to the area stated, "But this is nothing compared to the days before the dam." Prior to the dam, floods three to four times the strength of the artificial flow came through with each spring's snow melt. How successful this dam management program will prove to be, only time will tell. It cannot completely replicate the natural regimen of the river before the dam was built.

The silting of reservoirs is both a major long term and short term problem. In India, in eight large reservoirs, the actual rate of sedimentation ranged up to 16 times as much as the projected rate, with the lives of the reservoirs reduced accordingly.(20) It has been estimated that Lake Mead, the reservoir behind Hoover Dam on the Colorado, will be filled with sediment in about 400 years.(66) A raft trip through the Grand Canyon shows benches of sediment which have already built up in the lower end of the canyon. In pursuing this matter of reservoir life with the governmental agencies concerned, the gist of the reply to the question of what happens when the reservoirs are filled with mud was, in effect, "let the future generations worry about that."

The impoundment of sediment behind dams not only eventually fills the reservoir, making them useless, but it also prevents the sediment from going downstream as it normally would to maintain and build out deltas and marshlands which are home to myriad forms of wildlife. Instead, the lack of these stream-borne sediments results in deltas being cut back by wave and current action and diminished in size. Marshlands are invaded by the sea and destroyed.

In the United States, the era of building big dams is largely over, mostly because there are only a very few such sites left, such as the Grand Canyon and the lower Hell's Canyon on the Snake River, both highly sensitive environmental areas. But in other large parts of the world, South America, Africa, Asia, and even in Europe large dam construction projects are either underway or contemplated.(27) Canada seems to be in pause in this regard at the moment, giving the matter of huge dam construction more thought.

This is the age when technology has harnessed nature in a great many ways, some of which probably cannot be repeated. Dams and their power plants are marvelous examples of modern engineering which presently benefit us greatly, but they are structures which have a finite life. And it should be noted that a dam site is not a renewable resource. Once used and filled with sediment it is gone.

In the United States, and the world as a whole beginning in the 1930s, governments began building large dams (those over 50 feet high). There were 5,000 such dams in 1950. Now there are 38,000 and many more are either in the planning stage or already under construction.

We are now enjoying the use of the early and most useful part of the lives of these relatively recently constructed big dams. We enjoy the short term benefits, some of which, however, are already becoming costly in terms of such things as lost salmon runs. But the long term results when the reservoirs are filled with silt, may not be so useful. But that will not be our problem. We have willed those difficulties to future generations.

Water Use and the Environment

Needed the world over by everyone, and with soil, the two most valuable resources of all, water when used by humans may have a variety of environmental impacts. As a medium to support human life directly, and through agriculture it has a very positive effect. However, there can be and are some negative aspects to water use.

Irrigation waters

All water, except precipitation from the atmosphere, contains minerals in solution. Water diverted from rivers, or obtained from wells to irrigate land contains minerals which remain in the soil. If these minerals are not flushed out in some fashion eventually the soil becomes too salty for agriculture. In the Imperial Valley of California which lies below sea level, these minerals cannot be flushed out to streams which reach the ocean, they flow instead into the Salton Sea, the lowest point in the area. This is causing an increase in salinity of that body of water with a negative effect on the organisms which depend on it. In other parts of the world, soils have lost their agricultural capabilities by the accumulations of salts from irrigation water.

Groundwater Mining and Land Subsidence

When groundwater is pumped out faster than it can be naturally recharged, the water table is lowered. If this trend is not checked, the water level may drop below the ability of pumps to economically raise the water to the surface and the supply of water is lost. But there is also another problem which may result from pumping water from the ground.

Land subsidence

When fluids — oil and water — are taken out of the Earth the ground may sink if the strata are not highly indurated — that is, fairly rigid. If the aquifer's materials are very well sorted, as for example, in the St. Peter Sandstone, further compaction cannot take place. In other places the rocks overlying, or part of the aquifer system, are so solid that they will not compact when fluid is withdrawn. In central and eastern Oregon, water is pumped from fractured lava flows or sands and gravels interbedded with the lava flows. The basalt lava flows are very rigid and no ground subsidence takes place. Also the groundwater levels are monitored so that the rates of pumping are held equal to the natural recharge, and water table levels are maintained. However, in other areas land subsidence due to water withdrawal has been and continues to be a problem. Land subsidence and fissures related to groundwater withdrawal in the United States has now affected an area of about 8,500 square miles.(45)

San Joaquin Valley

Due to excessive withdrawal of water in the San Joaquin Valley of California more than half the area (or about 5,200 square miles) has subsided more than one foot, and in the period from 1925 to 1977 at one location the ground subsided 29 feet.(6,45) All this subsidence has resulted from over pumping. On the west side and in the southern end of the valley, the drop in water levels in wells has been as much as 400 to 500 feet, and land subsidence is evident in several areas.

The three principal areas of land subsidence in the San Joaquin Valley are the Los Banos-Kettleman City area, the Tulare-Wasco area, and the Arvin-Maricopa area. Subsidence in these places for several years ranged from about one-half foot to a foot and a half per year. Pumping of groundwater had to be curtailed to prevent further subsidence.

Santa Clara Valley

Land subsidence in the central part of the Santa Clara Valley of California has been going on for more than 50 years, being first noticed when a detailed series of re-leveling studies were done in 1932 to 1933. A level line established there by the Coast and Geodetic Survey in 1912 showed about four feet of subsidence in the San Jose area. A subsequent study by the U.S. Geological Survey published in 1988 showed that since 1933 the San Jose land surface had subsided a total of about 13 feet.(60) This large amount of subsidence is presumably due to the rapid and intense development of the San Jose area since 1933. It was concluded that the principal cause was the continual excessive pumping of groundwater. This same study showed that the artesian water level had dropped as much as 200 feet since 1916. Now, however, due to imports of surface water and a decrease in groundwater withdrawal, the water level has recovered 100 feet or more, but the land subsidence remains.

The California Central Valley in total

This includes both the San Joaquin Valley in the south and the Sacramento Valley in the north. It has more than 100,000 irrigation wells now drilled, and groundwater pumpage greatly exceeds the natural recharge rate.(6) It has the largest volume of land subsidence in the world caused by human action, mostly due to excessive groundwater pumpage. Some, however, is caused by drainage of marshlands for agriculture in the Sacramento River delta area where about 450 square miles has subsided, in some places more than 20 feet.

Permanent loss

When an aquifer collapses, its water-bearing ability is either greatly reduced or may be entirely destroyed. In either case, it has suffered irreparable damage, and cannot be restored. With the great importance which the Central Valley is to the agricultural production of both California and the entire United States, the diminishing ability of aquifers to supply groundwater for irrigation is a serious loss. An agricultural economy built on a groundwater resource which cannot be renewed to what it was once was will either have to contract or find some other source of water. California's Central Valley has been one of the most productive and most valuable pieces of agricultural land in the world. Because half of the irrigation water used in the Central Valley has been groundwater, its depletion is of major economic concern. With groundwater, on which this economy was partly built, now being partially lost, the economy of the Valley will have to adjust accordingly. Surface water brought in from considerable distances by canals is one solution being used. However, this source is limited by the increasing competition from the cities for these finite water supplies, and from the fact that the northern areas of California from which much of this water is obtained are becoming increasingly unwilling to continue to export their water.

Las Vegas Valley

In the Las Vegas, Nevada valley area, excessive pumping of groundwater has caused land to subside as much as five feet. Some 400 square miles of this valley now are showing subsidence due to groundwater withdrawal. Visible displays of this can be seen by the well casings and wellheads which now stand as much as four feet above the ground level at which they were originally placed. It is estimated, also, that excessive pumping has reduced groundwater levels as much as 180 feet locally around the Las Vegas Valley Water District well field. The resulting damage to buildings and roads has already amounted to several million dollars.

Houston-Galveston area

In the Houston-Galveston area of Texas, excessive withdrawal of groundwater has caused the land in places to subside as much as eight feet. The area affected covers several hundred square miles and damage from this subsidence is now in excess of a billion dollars.

Arizona

In Arizona, large land areas in the southern part of the state have been slowly subsiding. Great quantities of groundwater have been pumped out here since 1900; more has been withdrawn than has been replaced by natural recharge. Subsidence was first noted in 1948 near Elroy in south-central Arizona. A 1977 re-survey showed that since 1948 the ground subsidence has been more than 12 feet in places, with the maximum annual subsidence rate being about five and a half inches. The damage to roads and buildings has now run into the millions of dollars.

In Arizona's Salt River Valley, where fields have been irrigated by groundwater for many years, the water table has dropped more than 300 feet in places. This loss of water in the ground has caused subsidence of the surface in a number of areas. The Central Arizona Project aqueduct which brings in Colorado River water to Arizona had to be especially constructed to compensate for continual land subsidence near Apache Junction. Elsewhere, in Paradise Valley, a residential complex northwest of Phoenix, a 400 foot long fissure opened as an apparent result of groundwater withdrawal. Many more fissures are now known with hundreds occurring in the basins of parts of Cochise, Maricopa, Pima, and Pinal counties — all areas of groundwater pumping for agriculture.

Groundwater Withdrawal/Land Subsidence a Worldwide Problem

Land subsidence due to excessive groundwater withdrawal has become a worldwide concern. At Osaka, Japan, subsidence has been about 10 feet resulting in about 50 million dollars in damage. The Tokyo area has undergone subsidence in places in excess of 12 feet, and the damage cost between 1957-1970 amounted to an estimated 225 million dollars. Thailand (Bangkok area), Italy, and England (the London area) have also experienced land subsidence due to excessive groundwater pumping.

Mexico City

Probably the most spectacular area of ground subsidence has been in the Great Valley of Mexico, now the site of Mexico City with more than 20 million inhabitants. Here the land has dropped in places as much as 28 feet, and the large lake which once was there, has been reduced to a relatively small pond. Buildings continue to sink and the damage is now greater than 500 million dollars.

Various and costly effects

The results of ground subsidence are becoming costly in many ways. Flow of sewers may be impeded and in some cases even reversed. Bridges, tunnels, railroad lines, power lines, and highways are all adversely affected. Cracks in foundations of buildings occur. A railroad derailment was caused by land subsidence and shifting of the rails. In total, the mining of groundwater has become in many areas a substantial economic liability. And the subsidence and damage continue to grow, although in some places regulations now have prevented further excessive groundwater removal.

Groundwater Mining and Salt Water Intrusion

There is an additional negative effect of mining groundwater. Along the coastal plain of southeastern United States several excellent freshwater aquifers dip gently toward and ultimately into the sea, as, for example, in Virginia. But because they are open-ended in salt water, keeping the salt water out depends on a steady flow of fresh water from the higher level of the land, and toward the ocean. If fresh water is drawn out excessively, the hydraulic pressure of the fresh water is reduced, and there is a landward invasion of salt water into the aquifer. This has been a problem in the Norfolk, Virginia, area, and also occurs elsewhere in the Atlantic coastal plain.

Salinas Valley

On the Pacific coast in the highly productive Salinas Valley of Monterey County, California, over-pumping has caused the aquifers to be invaded by salt water. As of 1993, salt water had destroyed about 100 wells in the valley, and sea water had moved more than seven miles inland in some places. William Hurst, general manager of the Monterey County Water Resources, stated: "If nothing happens to slow this down, you won't be able to grow much of anything in this valley, and people are going to have to get drinking water from somewhere else."(61)

Particularly delicate fresh water-salt water relationships exist in some of the Pacific islands where thin lenses of fresh water sit on top of salt water in the very porous coralline limestones. If the fresh water is withdrawn beyond its recharge rate, the salt water rises and contaminates the wells. Reversing this situation takes considerable time. The development of tourism in some of these islands has increased the demand for fresh water and exacerbated the problem of salt water intrusion.

Fossil Fuels and the "Greenhouse Effect"

About 100 years ago the Swedish chemist Svante Arrhenius predicted that burning fossil fuels would cause carbon dioxide to build up in the atmosphere and trap heat. The present prodigious burning of fossil fuels has now raised a concern over the possible "greenhouse effect" which this may have. The "greenhouse effect" is the common term used to denote presumed influence of the retention of the Sun's heat in the atmosphere by the presence of carbon dioxide.(65) The effect of this is called global warming. A certain percentage of the Sun's heat is reflected back to outer space, but carbon dioxide in the atmosphere trends to prevent this to some extent. This raises the temperature of the Earth's atmosphere much as in a greenhouse, where the short light waves can come through the glass but they are changed to long infrared heat waves upon impact. These long waves cannot get back out through the glass, thus retaining the heat in the greenhouse. When fossil fuels, coal and petroleum, are burned one of the combustion results is carbon dioxide.

Numerous studies suggest that the carbon dioxide content of the atmosphere is increasing. On the Arctic island of Spitsbergen, plant life is spreading dramatically. In the Antarctic, a 65 kilometer crack in the Larsen Ice Shelf has developed. Samples of air from decades ago can be obtained from bubbles in the Greenland ice cap. Comparison with present air samples indicates a slight rise in CO_2 content. Certainly a reasonable conclusion would be that this huge — almost instantaneous in terms of geologic time — burning of millions of years of carbon accumulations will have some effect.

Climate modification

If the greenhouse effect resulting in global warming is growing, the climates of the world will be modified in a variety of ways, the net effects of which are hard to predict. One probable result, however, would be an increased rate of melting of the ice caps in the polar regions causing a rise in sea level. Large portions of the world's population live close to sea level in cities like Miami, lower New York City, Copenhagen, Stockholm, Tokyo, and Shanghai. Much of The Netherlands is already below sea level, protected by dikes. A rise in sea level would have profound effects around the world. If just present trends continue, the Marshall Islands in the Pacific, and the Maldives off southern India as well as many other low-lying ocean island communities will disappear in only a few centuries.

There is yet a continuing debate over the existence of the greenhouse effect from the burning of fossil fuels, and there are competent people on both sides of the argument. The climatic record of the past 10,000 years shows both warming and colder trends, with an overall trend toward warming as we emerge from the most recent glacial times. Some people suggest that the current warming is simply part of the post-glacial over-all warm trend. Yet, accumulating evidence backed by a definite measurable rise in the carbon dioxide content of the air suggests that the greenhouse effect is real and here.(52) Analyses of air trapped in bubbles in glaciers show the carbon dioxide content in 1860 was about 260 parts per million; today it is 346 parts per million.(11)

A conference on Global Warming was held in Berlin in March 1995, with a United Nations panel of 300 scientists. Sir John Houghton, conference chairman, told the London *Guardian* at the time, "There is no doubt that global warming is happening. It is inevitable. The question is whether we can slow it down enough to avert the worst effects."

The fear is, however, that if the greenhouse effect is proceeding now, then it may already be too late to stop it. To do it takes international cooperation. China was a key player at the Berlin Conference, because it is developing so fast that it could on its own destroy the efforts of other nations to reduce total world CO_2 emissions. China, in 1995, announced a program to emphasize nuclear power as an alternative to coal to meet the energy demands of an increasing population, and its rapid industrialization. China is also proceeding with the huge Yangtze River dam. Unfortunately, however, China will also have to continue to increase the use of coal. At present, coal supplies China with three-fourths of its commercial energy, and the government plans to double coal use in the next two decades.(15) Alternative less polluting energy sources cannot be obtained in volume and soon enough to halt the trend, which means more air degradation. Coal is one of the largest contributors among energy sources to CO_2 atmospheric pollution.

Western nations have tentatively established a year 2000 target of reducing emissions to the 1990 levels. The United States, currently the world's largest carbon dioxide producer because it burns the greatest amount of fossil fuels, has per capita carbon dioxide emissions nearly twice as high as in Western Europe, and five times the world average.(35) With present trends, the U.S. will go over the year 2000 target limit by 11 percent. Without China's cooperation, the total of world carbon dioxide emissions will expand much more.

CO_2 from fossil fuels

A summary of the effect of using various fossil fuels states:

> "Coal, which is nearly pure carbon, is almost completely converted to carbon dioxide when it is burned; oil and gas contain hydrogen in addition to carbon, hence both carbon dioxide and water are released when they are burned. For the same amount of heat produced, oil and gas produce less carbon dioxide than does coal, since some of the energy released from oil and gas comes from the reaction of hydrogen with air. The synthetic fuels, on the other hand, require extensive processing and as a result produce even more CO_2 for a fixed heat output than does coal. Natural gas is the lowest CO_2 producer (25 percent less than oil), while synthetic gas from coal produces 80 percent more than fuel oil."(70)

Half a billion years accumulation in an instant

We are in the age of fossil fuels, and we are, in a relative geological instant of time, throwing into the atmosphere great volumes of hydrocarbons. These have accumulated in the Earth during hundreds of millions of years, some as far back as the Cambrian Period, some 500 million years ago, and perhaps even a little earlier. The "sink" for absorbing the carbon dioxide produced by this activity is the vegetation on land, the floral plankton of the oceans, and the ocean water itself. Whether or not these can absorb the great increase in carbon dioxide emissions and preserve the balance of carbon dioxide in the atmosphere which we have known is the question. Weber and Gradwohl state: "More recent studies estimate that the oceans absorb between 30 and 50 percent of the carbon dioxide generated by human activities. Today, the role of the oceans in absorbing and storing carbon dioxide remains one of the largest unknown variables in the models climatologists use to predict the extent and effects of climate change."(91)

If the greenhouse effect is real and increasing, the question then becomes can we effectively cope with it in various ways?(17) Some suggestions have been made such as extensive reforestation, but too little is yet known about the whole problem to offer definite solutions, if, indeed, the problem exists. There are firm statements on both sides of the question of global warming. Some claim that no definite conclusions can yet be drawn.(51)

However, the evidence increasingly tends toward the conclusion that the "greenhouse" effect is occurring.(51) In 1995, the United Nations sponsored Intergovernmental Panel on Climate change which represents the consensus of the scientific community issued a draft report stating that the global warming which has taken place in this century, "is unlikely to be entirely due to natural causes." Vogel states: "Climate detectives are finally beginning to see the fingerprint of greenhouse warming on the planet...climate models clearly show that we cannot continue to pump such a large amount of carbon dioxide and other greenhouse gases into the atmosphere without having some effect."(87)

Other air pollution

The effect of burning coal and gasoline in producing other forms of pollution besides carbon dioxide is well established. The smog from vehicle emissions in the Los Angeles area and other large cities testifies to that, and the effect on the health of the citizenry is surely negative. In large metropolitan areas of China, notably in Beijing and Shanghai, the air quality resulting from coal burning, which at present is China's chief source of energy, is very bad. It no doubt seriously affects the health of those populations.

Oil Spills

Increasingly, much of the world's oil is moved by ship. Up until the time the United States' self-sufficiency in oil ended in 1970, most of the oil used in the U.S. was transported within the country by pipeline. However, as more and more oil is now imported, and also a relatively larger amount comes from offshore by ship, transport by sea is increasing. As a result, oil spills at sea occur. The Exxon Valdez spill in Prince William Sound, Alaska, is one of the more infamous of such disasters. Numerous other spills have occurred. In 1996, the tanker Sea Empress ran aground off the coast of Devon in southwest England. Some 20 million gallons of oil were spilled, about twice the amount of the Exxon Valdez spill. The spill encircled Lundy Island, a marine nature reserve, and ranked as one of the 10 worst oil spills in the world.

Accidents are an unfortunate fact of life; vehicle accidents occur on highways simply because people drive on highways. About 40 percent of them are probably preventable as they are alcohol-related, as may also have been the case the Exxon Valdez spill. Other marine accidents, even as are accidents on land, may be caused by storms, fogs, and other weather problems. They are largely unavoidable; and not predictable. Mechanical failures also cause accidents. As long as people want to drive gasoline-powered cars, and oil is transported by any means, oil spills will occur. Efforts can be made to minimize them, but they are not likely to be entirely eliminated. The way to completely avoid all oil spills is not to transport or use any oil. Even as the human race is not perfect, oil spills, just as car accidents, are an inevitable part of our civilization as we know it today.

Weber and Gradwohl state:

> "Although estimates vary, tankers and freighters are the source for 42 percent of the estimated 25 million barrels of oil entering the oceans every year — about 100 times the entire spill from the *Exxon Valdez* in 1989. Two thirds of the oil from marine transportation comes from routine operation of vessels including the discharge of oil in ballast water, of oil washed from tanks for storing oil, and of the sludge from fuel oil used to power these vessels. On average, spills from tanker accidents release nearly 3 million barrels of oil into the oceans each year.
>
> "Discharges of oil from sources on land account for about 32 percent of the oil entering the oceans annually. Of this, 5.1 million barrels or more than half the total comes from municipal sewage treatment plants, and about one-quarter comes from refineries and other industrial activities. Offshore production of oil and gas releases an estimated 360,000 barrels of oil in routine and accidental discharges each year. Winds carrying wastes released into the atmosphere in the refining and combustion of oil deposit another 2.2 million barrels of oil onto the oceans."(91)

"Citizen oil spills"

It is interesting to note that whereas the Exxon Valdez spill was about 10 million gallons, and there was worldwide outrage over this, in the United States 20 to 30 times as much oil is annually dumped onto the ground and into storm drains by the do-it-yourself oil changers. Even in the notably environmentally conscious State of Oregon, do-it-yourself oil changers discard about 2.5 million gallons of used motor oil a year. Of that, only about 600,000 gallons, less than 25 percent, is collected and recycled. The rest, it is believed, is disposed of improperly.

Natural oil seeps

In petroleum provinces, oil seeps out of the ground naturally. Locally, seeps may exude large amounts of oil continually over many years. Nearly all major oil fields including those beneath the sea, exhibit some surface seeps. In total around the world, it is estimated that about 1.8 million barrels of oil annually are discharged into the sea from natural oil seeps. At Coal Point west of Santa Barbara, California, a large shallow underwater oil seep continually leaks oil into the ocean estimated to be as much as 150 barrels a day. A recent study showed that the western part of the Santa Barbara channel is not well developed for oil production because of the high tar content of the oil, and that much of the oil and tar on the beaches as well as gaseous releases and air pollution result from the enormous volume of natural seepages in the area. If the area was drilled and oil produced, these naturally caused polluting effects would be lessened.(59)

Alaska Pipeline and the Environment

One of the most bitterly fought battles regarding the perceived danger to the environment from a mineral resource development, was concerning the construction of the 800-mile pipeline from the Prudhoe Bay Oilfield on the Alaskan North Slope, to the shipping terminal at Valdez on the southern Alaskan coast. The Prudhoe Bay Field was discovered in 1967. Subsequent drilling by 1970 proved it to be the largest single oilfield ever found in the United States. The only feasible way to get the oil out of Prudhoe Bay was by means of an 800 mile pipeline across large areas of permafrost, and over three major mountain ranges, including the Alaskan Range with the highest mountain in North America. It was one of the largest engineering projects ever undertaken anywhere.

Perceived environmental impact

Opposition to the pipeline construction was intense. Articles and books were written predicting numerous catastrophes from earthquake rupture of the line, to the prevention of the caribou from migrating.(5) It was projected by environmental foes of the pipeline that the warm oil pumped through the pipeline would cause all sorts of damage by thawing the permafrost. It was further suggested that wildlife all along the pipeline route would somehow be markedly affected. Environmental protests and related legal proceedings delayed the building of the line for five years.

Actual impact

The pipeline, however, was constructed. To see what effects the pipeline actually had, I visited Prudhoe Bay in 1992, and followed the pipeline from Prudhoe Bay to the terminal at Valdez. None of the predicted negative environmental impacts was apparent. Where the pipeline crosses the permanently frozen ground (permafrost) from Prudhoe Bay to a short distance south of Fairbanks, the pipeline is elevated nine feet above the ground. The supports each have miniature self-circulating refrigeration systems in them which prevent the steel supports from conducting heat into the ground and thawing it. This is necessary to keep the pipeline stable.

In the Prudhoe Bay Field where the greatest activity is, the caribou herd has grown from approximately 3,000 at the time the field was discovered to more than 15,000. The caribou graze extensively over the field and unconcernedly move beneath the pipeline, which offers no barrier whatsoever to their movement. The commercial airline, MarkAir, offers visitors a delightful, reasonably priced trip, starting at Fairbanks, to Prudhoe Bay, and includes a tour of the oilfield.

All along the pipeline where it is elevated, animals, be they moose, caribou, bear, or wolves, pass freely beneath it. Where there is no permafrost, the pipeline is buried. There is a major highway which follows the general route of the pipeline south of Fairbanks to Valdez. As you travel the highway it is hard to see where the pipeline is buried. The only readily visible evidences are the several pumping stations built along the way. Helicopters and ground personnel maintain a 24-hour watch on the entire length of the line.

Importance to U.S. of Alaska Pipeline

At the present time, one-quarter of all the oil produced in the United States flows through that pipeline, and is a significant part of the U.S. economy. If it had not been built so that the

Prudhoe Bay oil could be used, this would have had a markedly negative effect on the U.S. balance of payments, and further exacerbated the U.S. currency problems. Although it did not, and does not solve the problem of ultimate greater U.S. dependency on foreign oil, it has at least mitigated the problem for the time being.

Area involved in oil production

The North Slope of Alaska is that area between the foothills of the Brooks Range to the south, and the Arctic Ocean on the north, an area of approximately 69,000 square miles. Oil development, including Prudhoe Bay and some nearby smaller fields, has occupied about 400 square miles or slightly less than 6/10 of one percent of the total area of the North Slope. The actual area involved in various installations is, of course, much less than that. Wells are drilled directionally from gravel drill pads, with 16 to 40 wells per pad. The drill pad occupies an area of 20 acres or less, but the wells drilled from this pad produce from an underground area of up to six square miles. Using a drill pad and directional drilling procedures greatly reduces the need for roads connecting well sites. All together, in the 400 square miles of the oil field less than 15 percent has any road or installation of any sort on it. This is less than 1/10 of one percent of the North Slope. The total area of the Prudhoe Bay oilfield is about 130,000 acres (that is the total size of the underground oil reservoir). To tap this, about 5,500 acres have been disturbed, less than five percent of the total area. The caribou graze extensively over the field.

Arctic National Wildlife Refuge (ANWR)

This area is of great environmental concern. About 92 percent of ANWR is already declared a wilderness area, and off-limits to development. The rest of the refuge was specifically set aside by Congress for possible oil exploration. But the environmental discussion over even this eight percent has continued. With the U.S. now importing more oil than it produces, and with the resulting negative effect on the balance of international payments and on the dollar, the matter of ANWR and its possible oil is a relevant topic. The commonly held impression, fostered by environmental interests, is that oil exploration would take place over the entire refuge. This is not true. It is useful in any discussion to know the facts.

The facts of ANWR and oil

The entire Refuge covers about 19 million acres. The coastal plain, which is the only area of oil interest, is 1.5 million acres, which is eight percent of the Refuge, and is that part set aside by Congress for oil exploration. Oil company estimates are that at full development of what are believed to be the petroleum resources there, no more than one percent of the Coastal Plain would be involved. In terms of the entire Refuge, this means that oil development would involve 15,000 acres at the most. Therefore, out of the total of 19 million acres in ANWR, oil development would use less than eight one-hundredths of one percent. The U.S. Office of Technology has estimated that the "footprint" of human activity on ANWR, if oil is discovered, would actually cover only about 7,000 acres, which is less than the area of Dulles International airport in Washington, D.C.

The coastal plain of ANWR is now the best prospective land area for a major oil discovery in the United States. It has been given rather thorough study both by oil company geologists and by the U.S. Geological Survey.(7)

The Prudhoe Bay Field has begun an irreversible decline, and the nearby second largest field in North America, Kaparuk River, is near its peak, and will be in decline in only a few more years. The Alaskan North Slope currently produces about 25 percent of all U.S. domestic oil supplies. The Alaska Pipeline will not be economical to run sometime before the Prudhoe Bay and Kuparuk fields are exhausted. It will need more oil to keep it going. The ANWR area is about 70 miles east of Prudhoe Bay and can easily be tied into the Alaska Pipeline. By 2010 without the estimated ANWR production, North Slope production will decline to about 500,000 barrels a day. With ANWR production, it is estimated to be about two million barrels a day.

The caribou

The main environmental concern voiced about ANWR seems to be that the Porcupine River caribou herd which comes to that region in the summer would be hurt by the oil development. At Prudhoe Bay, as already noted, the caribou herd from the time of the beginning of oil development to the present has increased from 3,000 to more than 15,000. There is every reason to believe, based on past experience, and the small area actually occupied by roads and facilities over the field, that the ANWR caribou herd would not be harmed.(1)

In a human value approach to this oil/environment controversy, Henry Schuler, energy security specialist with the Center for Strategic International Studies in Washington, D.C., has commented from the perspective of the Persian Gulf War, "We're willing to risk lives in the Persian Gulf, but not the lives of caribou in Alaska."(82) All evidence is that the lives of the caribou would not be at risk.

In regard to the stated determination in 1995 of the federal officials not to issue oil leases in ANWR, it is interesting to note that at the same time plans were proceeding to offer ocean areas for lease just offshore from ANWR, which would be many more acres than the approximately 6,000 acres of ANWR land. These offshore areas are probably more environmentally sensitive than is the ANWR acreage, being the year-around home of seals and polar bears. In the summer the endangered bowhead whale migrates through the area. Also, leases were planned to be offered in the Cook Inlet area of southern Alaska which includes territory used by six endangered species of whales. (It is a region where petroleum has been produced for many years.).

In the case of ANWR oil lease acreage, no threatened or endangered species call it home. From this it might be concluded that ANWR is being used by elected government officials as a highly visible symbol of their concern for the environment and by that means enlist environmental support for the party in power. Environmental politics exist also.(88)

Some Good Environmental Trends

Recent assessments of environmental trends present somewhat encouraging facts.(29) Acid rain from burning coal is declining. The 23 million tons of sulfur dioxide put into the atmosphere in 1970 by U.S. power plants is now more than cut in half. Improved jet airplanes burn about 30 percent less fuel than the previous generation of planes. Greatly improved water treatment facilities have made the use of water supplies much more efficient. A very important positive trend is increased access to safe drinking water. Between 1980 and 1990, Mexico increased such supplies from 73 to 89 percent, Zimbabwe from 52 percent to 84 percent, India from 42 percent to 73 percent, Myanmar (Burma) from 21 percent to 74 percent, and Nepal from 11 percent to 37 percent. Similar large gains were made in a number of other countries.(94) Because water is the chief carrier of disease in many countries, the increase in safe water supplies is a major step in improving the health of these areas.

The notorious air pollution in Southern California caused by motor vehicle emissions has begun a modest reversal in trend. Although still severe, the use of reformulated gasolines, has helped to eliminate 98 percent of hydrocarbon emissions compared with the vehicles of 25 years ago. All gasoline sold in California now must meet new emission standards which has prevented an estimated 3,800,000 pounds of pollution from reaching the atmosphere each day.

Excess optimism

Some gains have been made, but even as environmental pessimists have frequently seen only the dark side of a situation, so too environmental optimism may also be excessive. Statements, for example, which suggest that the arid Southwest United States may be among the most appropriate places for large populations of people to live because solar technology will make it all possible ignores reality.(29) Solar energy cannot create water, and even if solar power could energize the pumps for groundwater sources, such sources in the arid Southwest are simply not adequate to sustain any large population.

The optimism for greatly expanded hydroelectric power is reflected in the charge of "damophobia" with regard to those who do not endorse dams.(29) This ignores the fact that all reservoirs eventually fill up with sediment. And it does not recognize the detrimental effects dams have on the regimens of rivers, or, in the case of the Columbia River system, of the extinction of some salmon runs and near destruction of others.

The suggestion that methanol could be an alternative to "petroleum" (in this citation the author apparently means oil), and be produced from "trees, small woody plants, and cane plants" does not face the fact of the huge volumes which would be involved to make any significant difference in the world oil supply situation. Also, studies indicate that production of methanol from trees and woody plants is a net energy deficit. Methanol can be produced with a net positive energy recovery from coal, but not from plants as suggested.(29)

The statement made in favor of environmental optimism that, "If the majority of U.S. automobiles were pushed to a 45 mpg standard, petroleum imports could end altogether" is unrealistic.(29) It ignores the continuing decline of U.S. oil production, combined with the increasing number of cars on the road, and the fact that automobile consumption of oil accounts for only about 40 percent of the total use of oil in the United States.

It is unfortunate that such misleading statements appear. The general public has little ability to evaluate them and is given a false sense of security by them. On both sides of the environmental debate, facts and realism are in order.

Society's Choices and Priorities

There is no doubt that development of mineral and energy mineral resources has an impact on the environment. Mitigating these impacts can be done in various degrees, but the question becomes two-fold. One is the cost which would be involved in completely repairing the landscape. At the present time some jurisdictions require that open pit mining operations restore the land to its approximate original contour. This can be done more easily in some mining operations than in others. In the case of coal mines, the thickness of the bed of coal is generally less than 50 feet and therefore the amount of material taken from the ground is not large. It is therefore possible to restore the landscape quite easily. The strip mine operation moves along, and the overburden of the coal seam, as it is removed, can simply be conveniently and economically thrown over into the previously mined area. When the mining operation is completed the land can usually be graded into relatively smooth contours at modest cost. In some areas leaving the pits as they are has provided ponds and recreational facilities, and valuable wildlife habitat.

However, in open pit metal mining operations, the entire rock material containing the metal is taken out. In the case of copper ores in the United States, as little as $\frac{4}{10}$ of one percent copper or less exists in the ore. Although this is a very small part of the rock, the entire mass has to be transported to the mill usually some distance away where it is crushed, and the copper is removed. Moving all the waste material back to the mine is very costly, but it could be done. But are consumers willing to pay the cost?

Do people want to use the materials?

In considering environmental impacts of mineral resource production, the first question is simply, does society want to have use of the materials which mining for metals, non-metals (e. g. sand and gravel), and coal, or oil and gas that drilling provides, although such activities in some cases do locally scar the landscape? Or, to put it another way, do you want to live in houses, keep warm in the winter, cool in the summer, and drive automobiles made of steel and powered by gasoline? Do you want to have electricity in your home and office? It is produced by coal-fired, oil-fired, or uranium-fired power plants, or by dams. Producing these things which are demanded by people have an environmental impact. Obviously people want to use Earth resources.

Environmental economics — who will pay?

The second question is whether society at large is willing to pay all the added costs which may be involved in mitigating the environmental impact of mineral production. In some cases, such as in certain coal mining operations, the answer has been yes. But in many metal mining operations, restoring the landscape would be so costly as to clearly make the project uneconomic. Then it is a choice of leaving the open pit as it is, and having the economic advantage of having the domestic production of a useful metal, together with providing employment, or not producing the metal, and having to import it with an increased balance of payments problem, and not providing employment.

Land use priorities

To accommodate what is generally termed "economic development" which is basically a reflection of growth of population, annually about one million acres of land in the United States are paved over or used up in various developments and cannot easily be reclaimed. This includes housing projects, shopping malls, factories, stadiums, golf courses and other recreational facilities, and roads. All these things are demanded by a growing population. Yet none of these things could be built without the use of minerals and energy supplies which use up far smaller land areas in their production, many of which can be and are reclaimed essentially to their original form.

In the case of mining, that industry has disturbed less than one percent of the land area of the United States to produce all the minerals since 1776. About one-third of that area has now been reclaimed. Keep in mind this includes all the iron, copper, lead, zinc, molybdenum, and other metals, and coal, all of which have been absolutely indispensable in building the U.S. economy to what it is today.

Also, in general, land used for houses, factories, shopping malls, and golf courses is relatively flat, commonly fairly good land which could be used for agriculture. This is in contrast to many mining areas which are in mountainous areas not generally suited for crops.

Many oil fields are in fairly rugged terrain, desert regions, or offshore — hardly farming country.

Yet, the hue and cry over mining and drilling operations which are essential to support various economic developments tends to be much greater than the complaints about housing projects and shopping malls (although there are some voices beginning to be raised about that also). The proliferation of golf courses seems to be quite popular. Society's current priority is for an expanding population, which needs more space on which to live, increased recreational facilities and areas set aside for recreation, and correspondingly increasing use of energy and mineral resources. Inevitably this has a marked environmental impact, even beyond what would be needed simply to maintain a stable-sized population. People must recognize the choice they are making. Today, worldwide, the choice is for more people, and a corresponding exponential growth in use of resources. This cannot continue much longer.

Individual lifetime environmental impact

If one is concerned with the environment, the place to start is with birth control. Hall, et al. have cited some interesting facts concerning the lifetime environmental impact of one baby born today in the United States. The child from birth to death will generate 13 tons of waste paper, 10,355 tons of waste water, 2.5 tons of waste oil and solvents, 3 tons of waste metals, and 3 tons of waste glass. From manufacturing processes, mining, and agriculture used to support this individual, there will be 83 tons of hazardous waste, 419 tons from mining (not including coal mining), 197 tons from manufacturing in general, 1,418 tons of carbon dioxide, and 19 tons of carbon monoxide. Consumption of materials during a lifetime will include 1,870 barrels of oil, and 260 pounds of pesticides used to produce the food to sustain the individual.(42) Given the present concern about environmental impacts of various endeavors, it has been suggested, only partially in jest, that before a child is conceived, an environmental impact statement should be required to be filed.

Treat the cause, not the symptoms

Organizations concerned with the environment would do well to allocate more funds toward birth control rather than using their money to fight the symptoms of the problem, which are increased resource development demanded by an ever-growing population.

Treating the symptoms of a problem rather than the basic cause is ultimately a futile effort. Preserving natural regions can be a worthy objective, but if the pressure of increased population is not stopped, wilderness and other such set-aside areas will inevitably be invaded. In 1993, and again in 1995, entrance to Yosemite National Park had to be closed for a time because of the huge press of cars and people wanting to come in. This has never happened before to any national park, and is an ominous sign for the future. In Africa, the pressure of population impinges on designated wildlife preserves, and hungry people poach the animals.

The basic problem is also illustrated in resources other than minerals and energy. The vigorous debate, and often even physical confrontations, over the cutting of timberlands is an example. The fact is that if you want to have a piece of tissue paper with which to blow your nose, or put it to any other end use, it is necessary to cut down a tree or some other type of vegetation to supply the raw material. More and more noses equate to fewer and fewer trees. This is also true with metals and many kinds of fuel supplies. More people will need more of these resources, and their production has inevitable environmental impacts.

In the final analysis, two things combine to degrade the environment: high individual resource consumption, and population growth.(9,28) Both are illustrated by the view stated by Tobias: "Today, countries are divided into two types: those exploiting their domestic environs in order to meet the survival needs of their expanding populations, such as India, and those impacting their environment strictly to meet their increasing expectations, such as the United States."(84) One might make the observation, however, that the impact on the environment in the past in the United States may have been largely for the purpose of meeting rising expectations, but the situation now has also to a considerable extent reached the stage of simply trying to take care of a growing population in its present standard of living. If current trends persist, the population of the U.S. will double by 2050, but the physical standard of living may not continue to rise and may decline.

Brown and Kane, in a recent book which should be required reading for all and especially for statesmen, provide ample evidence that as far as the world as a whole is concerned we now have "full house."(14) Pimentel and Giampietro arrive at the same conclusion: "This brings us to the present situation in which the world is full. The exponential increase in the demand for natural resources, due to demographic and economic growth, is rapidly eroding resource stocks and national food surpluses all over the world. As a result the assumptions typical of the 'empty-world development paradigm' are no longer valid."(68)

Everyone an environmentalist

To say that some people are environmentalists thereby implying that others are not, is not a fair statement. Surely everyone wants a clean environment in which to live. Geologists and others concerned with production of energy minerals such as coal, oil, and uranium, and the various metals are commonly drawn to their professions by an early interest in and affection for the out-of-doors. They find employment in the out-of-doors where these resources exist. They dig mines and drill holes in the Earth to produce materials which society at large demands. It would be nice if all mineral and energy resources could be obtained in some fashion that did not disturb the Earth, but that is not possible. The controversy comes about in regard as to how and where the Earth shall be disturbed, and how much can be restored to its original form and at what cost. The controversy must be resolved in a framework which recognizes the fact that most of the resources we use come from the Earth, and as we each use some of them every day, we each have an environmental impact on the Earth. It is unavoidable, and the more of us there are, the greater the total impact, and now about 250,000 more of us arrive each day.

Summary

With today's technology we are, in a geological instant, using the mineral and energy resources which have accumulated over hundreds of millions of years by slow geological processes. This has enabled the world's population to greatly increase. But, as each person has an environmental impact every day from birth to death, the total environmental stress on the Earth is increasing rapidly. The more people, the greater the impact. Treating the problems of the environment can only serve to buy time to develop a sustainable economy based on renewable resources, and adjusting population size to fit the resources available in that kind of an economy, at a reasonable standard of living.

At present, we are living on a great mineral resource inheritance. We must begin to live on current income, and on that basis, recognize that the world is probably now beyond the population size which can be maintained on a day by day renewable resource availability. Wackernagel and Rees conclude: "We have shown that current human consumption of agricultural products, wood fiber, and fossil fuel have an Ecological Footprint that exceeds ecologically productive land by close to 30 percent. In other words, we would need an Earth 30 percent larger (or more ecologically productive) to accommodate present consumption without depleting corresponding ecosystems."(89)

If the present population is beyond sustainable size, even with many people now still living a substandard existence, it is futile to try to solve environmental impact problems without also addressing the underlying cause which is excess population. All of society, and especially organizations which have the environment as their primary concern, must recognize this.

BIBLIOGRAPHY

1 ANONYMOUS, 1991, Oil and Caribou Can Mix: The Wall Street Journal, January 9.

2 ANONYMOUS, 1995, Oil Rigs as Artificial Reefs: The Futurist, May-June, p. 53-54.

3 BARKER, TERRY, et al., (eds.), 1995, Global Warming and Energy Demand: Routledge, Inc., New York, 352 p.

4 BARTELMUS, PETER, 1977, Environment, Growth and Development. The Concepts and Strategies of Sustainability: Routledge, Inc., New York, 163 p.

5 BENNETT, C. F., Jr., 1975, Man and the Earth's Ecosystems. An Introduction to the Geography of Human Modification of the Earth: John Wiley & Sons, New York, 331 p.

6 BERTOLDI, G. L., et al., 1991, Ground Water in the Central Valley, California — A Summary Report: U.S. Geological Survey Prof. Paper 1401-A, Washington, D.C., 44 p.

7 BIRD, K. J., and MAGOON, L. B., (eds.), 1987, Petroleum Geology of the Northern Part of the Arctic National Wildlife Refuge, Northeastern Alaska: U.S. Geological Survey Bulletin 1778, 329 p., 5 pls.(maps).

8 BOUGHEY, A. S., 1971, Man and the Environment: The Macmillan Company, New York, 472 p.

9 BOUVIER, L. F., and GRANT, L., 1994, How Many Americans? Population, Immigration, and the Environment: Sierra Club Books, San Francisco, 174 p.

10 BROWER, MICHAEL, 1992, Cool Energy. Renewable Solutions to Environmental Problems: MIT Press, Cambridge, Massachusetts, 219 p.

11 BROWN, L. R., et al., 1987, State of the World 1987: Worldwatch Institute, Washington, D. C., 268 p.

12 BROWN, L. R., et al., 1991, Saving the Planet. How to Shape an Environmentally Sustainable Global Economy: W. W. Norton & Company, New York, 224 p.

13 BROWN, L. R., et al., 1994, State of the World 1994: Worldwatch Institute, Washington, D. C., 265 p.

14 BROWN, L. R., and KANE, HAL, 1994, Full House. Reassessing the Earth's Population Carrying Capacity: W. W. Norton & Company, New York, 261 p.

15 BROWN, L. R., et al., 1995, Vital Signs. The Trends That are Shaping Our Future: Worldwatch Institute, Washington, D. C., 176 p.

16 BROWN, TOM, 1971, Oil on Ice: Alaska Wilderness at the Crossroads: Sierra Club, San Francisco, 159 p.

17 CARGO, D. N., and MALLORY, B. F., 1974, Man and the Geologic Environment: Addison-Wesley Publishing Company, Reading, Massachusetts, 548 p.

18 CARTER, L. M. H., (ed.), 1995, Energy and the Environment: U.S. Geological Survey Circular 1108, 134 p.

19 CARTLEDGE, BRYAN, (ed.), 1993, Energy and the Environment: Oxford Univ. Press, Oxford, 170 p.

20 CLARKE, ROBIN, 1993, Water: The International Crisis: MIT Press, Cambridge, Massachusetts, 193 p.

21 COATS, D. R., 1985, Geology and Society: Chapman and Hall, New York, 406 p.

22 CORSON, W. H., (ed.), 1990, The Global Ecology Handbook. What You Can Do About the Environmental Crisis: Beacon Press, Boston, 414 p.

23 CRAIG, G. M., (ed.), 1993, The Agriculture of Egypt: Oxford Univ. Press, Oxford, 516 p.

24 DEVINE, R. S., 1995, The Trouble With Dams: Atlantic Monthly, August, p. 64-74.

25 DEVINS, D. W., 1982, Energy: Its Physical Impact on the Environment: John Wiley and Sons, Inc., New York, 572 p.

26 DOLAN, R., HOWARD, A. D., and GALLENSON, A., 1974, Man's Impact on the Colorado River in the Grand Canyon: American Scientist, v. 62, p. 392-401.

27 DURBIN, KATHIE, 1993, Slovakia Builds Dam —Let the Consequences be Damned: The Oregonian, Portland, Oregon, October 25, p. A6.

28 DURING, A. T., 1992, How Much is Enough?: W. W. Norton & Company, New York, 200 p.

29 EASTERBROOK, GREGG, 1995, A Moment on the Earth, The Coming Age of Environmental Optimism: Viking, New York, 745 p.

30 ECKHOLM, E. P., 1982, Down to Earth. Environment and Human Needs: W. W. Norton & Company, New York, 238 p.

31 EGGERT, R. G., (ed.), 1994, Mining and the Environment — International Perspectives on Public Policy: Resources for the Future, Washington, D. C., 172 p.

32 EHRLICH, P. R., et al., 1977, Population, Resources, Environment: W. H. Freeman and Company, San Francisco, 1051 p.

33 EHRLICH, P. R., and EHRLICH, A. H., 1991, Healing the Planet. Strategies for Resolving the Environmental Crisis: Addison-Wesley Publishing Company, Inc., Reading, Massachusetts, 366 p.

34 ELWELL, H. A., 1985, An Assessment of Soil Erosion in Zimbabwe: Zimbabwe Science News, v. 19 (3/4), p. 27-31.

35 FLAVIN, CHRISTOPHER, and TUNALI, ODIL, 1995, Getting Warmer. Looking for a Way Out of the Climate Impasse: World Watch, March/April, p. 10-19.

36 FLAWN, P. T., 1970, Environmental Geology. Conservation, Land-use Planning, and Resource Management: Harper & Row, Publishers, New York, 313 p.

37 FOWLER, J. M., 1984, Energy and the Environment: McGraw-Hill, Inc., New York, 655 p.

38 FREUDENBURG, W. R., and GRAMLING, R., 1994, Oil in Troubled Waters. Perceptions, Politics, and the Battle Over Offshore Drilling: State Univ. of New York Press, Albany, New York, 179 p.

39 GOLDSMITH, EDWARD, et al., 1990, Imperiled Planet. Restoring Our Endangered Ecosystems: MIT Press, Cambridge, Massachusetts, 288 p.

40 GOUDIE, ANDREW, 1986, The Human Impact on the Natural Environment: MIT Press, Cambridge, Massachusetts, 338 p.

41 GRIGGS, G. B., and GILCHRIST, J. A., 1977, The Earth and Land Use Planning: Duxbury Press, North Scituate, Massachusetts, 492 p.

42 HALL, C. A. S., et al., 1994, The Environmental Consequences of Having a Baby in the United States: Population and Environment, v. 15, n. 6, p. 505-524.

43 HARDIN, GARRETT, 1968, The Tragedy of the Commons: Science, v. 162, p. 1243-1248.

44 HOLDREN, JOHN, and EHRLICH, P. R., (eds.), 1971, Global Ecology. Readings Toward a Rational Strategy for Man: Harcourt Brace Jovanovich, Inc., New York, 295 p. (Note: Includes a reprint of Garrett Hardin's article "The Tragedy of the Commons.")

45 HOLZER, T. L., (ed.), 1984, Man-induced Land Subsidence: Geological Society of America, Reviews in Engineering Geology, v. 6, Boulder, Colorado, 221 p.

46 HOUSER, F. N., and ECKEL, E. B., 1962, Induced Subsidence: Geotimes, v. 11, n. 2, p. 14-15.

47 HUBBARD, H. M., 1991, The Real Cost of Energy: Scientific American, April, p. 36-42.

48 JACKSON, WES, 1971, Man and the Environment: Wm. C. Brown Company Publishers, Dubuque, Iowa, 322 p.

49 JOHNSON, W., and MILLER, G. C., 1979, Abandoned Coal-mined Lands; Nature, Extent, and Costs of Reclamation: U.S. Bureau of Mines, 20 p.

50 KELLER, E. A., 1979, Environmental Geology: Charles E. Merrill Publishing Company, Columbus, Ohio, 584 p.

51 KERR, R. A., 1995, Is the World Warming or Not?: Science, v. 267, p. 612.

52 KERR, R. A., 1995, U.S. Climate Tilts Toward the Greenhouse: Science, v. 268, p. 363-364.

53 KERR, R. A., 1995, Studies Say — Tentatively —That the Greenhouse Warming is Here: Science, v. 268, p. 1567-1568.

54 KESLER, S. K., 1994, Mineral Resources, Economics and the Environment: Macmillan and Company, New York, 391 p.

55 KIESSLING, K. L., (ed.), 1994, Population, Economic Development, and the Environment: Oxford Univ. Press, New York, 312 p.

56 LEA, VANESSA, 1984, Brazil's Kayapo Indians. Beset by a Golden Curse: National Geographic, v. 165, n. 5, May, p. 675-694.

57 LEVINE, J. S., 1991, Global Biomass Burning. Atmospheric, Climatic, and Biospheric Implications: MIT Press, Cambridge, Massachusetts, 569 p.

58 LOFGREN, B. E., and KLAUSING, R. L., 1969, Land Subsidence Due to Ground-water Withdrawal in the Tulare-Wasco Area, California: U.S. Geological Survey Prof. Paper 437-B, 101 p.

59 MAJOR, M. J., 1995, Santa Barbara Takes Drilling Step: AAPG Explorer, American Association of Petroleum Geologists, Tulsa, Oklahoma, April, p. 16-17.

60 McARTHUR, SEONAID, (ed.), 1981, Water in the Santa Clara Valley, A History: California History Center, DeAnza College, California, Local History Studies v. 27, 92 p.

61 McCOY, CHARLES, 1993, Future of big Vegetable Growing Area in California Threatened by Salt Water: The Wall Street Journal, July 1.

62 MILLER, G. T., Jr., 1982, Living in the Environment (Third Edition): Wadsworth Publishing Company, Belmont, California, 500 p.

63 MORAN, J. M., et al., 1980, Introduction to Environmental Science: W. H. Freeman and Company, San Francisco, 658 p.

64 NORTHWEST POWER PLANNING COUNCIL, 1995, Northwest Energy News: v. 14, n. 1, Portland, Oregon, 34 p.

65 OPPENHEIMER, MICHAEL, and BOYLE, R. H., 1990, Dead Heat. The Race Against the Greenhouse Effect: Basic Books, Inc, Publishers, New York, 268 p.

66 PELLANT, CHRIS, (ed.), 1989, Earthscope: Tiger books International, London, 208 p.

67 PIMENTEL, DAVID, (ed.), 1993, World Soil Erosion and Conservation: Cambridge Univ. Press, Cambridge, 349 p.

68 PIMENTEL, DAVID, and GIAMPIETRO, MARIO, 1994, Food, Land, Population and the U.S. Economy: Carrying Capacity Network, Washington, D. C., 81 p.

69 POLAND, J. F., 1978, Land Subsidence in the Santa Clara Valley: Water Spectrum, Spring, p. 11-16.

70 PORTNEY, P. R., (ed.), 1982, Current Issues in Natural Resource Policy: Published for Resources for the Future, Inc., Washington D. C., distributed by The Johns Hopkins Univ. Press, Baltimore, 300 p.

71 POSTEL, SANDRA, 1995, Where Have all the Rivers Gone? World Watch, May/June, p. 9-19.

72 POTTER, L. D., and DRAKE, C. L., l989, Lake Powell. Virgin Flow to Dynamo: Univ. New Mexico Press, Albuquerque, 311 p.

73 RAMSEY, WILLIAM, 1979, Unpaid Costs of Electrical Energy: Johns Hopkins Univ. Press, Baltimore, 180 p.

74 REISNER, MARC, 1993, Cadillac Desert. The American West and its Disappearing Water: Viking Penguin, Inc., New York, 582 p.

75 ReVELLE, CHARLES, and ReVELLE, PENELOPE, 1981, The Environment. Issues and Choices for Society: Willard Grant Press, Boston, 762 p.

76 SCHNEIDER, S. H., 1989, Global Warming: Sierra Club Book, San Francisco, 317 p.

77 SIMMONS, I. G., 1974, The Ecology of Natural Resources: John Wiley and Sons, New York, 424 p.

78 SOUTHWICK, C. H., (ed.), 1985, Global Ecology: Sinauer Associates, Inc., Publishers, Sunderland, Massachusetts, 323 p.

79 STANLEY, D. J., and WARNE, A. G., 1993, Nile Delta: Recent Geological Evolution and Human Impact: Science, v. 260, p. 626-634.

80 STROHMEYER, JOHN, 1993, Extreme Conditions: Big Oil and the Transformation of Alaska: Simon & Schuster, New York, 287 p.

81 STROSS, FRED, 1995, Personal communication: Chemistry Digs the Past. The Nile: Then and Now.

82 SULLIVAN, ALLANA, 1990, Energy Options. It Wouldn't Be Easy But U.S. Could Ease Reliance on Arab Oil: The Wall Street Journal, August 17.

83 TESTER, J. W., WOOD, D. O., and FERRARI, N. A., 1991, Energy and the Environment in the 21st Century: MIT Press, Cambridge, Massachusetts, 1006 p.

84 TOBIAS, MICHAEL, 1994, World War III. Population and the biosphere at the End of the Millennium: Bear & Company Publishing, Santa Fe, New Mexico, 608 p.

85 TURK, JONATHAN, et al., 1984, Environmental Science (Third Edition): Saunders College Publishing, Philadelphia, 544 p.

86 VILLEE, C. A., Jr., (ed.), 1985, Fallout From the Population Explosion: Paragon House Publishers, New York, 263 p.

87 VOGEL, SHAWNA, 1995, Has Global Warming Begun? Earth, December, p. 24-34.

88 WALL STREET JOURNAL, 1995, Playing Arctic Politics: November 25.

89 WACKERNAGEL, MATHIS, and REES, WILLIAM, 1996, Our Ecological Footprint. Reducing Human Impact on the Earth: New Society Publishers, Gabriola Island, British Columbia, 160 p.

90 WALTHAM, A. C., 1989, Ground Subsidence: Blackie and Son/Chapman and Hall, New York, 224 p.

91 WEBER, M. L., and GRADWOHL, J. A., 1995, The Wealth of Oceans: W. W. Norton & Company, New York, 256 p.

92 WILLIAMSON, A. K., et al., 1989, Ground-water Flow in the Central Valley, California: U.S. Geological Survey Prof. Paper 1401-D, p. D1-D127.

93 WINNINGHOFF, ELLIE, 1994, Where Have All the Salmon Gone?: Forbes, November 21, p. 104-116.

94 WORLD RESOURCES INSTITUTE, 1994, World Resources, 1994-1995. A Guide to the Global Environment: World Resources Institute, Oxford Univ. Press, New York, 400 p.

95 YOUNG, J. E., 1992, Mining the Earth: Worldwatch Paper 109. Worldwatch Institute, Washington, D. C. 53 p.

CHAPTER 23

Efficiency and Conservation — To What Purpose?

With our increase in technology and the growing evidence of the need to make better use of the raw materials and energy which we have, there has been a growing emphasis on efficiency and conservation. The two oil crises of 1973 and 1979 in the United States served the useful purposes of getting people to think about efficiency and conservation. Efficiency and conservation are much the same thing. Efficiency seeks to make better use of what raw materials are available, and conservation seeks to save raw materials for the future. They each result in stretching out raw material supplies, be they energy, mineral, or agricultural resources.

The dictionary defines efficiency as "effective operation as measured by a comparison of production with costs as in energy, time, and money." Conservation is defined as "a careful planned management and protection of something, especially planned management of a natural resource to prevent exploitation, destruction, or neglect." Both of these are laudable goals, and should be pursued.

Efficiency

Houses can be made more efficient in energy used to heat or cool them by means of proper construction, particularly insulation. Cars can be made more efficient by using lighter weight materials in their construction, and by being smaller. However, this reduces safety, for a small light vehicle comes out second in a collision with a larger, heavier car. Even if all cars are small and light they are likely to be more severely damaged in a variety of accidents than are heavier more sturdily built vehicles, so there is a reasonable limit to how small and light vehicles should be made.

All sorts of equipment and appliances can be made more efficient in their use of energy. The transistor has surely helped in reducing electricity consumption in many things we use. Fluorescent lights are more efficient than filament lights. The refrigerators and water heaters now manufactured use less electricity than they did a few decades ago. Yet, all these things must use some energy.

There is a valuable practical aspect to efficiency, from both a business and a personal point of view. For business, making less do the same or more means lower costs and becoming more competitive. Factories hire engineering efficiency experts. For individuals it means containing or perhaps cutting the cost of living, or at least reducing the rate at which it would increase if efficiency were not employed. In general, increases in energy costs have an adverse effect on the economies of all countries, except for the energy exporting nations, which currently means chiefly the oil producers. As energy is a basic part of all manufacturing processes, energy efficiency will always be an important goal because the cost of energy enters into almost everything we use.

Conservation

Conservation implies preserving something for the future. Conserving, therefore, involves looking ahead for perhaps many generations. This can be conservation of wildlife, forests, a wild river, or a particularly scenic area. In minerals, it could mean saving coal, oil, gas, or metals for future generations.

In the case of wildlife, forests, a wild river, or a scenic area, the rationale is fairly clear as to conservation. But with minerals and energy mineral resources the matter is much more complicated. Unlike wildlife or trees which can reproduce, or in the case of scenery which can be "used" for viewing but continues to survive indefinitely, minerals and energy minerals do not reproduce. They can be used only once.

The question comes then as to how we might rationalize saving minerals or energy sources such as coal. These are all one crop products. Can enough be saved for future generations to make any difference? And for how long would it make a difference? An example of this is oil. The U.S. is already using more than twice as much as it is producing. Does the United States reduce consumption of oil from its own oil fields in order to save some for the future, while it imports more oil to make up the difference?(2)

These are national questions of choice, with which neither the United States nor any other nation has come to grips. People have suggested that as a continued national policy we should save "our" oil and use "theirs." But the international payments problem of buying more goods, including oil, from abroad than are sold as in the case of the United States is becoming acute. This deficit in balance of payments is huge and growing and there are indications that this situation is major factor in the decline and international loss in prestige of the dollar. The oil import bill is the largest single item in the U.S. deficit in balance of payments.

Also, if Earth resources are not developed and also processed domestically, the industries which do these things do not survive, along with the many suppliers to these activities, and unemployment results. One job in the oil fields has generally meant two jobs in related services. The United States' oil production has been declining for a number of years, and in the process nearly half a million jobs have been lost.(3)

Delaying the need

We "conserve" electricity by increased efficiency in our homes and factories. This helps in that we can delay the necessity for the construction of new power plants. We can buy smaller cars and drive less, delaying the need to import more oil or explore for more domestically. But, as we conserve to delay the need, we have creeping up behind us a continued growth in population. Thus, the problem of facing the need is only delayed, not solved.

Growth — the over-riding factor

Growth in population puts a tremendous impediment to any program to conserve the world out of the energy problem. The nearly 100 million people now being added each year to the world equal more than the combined populations of Canada and France. Some states including Oregon have Energy Departments, or some similar agency designed to encourage efficient use of energy supplies, particularly electricity.

Oregon, on the west coast of the United States is one of the faster growing parts of the country, caused by the natural increase of the people already there together with immigrants coming from a variety of places both within and outside the U.S. The Oregon Department of Energy has an energy conservation program which takes various forms. The question was raised as to whether or not the amount of energy saved each year by these programs was equal to the additional amount of energy required to take care of the needs of the immigrants into Oregon. The answer was "definitely not."(9) With a stable population, the energy efficiency efforts would result in a reduced need for power, but when the needs of an increasing population are included, it is a losing effort. Admittedly, it would be even worse if these energy programs were not in effect, but the over-riding fact is that more, not less energy, is needed each year. As no more electrical energy can be developed from the now fully harnessed Columbia River (Bonneville) power system, additional electric power will have to either be imported — which places the problem in someone else's backyard — or fossil fuel or nuclear plants will have to be built, which again involves importing fuel from some other area to supply these plants. Perhaps sunlight and/or wind power could make up the difference, but such projects as have been tried so far have achieved only modest results. If the population keeps growing indefinitely, there is no way to keep up with energy demand. The physical standard of living will decline.

Against an ever-increasing population it is not possible to conserve yourself out of the energy problem related to the non-renewable energy sources which dominate our economy today. It is akin to the fable of the tortoise and the hare. The tortoise is the ever-increasing population. The hare is the technology which is able to develop more mineral and energy supplies. But, at some point, it is not possible to continue to expand production of non-renewable resources, and in fact, the production trend will turn down. The tortoise of continuing population growth catches up with the hare that is no longer able to come up with more non-renewable resources.

With a stable population, the hare is the population which continues to use up non-renewable resources, and the tortoise is the approaching end of these resources which continues to creep up on the hare. Thus even with a stable population it is not possible to conserve ourselves out of the energy problem in our present circumstances. It will take both a fixed number of people and a renewable resource base large enough to support that number of people to solve the problem. Neither of these necessary factors is in sight at the present time.

If the population is adjusted to a stable size which can satisfactorily live on a soundly established renewable energy resources, then an energy allotment to each person might be made. However, as long as we draw on non-renewable energy sources, efficiency and conservation can simply try to buy time to allow a sustainable renewable energy resource economy to be put in place.

Conservation: an energy source?

Unfortunately, there frequently is heard the mistaken implication that somehow conservation is an energy source. In a widely acclaimed national best seller there is a chapter headed "Conservation: The key energy source." There it is stated, "There is a source of energy that produces no radioactive waste, nothing in the way of petrodollars, and very little pollution. Moreover, the source can provide the energy that conventional sources may not be able to furnish...The source might be called energy efficiency, for Americans like to think of themselves as an efficient people. But, the energy source is generally known by the more prosaic term *conservation*."(10) To call conservation an energy source is to confuse the issue. Reducing the amount of energy used, so that not as much is withdrawn from the source during a given time period is efficiency, but neither conservation nor efficiency produce any energy. Also, there is a limit to the amount that energy consumption can be reduced and still do what is absolutely necessary. Energy supplies are continually needed. Conservation or efficiency are not an energy source. These are, however, laudable actions.

Non-renewable resources

If use of non-renewable resources continues, at some point there is an end. In the United States that end for coal is perhaps two or three centuries away. In the case of oil, the United States, has not been self-sufficient since 1970, and now imports more oil than it produces. With the large influx of immigrants into the U.S. at the present time combined with the natural growth in population it is likely that the trend to import more and more oil will continue. Efficiency can slow the need for increased imports but it probably cannot reverse it. Higher oil prices may make some domestic deposits economic which are not now being produced. But over any five to ten year period in the next several decades and indefinitely beyond, U.S. oil production will continue to decline, and more and more oil will have to be imported. The large foreign oil bill will get larger.

Efficiency and conservation — why?

There are two things which efficiency and conservation can do with regard to Earth resources. These actions can delay the further environmental impact on the Earth which the additional energy and mineral resource developments would have had if they had been built.

The second thing that efficiency and conservation can do is related to the first effect. The delay buys time for nations and individuals to make important basic decisions, and implement them for the future. Among these would be:

1. Attain a size of population which can be sustained in a reasonable standard of living based on the development of a truly renewable resource economy.

2. Develop a renewable resource supply which can match the need of the population size which is regarded as desirable.

3. Make the social and moral decision as to what is acceptable as a "reasonable standard of living."

A match between the size of the population and the sustained carrying capacity of the land must be achieved. Items one and two might be reversed, and they are to some extent the same. Whatever form a truly renewable resource economy will take will probably determine how many people would be desirable or could exist in a reasonable standard of living. How to eventually come to a genuine renewable resource economy and at the very same time achieve a population size which that economy will reasonably accommodate is hard to visualize. Nature in the past has always done so by the imposition of the harsh realities of lack of food. The process is called famine. Can intelligent humanity fit the population size to a sustainable economy? Johnson thinks it may be accomplished by the process he has called "muddling toward frugality."(5) How intelligently, orderly, and with no harsh circumstances this "muddling toward frugality" can be done is not certain.

Unfortunately, at the present time there is little evidence that much progress is being made toward matching population to a sustainable economy. One nation might presume to try to do this, but if its borders are not completely closed the endeavor will not succeed because people from areas where population exceeds resources will migrate in. Maintaining totally closed borders does not seem to be a realistic possibility. Accordingly, the project would have to have a world-wide consensus to be accomplished. This, too, seems difficult, and nowhere in sight.

However, for the present, efficiency and conservation remain worthwhile goals. They will help to try to maintain the standard of living as we now know it. Conservation and efficiency can help in the short run. But, if population continues to increase, that increase will ultimately use up all the savings from conservation and efficiency, and will eventually move ahead to make even greater demands on resources and the environment. Conservation and efficiency only provide a brief window of opportunity in which to make more significant and permanent decisions. That opportunity must be seized or current efforts of conservation and efficiency are largely for naught.

Crises Cause Change

Unfortunately, it seems that only when a crisis occurs is change made. It would be better to try to anticipate the problem, but we do not seem to have evolved to that point in society. The oil crises of 1973 and 1979 provided an example of crisis-forced change.

The oil crisis was, for a time, damaging to the U.S. automobile industry, which had largely ignored the matter of efficiency and instead had been promoting bigger and bigger cars. But with the gasoline shortage, the Japanese cars and some of the smaller German cars suddenly became popular. The Japanese in particular, not having any significant domestic oil supplies, had developed quite efficient cars. When the U.S. car market needed them, Japan was ready and able to supply them. It took Detroit a number of years to redesign and retool to make a product to fit the new realities, and compete with the Japanese.

Recent change in ethics

The concept of endless supplies of resources which tended to dominate the thinking of earlier times in the United States has given way to a more rational view that we have to make the most efficient use of what we do have. This ethic has been in Europe for many years, but because of the abundance of natural resources in the United States, and the fact that the U.S. was settled so much later than was Europe, that attitude has only recently arrived here. It is welcome, but still only partially adopted. Although there has been much lip service given to efficiency and conservation, in actual practice the adoption of these programs is by no means universal in the United States. Some people conscientiously recycle their refuse. Others continue the throw-away culture.

Automobiles

This is an area where the different attitudes among people is perhaps the most conspicuous. For a time during the gasoline shortages, large automobiles with their relatively poor miles per gallon performance were not popular. But, the American public has a short memory, and with the apparent abundance of cheap gasoline, large cars are back. A particularly conspicuous aspect of this are the huge motor homes which now fill the highways. Weighing several tons, many of these vehicles get less than 10 miles to the gallon. The fact that the United States now imports more oil than it produces, resulting in a substantial part of our international deficit in balance of payments, does not seem to be of concern to many people. While waiting for passage through a road construction area, I found myself behind one of these large motor homes. On the back it had a bumper sticker which read "Ban Offshore Drilling." There was an ironic contradiction in that situation, as the license plate on this large vehicle was from California. Where does the owner of that vehicle propose that his fuel be obtained, if he doesn't want it produced in his state? The only substantial prospective oil producing areas left in California are all offshore where the State of California has banned drilling.

As we look at social attitudes, one might also remark upon what is common in some places, the Saturday night ritual where younger people do what is called "dragging the gut." This involves getting into a car and driving for several hours, with many others, in a more or less circular route through the town. It is a way of socializing, but it also creates air pollution and burns up imported gasoline. One would think there would be better ways of using time and energy. This sort of thing tends to justify the thought that Americans have been living in a "fuel's paradise." Some have realized this and have made rational adjustments in their lifestyles. Others have not.

Collectively, these attitudes do affect the course of a nation. They affect its economic position and its ability to continue to survive in a world which has a continually expanding population drawing on a diminishing resource base. The attitudes of the population toward the use of mineral and energy mineral resources will go far toward determining its future.

The Third World

The foregoing remarks apply chiefly to the industrial world and especially the United States, but what of the relatively undeveloped nations, many of which already are in a very marginal resource situation? To these peoples, conservation and efficiency are an immediate concern just for survival. Whatever low cost and relatively simple technologies can be developed locally or supplied by the industrialized nations will be very useful. Solar cookers used in India are one such example, and can reduce the demand for firewood which has been deforesting the country.

In countries such as Haiti and India, where population demands already exceed available domestic resources to provide an adequate standard of living for many people, the concern for saving resources for the future is largely lost by the need to simply exist from day to day. The efficiency ethic is one which can be immediately appreciated to be used to make better use of what resources exist. How to implement a conservation program is more difficult. The chief resources in these regions are water, land, and vegetation. Significant conservation of metals and energy resources beyond just firewood is not germane to these situations of marginal existence, and even potential future supplies of firewood are used for the needs of the day. A reduction in population would probably be the most useful of all programs to implement.

Ethics of Mineral Use

The question of conserving mineral and energy resources in any country is one which cannot easily be answered. We conserve these resources for whom? And for how long?

To whom do resources rightfully belong?

When considering conservation and efficiency in order to save resources, there is the philosophical question as to whom the resources belong. At present, we are operating under the law of the "right of capture." Those who can capture the resources, either within their own borders or by buying them can have them. We now have the ability to produce oil and minerals in quantities which generations past did not. So we are producing them and using them. But if we did not use them now, who in the future would have the ethical right to use them — the next generation, the people who will live 100 years from now, or, those alive 500 years hence? Who?

Should we use all the resources we want to use now in any manner we chose, or just for basic daily needs? Or, should we also use substantial quantities simply for recreation? Do we ration resources to the consuming public as has been done in wartime? Or, do we assume for certain that scientists in coming generations "will think of something." That is a tenuous assumption, that is widely expressed and it allows us to enjoy resources used today without being burdened with ethical concerns for the future. Then there is the fact that even today large segments of world population are consuming and enjoying very few resources, and lead a very marginal existence.

Obtaining resources from other lands

There is also the fact that currently if one country does not have the resources it will buy them from other lands. Some feel that industrialized countries are, in effect, stealing the resources from other lands. But it may be pointed out that the Persian Gulf countries, for example, are not required to sell oil to anyone. Nor are Chile nor Zambia required to sell copper. These countries sell their energy and mineral resources because they want the benefits of what the money from the sale of these resources can bring, and they want these benefits now.

For the countries which now lead a marginal existence, if they have resources which can be sold, this money presumably helps to prevent their situation from being even worse.

Today and tomorrow

For the most part, the world view seems to be that resources are for use by the present generation. What might remain goes to future generations . . . who may well have the same point of view. The higher grade energy and mineral resources tend to be used first. The lower grade mineral resources which generally require more energy to process remain. In the future, then, unless technology can mitigate the situation, this will have the pyramiding effect of causing higher and higher costs for Earth resources for future generations to pay.

The answer?

If one would save resources now for the next generation, it might be asked, why not for the 10th generation beyond? Do they have less right to resources that exist now than the first generation beyond us? This is a huge philosophical question and it is surely not presumed to be answered here. It is a large area for discussion. Realistically, what is occurring is that each generation uses the least expensive and highest quality resources it can obtain, leaving future generations to shift for themselves, with the hope that technology will come to their aid. But one might wonder if the dual problems of lower and lower grade resources demanded by what apparently are going to be larger and larger numbers of people can be solved by technology? Can current living standards be maintained for the affluent societies, much less be raised for the many depressed populations, against these realities? Future generations face large challenges.

Summary

Efficiency and conservation are neither energy nor mineral sources. They will simply buy time to allow for population to be stabilized at a size which renewable resources can sustain. Efficiency and conservation practices at present are probably being more than cancelled out by the increased demands caused by a growing population. Trying to solve the problem of energy and mineral demands by efficiency and conservation is simply treating the symptoms of the problem rather than the cause and ultimately will prove futile. The cause is population growth, and the inability so far to establish a stable population of the size which can be adequately supported by a renewable resource economy.

The ethical question as to whom Earth resources belong is one to which there is no apparent answer. The reality is that each generation uses the highest quality most inexpensively recovered resources available, leaving what remains for future generations. The future, given present trends, involves more and more people trying to exist on lower quality and more expensive, and in many cases, diminishing resources, with the hope that technology can solve the problems by ultimately achieving a completely recyclable sustainable economy. Can this be accomplished, or will other basic decisions have to be made as to population size and what are acceptable standards of living? And will these decisions be voluntary or forced, either by government edict or nature's harsh realities?

BIBLIOGRAPHY

1 CHANDLER, W. U., et al., 1988, Energy Efficiency. A New Agenda: The American Council for an Energy Efficient Economy, Washington, D. C., 79 p.

2 DE-SHALIT, AVNER, 1995, Why Posterity Matters. Environmental Policies and Future Generations: Routledge Inc., New York, 176 p.

3 HODEL, D. P., and DEITZ, ROBERT, 1993, Crisis in the Oil Patch: Regnery Publishing, Inc, Washington, D. C., 185 p.

4 HU, S. D., 1983, Handbook of Industrial Energy Conservation: Van Nostrand Reinhold Company, New York, 520 p.

5 JOHNSON, WARREN, 1978, Muddling Toward Frugality: Sierra Club, Shambala Publications, Inc., Boulder, Colorado, 252 p.

6 MEYERS, R. A.,(ed.), 1983, Handbook of Energy Technology & Economics: John Wiley & Sons, New York, 1089 p.

7 MUNASINGHE, MOHAN, and SCHRAMM, GUNTER, 1983, Energy Economics, Demand Management and Conservation Policy: Van Nostrand Reinhold Company, New York, 464 p.

8 SAWHILL, J. C., (ed.), 1979, Energy Conservation and Public Policy: Prentice-Hall, Englewood Cliffs, New Jersey, 259 p.

9 SIFFORD, ALEX, 1993, Oregon Department of Energy, Letter dated November 12.

10 STOBAUGH, ROBERT, and YERGIN, DANIEL,(eds.), 1983, Energy Future: Random House, New York, 459 p.

CHAPTER 24

Minerals, Politics, Taxes, and Religion

This may appear to be a rather mixed group of topics, but in fact they are quite closely related. In some countries religion and politics are separate — the separation of church and state. In other countries religion and government are closely entwined, and religious attitudes influence the development and free flow of mineral resources. An example is the Arab embargo of oil to the United States in 1973, imposed because the United States supported Israel in its war with Egypt.

Taxes on minerals since the time of the ancient Egyptians who levied a tax on the commodity everybody needed, salt, have been and continue to be a major source of government revenue. More recently, another item in common use, petroleum products, and especially gasoline, are the object of moderate to, in some places, very high taxes, levied not only for the purpose of building roads, but in many countries as well as in some states of the United States, as a source of general government revenues. In 1993 the U.S. federal government imposed a 4.3 cent increase in the gasoline tax which simply went into the general treasury account. Political bodies recognize that taxes placed on things which people need and use will yield rich revenues.

Perversely, in some places, although mineral industries and their products are a source of substantial employment and government revenues, politicians denounce and assail these enterprises in the presumed interest of their constituents. This is commonly done to gain general popular political favor. Oil companies are among the favorite targets for these attitudes. Because petroleum products, especially gasoline, are commodities which are used by nearly every one, assailing the companies which sell these products touches the interests of a large portion of the population of industrialized countries.

Together, the matter of minerals, politics, taxes, and religion form a large area of discussion. Some examples of where these have an influence on the future of nations and individuals are briefly cited.

Politics

The politics of minerals, particularly oil, has filled many books, among which is Yergin's epic volume, *The Prize*.(35) Only a few salient points can be addressed here. There is no doubt that natural resource companies have engaged, as do all other economic enterprises, in influencing politicians for various ends. This ranges from matters concerning taxes on minerals, obtaining mineral leases (particularly from certain foreign governments where corruption is a way of life), rights of way through public lands, workers rights, and a host of other considerations.

Domestic politics

The mineral industry, and particularly the oil industry, is involved in politics in another way. Gasoline, especially in the United States, is such an important everyday commodity in the lives of nearly every citizen that everyone is keenly aware of its price. As more and more oil has to be imported, oil companies domestically have less and less control of the price of the crude oil which they process. In the past two decades, crude oil prices have fluctuated markedly from as low as $10 a barrel to as much as $40 a barrel. Retail prices of gasoline have gyrated accordingly. In the presumed role of protecting the public, politicians sometimes base part of their campaigns on the theme that it is every American's God-given right to always have cheap gasoline. Ignoring the fact of the large international variations in the price of crude oil over which companies have no control, politicians assail the oil companies for exploiting the public. The politicians cast themselves in the role of public defenders. It can win votes.

In the case of the United States, the fact is that its citizens continue to enjoy some of the world's cheapest gasoline prices. Furthermore, less than half the cost of gasoline at the service station is the basic cost of the gasoline. The majority of the cost comes from transportation, the cost of wages of the employees at the retail level, and a reasonable profit margin for the service station (which itself has to pay local real estate taxes) — and state and federal gasoline taxes. Federal and state gasoline taxes in the United States are commonly equal to or more than the basic cost of the gasoline. Yet, to find a campaign issue which has broad appeal, politicians have frequently used the oil companies as their targets.

During the oil crises also, the political body has not admitted to the fact that the U.S. since 1970, has not been self-sufficient in oil, but has used the oil crises to blame the oil companies and in turn appear as a protector of the public interest. Stobaugh and Yergin have described this situation very well:

> "Even opinion polls taken in 1979 and 1980 found that half the American people were unaware of the high level of U.S. oil imports.
>
> "This simple ignorance, in turn, contributed heavily to the search for a domestic villain on whom to blame the problem. The adversary character of the American political system, the never-ending welter of charge and countercharge, the regulatory confusion, the cacophony of contradictory expert opinions — all reinforced the tendency to embark on an energy witch hunt. Various groups argued that the villain was the government, environmentalists, or both. But much more pervasive was the belief that the 'oil companies' were the villain. More than half the people in a poll in October 1979 asserted that the oil companies had fabricated the energy shortage, and one-quarter thought that the government should take over the companies and run them. Here too was an example of time lag — the belief that the oil companies were as powerful in 1980 as they were in 1960, when in fact one of the causes of the problem we face today is the waning power of the oil companies in the face of the growing assertiveness of OPEC producers."(30)

This problem has only grown. Increased nationalization of oil companies abroad, and the continued movement abroad of U.S. oil companies because domestic exploration prospects are diminishing, have made U.S. companies more and more hostage to foreign governments. Oil companies do not own the oil, they simply have lease arrangements for developing those resources. The foreign governments own the oil and have control. In turn, the American oil consuming public is hostage to foreign governments. When the United States could no longer produce enough oil to supply its own needs, it lost an important part of its economic sovereignty.

Oil company profits

In terms of return on investment, which is a basic measure of profitability, the oil industry historically has had a return on its investment which has ranged between twelfth and seventeenth among the various industries and businesses. Tobacco, liquor, and the media (which frequently joins in the chorus against the oil industry) have commonly ranked one, two, and three. Mining companies generally have had an even lower return on investment than oil companies. Both oil and mining companies are involved in high risk exploration ventures. Much money is spent on things which do not pay off, and when a property is found which has some potential, much more money has to be invested before there is any return. More of this is taken up in Chapter 26, *Mineral Economics.*

International Politics

The area of international relations regarding mineral resources has seen great changes the past half century. There are two important trends, one is the rise of nationalism following the colonial period. The other, related considerably to the first, is the increasing dependence of the industrial nations on mineral and energy mineral supplies they do not control, and how this affects their foreign policies. International politics also are modified by what strained relationships may exist, as, for example, the dislike of Iran for the United States, and reciprocated by the United States, dating back to when Iran took U.S. Embassy officials hostage for more than a year. Expressed in terms of minerals, the United States has cut off all purchases of oil from Iran, and forbidden oil companies to do business with Iran.

Restructuring of nations such as occurred in the break-up of the Soviet Union with some of its oil resources now in control of the independent republics requires that oil consuming nations establish new relationships in that region.

Nationalization of mineral resources

One of the advantages which Great Britain had in both World War I and World War II was the ability to draw mineral resources from its widespread colonial holdings. The main reason Japan went to war in World War II was because it was thwarted in trying to expand its colonial position to obtain natural resources. The era of colonialism came apart after World War I for Germany. It completely ended after World War II when the colonial holdings of Britain, France, The Netherlands, Belgium, and Italy began to assert their independence. It was said that the chief international sport became "twisting the British lion's tail." India and Pakistan were among the first to do it, in 1947.

Almost immediately thereafter there was a further move toward establishing a total national identity by taking over mineral developments formerly held by the colonial powers. This then spread to almost all foreign mineral holdings in countries which were not previously colonies. United States held no colonies, but U.S. corporations had large international mining and oil interests. Soon the international sport became "yanking the tail feathers out of the American eagle."

The American copper companies, Kennecott and Anaconda, were completely nationalized in Chile. Zambia and Zaire took over all multinational copper operations there. "American" was rubbed out of the name Arabian American Oil Company on the desert sands of Saudi Arabia. All foreign interests in Iran and Iraq were taken over. Kuwait nationalized Gulf Oil's interest there. Venezuela nationalized Creole Petroleum Corporation, a division of what is now Exxon Corporation, and the company which had developed the great oil deposits of the Lake Maracaibo Basin. Peru took over International Petroleum Company, also an Exxon affiliate. This was done with no compensation whatsoever, and was done not long after Exxon had invested large sums in rebuilding the oil camp and related facilities, including a modern hospital free to all employees and their families, building the safest water supply system in the entire country, and even a fine large church.

"For the people"

The stated reason one government gave for the nationalization of a foreign company and its mineral holdings was that "we wanted to have a gift for the people." Whether such a gift ever reached the people is debatable. In some countries nationalization has meant some increased affluence for the general citizenry. In other nations it has only meant more money which has benefited a relatively few through graft and corruption. The initial attraction for nationalization by the government officials in power at the time was to get the revenue flow. What happened after that has been a mixed bag. In general, the countries of the Persian Gulf have done better in spreading the mineral wealth through the general population than have most other nations.

Military regimes and oil

In some countries, Nigeria being a recent notable example, oil revenues have served to support governments which have been oppressive. This is an unfortunate situation for which the oil companies have occasionally been severely criticized, but over which they have little or no control. Interference in the internal affairs of a country is not feasible for a foreign company. The options are limited to either continuing operations or leaving. These situations put the companies in a difficult position.

In Nigeria in late 1995, a particularly outspoken critic of the government complained that the oil income was being taken by a relatively few government officials and not being fairly distributed to the impoverished people in the delta area where the oil was being produced, with some environmentally negative effects. The Nigerian government's response to this, in spite of international condemnation, was to execute this individual. Oil income to governments can be so lucrative as to spawn widespread corruption. This is a continuing problem to a greater or lesser extent in several countries. About the only thing which can be done is for the international community to try to support the democratic elements in such countries as best they can. This has been done to a limited extent in the past (for example, the oil embargo against Libya by the USA), but these tactics have met with limited success. Foreign oil and mining companies really cannot afford to get involved in the domestic politics of a nation. From a practical point of view, if they back the political element which comes out the loser, the companies have much to lose themselves.

After nationalization?

Some countries, after nationalizing their minerals, discovered they did not have the technical expertise to run the nationalized operations. Also, in some cases, so much money was drained from the operations and into political and social pockets and causes, that there was not enough capital to continue to maintain and develop the resource facilities. Also, some countries did not have the capital even from the mineral income to invest in developing new resources, so that money had to come from abroad. Therefore, many countries have now invited foreign companies to come back, under various financial arrangements. In a two page ad in 1995, Zambia stated they were privatizing the government monopoly of copper mining, and asked for foreign capital to come in and help them.(1) On January 1, 1976, Venezuela took over all foreign oil interests. But in 1995, Venezuela was in need of help to run its oil operations, and made arrangements to begin to auction off some exploration rights in various prospective areas to foreign oil companies. However, the terms include taxes that will take from 71 percent to 88 percent of the profits from these ventures.(13) It should be noted that Venezuela is not risking any money. If the leases are unproductive the companies will lose all their lease and exploration costs. Venezuela has no financial exposure.

Foreign Policies and Resource Dependency

If an industrial nation would be self-sufficient in all necessary mineral and energy resources, it could act independently of other countries. However, no nation is entirely in that position. Russia comes the closest to it. Japan is the most dependent on foreign mineral and energy supplies. This is probably the single most important factor in guiding Japan's foreign policy.

The United States for many years could be quite independent of the rest of the world. But when, in 1970, the United States could no longer supply its own oil needs, its foreign policy had to begin to be modified, especially toward the Persian Gulf nations. The United States' strong support of Israel brought on the oil embargo and crisis of 1973. Since that time, because of the increasing U.S. dependence on Arab oil, U.S. foreign policy has been adjusted to try to accommodate both the Arab and Israeli positions. Britain, Germany, France, and Italy have similar foreign policies for the same reason.

Minerals and military alliances

Mineral resources can greatly influence international military alliances. The Persian Gulf countries are a prime example. The Gulf War clearly showed the Gulf nations that they were in need of alliances which would preserve their integrity. They have since sought and generally developed alliances with the United States, Britain, France, and Russia, more or less in that order. As both the U.S. and France are increasingly dependent on Persian Gulf oil, and Great Britain's oil fields are peaking out, their interest in Persian Gulf affairs is obvious. Russia being the closest major military power to the Gulf also had to be considered, and Russia is reaching the point where domestic oil demands equal the supply, and may soon exceed it.

U.S. oil industry and Iraq

The Gulf War, however, has had a perverse effect on the U.S. oil industry. Although the United Nations imposed sanctions against Iraq by not allowing their oil to be sold on the world market, there were no restrictions made on doing business within Iraq. However, the United States, unilaterally, decreed that no U.S. oil company can do business with Iraq. No such rules have been made by the governments of France, Great Britain, or Italy. The result is that French, British, and Italian companies were able to negotiate with Iraq on future development of what appear to be huge oil fields in southern Iraq (potential 300,000 to 500,000 barrels a day). Being prevented from getting in on this rich oil province dismays the U.S. oil companies. "Even oil executives from outside the U.S. — whose governments allow them to negotiate, though not sign deals with the Iraqis — say the Americans have a right to be angry. International politics is creating an uneven playing field for U.S. companies and could lead to a major power shift away from the U.S. industry, they say. Indeed, the companies that win the rights to develop Iraqi fields could be on the road to becoming the most powerful multinationals of the next century."(21)

The fields were discovered but not developed during the eight year period of the Iraq-Iran war. The French companies, Elf Aquitaine and Total SA, and the Italian Company Agip, all were in Iraq doing negotiations while U.S. companies sat on the outside looking in. The chief executive of British BHP Petroleum Ltd. of Australia was quoted: "Non-American companies are being given new opportunities by the U.S. government. One day we'll wake up and say 'Good Lord, Total and Agip are the largest oil companies in the world.'"(21) Present United Nations sanctions prevent Iraq from freely selling oil on world markets. When this embargo is lifted, the United States will be largely left out of further oil ventures there.

Oil and the moral dilemma

The United States now imports more oil than it produces. However, this economic disadvantage is modified somewhat by the fact that U.S. oil companies increasingly are shifting their operations overseas where large oil reserves still remain. Profits from these operations are repatriated by these companies to the United States and appear as dividends paid to shareholders which in many cases are insurance companies and pension plans holding oil company stock. Thus many U.S. citizens benefit. However, if U.S. companies are barred from developing rich oil fields abroad, the economic position of the United States with regard to oil is worsened, and the balance of payments problem increases.

In 1995, Conoco (a division of DuPont) had made arrangements with Iran for development of some offshore oil fields, but, because of restrictions imposed by the U.S. government on dealing with Iran (an on-going aspect of the strained relations which date back to the hostage crisis with Iran in 1979), those negotiations were canceled at U.S. government request. If United States oil companies lose their strong position as world oil producers, the economic consequences will be unfavorable for the USA. The on-going shifting of the international balance of power coming from a barrel of oil will surely be important to the individual nations. Thus it may be that the United States, with its allies, won the Gulf War, but the winners will not ultimately include the United States.

It is a moral dilemma. The U.S. chooses to prevent its own companies from doing business in countries which have policies and politics of which it does not approve. But other countries allow their companies access to prospective oil territories in the countries where the U.S. voluntarily refrains from operations.

Breakup of the USSR and the new politics

With the break-up of the Soviet Union, a new set of politics caused by oil appeared in that region. Large oil deposits exist in several of the countries which split away from the USSR. The extensive Caspian Sea area oil is now held by Turkmenistan, Azerbijan, Iran, Russia, and Kazakhstan. Some oil, an estimated 4.1 billion barrels, is located in nearby Ukraine.(22) Kazakhstan — five times as large as France and larger than all the other former Soviet Republics put together, excluding Russia — is reported to have as much as three to

10 times as much oil as in Alaska's Prudhoe Bay Field. Foreign companies and American oil companies (in particular, Chevron) negotiated to develop some of these resources.(9) The continued tendency of Russia to meddle in the affairs of these new republics has caused the United States to modify its support for Russia, and to try to achieve a balance in its foreign policy between Russia and the other republics.(23) "Would-be investors in Kazakhstan must contend with Russia's reassertion of its regional power, haggles over pipeline routes, and rising Kazakh nationalism."(24) This, for U.S. foreign policy, has meant a somewhat lesser support for Russia and in some instances a statement of support for the new republics against Russian influence. Oil causes a difficult international political balancing act for the United States as well as for other industrial nations.

In the rich long-time oil-producing Baku region, now part of the independent Republic of Azerbaijan, and in the neighboring republics of Turkmenistan and Kazakhstan, Thomas Hamilton, president of Pennzoil Exploration & Production Company, stated, "We think the potential of the whole region is between 100 billion and 200 billion barrels. That is why there's such enormous interest in the industry."(9)

Galuszka writes: "Under pressure from such oil heavyweights as Amoco, Mobil, Exxon, McDermott, Brown & Root, Bechtel, and Chevron, the Clinton Administration is increasingly being pushed to alter its pro-Russian policy and start backing the republics."(9) From this region it is estimated "You're going to have at least 1 million barrels a day of oil coming out of [there] before long."(9)

Oil is now, and will be for decades to come, an integral part of U.S. diplomacy. As oil production continues to decline in the United States, oil sources abroad accordingly will gain in importance, and will be one of the strongest influences on U.S. foreign policy.

Taxes

The "salt tax" was collected back to the time of ancient Egypt and Israel. More recently in the Nineteenth Century both China and Mexico imposed a salt tax. At various times it has been a favorite tax for the French government.(19) In the United States in 1654, Father Simon Le Moyne, a Jesuit missionary, discovered salt springs in the vicinity of what is now Syracuse, New York, the original name of which was Salina because of the salt. Salt works were set up in 1788, and salt manufacturing flourished until the 1860s. For many years the tax on salt was New York State's chief source of revenue.

Gasoline tax, currently the world's favorite

The cost of gasoline in the United States is among the lowest in the world, in large part because the tax on gasoline is low. Attempts also have been made by the U.S. Congress to raise gasoline taxes in an effort to reduce consumption, and encourage automobile efficiency. But all such actions have been strongly resisted by both liberals and conservatives for diverse reasons.(23) In comparison with the other industrialized countries, gasoline taxes in the U.S. are the lowest. In many other countries, especially European countries, the price is three to four times as high due to taxes, and the taxes on gasoline may exceed the basic cost of the product by several times. There are several reasons why governments do this.

Reduce import bill

One reason for high gasoline taxes is that a high price tends to cut down usage of gasoline which has to be imported. This reduced gasoline consumption is important to a country in terms of its balance of trade. Some countries simply do not have the money by which to import much crude oil or refined petroleum products.

Environmental protection

High prices on gasoline and other fossil fuels are also favored by some environmental organizations as a way of cutting down the emissions which result from burning fossil fuels. Several governments have considered a "carbon tax" but have failed to enact it so far. The idea is that this would reduce the use of fossil fuels which pollute the atmosphere, carbon dioxide is the chief pollutant and a possible cause of the perceived rising problem of the "greenhouse effect."

General government revenue source

Increasingly the reason why gasoline is so expensive in some countries is that these nations put higher and higher taxes on it as a general source of government revenue. In the United States there is the concept that taxes on transportation fuels, gasoline, and diesel, are to be dedicated to the building and maintenance of the road system. The public is under this impression. However, a study by the American Petroleum Institute showed that in 1992 road user taxes and fees subsidized $38 billion worth of other government activities. At the present time annually in the United States, the federal excise tax on gasoline collects about $20 billion. State and local government taxes add up another $25 billion.

Severance tax

Taxes on nearly all mineral production are levied in all countries by various governmental agencies, usually in the form of what is called a "severance tax." The concept is that when the mineral is severed from the ground there is a loss to the country and this should be compensated for by a tax. The enactment of these taxes and how much they are is a political matter and can be and has been a source of much political maneuvering, as huge sums of money are involved. In addition, mineral resource companies also pay taxes on their general income, which taxes are again set by politicians.

Religion

In most of the world, governments and religion are separate, but in the important Persian Gulf region this is not the case, a fact which has been important in the past and may be so in the future.

Religion and the Persian Gulf oil bonanza

Some 60 percent of the known world oil reserves lie in the region of the Persian Gulf, an area which is almost entirely Muslim. These countries include Saudi Arabia, the country with the world's single largest oil reserves, Iraq, Kuwait, Iran, Bahrain, Qatar, Oman, and the United Arab Emirates. In these nations, religion commonly permeates the governmental structure, and affects how governments deal with the rest of the world.

Iran and Muslim fundamentalists

There is at the present time an apparent rise in religious fundamentalism in the Muslim countries. In part it is an aversion to what Muslims regard as rather loose lifestyles of the western nations, and in this regard one may concede in some respects they may have a point. A considerable number of fundamentalists have been educated in the West, so they have seen and know the culture. "Many fundamentalists were impressed by Western scientific and economic prowess but were repelled by aspects of Western culture —pornography, alcoholism and drug abuse, materialism and old age homes."(14)

But for whatever reasons these attitudes arise, the fact is that it seems to be a growing movement. In 1979, the Muslim fundamentalists who seized control of Iran, also took American hostages. This resulted in the United States cutting off oil trade with Iran and caused a brief oil supply problem. Strained relations still continue as the American flag is frequently burned in Iran, and the United States is referred to as "The Great Satan." The United States has now arranged for other oil sources and does not import any Iranian oil, nor allow its oil companies to trade with Iran.

These fundamentalist religious groups do not agree even with the more moderate Muslims of other nations, and are suspected of being the cause of terrorist attacks in other Gulf nations. In 1994, such attacks were made in Saudi Arabia, Bahrain, and Oman (most of which is not strictly on the Gulf but lies just beyond the Gulf at the lower end of the Arabian peninsula). During the 1990s in Algeria, Muslim extremists, believed to be supported by Iran, engaged in a variety of terrorist activities against foreigners and particularly foreign oil operations.

In December 1994, terrorists hijacked a French airliner in Algiers, and killed two passengers. The group claiming responsibility said it sought to drive out foreign expertise and capital to cripple the oil industry which is a major support to Algeria's economy. There is a major gas pipeline which crosses the Mediterranean from Algeria to France that is important to the economies of both countries. These and other terrorist acts forced the withdrawal of most of the personnel of western companies. Anadarko Petroleum of Houston, Texas was among those so affected.

In November 1995, a bomb blast in the Saudi capital, Riyadh, killed five American military men.(33) Subsequently it was determined that the blast was the work of Muslim fundamentalists and four were subsequently beheaded. This was done amidst threats of further terrorist activities if these four were executed. On June 25, 1996, a truck-based bomb blast in Dhahran, the city which is the oil capital of Saudi Arabia, shattered the building housing U.S. Air Force personnel.(26) The final toll was 19 Americans killed and more than 200 injured. The influential Council of Islamic Scholars issued a statement that this act was against Muslim rules. But other Muslims stated that similar attacks would continue until all American and other foreign troops "occupying the holy Saudi land" were expelled. Oil, in effect, has split the Muslim faith into moderate and extreme elements which becomes a very unsettling factor dividing this strongly religious nation.

Supplies of oil to western nations are at risk with any major internal Saudi strife. Indications are that the presence of non-Islamic troops to ensure the continued flow of oil to the industrialized nations will remain a major source of discontent in parts of the Muslim community. The avowed determination of some Muslims to eradicate foreigners and their influence from Saudi Arabia, which holds Islam's most sacred shrines, cannot be taken lightly. It is probable that there will be more efforts in this regard for religious zeal historically knows few bounds. The cost of the West's addiction to oil is likely to go up in ways other than simply in money.

That Iran appears to be the base for many Muslim extremists further complicates the problem of Middle East politics, for although Iran is Muslim it is not an Arab nation. Historically Iran (Persia) has been at odds with the Arab nations, particularly with what is now Saudi Arabia, which Persia in earlier times had vowed to invade and take over. And there is a religious power struggle within Iran among the various mullahs who vie for power within the Muslim framework. There are some more moderate elements in Iran, but there is likely to be continued internal unrest while all this is being resolved, and that resolution will take considerable time.

This is an unstable situation, but there is another growing trend in Iran which may cause further internal turmoil. As of the most recent census (1986), Iran had a population of 49.8 million, with an annual increase of 3.5 percent. Each year the population increases by more than 1,750,000 people, and the average age is 17. Iran's living standards even now are declining. Greatly increased population pressure will cause continued unrest, some of which may be exported. Iran is probably even more of a threat to Middle East peace than is Saddam Hussein, who is temporary. Iran's problems and the impact they will have in the Gulf region, will probably continue to grow over the longer term with the increase of population. The death or deposition of one leader will not change the course which it is now on. The leadership, in fact, is not well defined as several religious/political groups vie for power, and the President of Iran has no clear control of the country.

In spite of the need to address the domestic basic needs of the nation, Iran continues to invest a substantial part of its oil revenues in arms. Recently Iran has placed missile batteries and other military hardware on their side of the strategic Strait of Hormuz, at the mouth of the Persian Gulf.

Muslim nations: oil

But whatever the politics and religious attitudes may be, the fact is that Muslim nations have been fortuitously destined by geological events to control most of the oil reserves of the world. Oil deposits under control of the western and more highly developed industrial nations are smaller. Geology has determined that in the future more and more oil production will have to come from the Persian Gulf countries, within the realm of Islam. Other petroleum deposits, however, are also in Muslim nations, including Libya, Algeria, Brunei, and Indonesia, but the total proved oil reserves of all these countries combined is less than those of Saudi Arabia.

Historically, Muslims and Christians have had great conflicts dating back to the Crusades. In the future the question is whether the economic interdependence of oil-rich Muslim nations with the oil-dependent major industrial nations most of which are Christian, will prevail over religious fundamentalism. The prospect is that it will, but with some rough spots along the way. However, as more and more of the world's oil production becomes centered in the Middle East, religious extremists have the potential of being an increasingly important international factor.

The United States, among all the countries in the world, is by far, the largest per capita consumer of oil. Each day, California uses more oil than does either Germany or Japan. The rest of the world, also, is enjoying a rising oil consumption, and oil remains a relatively low cost commodity, considering what it can do. Flavin and Lenssen have made the comment with regard to the concentration in the Persian Gulf area of the oil supplies for export that "Not only is the world addicted to cheap oil, but the largest liquor store is in a very dangerous neighborhood."(7)

Religion, population, and resources

Attitudes toward family size, and all that relates thereto such as appropriate age for marriage, and use or rejection of birth controls, natural or otherwise, are closely bound up with the several religions. There are also ethnic considerations whereby the suggestion of the implementation of birth control is sometimes regarded as a form of genocide.

One theme throughout this discussion of mineral resources as they may affect the destinies of nations, is the effect of population on resource use. Therefore, in a very real sense the religious attitudes toward birth control and population size affect national futures. This is a very controversial subject and it is not the intent of this volume to pursue the matter which has many ramifications. Suffice it to say that if religious precepts result in a population size beyond the carrying capacity of a country, the nation will suffer and the quality of its future is diminished. In this regard, the observations of Jackson are pertinent:

> "It needs to be said over and over again that the bringing of surplus children into this world, whether from personal desire of from religious edicts, destines not only some of these children but many others to a premature death...The 'morality' of birth control in today's burgeoning human population has taken on an entirely new aspect. God clearly never meant for man to overpopulate this earth to the point where he would destroy many other forms of life and perhaps even himself."(11)

The edict, "Be fruitful and multiply" must be tempered with the fact that to survive those who are "multiplied," which presumably includes all of us, must also divide. We divide among ourselves the basic resources of our existence. As they are unevenly distributed by nature, and then by various human factors are further unevenly divided, this is a cause of conflict.

Be fruitful and multiply... Now, divide.

Figure 7. "Be fruitful and multiply . . . Now divide"

(Source: Clay Bennett, North America Syndicate, by permission)

Summary

Mineral and energy resources are political pawns in several ways. They are widely used as a source of tax income, the spoils of which may or may not be shared with the general population. Mineral resource companies, particularly the oil companies, are used at times by political elements as a way of manipulating public opinion to gain votes.

The nationalization of mining and oil companies has further complicated the political picture. Inability of some countries to manage the nationalized properties has resulted in various kinds of joint government/private company arrangements which, however, are subject to unpredictable political events.

With petroleum providing so much money and so many jobs to so many diverse groups around the world, and with the importance of petroleum in running the world's economy, the political maneuvering with regard to this particular resource involves very high stakes. As a result governmental pressures can be severe, and are used for a variety of purposes including religious objectives.

The Muslim nations now control more than half the world's oil reserves, and the geology of world oil occurrence is such that this concentration is likely to increase as other areas of the world which have in total much smaller petroleum reserves deplete their deposits. In these Gulf region Muslim countries, religion and the state are much more closely linked than in the western world. This influence was clearly demonstrated in the 1973 oil crisis when oil was used by the Arabs as a weapon against those nations supporting Israel in the war against Egypt. Whether or not economic interdependence between Muslim oil-producing nations and the non-Muslim industrialized West will prevent such problems in the future is questionable. Minerals, politics, taxes, and religion will be a continuing volatile mixture.

BIBLIOGRAPHY

1 ANONYMOUS, 1995, Mining. The Business of Copper: (advertisement) Fortune, July 24, p. 144.145.

2 BASIUK, VICTOR, 1977, Technology, World Politics, and American Policy: Columbia Univ. Press, New York, 360 p.

3 BILL, J. A., 1988, The Eagle and the Lion: The Tragedy of American-Iranian Relations: Yale Univ. Press, New Haven, Connecticut, 520 p.

4 deWILDE, J. C., 1936, Raw Materials in World Politics: Foreign Policy Reports, v. 12, September 15, p. 162-176.

5 EDEN, RICHARD, 1981, Energy Economics: Growth, Resources, and Politics: Cambridge Univ. Press, Cambridge, 442 p.

6 FEDUSCHAK, N. A., 1993, Kazakhstan's Oil Estimates Spark Rush to Risky Region: The Wall Street Journal, June 17.

7 FLAVIN, CHRISTOPHER, and LENSSEN, NICHOLAS, 1990, Beyond the Petroleum Age: Worldwatch Paper 100, Worldwatch Institute, Washington, D. C., 65 p.

8 FLAWN, P. T., 1966, Mineral Resources. Geology, Engineering, Economics, Politics, Law: Rand McNally & Company, New York, 406 p.

9 GALUSZKA, PETER, et al., 1995, The Great Game Comes to Baku. Who Will Tap — And Transport the Caspian's Sea of Oil?: Business Week, July 17.

10 GREEN, STEPHEN, 1989, Soviets Running Out of Oil. Moscow Involvement in Middle East Politics an Indicator of Domestic Oil Production Woes: The Oregonian, Portland, May 7, (reprinted in part from The Christian Science Monitor) April 6, 1989.

11 JACKSON, WES, 1971, Man and the Environment: Wm. C. Brown Publishers, Dubuque, Iowa, 322 p.

12 KHAVARI, F. A., 1990, Oil and Islam. The Ticking Bomb: Roundtable Publishing, Inc., Malibu, California, 277 p.

13 KNIGHT, JANE, 1995, Big Oil Can't Resist This Slick Offer. Venezuela's Oil Patch is up for Action — Strings Attached: Business Week, August 14.

14 LIEF, LOUISE, 1991, Battling for the Arab Mind: U.S. News and World Report, January 21.

15 LEITH, C. K., 1931, World Minerals and World Affairs. A Factual Study of Minerals in Their Political and International Relations: McGraw-Hill, New York, 213 p.

16 LEITH, C. K., 1940, Peace — Its Dependence on Mineral Resources: Speech given at National Convention of League of Women Voters, New York City.

17 LEITH, C. K., FURNESS, J. E., and LEWIS, CLEONA, 1943, World Minerals and World Peace: The Brookings Institution, Washington, D. C., 254 p.

18 MELLOAN, GEORGE, 1996, Five Years On, The Persian Gulf Still Simmers: The Wall Street Journal, January 15.

19 MULTHAUF, R. P., 1978, Neptune's Gift. A History of Common Salt: The Johns Hopkins Univ. Press, Baltimore, 325 p.

20 REES, JUDITH, 1990, Natural Resources. Allocation, Economics, and Policy: Routledge, New York, 499 p.

21 REIFENBERG, ANNE, and TANNER, JAMES, 1995, U.S. Oil Companies Fret Over Losing Out on Any Jobs in Iraq: The Wall Street Journal, April 17.

22 RIVA, J. P. Jr., 1993, Oil and Natural Gas in Ukraine and Byelarus: Congressional Research Service, The Library of Congress, Washington, D. C., 14 p.

23 ROSENBAUM, W. A., 1987, Energy, Politics, and Public Policy: CQ Press (Division of Congressional Quarterly, Inc.), Washington, D. C., 221 p.

24 ROSSANT, JULIETTE, and GALUSZKA, PETER, 1994, Why The West May Come Up Empty in a Monster Oil Patch: Business Week, May 30, p. 57.

25 ROSTEN, K. A., 1991, Kazakhstan's Vast Potential. Natural Resources Attract Investors: San Francisco Examiner, December 15.

26 SHENON, PHILIP, 1996, Saudi Truck Bomb at a U.S. Complex Kills 11, Hurts 150: The New York Times, June 26.

27 SHIRLEY, KATHY, 1993, Kazakhstan Tops Energy Agenda: AAPG Explorer, American Association of Petroleum Geologists, Tulsa, Oklahoma, November, p. 6, 8.

28 SMITH, G. O., 1927, Raw Materials and Their Effect Upon International Relations: Carnegie Endowment for International Peace, 69 p.

29 SPANIER, JOHN, (ed.), 1991, American Foreign Policy Since World War II: Congressional Quarterly Press, Washington, D. C., 441 p.

30 STOBAUGH, ROBERT, and YERGIN, DANIEL, 1983, Energy Future: Vantage Books, Random House, New York, 459 p.

31 SUTLOV, ALEXANDER, 1972, Minerals in World Affairs: Univ. Utah Printing Services, Salt Lake City, 200 p.

32 TIPPEE, BOB, and BECK, R. J., 1995, Politics Has a Strong Say in Who Benefits From Oil: Oil & Gas Journal, April 3, p. 41-48, 51-53.

33 WALDMAN, PETER, 1995, Saudi Bombing Prompts Fears for Nation: The Wall Street Journal, November 14.

34 WILLRICH, MASON, 1975, Energy and World Politics: The Free Press, Macmillan Publishing Company, Inc., New York, 234 p.

35 YERGIN, DANIEL, 1991, The Prize: Simon & Schuster, New York, 877 p.

CHAPTER 25

Minerals, Social, and Political Structures

In our daily lives we fail to recognize how the mineral resources we have available to us in very large measure control our lifestyles and the structure of society. Until recently, and still today in isolated groups, there are people living much as they did thousands of years ago. But for most of the world's population, the discovery and/or the access to mineral resources has set societies on courses which are vastly different from what they were only a few hundred years ago.

Gold is one example of how a mineral resource has affected social and political structures. The search for gold by the Spanish and others in the New World eventually caused the destruction of entire native cultures. Today in the highlands of New Guinea and the jungles of Brazil, gold mining is markedly impacting age-old native cultures, and parts are being lost.(12)

But much beyond gold, other minerals, especially iron, coal, and oil have had a huge impact on social and political organizations. The Industrial Revolution, dependent largely on iron and coal, resulted in a great transformation of the way many societies were organized. This is continuing as more and more countries enter the industrial age, based on access to mineral and energy resources. Among other things, industrialization causes mass movements of population from rural areas to manufacturing regions, and a concordant change in the way of life for many people. Oil has transformed societies in the Persian Gulf region. Water in the past has been the basis for developing social and political structures, as, for example among the Native Americans in the semi-arid southwestern United States. When water supplies declined, social and political structures crumbled, and people moved away.

Mobility

In the industrialized world today, one of the most visible effects of mineral resources on individuals' daily lives is mobility. The automobile represents freedom for the individual far beyond what could have been imagined even in quite recent times past. The economies which have grown up around this mobility are huge. These include motels, hotels, resorts, and other facilities related to travel. Recreation in many aspects from skiing to salt water fishing are dependent upon ready and fairly cheap mobility. The world's huge road systems, both local and transcontinental with the attendant commerce which moves by these systems is one aspect of the mobility which one particular resource, oil, has given to society. About 25 percent of the U.S. economy is one way or another based on the automobile — a machine which was first made in 1886.

The option to take the family to the mountains, to the coast, or on an extended trip to national parks and other recreational areas comes only as a result of ability to command mineral resources. When one considers that travel either by plane or ship to the far continents is now available to major segments of the world's industrialized societies at reasonable cost, the situation is truly astonishing. One need only to be part of the present older generation to appreciate that.

In lesser developed countries where individual ownership of motorized vehicles is still uncommon, mass transport by bus, train, or truck is available to many people which again represents a much different lifestyle than before the advent of oil. At the other end of the spectrum of travel made available by oil is intercontinental air transport, which has essentially only existed since World War II. As a child, I never dreamed I would see China, much less fly across the Pacific ocean in 11 hours to get there, which I, and many others now have done. It all is related to the availability of the cheap and currently abundant energy, oil, which has had such a profound effect on the world.

The mobility which oil offers to people on an individual basis through the private automobile has surely changed the social structure. When the younger generation got "wheels" the family fabric began to be stressed. The closeness of families in the past when weekends were times of local gatherings of clans was replaced by diverse activities of the several family members, frequently in different directions and at considerable distances. It is no remarkable event to travel a total of three hundred miles or more on a weekend to visit some point of interest or engage in a recreational activity. Two hundred years ago this was not possible. You could not travel 300 miles in a week. Now it is possible to travel around the world in less than a week by commercial airliner.

All this mobility has markedly altered the lifestyles of many millions of people, and illustrates the fact that we individually have our daily routines markedly determined by mineral resources. If it were not for oil, aluminum, and iron, we would each day lead vastly different lives than we do now. For one thing a great many more of us would be working on farms than we are today.

Making the world a smaller place

Mineral resources in the form of energy to transport people, or simply to put them in communication by means of telephones, radio, and television have made the world a much smaller place than it was less than two centuries ago. By means of these communication systems, mineral and energy resources have also made the world intellectually much larger, and ideas are now able to be immediately exchanged around the world. There are both immediate and longer term consequences of this, the results of which cannot readily be predicted except to say that they have surely changed and will continue to change societies.

Interchange of people

Several hundred years ago, only a privileged few people from each nation could visit other countries, and usually those which were nearby. Now students are transported from Middle East countries, from China, and from other distant lands, to North America and Europe where they study and see different ways of life afforded by the higher standard of living.

When these young people return to their home countries, they bring back impressions and frequently retain in their own lives and convey to their families, bits of the culture in which they lived abroad. They will influence how things are done at home from these experiences. Some of them will become their nation's leaders. Economical air transport or rapid ship transport made possible by oil does this. In the case of the students from Saudi Arabia and other Persian Gulf nations, it is their oil riches which allow these cultural influences and interchanges to occur. It is a facet of *geodestinies.*

Cheap energy and the mass meeting of cultures

The availability of cheap energy has made it possible for literally millions of ordinary citizens of many countries to visit other countries. There is an inevitable interplay of cultures, as people meet and visit with one another, and see how others live. There is the story of the Russian visitors during the time of the former Soviet Union, who came to Washington, D.C. and were taken to several local supermarkets. It was with some difficulty that they were finally persuaded that these were really the way things were in the United States, and not simply show places set up for their visit. Cheap energy allows global travel for millions of people, and a much better understanding of the world.

The gap

Those who do not have access to these resources for mobility lead quite different lives. This includes perhaps half or more of the world. The average Chinese or Indian, and many people in more remote areas of Latin America and Africa do not have access to this mobile lifestyle, nor do they have the ability to command for themselves the energy resources which other people can. The result, among other things, may be plowing with a horse, an ox, or a camel, riding a donkey, and pumping water by hand. Their daily lives are quite different from those in the industrialized world. Ability to use oil, or not use it, has markedly divided cultures. In the Stone Age, everyone lived pretty much at the same level — perhaps those who had access to good deposits of flint or obsidian may have done slightly better than others. But now the ability of each individual in the industrialized world to command large amounts of energy in the form of petroleum, compared to someone in rural India or Africa having to depend on himself or herself alone for energy creates a huge gap in lifestyles never before seen in this world.

Television and aspirations

Related to the gap between the industrialized and third world countries is television's influence on the future. Even in remote areas TV has reached people who now see the affluence of the industrialized world in contrast to their own situation. For third world areas, seeing how other parts of the world live at first generates curiosity. Ultimately it raises hopes and expectations of what things might be for the lesser developed peoples. Availability of energy, usually in the form of gasoline or diesel powered generators, to bring the one world to the other by television has given many people windows on a world they would not otherwise be able to imagine. It may be noted that the television set itself is a complex of minerals, including such things as the rare earth, yttrium, which makes the colored television screen possible. Being able to buy this mineral complex called a television set has changed the view of the world for many people. Through television, wants are generated, and as nations, particularly in Asia, increase their industrialization and their exports, their buying power increases. This is certain to increase the demand for energy supplies, of which oil is now the main source. Increased energy consumption and a rising standard of living are closely related. As more energy becomes available to the lesser developed countries the competition will increase for raw materials from which, by means of energy, to produce the material things of the good life. Great and important stirrings in this regard are now occurring in Asia, where half of the world's people live.(5)

Oil and its effect on Persian Gulf area culture

Culture is defined by Webster as "a particular stage of advancement in civilization." This would include personal lifestyles, ways of earning a living, political structures, degree of industrialization, and religious practices and beliefs.

The change in culture by the possession and development of a mineral resource — in this case petroleum — is nowhere more striking than in the nations bordering the Persian Gulf. Bulloch, in a perceptive volume based on many years' observations, has described what oil has done:

> "For the pace of progress in the Gulf has been faster than anywhere else on earth: money can work miracles, and has; in something less than two decades Kuwait, Qatar and the Emirates have been transformed from quiet, unknown backwaters into the most modern cities, banking centers, and world financial powers. Traders who ran little hole-in-the-wall stores now control international corporations; boys who could look forward to a lifetime tending their family flocks now write computer programmes; men who thought it an adventure to cross the Gulf by dhow now casually take the daily air shuttle between Bahrain and the Saudi mainland...In every State of the Gulf, oil has worked its magic..."(4)

In the early stages of oil development in the Persian Gulf area, some people thought the old order could remain. But the huge impact of oil which brought vast sums of money and modern technology eventually affected everyone. The ways of the past were overwhelmed. A few clung to their old lives, but they were soon washed over by the wave of oil and all that came with it.

One of the marked features of Islamic society has been the traditional isolation of women from the broader economic scene. But the impact of petroleum wealth on these countries which has caused much contact with the West, has begun to modify the status of women. Still keeping women separate as they have traditionally been in Muslim countries, all-women banks have been established. Women have begun to work outside the home. In Kuwait an all-women factory was built, making small electrical components which in their assembly make use of the dexterity of women's smaller fingers. It has been a success.

In Muslim countries religion is much more integrated into the economic and political life than it is in western nations. A bank in Kuwait specializes in supporting ventures which promote Islamic objectives, and a special insurance company was set up to take care of Muslim interests.(4) Thus oil money promotes religious objectives.

Incipient democracy

Another change which petroleum has brought to the Gulf is the erosion of the former desert chieftain or tribal rule form of government. It is no longer a one-man government in some countries, as it was but is no longer in Abu Dhabi.(4) In others, there still remain the ruling families. Saudi Arabia is at the present time not a democracy, it is a kingdom. There is no constitution spelling out individual rights, and it has no elections.(14) But even there, some evidence of a more democratic approach is becoming apparent, chiefly caused by the rise of a middle class, based largely on the oil industry and related enterprises.

With the oil riches and the need to manage both the oil resources and all the economics related thereto, Saudi Arabia the past 25 years has sent more than half a million of its people to be educated abroad and exposed to outside influences. When they return they have seen democracy. Being educated they want to have a greater influence on the future of their country than they have had in the past. The ruling House of Saud remains intact, but recognizing the winds of change, in 1990, King Fahd announced a new "Basic Law" which, was, in effect, a constitution. Its stated purpose was to "ensure security for all its citizens and residents. No person can be arrested, or jailed, or have his actions restricted except under the law." In part this was to restrict the activities of the religious police who had begun to extend their activities beyond Islamic matters to enforcing conservative cultural customs(1).

The great influx of non-Muslims into Saudi Arabia occasioned by the deployment of foreign troops for the Gulf War into Saudi Arabia disturbed the very zealous Muslims. When most of the foreign troops departed, these elements increased their efforts to retain the more conservative Muslim customs, and particularly those related to women and their dress habits. Business dropped off because women would rather not appear in public and be harassed. Ultimately, the country's mainstream religious leaders turned against the extremists, and thus the course of Muslim religion was set in a more moderate direction. The presence of the foreign troops brought in by the war had precipitated this confrontation. Oil was a factor in defining religious direction.

Autocracy and democracy

For the most part the Gulf nations are autocratic. Only Kuwait has an elected parliament, to the extent that it is elected by the some 80,000 male Kuwaitis who can trace their Kuwait ancestry back to before the 1920s. This is about 10 percent of the native population, but it is, at least, a start on democracy.

But other countries remain under various forms of autocracy, and in the spring of 1995, riots broke out in Bahrain. Cars were burned and electric substations set afire. The large disparity between the wealthy and the poor was part of the cause.

> "There are two worlds in Bahrain. One is home to the gated compounds of diplomats and Western bankers who help make Bahrain, in terms of assets, one of the biggest banking centers in the world. Here are the beach resorts of wealthy Arabs, who come to drink alcohol, visit their money, and be waited on by about 250,000 foreign workers. But the other world where a large share of the 350,000 native Bahrainis live, is a parched island of mud huts and poverty".(15)

The other cause of the civil uprising was lack of democracy. At the same time as the riots in Bahrain, both Bahraini and Saudi dissidents held a joint meeting in London, hosted by the House of Commons, to express complaints. Some 25,000 Bahrainis signed a petition for the restoration of Parliament and other rights. The Emir of Bahrain, Sheik Isa Bin Salman Al-Khalifa, would not accept it.(26)

The possible if not the probable growth of democracy caused by a growing westernized educated middle class (education paid for by oil money, and in many cases in American and other western universities), is one of the most important influences which petroleum has brought to the Gulf. How it plays out will determine much about the future of the region. In the meantime, there is continuing pressure on the tribal monarchies to give up some of their absolute power and share political and economic decision-making. In the past, to a large extent, these ruling groups have simply bought their way out of trouble when faced with internal dissent. Lamb, a long-time observer of the Persian Gulf region, has observed,

"...nothing ensures political stability better than prosperity across the board and rising expectations that are being fulfilled."(11) But now with the national budgets equaling or in some cases exceeding oil income (and deficit spending is actually occurring in some countries including Saudi Arabia), the cushion of petroleum wealth is no longer available against dissent. Complaints will have to be addressed more directly, which means more democratic conduct of national affairs.

Saudi Arabia is the only country which is named after a family. (Liechtenstein is named for a family but is a principality). The country was welded together by the House of Saud and this group with its several thousand Saudi princes has run the nation very much like a family business.(11) Ministerial positions can be held by commoners but the key posts are held by Saudi royalty. But the need for educated Saudi nationals to run the increasingly large and complex oil industry now under control of the government has fostered an expanding middle class. This combined with more stringent economic times is forcing the House of Saud to reconsider its role in Saudi Arabia's future. The huge oil reservoirs beneath the Kingdom are bringing changes to the desert's social and economic structure. Incipient signs of democracy are emerging. It is a course which was destined by geological processes millions of years ago.

Iraq invaded Kuwait because of its oil riches. As a result, the political and social structure of Kuwait has begun to be altered toward a more democratic form in order to develop and maintain more governmental support by the common people. The parliament, disbanded in 1987, has been reinstated. Also, a new military conscription law was enacted in 1992 making the safety of the country a concern and responsibility for everyone. During the Gulf War, it was widely reported that many young Kuwaitis simply went abroad and enjoyed the good life, while the U.S. and its Gulf War allies prepared for, and eventually did the fighting. Now Kuwait is beginning to use its citizens much more extensively in expanding its armed forces, making apparent to the population the need to recognize the obligations which go along with oil riches.

Clearly, the wealth created by oil is causing unrest and a demand for political and economic reform. This is altering the future course of all the nations of the Gulf region.

Cheap energy and rise of the middle class

The ability to use other than human energy to do many things, created the industrialized civilization we know today. It was the basis for the growth of cities as solitary farmers and scattered rural people changed to become social work groups in larger communities. The entire social structure has been reorganized because of cheap and abundant energy. One of the most important features has been the development in many nations of a middle class which is a markedly different situation from the feudal system of a few rich and many poor. Much of what we know as democracy today in the western world came from the development of a middle class which owes its existence to the availability of cheap energy. This energy, chiefly in the form of oil, allows a relatively small part of the population of industrialized countries to produce the basic food for the rest of the population. These people, in turn, do not have to spend their time in manual labor in the fields, but can be part of industrial complexes which produce a host of products, the manufacture of which again is dependent on abundant, cheap energy. Freeing much of the population from routines of agriculture, also allows for the development and staffing of institutions of higher education, and of great research facilities, all of which further establish a broad-based, educated middle class.

Petroleum and job creation

In many countries, mineral resources and particularly petroleum, have greatly broadened the work spectrum. This ranges from the automobile, aircraft, steel, aluminum, and petrochemical industries, to all the facets of the tourist industry. Tourism, facilitated by cheap oil and air transport, has built large and important economies in many lesser developed countries, and in some cases it has become the principal part of the economy (for example, many island nations in the Caribbean).

As the population has increased, so have the variety of employment opportunities in fields which did not exist before this age of the huge use of minerals and energy minerals. In the industrialized world, the demands of physical labor have decreased, and in some places the work ethic has changed because of this, as apparently has happened in Brunei.(3)

Prosperity on a Non-renewable Base

One of the great challenges with respect to the development and use of non-renewable resources which minerals are, is how society will cope with the scarcity or loss of these resources. In a micro-way, this can be seen in the abandoned mining camps across the western United States, in Australia, and Canada, and the beginnings of the decline of petroleum and hard mineral resources in parts of Texas, California, Nevada, Alaska, and elsewhere.

In other parts of the world, entire nations have economies largely or wholly dependent on a non-renewable resource base. The oil-rich countries of the southern shore of the Persian Gulf are a classic example of this situation. To these arid areas previously inhabited by nomadic tribes and scattered villages, oil has brought a colossal transformation of the economies, but all built on a non-renewable resource. This is a *non-renewable prosperity*.

Kuwait is a country nearly 100 percent dependent on oil income. At the beginning of the Twentieth Century the population was less than 50,000. Now the population is approximately two million. At Kuwait's current six percent annual rate of increase, the population will double in twelve years.(20) The 1962 Kuwait constitution guaranteed employment for all Kuwaitis. At present 97 percent of them work for the government at one level or another, and the government is almost entirely supported by oil. No doubt a higher standard of living including better sanitary facilities, improved nutrition, and greater access to medical care paid for by oil revenues have accounted for much of the increase in population. Will oil sustain twice the present population in just twelve years: And when oil income begins to decline...? Another aspect of current oil riches and the future is noted by Reed who reported the concern by some in Kuwait that the present affluence will adversely affect the ability of the nation to adjust to the coming more competitive times. Even now Kuwait's spending is more than its income, and the investment portfolio which has been set aside against the time the oil is gone is now being drawn upon.(22)

Qatar, with 6.5 percent has an even higher population growth rate than Kuwait. The United Arab Emirates' rate is 7.3 percent, which will double the population in less than 10 years. When oil income shrinks and finally disappears, what happens to these areas? Social benefit programs may not be sustained. The rising oil income which bought political stability, jobs, and harmony, and a valid hope for a better future for all, will start to decline. Instead of being able to continually look forward to a better life, the outlook may begin to deteriorate. As long as there is more and more to divide to raise the standard of living of a population or at least keep it steady against a rising population, people are reasonably content. But what happens when the inevitable day arrives that there is less and less to divide among more and more people?

Even such a rich oil country as Saudi Arabia may be facing problems. Chandler writes:

"Although much of the oil windfall of the 1970's was invested wisely in Saudi Arabia — on hospitals, roads, and bridges, seaports, power plants and the like — a huge proportion was diverted to social programs that can't possibly be sustained in a nation whose population is growing at a rate of nearly 4 percent a year, one of the highest rates in the world. Government subsidies touch every aspect of Saudi life.

"'The problem is that we have all been spoiled for 20 years,' said a prominent Saudi prince who plays an active role in policy making. 'We have become too accustomed to receiving help from the government. Sometimes I wonder if it will be possible for us to get used to life as an ordinary economy.'"(7)

In August 1995, there was a wholesale shake-up of the Saudi Arabian cabinet caused by problems in the Kingdom due to the combined costs of paying for the Gulf War and continuing to finance the many social programs. Both the oil minister and the finance minister lost their jobs. All together, sixteen ministers were replaced and two swapped jobs. The five royal family members of the House of Saud retained their ministerial positions, as did five other ministers.

The Saudi family sits atop the world's largest and richest family business, and the country has been run largely as a family enterprise when at one time "each Saudi prince got a minimum monthly allowance of $20,000, even as the number of princes swelled to 6,000."(23) To keep its increasingly restless society from upheaval, Saudi Arabia must pump all the oil it can sell without unduly depressing the price. But the demands of the social programs now in place together with the rapidly rising population receiving these benefits cannot be met by the current oil revenues. "So here is the picture: Despite all that oil, the Saudis are running out of money."(23) There are more and more hands out for a limited amount of funds, and Saudi Arabian per capita income is now falling.(7)

Reed and Rossant reported that:

"A population explosion has also helped sharply erode per capita gross domestic product from more than $12,000 in 1982 to little more than $7,000 today [1995]. Some 3 million Saudis — 44% of the labor force — work in the public sector where salaries have been frozen for almost a decade. This year, in a huge departure from traditional largesse, King Fahd is more than doubling the fees charged residents for electricity, water, and other services...Such erosion of the desert welfare state sorely strains the paternalistic social contract between the ruling Al-Saud clan and the population."(21)

It is the population explosion cited by Reed and Rossant which is the critical and on-going factor in the unrest in the general populace. Initially, when oil was first being produced in substantial quantities, the bounty was spread over relatively few people. With subsequent continued increases in oil production and attendant rise in national income, the citizenry could look forward to even better times. But as oil production begins to level off and not increase on a per capita basis as fast as the population, the outlook changes. Today in Saudi

Arabia, half the people are younger than 15 years old, portending a continuing sharp rise in population. The time when money was available for almost any social demand is passing and the coming generation cannot look toward a continually rising income. And even the present generation now sees evidence that the time of subsidies, free services, and other elements of affluence which oil has brought is coming to an end. This is having an unsettling effect.(21,23,26,27,28)

"Experts are calling it the Gulf Disease...The roots of the problem are the same across the Gulf. The era in which ruling families could use seemingly endless oil revenues to buy the loyalty and silence of the population is coming to an end. Cash-strapped governments are cutting back social services, while the stream of rich contracts that helped oil the economy has dwindled to almost nothing."(21)

Melloan comments on Saudi Arabia that the "kingdom is heavily in debt despite its $45 billion annual oil revenues . . . [and there might have to be] unpopular cuts in spending on development, welfare, weapons and the lavish lifestyles of several thousand princes."(15)

The bomb explosions in Saudi Arabia are an unsettling event in a society which has long been among the most stable in the Middle East. If civil strife spreads throughout the Gulf, site of more than half the world's oil reserves, the implications for the industrial oil-short nations could be profound.

Elsewhere other nations have built up their economies on a non-renewable resource. Nigeria is an example. Cetron and O'Toole state, "Nigeria is afflicted with a disease one might call 'rising expectations'. Today, it is oil rich. Unless it can find massive new oil reserves that so far have not been tapped, Nigeria will not be oil rich in 10 years. If the day comes when Nigeria is not oil rich, it will become a land in chaos that will revert back to the bloody civil wars that marked the country 10 years ago."(6) As of today, this prediction appears to be beginning to come to pass, as a corrupt, mismanaged economy based largely on a single resource — oil — is coming apart.(10)

In Venezuela, in 1989, there may have been a preview of what could happen more widely when income from a depletable resource, such as petroleum or metals, declines. Oil income began to falter. The government had to change free spending ways based on abundant oil income from 1974 to 1979, when oil prices moved up very rapidly, to budget cutting as oil prices declined. When subsidized bus fares were raised along with previously cheap gasoline prices, riots erupted in Caracas and 17 other cities. More than 300 persons were killed, 2,000 injured, and several thousand arrested. The government had to rescind these increases. The inability of oil income to keep up with social demands was not well received.

In 1995, Venezuela, because of trying to continue to live off the oil revenues which are not sufficient to take care of all the social programs inaugurated earlier, began to encounter even more troubles. "Any lasting economic solutions would require asking Venezuelans to sacrifice government-provided privileges, something no one seems willing to contemplate. Consider university students. Thanks to their ability to mobilize in defense of their interests, colleges offer free five-year education as well as cut-rate hot lunches...Yet university students won't yield an inch to help out. This summer they threatened street protests if the government raised transportation prices by reducing the gasoline subsidy."(16) The present price of gasoline, with the government subsidy, is about 13 cents a gallon.

The Venezuelan President, Rafael Caldera, also deferred cuts in the swollen public employees sector which now employs about 15 percent of the total Venezuelan workers.

"Carmelo Lauria, the vice president of the [Venezuelan] congress says Mr. Caldera will continue running the country on oil and a prayer until painted into a corner by a budget crisis — which may not be far off. Most federal agencies have received only 30% of their budgets this year, just enough to pay salaries. If the government can't meet the payroll, Mr. Laurio says, it will face a dilemma: printing money and risking hyperinflation, or firing public employees and provoking social unrest."(16)

In 1996, because of the continually deteriorating economy, Venezuela made application for a $2.5 billion loan from the International Monetary Fund. As part of the conditions, the government subsidy on gasoline would have to be reduced or eliminated, and other adjustments in the Venezuelan economy would be required. These provisions caused fears of further social disturbances.(25) In 1996, with the continuing low gasoline prices, the average annual consumption of gasoline for each registered vehicle in Venezuela was the highest in the world at 1,445 gallons. This compared with 690 gallons per vehicle annually in Mexico where gasoline is also subsidized, and 591 gallons in the U.S. where taxes on gasoline are low relative to other industrial countries. In high gasoline tax countries, German vehicles annually use 270 gallons, Japan 199 gallons, and France 190 gallons. The present gasoline subsidy in Venezuela is about $500 million annually. Taking this away, and the resulting social outcry, is part of the destiny of Venezuela imposed by the Earth resources which have brought the country wealth but which ultimately cannot continue to meet rising export and domestic requirements. Over 70 percent of Venezuela's export earnings come from oil. Oil has brought a degree of affluence to Venezuelans which they are reluctant to lose. Oil, and eventual lack therefore, has caused and will cause profound social changes. It is inevitable.

Oil revenues in Venezuela have not kept pace with the growth in population and the related growth in the costs of the social services which were established in the earlier relatively more affluent oil income years. The people are outgrowing their resources, a situation which, in diverse ways, is happening in many other nations and affecting the lifestyles and habits of their citizens. The destinies of these nations are now being changed by the unalterable geologic fact that Earth resources which include petroleum, metal deposits, water resources, and soil on which the economies have been built are finite.

In the United States, the once abundant resource-based affluent life is now beginning to be eroded. The U.S. still has access to ample oil supplies through imports which allow its citizens to continue to live the pleasant oil-based life which has become so important. But the fact that the United States now imports more oil than it produces makes the future much less certain than when the United States was self-sufficient in oil. Shades of the future were briefly evident in the minor social unrest during the 1973 and 1979 oil crises.

Renewable resources: fixed supply

Other resources such as surface water may be renewable but are in fixed supply at any given time. Against a rising demand from increased population, their availability or lack thereof, and the demands of cities against rural areas, are changing lifestyles and societies. Although modified by occasional years of more than normal rainfall, water supplies have become an increasing concern in a number of areas of the U.S. Water rationing has become a fact of life in many communities. "Water police" even patrol the cities in times of water shortage, ensuring that the citizens do not use water beyond their allotment. Fifty years ago with the same basic water supplies but many fewer people such a situation never occurred. The population is overtaking even the available renewable resources.

Abroad, water supplies are affecting international relations. The loss of water in some areas has been devastating, as in the once highly productive Aral Sea region of Kazakhstan which is now becoming a desert. Ways of life are being changed by the changes in distribution of water, and the upstream use of water by one country to the detriment of countries downstream on international waters. This is certain to have profound effects on the lives of many of the people of the Middle East, and, indeed this is already occurring.

Possession of Mineral Resources: Pro and Con

Easy to adjust to "more"

Having a continually increasing supply of mineral and energy mineral resources such as much of the world has enjoyed to date has been a pleasant experience. But with the exponential increase in populations (U.S. to double by 2050), drawing on finite and in many cases now clearly diminishing resources (e.g., oil in the United States, groundwater in the Middle East), how will societies make the adjustments? Will it be thoughtfully and peaceably or will it involve social and political chaos, elements of which seem to be already appearing in Nigeria and Venezuela?

The possession of valuable mineral resources should be helpful to a country. It may provide needed foreign exchange. It can enrich national treasuries which can distribute the money into useful domestic enterprises in many ways including education, roads, communications, and sanitary and health facilities. Developing mineral resources can provide much needed local employment. The money can also be invested wisely so that there is a continual stream of income when the resources are depleted. To a greater or lesser extent, these things have been done in some countries. But is there a negative side to possession of mineral wealth?

A destabilizing influence?

Mineral wealth can also be a destabilizing influence in a country, as just noted by the situation in Venezuela where population is outrunning mineral income, and the future looks less bright. In Mexico, in 1996, the Democratic Revolutionary Party (PRD) in the oilfield areas of the southeast coast set up blockades at oil installations protesting the actions of Pemex, the Mexican government oil monopoly. The PRD said Pemex had contaminated farm land, had not cleaned up oil spills nor raised living standards as promised, but had diverted the money to political ends and politicians pockets.

In Nigeria it has been rather clear that oil money has been used to support a corrupt and highly dictatorial and oppressive government. The affluent lifestyles of the Saudi princes has encouraged groups of Saudi citizens to move outside of Saudi Arabia and engage in efforts to change Saudi Arabia government structure. This, incidentally, has created a problem for foreign governments such as Britain which find themselves host to these dissident groups but wish to keep good relations with other countries such as Saudi Arabia.

Mineral wealth properly employed can be a stabilizing influence in a country if it is honestly used to help the general populace. But if it is used to support lavish lifestyles by a few individuals or the money is taken by a few in various forms of corruption, and to support oppressive regimes, this ultimately plants the seeds for civil unrest and possible revolt.

If mineral wealth has encouraged a large increase in population used to looking toward better times, and the population continues to increase when the resource income levels out and then declines, the inevitable decrease of social spending can destabilize social structures.

Managing this problem is becoming pertinent in some countries already and will soon be of concern in an increasing number of nations. Thus when the complaint is heard that oil is frequently found in politically unstable countries, it may be that the presence of oil or other mineral riches is what makes the country unstable. One example may be equatorial Guinea. Following the recent discovery of oil, the government promised the country will become the "Kuwait of Africa." However, critics of the repressive state doubt that the oil income will go anywhere but to the President and his ruling clique. The Mayor of the Capital, Malabo, predicted that the oil money will reinforce the privilege of a regime built around a clan from the president's home town. He states "The country will continue in poverty and misery. They'll keep arresting and torturing people and seeing coup plots because all that [oil] money will make them even more paranoid."(24)

Planning Ahead in Time?

Can foresighted governments or business and industrial interests, or simply citizen movements, or a combination of all, devise economies which can be put in place early enough to allow presently resource-rich populations to allow a smooth transition to other equal sources of revenue? Or can an orderly adjustment be made in the standard and modes of living which will accommodate a less-wealthy situation in the future? Against a growing population which has had rising expectations, either of these alternatives will be difficult to implement.

Isolated violence did occur in the gasoline lines in the U.S. during the 1973 and 1979 oil crises. This suggests that civilization — even in the presumably highly civilized United States — is a perilously thin veneer. The British social critic C. P. Snow wrote, "Civilization is hideously fragile and there's not much between us and the horrors beneath, just about a coat of varnish." That thin coat of varnish is basically made up of the current availability of the basics of life — food, shelter, and clothing. The ability to produce and distribute them in the huge quantities in which they are in demand today is made possible only by the use of mineral and energy resources. Increasing competition for Earth resources is now and will increasingly affect the stability of social structures.(13)

How energy and mineral resources are managed, and how growing populations can be accommodated by these resources, many of which are non-renewable, will determine the future of civilization.

BIBLIOGRAPHY

1 AMOS, DEBORAH, 1992, Lines in the Sand. Desert Storm and the Remaking of the Arab World: Simon & Schuster, New York, 223 p.

2 BECHT, J. E., and BELLZUNG, L. D., 1975, World Resource Management: Key to Civilizations and Social Achievement: Prentice-Hall, Englewood Cliffs, New Jersey, 329 p.

3 BEGAWAN, B. S., 1989, Brunei's Work Ethic: What, Me Worry?: The Economist, February 25, p. 66-67.

4 BULLOCH, JOHN, 1984, The Persian Gulf Unveiled: Congdon & Weed, New York, 224 p.

5 CARR, EDWARD, 1995, Energy: The Economist, June 18, p. 3-18.

6 CETRON, MARVIN, and O'TOOLE, THOMAS, 1983, Encounters with the Future: A Forecast of Life Into the 21st Century: McGraw-Hill, New York, 308 p.

7 CHANDLER, CLAY, 1994, Desert Shock: Saudis Are Cash Poor: The Washington Post, October 28.

8 COLOMBO, BERNARDO (ed.), 1996, Resources and Population: Oxford University Press, New York, 384 p.

9 HUNTER, BRIAN, (ed.), 1992, The Statesman's Year-book 129th Edition, 1992-1993: St. Martin's Press, New York, 1702 p.

10 KAHN, S. A., 1994, Nigeria. The Political Economy of Oil: Oxford University Press, New York, 284 p.

11 LAMB, DAVID, 1987, The Arabs. Journeys Beyond the Mirage: Vantage Books, New York, 333 p.

12 LEA, VANESSA, 1984, Brazil's Kayapo Indians: Beset By a Golden Curse: National Geographic, v. 165, n. 5, May, p. 675-694.

13 LINDAHL-KIESSLING, KERSTIN, (ed.), 1994, Population, Economic Development, and the Environment: Oxford University Press, New York, 312 p.

14 LINDSEY, GENE, 1991, Saudi Arabia: Hippocrene Books, New York, 368 p.

15 MELLOAN, GEORGE, 1996, Five Years On, the Persian Gulf Still Simmers: The Wall Street Journal, January 15.

16 MOFFETT, MATT, 1995, Venezuela is Suffering, Its Economy Strangled by Too Many Controls. Its Oil Riches Spur Demands for Government Services and Block Reform Moves. Drifting Into Social Chaos: The Wall Street Journal, August 16.

17 MORGAN, M. G., (ed.), 1975, Energy and Man: Technical and Social Aspects of Energy: The Institute of Electrical and Electronics Engineers, Inc., New York, 519 p.

18 MURDOCH, W. W., 1980, The Poverty of Nations: Johns Hopkins Univ. Press, Baltimore, 382 p.

19 O'TOOLE, JAMES, et al., 1976, Energy and Social Change: The MIT Press, Cambridge, Massachusetts, 185 p.

20 PERKES, DAN, (project director), 1988, Associated Press World Atlas: 184 p.

21 REED, STANLEY, and ROSSANT, JOHN, 1995, A Dangerous Spark in the Oil Fields. Terrorism in Riyadh Ratchets up Tension as the Desert Sheikdoms Trim Largesse: Business Week, November 27.

22 REED, STANLEY, 1996, Kuwait Looks at its Soul —And Isn't Happy. Has the Welfare State Destroyed the Competitive Spirit?: Business Week, January 29.

23 ROGERS, JIM, 1995, The Tent of Saud: Worth Magazine, November, p. 35-40.

24 SHELBY, BARRY, 1996, National Review, p. 24.

25 VOGEL, T. T., Jr., 1996, Venezuela To Increase Gasoline Prices, Rising the Backlash to Secure IMF Aid: The Wall Street Journal, April 15.

26 WALDMAN, PETER, 1995, Restive Sheikdom. Riots in Bahrain Arouse Ire of Feared Monarchy as the U.S. Stands By. Gulf Rulers Clamp Down on Dissidents Demanding Democracy and Work: The Wall Street Journal, June 12.

27 WALDMAN, PETER, 1995, Saudi Bombing Prompts Fears for Nation: The Wall Street Journal, November 14.

28 WALDMAN, PETER, PEARL, DANIEL, and GREENBERGER, R. S., 1996, Terrorist Bombing is Only the Latest Crisis Facing Saudi Arabia: The Wall Street Journal, June 27.

CHAPTER 26

Mineral Economics

Economics of mineral and energy exploration and production are exceedingly complex. They are treated here only in very broad generalities, emphasizing a few factors common to most activities.

It is important that the public understand and appreciate some facts of mineral economics so that they can, especially in a democracy, properly relate to mineral resources and the organizations which produce them. Individuals must be knowledgeable consumers and informed policy makers with respect to the resources which are the cornerstones of modern civilization. Lack of knowledge and appreciation of mineral economics and the general importance of minerals to the public welfare has led and continues to lead to antagonisms and mis-directed hard feelings many times between the consuming public and the resource producers. This is not in the best interests of either group.

Cameron has reviewed the public perception of mineral resources:

> "The result of changing social attitudes is a new set of priorities for the use of natural resources. Development of mineral resources has a much lower priority than in the past and often fares poorly in competition with use of land for wildlife refuges, wilderness areas, wild and scenic rivers, and other scenic and historical attractions. Perhaps the new set of priorities has found its ultimate expression in the massive withdrawals of lands in Alaska, the last great wilderness frontier but also the last great frontier of mineral exploration in America.
>
> "To those who are concerned over the deteriorating mineral position of the United States, the new priorities are difficult to reconcile with the continuing large-scale use of minerals in the national economy, with our growing dependence on mineral supplies from abroad, with our growing deficits in international trade, with the loss of some of our mineral-based industries, with the related weakening of American power and influence in international affairs...In the early part of this century, development of mineral resources was considered essential to the economic welfare of the United States. In the 1980's the development of mineral resources seems to be viewed by the general public (with the active encouragement of environmental groups) primarily as a source of quick profits to large mining companies, as a source of disturbance of the environment, and as an activity in

conflict with more attractive pursuits. The fact that mining industry is basic to the American economy (and the world economy) is not widely perceived. The lessons of the mineral shortages of two world wars are long since forgotten. The 'energy shocks' of the 1970s have not even led to design of a national energy policy.

"This is not a healthy situation. Agriculture, energy and mineral raw materials are still the pillars of the economic structure of a nation."(3)

To continue to enjoy the affluence which the United States and the rest of the industrialized world now has, requires mineral and energy supplies. As the world population continues to increase and large segments of today's population aspire to a better standard of living, the demand for these raw materials grows even faster than the population. Rising consumer expectations and the increasing economic development of China and Southeast Asia is an example.

Today's growing focus on environmental considerations makes it even more important for the general public and national policy makers to know something of mineral economics. What are the monetary and inevitable environmental costs (mitigated though they might be) in supplying to the world the basic materials for human survival?

To some extent each mineral resource has its own set of economics. Some are natural in origin, and some are the result of various governmental regulations and tax policies. However, there are some economic factors which apply to all.

Seven Factors in Common

Seven factors which are common to all mineral exploitation are:

1. The economics of location
2. High costs
3. Risks of unsuccessful exploration
4. The large amount of time involved in searching for and developing the resource before any income can be generated
5. The fact that mineral deposits are non-renewable, and if new discoveries are not made, the company gradually puts itself out of business
6. Political risks: the instability of some governments whereby agreements made by one regime can be repudiated by another regime, and the economic treatment of resource companies which change with political winds in any country
7. Taxes

Location economics

Mineral resources must be developed where they exist. Writing about the United States, Lovering makes the point:

"The mineral resources of a country are fixed as to location; they must be exploited where they occur or left to the future. Even such abundant and widespread deposits as coal and oil underlie only a very small fraction of our land. Although they are relatively common in some sections of the country, they are entirely lacking in great areas. The proportion of the earth's surface that is underlain by many other important minerals is infinitesimal. The molybdenum deposit at Climax, Colorado, is included within less than a square mile, but for many years approximately 85 percent of the world's molybdenum came from the Climax mine."(7)

Some mineral deposits may be in convenient locations. Others may be remote, and have difficult access. In general, the convenient, low cost, readily accessible mineral deposits are already developed. Some have been depleted. Disregarding the different geological conditions which cause marked differences in initial production costs, the cost of bringing a barrel of crude oil to the refinery from a well at Prudhoe Bay is considerably different than bringing a barrel of oil from a well a few miles offshore in Galveston Bay to a refinery in Houston less than 20 miles away.

At Prudhoe Bay the ground is permanently frozen, temperatures drop to 60 below zero F., wind-chill can approach minus 100 degrees F., there is little or no daylight for six months of the year, and it is 800 miles to the shipping point through a pipeline which has to be constantly patrolled and maintained. From the shipping point, it is another thousand miles or more by ship to a refinery. Supplies must be barged in through Bering Strait during a few months of open water, or hauled several thousand miles by truck from the lower 48 states. On-site housing must be provided with electric power and other utilities.

The well in Galveston Bay enjoys a mild climate, supplies are close at hand, a remote, isolated community does not have to be built, and many of the oilfield workers can go home every evening. The oil's journey to the refinery just a few miles away is relatively inexpensive.

Each mineral deposit, depending on its location relative to where it will be processed, is produced subject to a unique framework of economics. When mineral and energy resources are to be developed, the ruling governmental entities involved need to understand the special differences, and treat the projects accordingly.

High costs

A second common factor is cost. Any mineral development of consequence is expensive. Prior to both petroleum and mining development, an initial study of the geology of the areas must be done. Before one area of interest is located, many areas must be evaluated, almost all of which will prove to be of no value. Many under-financed companies have never survived beyond paying the cost of geological studies. They went broke before they found anything.

Once an area of interest is located, the land must be obtained, sometimes by competitive bidding wherein some companies may lose out in spite of money they have already expended for exploration. Then the prospect must be evaluated to determine if an economically recoverable resource exists. This is done by further geological exploration, usually aided by drilling. Drilling, whether for metals, or oil is expensive.

The easy oil and other minerals have been found. Most of the surface of the Earth has been mapped geologically and mineral resources that were readily apparent have been developed for the most part. The early prospectors quickly found the rich mineral deposits which were exposed at the surface. In many cases these were small operations, but the quality of ore was such that relatively little work could yield great wealth. And so the easily found, rich mineral deposits were quickly exploited.

In the case of petroleum, oil and gas seeps indicated the presence of easily discovered and inexpensively drilled shallow oil fields. Big up-folds in the Earth's crust called anticlines were early known to be common sites for commercial oil accumulations. Some of these, as in the hills of Southern California, were clearly evident even to persons with no geological training. They were quickly drilled with little or no cost of exploration. Now the oil industry must find anticlinal oil traps at great depth, or find much more subtle oil traps such as a lens of sand representing a delta finger at a depth of 10,000 to 15,000 feet, or locate a buried coral reef which has no surface expression.

The same circumstances apply to metals and other hard minerals. Native copper in the Upper Peninsula of Michigan was the beginning of the great copper industry of the United States. Anyone could see this copper which occurred in large quantities in the "Great Conglomerates." Native Americans in prehistoric times dug more than 10,000 pits to produce this copper which was a major trade item up and down the entire Mississippi River Valley. It took no skill to find the copper, and, because it was pure copper, it did not need to be refined in any way.

Now, the copper of the United States is produced from a tough, fine-grained igneous rock (commonly quartz monzonite) with only small specks of a copper mineral, not pure copper, in it. And, the quality (tenor) of the ore is down to as low as 4/10 of one percent. This means that a ton of rock has to be blasted out, crushed, milled (upgraded by various processes), smelted, and ultimately put through an electrolytic process to obtain just eight pounds of copper. It takes a very efficient huge mining, transporting, milling, and smelting operation to do all this.

If the mineral deposit is in veins and too deep to mine by the open pit method, shafts must be sunk. Underground operations must be pumped free of groundwater that continually comes in, huge fans must be installed to ventilate the mine, and the mine must be electrified. A variety of mine safety devices must be installed. Trying to follow veins of ore through the various complex rock structures which control the ore is difficult and expensive. Because of all these factors, it is estimated that it takes the work of seven men underground to produce the same amount of ore as one man working a surface mine. Today mines more than two miles deep are in operation. At such depths, the natural heat gradient of the Earth necessitates that the mine be air-conditioned. Also, because of the great pressures of the overlying rocks, violent rock bursts may occur. Rocks simply burst out of the sides of the mine because of pressure from the overlying rock. Rock bursts are unpredictable. They smash ore cars and kill miners. Mining is hazardous and insurance and other costs are high.

In the case of underground mines, fluctuating metal prices become a special economic hazard. Unlike surface mining operations which can simply be shut down if the price of a metal temporarily drops below its production cost, an underground mine must be constantly maintained. The main problem is the groundwater which continually comes into the mine. It must be pumped out and the mine kept reasonably dry to prevent the hoisting and other equipment from being destroyed.

Risks of unsuccessful exploration

With the obvious, easily reached deposits of minerals already discovered and developed, the search for minerals today involves looking beneath the ocean floor to find an oil or gas trap many thousands of feet below the floor. Or it could be exploring for deposits beneath the muskeg swamps of Canada or under thick jungle cover in Brazil or New Guinea. If a mineral exploration company is formed, there is no assurance whatever that anything of value will be ever found. Too many consecutive dry holes have put many an oil company out of business. Imperial Oil of Canada, a partially owned subsidiary of Exxon Corporation, drilled 131 dry holes before striking the buried coral reef where their Leduc well number one was a major oil discovery. The story is told that when the discovery was made, the geologist on the well in Alberta wired the Imperial home office in Toronto, Ontario, saying, "Leduc Number l flowing 640 barrels." The return wire from Toronto asked, "Six hundred and forty barrels of what?" They had almost given up on the thought that their efforts would result in an oil discovery. It now takes courage and large amounts of money to get into the mineral development business. And, once in it, more companies fail than survive. The names of failed mineral ventures are legion. Walls have been papered with worthless stock certificates from failed mineral ventures. Mineral exploration is a high risk, expensive operation. Years of fruitless searches for oil, silver, copper, gold, or other minerals have exhausted the capital of many companies.

The Mukluk oil prospect located about 90 miles north-northwest of Prudhoe Bay was thought to be a prime candidate for another huge discovery. One company paid $300 million just for leases in the area. The exploratory well itself cost $130 million. The combined costs of all the participants in this venture which included the prior geological exploration, the leases, and drilling the test well were, in today's dollars, more than $1.5 billion.(2) The well was a dry hole. The prospect had no oil. These, and similar costs of other failed ventures, must be recovered from projects which are successful.

Length of time to begin to get return on investment

Another common factor to all mineral ventures is that it takes a good deal of time to realize any income even from a successful project. The time is almost always measured in years. In the case of the Prudhoe Bay Oilfield, it was twenty years from the time when the first exploration money was spent just to the time of drilling the discovery well in 1967. June 20, 1977, The *Anchorage Times* carried the headline, "FIRST OIL FLOWS (After 8 years, 4 months, 10 days)."(9) Actually, from the discovery it was nearly 10 years until oil was sent down the pipeline, and income could begin to flow from the wells. It was 30 years from the time when the first money was spent on exploration.

The huge Hibernia Oilfield project off the east coast of Canada has been almost two decades in preparation. One of the problems is to build a big enough and strong enough drilling platform to resist the icebergs which frequently come down "iceberg alley" in the path of which the offshore Hibernia Field is located. By the time the platform is in place and production can begin, the various participants in the project will have invested a total of more than six billion dollars (U.S.) which to that time have not earned a penny in return over a period of about 20 years.

Individuals do not have such large sums of money to invest, nor can they wait many years for a return on their investment. It requires large corporate structures to take on such long term ventures and carry them to completion. "Big oil" is necessary to put gasoline in the world's automobiles.

For mining operations, the average time from the discovery of the prospect to production is about seven years. Previously there were costs of exploration. It may have taken many years to discover the prospect. Then add seven years of costs of drilling, building the mills to crush the ore, and other processing facilities, and the increasing financial and time costs of environmental concerns. Time enters into this because until production is started, all the money invested earns nothing. Money has a time value. For example, if all costs from the beginning of exploration to bringing the mine to production would mean that $100 million has been invested for a net total average time of ten years, that $100 million must either be borrowed for ten years at the going rate of interest, or provided from retained earnings from other projects, or supplied by stockholders who buy the stock in hopes of eventually getting a reasonable return for their risk investment. And they may lose it all if the project fails. Many do.

The people who buy the stock, or the company itself could have invested the money into some income producing instrument such as a bank deposit or a bond and earned an income beginning immediately. Instead, the money was spent on trying to develop a mineral prospect which not only has to earn a current return on the investment but also make up for the years when the money earned nothing.

Cameron puts the present situation in perspective:

> "Part of the current American attitude toward mining is a carry-over from the 19th century, when there were spectacular successes in some districts of the West. Mining became identified as a quick source of easy profits. Those days are long since gone, although there was a brief revival during the uranium boom in the late 1940's and 1950's. Mining today is a highly competitive industry, in which profit margins are low. It is capital-intensive, yet the profit margins and the long lead times between discovery and first production make it difficult to attract capital funds in competition with other industries in which returns on investment are higher and can be realized in much shorter periods of time."(3)

Mineral resources are non-renewable

The mineral industry differs from other basic wealth-producing activities which are farming, fishing/hunting, and forestry in that minerals are non-renewable. The average mine life is seven to ten years. Oil may first flow from a well. Then it has to be pumped. The field during production usually has to be repressured by water-flooding or gas injection. Finally, all oil fields are abandoned. Each pound of copper produced and each barrel of oil produced puts the company involved a bit closer to being out of business unless some of the money made from current production is set aside to pay exploration costs to find more resources. A new crop of corn may be grown each year to replace the crop produced the year before, but oil wells and mines are a one-crop situation.

Political risks

These risks depend on the relative stability of governments and the politics within the countries where mineral operations are conducted. The risks can be very large, and range from the risk of civil war, such as has been going on in Angola for many years, the destruction of the resource producer's equipment, and even the expropriation of the company.

Contracts made by one political regime may be invalidated by a succeeding political regime. Or the same governing group may simply change its mind and not honor a contract. This can happen even in what are regarded as civilized countries. Without naming the country, a number of years ago, an oil company made an exploration and development agreement covering an ocean area which was difficult to explore, and which did not have any known oil deposits. It was strictly a wildcat venture and to explore this area was very expensive. The country which had the rights to that part of the ocean, had been paid a considerable sum of money for the lease by the company just for the opportunity to explore for petroleum. The country itself had no capability to do the job. The terms of the lease provided for a royalty to the country if petroleum was discovered.

The company spent a lot of money in exploration and finally made a major discovery, whereupon the country announced that it was changing the terms of the lease so that it would get a much larger royalty than initially agreed. When the oil company protested, and asked why the original more generous lease terms to encourage the high risk exploration in this area had been altered, the government replied, "When we sold you the lease we didn't think you would find anything."

Taxes

The seventh but very important factor in mineral development, and one completely in the control of people, is taxes. In general, companies are allowed to write off their exploration costs before their income is taxed. With the nearly universal inflationary trends, it is more useful for the companies to write off their exploration costs immediately with current dollar value, rather than over a period of years with depreciated dollars. However, some countries insist that the costs be written off over a period of years against current income, which means that the companies are forced to use the earlier more expensive dollars to reduce the current income of cheaper dollars, which is not to their advantage.

Another aspect of taxes is that companies are commonly taxed on their plant and equipment and also on proved reserves. This means that the tax bill gets higher if exploration to prove up reserves gets very far ahead of production needs. Taxing reserves discourages exploration.

Taxes on production of low grade mineral deposits may prevent their development. On Alaska's north slope, at Milne Bay, just west of the prolific Prudhoe Bay Field, there is a large deposit of heavy (thick) oil. This heavy oil is costly to pump and carry down a pipeline in the cold arctic climate. Developers are willing to drill some 230 wells at a cost of more than half a billion dollars if the State of Alaska would eliminate its 12½ percent royalty on this production for the first five years. Taxes now prevent this oil from reaching the market.

As each barrel of oil produced puts the company closer to being out of business, money must be set aside to discover replacement oil in the future. At one time, Britain levied taxes as high as 90 percent on the income which oil companies received from British North Sea oil production. This left little for companies to reinvest in further exploration in this high cost area, and firms began to reduce operations. Seeing this, the British government has since reduced its taxes on North Sea operations to some extend but still taxes them very heavily. There are also some smaller fields in the North Sea which could be found and developed if taxes were lowered. At the present time, only large fields which can be produced with relatively few wells are economic. As these fields are depleted Britain will have to make the decision of reducing taxes or having to begin to import more oil. To date, the North Sea oil fields have been milked very heavily by British taxes.

Metal mining also tends to be a "cash cow" for both federal and local governments, with total taxes commonly approaching 50 percent or more of the gross income. Local governments frequently expand their political boundaries in order to include mining and oil properties into their tax base.

Mining and Public Lands in the United States

There has been considerable recent controversy over the 1872 Mining Law by which public lands can be patented and become private property for the production of minerals. In earlier times, obtaining public lands by this method was easy and no doubt abused. Recently, however, requirements for patenting lands have become much stricter, and it has become much more difficult to patent public lands. In 1989, for example, only 43 patents were granted and most of them went to Native American tribes through land settlements in Alaska. At present, it is necessary to prove without reasonable doubt that a mineral deposit of value exists before the land can be patented. To do this an expenditure of between a half a million and a million dollars must ordinarily be spent on each claim. A claim is 600 feet by 1500 feet. A placer claim, one on sand and gravel deposits, is 660 by 1320 feet. Subsequent to obtaining a patent, many millions must be spent in developing the property. Also, no other industry in the U.S. is covered by more stringent federal, state, and local permitting, safety, reclamation, and environmental laws. In the past twenty years the American mining industry has spent more than $15 billion in compliance.

From these operations come the materials for building the things used by everyone — cars, trucks, roads, houses, factories, office buildings, home appliances, and myriad other products in every day use. The bottom line is that mining is an important part of the U.S. economy but even when public lands are patented to mining companies, that industry remains one of relatively low profitability. If public lands are to carry a royalty to the government on minerals produced, that cost ultimately will have to be borne by the consumer — the general public.

The Oil Industry and Land

Initially in the United States, most oil drilling was done on private lands where the mineral rights were held by the land owner. This is in contrast to the rest of the world where these rights are usually owned by the respective governments. And now, in the United States, increasingly, oil development is going offshore where the mineral rights are owned either by the federal or state governments.

Competitive bidding

Oil leases which are the right to explore for oil, with no assurance that any will be found, are put up for competitive bidding by the governments. The cost of buying these leases routinely runs into many millions of dollars. Not uncommonly leases are bought at high prices which ultimately prove to have no commercial oil. If a company is to survive, these costs have to be recovered from exploration projects which are successful.

Some countries have an additional way of getting money from the oil industry. They charge for copies of the lease forms which contain the provisions attached to each lease. In 1995, Ecuador offered nine leases in the Amazon basin, and charged a non-refundable $100,000 (U.S.) for copies of each of the lease forms. Thus it cost an oil company or an individual $100,000 just to find out what they would be bidding on.

Oil: A Highly Complex Expensive Enterprise

The oil industry in its entirety from the exploration, drilling, production, transportation, refining, petrochemical production, and marketing uses more varied technology than any other industry in the world. These include everything from space satellites, to a study of the highly complex science of organic chemistry. In between, seismology, directional drilling,

drilling in waters a mile or more deep, and putting vast quantities of steam into the ground may be involved, among many other activities. All this takes time and money, lots of both. The oil industry invests more money, by far, than any other industry in the world to conduct its operations. Return on investment, however, is modest. The average profit earned by an oil company on a gallon of gasoline is less than two cents.

Resource Development: Open to Everyone

The western world and increasingly the rest of the world are democracies and open economies. Anyone has the opportunity to raise capital and invest it in mineral development. If anyone believes that searching for and developing oil, copper, gold, silver, or any other mineral resource is the easy and quick road to riches, they are free to try. We need mineral exploration people with an entrepreneurial spirit as those in the past who have provided the raw materials by which to build our present civilization, and the high standard of living we enjoy. But those who try should know they must have a lot of time and money to spend before they can expect to get anything back. And they may get nothing.

Actually, it is not necessary to form one's own company to get into these enterprises. Almost all metal resources and energy resource companies are public corporations, with stock traded on various financial exchanges. Individuals can buy into these enterprises with only small amounts of money. There are mutual funds which specialize in owning mineral and energy mineral resource companies into which individuals can easily invest. Individuals, mutual funds, pension funds, and insurance companies offering life insurance policies and annuities are the major investors funding mineral resource development worldwide. In general, however, returns from natural resource companies are modest compared to investments in many other enterprises. But they are attractive investments because they produce vitally needed resources which will always be in demand, and because these are tangible things which are a hedge against inflation. Precious metal stocks are particularly attractive in that regard.

Basic and Useful

Much of the strident opposition to resource developments fails to consider that if these were shut down and did not exist, the human race would still be close to living in caves, and heating only with wood. Even today's solar energy devices take metal to produce. Bicycles are made of metal, as are knives and forks. It is important that companies which produce the raw materials for our present civilization and standard of living be allowed to continue to do so. It is also important that there be economic rewards for those willing to supply risk capital to do these things. Communist economic societies, even if they have huge natural resources, do not provide a good standard of living for their citizens, as the experience of the recently deceased Soviet Union has so amply demonstrated.

Printed Capital and Natural Capital

Finally, in considering economics in relation to natural resource supplies, mention should be made of an economic concept sometimes heard — that an increase in capital investment, that is, spending more money to obtain a commodity, will solve any shortage. A variation of this theme is the statement that, "...no mineral, including oil, will ever be exhausted. If and when the cost of finding and extracting oil goes above the price consumers are willing to pay, the oil industry will begin to disappear...The amount [of oil] extracted from first to last depends purely on cost and price."(6)

It is true that increased prices in some circumstances will bring into the economic stream more marginal resources. More investment combined with better technology has now made it possible to produce copper from deposits with as little as 4/10 of one percent copper. And this approach has worked for some other mineral resources as well. And it may work for a time but not indefinitely. There are limits to the problems which money will solve. And the statements made in the article(6) that "no mineral, including oil, will ever be exhausted" and that if the price of oil gets too high for consumers, "the oil industry will begin to disappear" ignore reality. If consumers can no longer afford oil, for all practical purposes oil supplies are then exhausted, even if some remain in the ground. Theoretical economical arguments tend to blur the important fact that Earth resources are finite, and if a given industry such as oil "begins to disappear," without an apparent equal substitute, large problems will arise which economics may not be able to solve, particularly with the ever-growing population demands on Earth resources.

Recently a few economists have demurred regarding more investment capital and price as a solution to any scarcity problem,(10) They suggest there are two kinds of capital, that represented by money and that in the form of natural resources — termed "natural capital." We are using this natural capital at a tremendous rate, and simply printing money will not make more natural resources than nature provides. The idea that a higher price is all that is needed to maintain or increase the supply of a commodity, taken further, would suggest that as long as more money is printed there would be no lack of oil or any other resource. Or suggesting that if the price gets too high, then the use of a given commodity will simply fade away, ignores the fact that the commodity may be basic to the economy and have no adequate substitute. Losing it could be a disaster which economics cannot solve.

Oil is a particularly good example to show the fallacy of the economic argument that price is the overriding factor. Energy is critical as it is the key which unlocks all other natural resources. But the economic argument ignores the fact that if it takes more energy to produce the energy obtained, then that becomes a futile situation with a net energy loss which cannot be continued. The amount of oil extracted does not depend simply on price, and if a given energy supply (of which petroleum is now the world's chief source) cannot be produced at a net energy profit, then all the things which energy enables to be produced, such as metals, are lost.

The article cited(6) also states that, "Increasing global oil scarcity is an illusion." This ignores the fact that the world discovery rate of oil peaked in the early 1960s and has fallen markedly ever since and the peak of oil production is in sight.(8)

Summary

The economics of the mineral industry are poorly and frequently incorrectly and unfavorably understood by the general public, the people who need and use the products of this industry.

The most important and seemingly obvious fact is that minerals must be developed where they have been placed by geological processes. Accommodation of this reality must be made with regard to environmental restrictions if the resource is to be developed. Additional costs of complying with environmental regulations must be added to the final price of the resource to the consumer.

Large amounts of money are required to find and then bring a substantial metal deposit or oil field into production. There is also a large time interval between the date of discovery and the time the discovery can begin to make a return on the investment.

Explorations for metals and petroleum are high risk ventures. Many projects fail. Once a resource is found and development started, more resources must be discovered to replace this deposit when it is depleted. Tax laws must allow the company to accumulate money for this purpose. If not, then the company will eventually be out of business, no additional resources will be discovered, and the consumer will not have those resources for use.

The mining and petroleum industries are commonly perceived to be large sources of wealth. They do produce great quantities of valuable resources benefitting nearly everyone, but for the producing companies and their investors, the returns on these high risk investments are modest. Because most major natural resource companies are publicly held, it is possible for anyone to become a part owner of mineral resources and to begin to understand firsthand the problems and risks and rewards of mineral resource exploration and production.

Mineral and energy resources are finite and have limits. One cannot simply print money to have more and more resources. Economic capital cannot increase the natural capital of the amount of water which falls on Israel, which is in increasingly difficult straits in regard to fresh water supply. Nor will higher prices cause more water to flow in the already overdrawn Colorado, Jordan, or Nile rivers. Topsoil is now being lost around the world faster than it is being formed (a process which takes thousands of years). No amount of money can produce more soil. There are limits to natural resource capital which printed capital cannot solve. Geology, not economics, ultimately controls the availability of Earth resources.

BIBLIOGRAPHY

1 ADELMAN, M. A., 1970, Economics of Exploration for Petroleum and Other Minerals: Geoexploration, v.8, n. 3/4, p. 131-150.

2 BRAY, RICHARD, 1995, personal communication, May 10.

3 CAMERON, E. N., 1986, At the Crossroads. The Mineral Problems of the United States: John Wiley & Sons, New York, 320 p.

4 EDEN, RICHARD, 1981, Energy Economics: Growth, Resources, and Politics: Cambridge Univ. Press, 442 p.

5 GRIFFIN, J. M., and STEELE, H. B., 1980, Energy Economics and Policy: Academic Press, New York, 370 p.

6 HANKE, S. H., 1996, Oil Prices are Going Down: Forbes, February 26.

7 LOVERING, T. S., 1943, Minerals in World Affairs: Prentice-Hall, Inc., New York, 394 p.

8 MacKENZIE, J. J., 1996, Oil As a Finite Resource: When is Global Production Likely to Peak?: World Resources Institute, Washington, D. C., 22 p.

9 STROHMEYER, JOHN, 1993, Extreme Conditions: Big Oil and the Transformation of Alaska: Simon & Schuster, New York, 287 p.

10 ZACHARY, G. P., 1996, A 'Green Economist' Warns that Growth May be Overrated: The Wall Street Journal, June 25.

CHAPTER 27

Myths and Realities of Mineral Resources

Although minerals and energy minerals are fundamental to our existence, the facts of these resources and of industries which produce these materials are subject to many myths and much misinformation. This is unfortunate for it clouds the ability of individuals in a democracy to make intelligent choices. Some of the distortions are deliberately made by political interests who play upon the fears and hopes of the electorate, and then in the role of the defender of the public interest against the oil or mining companies seek to obtain votes by this device. Some statements are made from ignorance, and some are made by people who have their own political and social agendas which they wish to perpetrate upon the public. Some are made by people who are a bit over enthusiastic about a particular resource and do not carefully examine the hard facts, or may not be aware of them. Some statements are made by promoters wanting to raise money for a particular mineral development, whether that development has a sound basis or not.

It is important that facts be sorted out from fiction if democracy is to be the form of government which makes our laws and guides our international, national, and personal affairs. The tax structure which is a political matter has much to do with the success or failure of many mineral ventures. However, what is fact must be carefully presented and any doubts about a statement be fairly noted.

Also, it is important for correct public policy that the basic geologic, economic, and technical facts be known about a given resource, so that there are no illusions by government leaders or the citizenry in general as to how important that resource is now or may be in the future. There are numerous glowing statements in print about what can be expected from things ranging from oil shale to mining the moon. It is indeed nothing short of amazing what claims are made, and what people may believe. This also applies to solutions to resource-based problems such as population pressures, where colonizing space has been suggested. Some of these myths are discussed here.

In part, this chapter is a summary of statements made in other parts of this volume with respect to particular resources. However, for the sake of emphasis with regard to some of the misunderstandings concerning mineral resources, the facts are here brought together in one chapter for convenience of review.

Myth: There is no oil supply problem in the United States

During the two oil supply crises of 1973 and 1979, in the U.S. the average citizen frequently stated the belief that no "real" oil shortage existed, and that the shortages were caused by the oil companies withholding oil from the market. But when the individual is asked two basic questions: How much oil does the United States produce each day, and how much oil does the United States consume each day, there usually is no reply. People are "experts" on the oil situation with no knowledge of the facts. This serves no useful purpose.

Reality

The United States passed the point of oil self-sufficiency in 1970, and has been an importer of oil ever since then. In 1996, the United States produced about 6.4 million barrels of crude oil a day but imported more than 7 million barrels of crude oil plus 1.7 million barrels of refined oil products. The U.S. is now importing more oil than it produces.

The United States is the most thoroughly drilled area in the world and there is no possibility that this nation will ever again be self-sufficient in oil in the volumes and ways in which it is now being used. In what are reasonably prospective available oil areas in the United States, there are very few undrilled places left large enough for a major oilfield to be discovered onshore. Offshore there are some prospects, but offshore drilling has been banned in many areas.

A major oilfield covers many square miles and almost all sizeable prospects have been drilled onshore U.S., except in a small corner of the Arctic National Wildlife Refuge, which has been under environmental limitations. Some prospective areas do exist in more distant offshore areas open to exploration but these are generally in increasingly deeper waters, and are difficult and expensive to drill. The amount of oil which would make the United States again self-sufficient in petroleum cannot be found and produced in what areas remain to be explored, either onshore or offshore. As one petroleum geologist put it, "Exxon has run out of real estate." This is true of all the major companies who have now had to mostly go abroad to obtain acreage prospects of worthwhile size.

Myth: Just drill deeper for more oil

The statement is sometimes made that deeper drilling would find more oil.

Reality:

Oil occurs in sedimentary rocks which are a fairly thin part of the Earth's crust. In the long-time oil-producing State of Kansas, for example, granite or something else besides sedimentary rock exists everywhere at depths of 15,000 feet or less. All over the world, at some depth, non-petroliferous rocks are encountered below which there is no oil. Where there are great thicknesses of sedimentary rocks, 16,000 feet is, with a few exceptions, the limit of oil occurrence. Below that depth, because of the temperature of the Earth, only gas exists.

Myth: Oil companies have capped producing wells to keep up the price of oil

This is one of the oldest and most persistent myths about the oil industry. The idea is that oil companies will drill wells and then cap them, thus withholding production from the market until the price of oil goes up.

Reality:

It is true that many wells are drilled and then capped. Almost all of them are capped because they are dry holes — that is, they are failures. Less than one in eleven exploration wells is successful. The law requires that failed wells be filled with cement at key points in the well to avoid groundwater contamination, and then capped. To the landowner who had great hopes for the well drilled on his property, the face-saving statement to the neighbors is that "they found oil but just capped the well." Only when the oil company drops the lease does reality arrive.

There are some wells which could produce oil which are temporarily capped. There are two common reasons for this. One is that there is no facility for transporting the oil from the well at the moment. Either a pipeline does not exist or it is too expensive to truck it out. Generally, if the well is a producer, other wells will be drilled in the area to establish the presence of enough recoverable oil to justify developing a transport system by which the oil can be brought out economically.

A second reason may be that occasionally it is true a well may be drilled, completed, and capped when the current price of oil is not high enough to pay for the expenses of producing the oil —the pumping costs and perhaps the problem of the disposal of the salt water which may be produced with the oil. However, capping a well and leaving it for a time is risky because sometimes the well cannot be restored to production.

Drilling a well is so costly that if the well is productive and capable of bringing a return on investment, the well will be produced. If a million dollars is involved in exploration, lease, and drilling costs — and one million is much less than many wells cost — then the cost of that money in lost interest which that money would otherwise bring, demands that the well be produced. No one can afford to tie up a million dollars, or many millions with no economic return. And it is not done.

Myth: Don't drill this prospective field. Only 90 days of U.S. oil supply there

One of the most misleading arguments used against drilling a particular area is the statement that it would only supply X number of days or months of U.S. oil demand. Yet to the average citizen this is one of the most "logical" reasons for not allowing drilling in a particular area. It is one of the most widely and most effectively used arguments against oil drilling. It appears frequently in numerous newspaper editorials and letters to the editor, and at public hearings.

With regard to the long-running debate about opening a portion of the Arctic National Wildlife Refuge in Alaska for oil exploration, in 1995, the president of a prestigious environmental organization said "...there may be at best only 90 days supply of oil for the U.S. There can be no justification to develop the arctic refuge."(27) Let us pursue this argument.

Reality:

At the present time the U.S. uses about 19 million barrels of oil a day. A 100 million barrel oil field is regarded in the petroleum industry as a "giant." They have been discovered only infrequently. Yet if one of these giant oil fields was used to supply U.S. oil demand, it would last less than six days!

To put this in further perspective, at the present time only 15 oil fields in the United States have produced as much as a billion barrels of oil. This is done, of course, over a period of many years. But if the argument is applied that the oil field would only supply oil for a given length of time in the U.S., it should be noted that a billion barrels from each of these 15 fields – if it could have theoretically been used alone at one time – would have only supplied the U.S., at its current rate of consumption of about 6.9 billion barrels a year, only about 53 days.

If the argument used by the president of the environmental organization was to be followed, there would be no oil drilling at all in the United States. These days, a ten million barrel oil field discovery is an important event in U.S. oil exploration. But that amount would last the U.S. less than 13 hours! The fact is, we are not discovering ten million barrel oil fields every 13 hours in the U.S. That is why our oil reserves are in decline. Prudhoe Bay, the largest oil field ever discovered in North America, would have lasted the U.S. less than two years if it alone had been used.

But it is not possible to produce all the oil out of Prudhoe Bay or in any other field in 90 days, or six months or two years. If one divides the number of producing oil wells in the U.S. into the total proven U.S. reserves, each well has a reserve of about 40,000 barrels. These 40,000 barrels of oil, if they could be immediately produced, would supply U.S. oil demand for about three minutes. On this basis, it might be argued that none of these wells should have been drilled, in which case the U.S. would have no oil production. But oil supplies are produced over many years from many wells which make up the total U.S. production.

Each well makes a contribution, and each discovery serves to stretch out domestic supplies a little longer. Individually most fields, with the notable exception of the huge Prudhoe Bay Field, and each well produces an insignificant amount of oil relative to total U.S. production. But taken together they add up to the 6.4 million barrels a day now being produced.

People who use this argument presumably drive to work in gasoline-powered cars. Where do they expect that gasoline to come from? People demand and use oil. With few significant prospective areas now still open to drilling in the U.S., where is the oil supposed to be obtained? Those who would curtail exploration first need to reflect on what is causing the huge and increasing demand on mineral and energy resources, and address that cause and not the symptoms of the problem. The cause is the resource demands of growing numbers of people, and the desire to continue to maintain the largely petroleum-based standard of living enjoyed by citizens of the industrialized nations. Use no oil and there is no need to drill. Otherwise drilling is necessary —somewhere. And each well and field are a necessary part of the total supply picture.

Politics and Oil

In the United States, and in many other countries, gasoline is the commodity which most touches individual lives every day. It has been politically popular to proclaim that it is the right of every American to enjoy cheap gasoline, and if this does not occur the politicians blame someone — usually the oil companies. It is "they" versus "us". "They" are the oil companies. "Us" is the public, and the public elects the politicians. When considering the problems of gasoline supply, people should simply look in the mirror to see the major part of the problem. The United States uses more gasoline per person than any other nation in the world, except Venezuela. In the Los Angeles area more people drive more cars more commuting miles every day than anywhere else on Earth. To a lesser extent this occurs in many other cities in the United States including greater San Francisco, Houston, New York, Chicago, Denver, and Seattle.

Myth: Gasoline is high-priced

When gasoline in the United States crossed the one dollar per gallon retail price there was a general public resentment of the oil companies. Gasoline was "too high priced."

Reality:

In the 1990s in the U.S. the basic cost of gasoline (before taxes) was less in terms of inflation adjusted dollars than anytime in the past 40 years. In fact, it was nearly as cheap as anytime in the history of the oil industry. It was also historically inexpensive in terms of how long the average wage earner had to work to buy a gallon of gasoline.

The cost of gasoline at the pump is the basic cost of exploring for, drilling, producing, refining, and marketing the gasoline together with the taxes which are placed on this commodity. Lesser costs are the cost of transporting and storing the gasoline enroute to the service station. Profit margins are spread all through this system, and are generally in line with average market returns on investment. The biggest single cost in the final price of gasoline at the service station is taxes. Gasoline is a favorite source of revenue for government. In 1993, for example, U.S. President Clinton signed a bill which increased the U.S. federal gasoline tax by 4.3 cents. This was not dedicated for the purpose of road building and maintenance, but went into the general U.S. Treasury, and was stated to be for the good cause of reducing the annual government deficit, which end result has since seemed rather elusive in practice. States and cities also impose gasoline taxes. In the United States, federal and state gasoline taxes on the average are in total equal to more than the basic cost of the gasoline at the refinery.

Based on constant 1967 dollars, *exclusive of taxes*, the retail price of gasoline in the U.S. in 1920 was 49 cents, in 1930 it was 39 cents, in 1950 it was 37 cents, in 1970 it was 30 cents, in 1974 it was 40 cents. The price in 1995 was 67.7 cents a gallon.(19) But this 1995 price is for a much improved quality of gasoline with additives for better engine performance, and also for reduction of air pollutants. The price is also for unleaded gasoline which was not available in 1974, and which costs more to produce than does leaded gasoline. This record of price stability is in marked contrast to the large increase in prices of virtually all other consumer items. The oil companies have done a remarkable job in supplying the world's largest consumer of gasoline, the U.S. citizen, with inexpensive high-quality gasoline without restrictions as to quantity.

However, because gasoline price touches so many people, the political posturing over gasoline prices in order to gain voter favor seems to be a continuing phenomenon. In the U.S. in the spring of 1996, gasoline prices rose about 10 to 15 cents per gallon. This was due to the fact it had been an exceptionally long, hard winter, and refineries had delayed their shift of refinery output emphasis from fuel oil to gasoline. There were also weather related problems in the North Sea and Mexico which interrupted oil shipments, and the world oil price rose from about $17 a barrel to $25. U.S. oil companies have no control over the price of world oil, from which now comes more than half the U.S. oil supply.

But both major U.S. political parties tried to make campaign advantage of the situation. The administration announced that the Justice Department would immediately look into the matter of a possible price conspiracy among the oil companies. The opposition in the Congress said it would try to repeal the 4.3 cent gasoline tax increase which the administration had pushed through in 1993. The media interviewed motorists at the filling stations who by and large were of the view that the oil companies were greedy, which view was widely echoed by cartoons, editorials, and radio and TV commentators.

Some of the media, however, had more informed observations. The syndicated columnist, Mike Royko, viewing oil prices both historically and currently, wrote some very direct comments about the 1996 oil price situation:

> "What I didn't hear any reporter say was: 'Of course, in this country, we pay far less for gasoline than they do in Canada, Europe, or just about any other developed nation.'
>
> "Nor did they point out that when you factor in inflation that the price of gas is less than it was 40 years ago.
>
> "If the broadcast hysterics took note of these few simple facts,there wouldn't be any talk of a gas pump crisis...
>
> "If CNN insists, every half hour, that helpless American motorists might suddenly be sputtering to a stop on the shoulder of the road, is the White House or Congress going to deny that we are suddenly fuel-starved? Is any self-respecting politician going to stand up and say: 'Hey, what's the fuss? You want to see high gas prices, go to Canada or Europe. What are you network magpies chirping about?'
>
> "Of course not. When the nation's broadcast babblers, from whom the majority of Americans get their news, say we have a crisis, it's time for the political speech writers to crank out something, even if it is something stupid.
>
> "That stupidity includes the instant-investigation into the vague possibility that the oil companies have somehow conspired to pick our pockets.
>
> "All that the investigation will show is that if there was a conspiracy, they've somehow conspired to give us the world's cheapest fuel for our cars."(21)

In terms of oil, American's live in a "fuel's paradise." A British observer on the scene has written, "...by European standards petrol [gasoline] is almost given away in the United States..."(28)

It should be noted that in other countries, the retail cost of gasoline without tax is about the same as in the U.S. That gasoline costs more than five dollars a gallon in some nations is due chiefly to taxes, and to a lesser extent to retailer's profit, which commonly is higher than in the United States. Also, in some countries the gasoline distribution system is less efficient than in the U.S. and it costs more to transport the gasoline to the retail outlets.

Myth: "They" own the oil companies

To gain popular favor, many politicians, frequently joined by the media, assert that oil companies are vague and distant entities owned by "they" and it becomes "they" versus "you." The oil industry is a favorite whipping boy for politicians seeking to gain votes. Because the average citizen is not well informed on these matters, political rhetoric often reinforces prejudices against the oil industry rather than dealing in realities.

Reality:

Who does own the oil companies? During the 1979 oil crisis I was invited to address a luncheon meeting of a State Employees Association in the State Capitol. The topic was the oil crisis. The oil industry was being widely blamed. I asked who among the State employees owned any oil company stock. Not a hand was raised. However, just prior to the meeting I had been in the office of the Public Employees Retirement System which administered the pension plan for all State employees. I had examined the holdings of the fund and discovered that the largest single industry holding in terms of dollar value, was oil company securities. The fact was that everyone in the room owned oil company stock. The conventional myth is that large oil companies are owned by some vague group distinct from the general public, the "they." The reality is that "they" are us.

And this is very broadly true. Insurance and investment companies place the funds of their clients in a variety of investments among which traditionally have been oil companies. Through life insurance, and other insurance policies, annuities, and mutual funds, the major oil companies as well as the mining companies are owned by the general public.

A recent study of ownership of stocks in the six largest oil companies in the United States disclosed the following: nearly 200 mutual insurance companies hold close to 16 million shares. Ninety-one colleges own these stocks, and about 1,000 charities and educational foundations in the United States are holders of these oil company securities. In direct ownership more than 2.3 million Americans hold stock in these six companies.

Many other Americans own interests in smaller oil companies. As to who produces U.S. oil, it should be noted that currently in the United States, excepting the North Slope Alaskan oil which is a very high cost operation and requires a very large investment, more than half of the oil produced in the U.S. is produced by small independent producers. It is the oil produced abroad in such high cost areas as the North Sea, where major oil producers are dominant. This is inevitable as expense of operations in these areas runs into billions of dollars, and are much beyond the financial and risk taking abilities of small independent oilmen. And, as noted, these larger companies are owned directly, and through pension plans, annuities, and insurance policies, by millions of citizens.

Myth: Some remote special group of people run oil companies

Here, also, people frequently believe that persons who are not part of the general public run the oil companies, just as they may believe that some distant remote group owns the oil companies.

Reality:

People who run oil companies just as those who own the companies are again not "they" but us. Geologists, engineers, accountants, and business administration majors make the oil companies function. They are our sons, our daughters, our neighbors. I taught petroleum geology at a state university. From my experience, which is typical, I cite two examples of who runs oil companies. One student had worked as a meat-cutter in a butcher shop in his small central Oregon hometown during his high school days to help out his family. He worked his way through college by various jobs and went on to graduate school and received a well-earned Ph.D. He worked his way up through the oil industry and is now vice president of a major U.S. oil company. He is based in London in charge of the company's North Sea operations. Another student worked as a clerk in his father's shoe store during both his high school and university days. After earning a graduate degree he held various positions in an oil company, and now represents one of the world's largest companies in examining oil prospects from Russia, to Norway, to Africa. Each of these men was the boy next door. The people who run the oil companies are us.

Myth: "Big oil" is bad

"Big oil" is a favorite expression frequently used in a derogatory manner by many in the media, and others who, for various reasons wish to turn the public against oil producers. The myth is that somehow "big oil" is bad.

Reality:

It is true that worldwide oil production is becoming a bigger and bigger business. The reason is that the easy to find, shallow oil has been found. Now, more and more significant discoveries have to be searched for in remote "frontier" areas (arctic, or jungle) or must be sought after in deep water offshore areas which involve very expensive exploration programs. Costly leases must be negotiated with foreign governments, and if the area of interest is offshore, huge drilling platforms which may cost half a billion dollars or more must be built. Oil exploration is being conducted offshore Greenland and in the frequently violently stormy North Sea. These are expensive areas in which to operate. Oil exploration and development in the areas east of the Andes Mountains in Peru, Ecuador, and Colombia means building roads and hauling equipment through difficult terrain. Ultimately pipelines must be built over the mountains. Oil companies must be big to do these things and deliver gasoline to consumers. Individuals, or small companies with small amounts of money cannot do it.

Myth: Oil companies own oil

Reality:

In a number of countries, including Saudi Arabia, Venezuela, Kuwait, Iran, Iraq, Peru, and Mexico, oil was originally discovered and developed by foreign companies with the expertise which the country itself did not have. Subsequently, with the rising tide of nationalism following the colonial period, oil company properties — oil fields, pipelines, shipping facilities —were taken over by the respective governments, at times with little or no compensation.

Most of the oil in foreign countries is owned by the governments, not the oil companies. Oil companies simply hold leases (abroad commonly called concessions) to develop the oil deposits. The companies are allowed to search for and produce what commercial oil may be found. Sometimes the oil companies can sell it themselves and sometimes they have to market it through state-owned companies. In a sense they own the oil they produce, but they never really own the oil in the ground. They only lease the right to produce it. This is an important point, because it means that U.S. companies or any other companies operating in a foreign country do not own an assured safe resource base.

In the United States, the mineral rights which include oil and gas usually belong to the owner of the land. The owner can sell these rights to a resource development company, so, in effect there can be more than one owner of a piece of land. The surface can be owned by one individual and the subsurface can be owned by someone else. Oil companies can buy the mineral rights to oil and therefore own the oil. However, even in the United States, more often than not, the oil companies have to lease the mineral rights. Offshore oil belongs either to the adjacent state, or beyond the state limits, to the federal government. Oil companies, for the most part, do not own much oil. Many own no oil. On the oil they do produce, they pay a royalty to the private owner, or royalties and taxes to the government. These costs range from 12½ percent to as much as 90 percent of the value of the oil.

In other countries, the government generally owns all the mineral resources which may be leased out to developers. But governments can change their minds about lease terms or cancel them with or without any compensation. Quite a few have done so — another severe hazard of the mineral resource business.

The existence of OPEC is obvious proof that oil companies do not own or control most of the world's oil.

Myth: Oil companies make big profits compared with other enterprises

The profits of oil companies are frequent targets of criticism by both the politicians and the media. Many people believe that mineral resource companies are excessively profitable relative to other enterprises.

Reality:

As pointed out in Chapter 26, *Mineral Economics*, the amount of capital which has to be invested in the production of oil is very large and it takes a long time, in some cases, many years, before any return can be realized on the investment, if indeed there is a return at all. Many smaller oil companies go bankrupt from a series of dry holes. One such example was a firm which drilled in the geologically rather unpredictable deltaic sedimentary complex in the Denver-Julesburg Basin of Colorado. The first well was a small producer. Subsequently four wells were drilled around the first well. All four were dry holes. The small amount of oil coming from the first well was insufficient to repay the bank loan which had been used to finance the drilling of the other four wells. The company went out of business.

Oil exploration and production is a high risk venture. Companies that do survive, earn a relatively modest return on investment. On records kept since 1968, the average return on stockholder investment in 30 representative U.S. oil companies has been 12.5 percent. In 1994, it was only 9.2 percent.(7) For 30 representative manufacturing companies, the return has been 13.1 percent.(1) The average return for oil companies is less than the average return for manufacturing industry in general.

Relating this to the gallon of gasoline which we buy, an editorial review of this matter stated:

> "No one needs to be reminded that gasoline prices have risen since the OPEC cartel began flexing its muscles. But oil industry analyses show that oil companies aren't exaggerating when they say they make a profit of only about two cents on every gallon of gasoline sold. In fact, only Exxon reports making that much. Standard of California, Phillips Petroleum and Texaco report making no more than $1\frac{1}{2}$ cents a gallon. The big winners in the gasoline sweepstakes are the federal and state governments, which collect six times as much in taxes per gallon as the companies earn in profits and some of the most spectacular increases in gas pump prices are attributable to state tax boosts."(2)

Although this editorial was written in 1975, the economics of the oil industry remain about the same today. In spite of intervening inflation, two cents a gallon is regarded by the oil companies as a very good profit on a gallon of gasoline. Adjusted for inflation since 1975, the profit is barely one cent a gallon.

At the upper end of the list of profitable segments of the economy are the so-called "sin-stocks", the tobacco and liquor companies. It is ironic that companies which produce products that are harmful to the health and welfare of the country are much more profitable than is the oil industry which produces a basic necessity and makes life for much of the world much more pleasant than it would be without this important energy source.

If anyone still believes that the oil business is very highly profitable, it should be noted that in the developed nations it is a free economy and anyone is welcome to form an oil company and get into the business, or simply buy stock in oil companies. Almost all major companies are publicly held, with their securities listed on both national and international stock exchanges.

Mining Companies

What has been said about oil companies in terms of huge capital costs, the risks of failed exploration efforts, and the long time from a discovery to when income is realized also applies to mining companies. Their economic returns are no better on the average than for oil companies, and in many cases are less. Mining company securities also may be bought on the stock exchanges of the world if one wishes to participate in this industry. Many other businesses show a better consistent and higher return.

Alternative Energy Sources

Alternative energy resources are those which could presumably replace the largest single conventional energy source which is oil. Because of occasional oil crises and the increasing dependence of the United States and almost all other industrialized nations, as well as most Third World countries on foreign oil supplies, the urgency for developing and using alternative resources is growing.

Well-meaning but uninformed people make a great variety of statements as to what alternative sources might do for the country. Unfortunately poorly founded statements are frequently picked up by the media who repeat them without any research as to what the facts might be. This in turn misleads the public.

There are three considerations when evaluating the worth and validity of alternative energy sources. One is the ability of alternative sources to really replace oil in the quantities we are now using oil. A second concern is how using alternative energy sources might affect and change current lifestyles. What would it really involve to change to a "solar energy economy" as is the popular concept among many alternative energy enthusiasts. The third consideration is the environmental impact of converting to alternative energy sources. These three factors with their myths and realities are briefly treated here.

Myth: Alternative energy sources can readily replace oil

This is the assumption made by many people who advocate alternative energy sources as an early easy solution to our dependence on imported oil, and the perceived negative environmental effects of burning oil.

Reality:

The facts relative to this myth are mixed. Alternative energy sources can replace oil in its energy uses, but in some uses much less conveniently than in others. Fuel oil used under steam boilers can be replaced by nuclear fuel, or coal. But replacing gasoline, kerosene, and diesel fuel for use in vehicles, airplanes in particular, by an alternative energy source will be much more difficult. At the present time, 97 percent of the world's approximately 600 million vehicles are powered by some form of oil. Going to another fuel source to meet this huge energy demand now met by the convenient, easily transported, very high grade energy source which is oil will not be easy.

The British scientist, Sir Crispin Tickell, has stated a sobering fact, "...we have done remarkably little to reduce our dependence on a fuel which is a limited resource, and for which there is no comprehensive substitute in prospect."(28) It is important to note that there is no apparent replacement for oil in the volumes and ways in which we now use it. The transition to a comparable energy source or sources will be difficult, and probably much less convenient than using oil. Even if it could be done it would markedly change the lifestyle of industrialized society as we know it today. This leads to the next and related myth.

Myth: Alternative energy sources can simply be plugged into our present economic system and lifestyle, and things will go on as usual

This also is a common assumption with regard to a transition to alternative energy sources, even to the major renewable energy source, solar. People do not appreciate the close relationship between the current energy sources, principally oil, and the control which energy forms have over the activities of their daily lives, and where and in what sorts of structures they live and work, and use for transportation.

Reality:

Conversion to a solar energy economy would involve vast construction projects installing huge collecting systems. Houses and factories would have to be redesigned to much more energy efficient standards. In transport, an electric economy means electric cars, and the facilities to generate huge amounts of power beyond what is presently being used. And the electric car, as far as can be visualized with reasonably foreseeable technology, would not offer the degree of mobility which gasoline powered vehicles do. This would markedly alter both the work and recreational habits of people. It would markedly affect recreational related economies.

Other energy sources, beyond oil, similarly would involve a restructuring of daily routines. Our activities are very much controlled by the energy forms which we use. Our standard of living is largely a function of how much and in what form we can command energy supplies. Changing from the energy form which is oil to other energy sources can and will have to be done, but lifestyles will be altered, as may also be the standard of living.

Myth: Alternative energy sources are environmentally benign

Advocates of alternative energy sources, commonly believe that these energy supplies have very little impact on the environment. Sunlight as a source of energy would seem to be an ideal energy source with virtually no negative environmental consequences. Or, converting a relatively more polluting source of energy such as coal into a less polluting liquid fuel appears to be a good exchange.

Reality:

Converting coal to some liquid fuel form which could be used in transportation is possible but to do so to the extent of replacing oil would involve the greatest mining endeavor the world has ever seen. It would require strip mining vast quantities of western land each year. If alternative energy considerations do not include coal, but rather are thought of in terms of solar energy, biomass, nuclear power, wind, hydropower, tidal, ocean thermal energy conversion (OTEC) or shale oil, they also have environmental impacts.

These have been discussed in more detail in Chapter 22, *Mineral Development and the Environment*, but some of the environmental problems are briefly summarized here. Solar energy collectors in numbers sufficient to be significant in our energy supplies would use very large amounts of land. Mining the materials used to make these collectors would have an impact. Because the collectors would not have an infinite life, there would be the continual problem of replacement, involving more mining operations.

The environmental impact of using biomass as a major source of energy would be huge, especially in terms of the degradation of the highly important mineral resource, soil. Nuclear energy from fission has the potential (and the reality, in the case of Chernobyl) of having a huge impact on the environment. Fusion nuclear power is relatively more safe but not entirely so. Wind power devices are unsightly, noisy, kill birds, and, like solar collectors, deteriorate and have to be replaced with more materials mined from the Earth. Tidal power, hydroelectric power, and OTEC have undesirable effects on aquatic environments. If oil shale is part of the energy alternative for the United States, the impact of developing that energy source on already scarce southwestern water resources would be large, and probably not sustainable.

In brief, as the saying goes, "there is no free lunch" in the use of any alternative energy source with respect to the environment. All make an impact. Eventually some or all of these sources will be used. The decisions to be made involve which sources have the least environmental effects and yet can meet the projected energy demands. With an ever-increasing world population requiring more and more energy, any energy source or combination of sources which will adequately meet this demand will inevitably have a large environmental impact by the sheer size of the operations.

Myth: Biomass — plants — can be a major source of liquid fuels

This myth comes up frequently, and it has been rather thoroughly explored through various projects and proven to be a myth. A variety of plants including greasewood in the arid Southwest U.S., sugar cane, sugar beets, trees in general, seaweed, and seeds have been cited as important possible sources of liquid fuel for the future. In 1979, an article in a widely read U.S. magazine states: "Myriad forms of natural organic matter can provide heat or be converted into gas, oil, or alcohol. Wood holds the most immediate promise."(9)

Reality:

In regard to wood as an alternative liquid fuel, a final report on a U.S. government-sponsored project on the conversion of wood to a liquid fuel stated as a conclusion: "Investigations to date have led the authors to be optimistic about the possibilities of oil from biomass. While difficulties in bringing the current facilities on-stream have somewhat limited information to date, it is felt that a vigorous activity in the future can eventually provide a new source of energy for the country in the form of oil from biomass."(6) A translation of this statement might be that "the project didn't turn out very well, but maybe in the future a lot of research could improve results." That may or may not be true. The project involved wood-to-oil conversion, and one conclusion was that "Information gained here should provide the means to be commercially competitive by approximately 1990."(6) The project was abandoned in 1981. No wood anywhere in the world is now being converted to liquid fuel.

There are several reasons why converting growing plants to oil will not be a significant substitute for oil obtained from wells. These have been touched upon in other chapters. Briefly they are:

- The energy conversion efficiencies are low, in some cases as with ethanol from corn, it is negative.
- The energy cost of harvesting and transporting the materials is high relative to the energy produced. In the case of wood, cutting the trees and loading and hauling them to a processing plant would be energy intensive even before processing into a liquid.
- The volumes of plant material available are not sufficient to yield large amounts of oil, given the low energy conversion efficiencies.
- The degradation of the land growing these materials by continuing harvesting without returning the fiber to the land is severe.
- If wood is considered, there is already a scarcity of wood in most of the world. In the form of wood waste (little is wasted now) there is insufficient raw material from this source to provide significant amounts of feedstock to convert to liquid fuel.
- The best land is now under cultivation for much needed human food supplies. If plants were used for raw material for liquid fuel conversion they would either have to displace food crops from present agriculturally developed land, or put marginal lands (thin soil, steep hillsides) into production which would greatly increase land degradation by erosion, and also have serious downstream effects, including silting up of reservoirs.

In final view, the Energy Research Advisory Board of the U.S. Department of Energy stated in 1981 (U.S. population then was 258 million compared with 267 now), that the 258 million Americans used 40 percent more fossil energy than the total amount of solar energy captured each year by all U.S. plant mass. Current annually available biomass volume is no significant replacement for the large storehouse of organic energy accumulated over millions of years in the form of coal and petroleum.

In summary, biomass, at least considering the size of world population today which has to be supported by crops, cannot be diverted from food supplies in significant quantities to be important as a liquid fuel, and at best energy conversion efficiencies from biomass to oil are low. The environmental impact of using biomass for conversion to liquid fuel on a large scale would be severe and unacceptable. Biomass is not a potential source of significant quantities of liquid fuel.

Myth: There are billions of barrels of oil which can be readily recovered from oil shale in the U.S.

As the United States has the world's largest and richest deposits of oil shale, the optimistic statements which sometimes arise from that fact are among the more commonly heard in regard to the U.S. energy future. An enthusiastic article about oil shale in the prestigious *Fortune* magazine is titled: "Shale Oil is Braced for Big Role." It concludes, "Shale oil is not the whole answer to the energy problem but it's one of the few pieces that is already within the nation's grasp."(19) The article was written in 1979. As of 1997 no oil from oil shale is being produced in the U.S. . . . or anywhere else.

Reality:

The supposedly great prospects for the production of oil from oil shale in the United States has been one of the most widely promoted and heard energy myths for many years. Statements even made by government agencies can be quite misleading. These arise perhaps because it is good government policy to take as optimistic view as possible toward any national problem. The statements also are due to a less than careful examination of the facts, and perhaps a bit of promotion for the agency involved. The statement is made by a U.S. government organization that "...using demonstrated methods of extraction, recovery of about 80 billion barrels of oil from accessible high-grade deposits of the Green River Formation is possible at costs competitive with petroleum of comparable quality."(12) This is a clear misstatement of the facts. At the time it was written (1981) there had been no demonstrated methods of oil recovery at costs competitive with oil of comparable quality, nor have there been any such methods demonstrated to this date. A variety of processes have been tried. All have failed. Unocal, Exxon, Occidental Petroleum, and other companies and the U.S. Bureau of Mines have made substantial efforts but with no commercial results.

A state government agency issued a pamphlet on oil shale stating, "The deposits are estimated to contain 562 billion barrels of recoverable oil. This is more than 64 percent of the world's total proven crude oil reserves."(29) The implication here is that the oil which could be "recoverable" could be produced at a net energy profit as if it were barrels of oil from a conventional well. The average citizen seeing this statement in a government publication is led to believe that the United States really has no oil supply problem when oil shales hold "recoverable oil" equal to "more than 64 percent of the world's total proven crude oil reserves." Presumably the United States could tap into this great oil reserve at any time. This is not true at all. All attempts to get this "oil" out of shale have failed economically. Furthermore, the "oil" (and, it is not oil as is crude oil, but this is not stated) may be recoverable but the net energy recovered may not equal the energy used to recover it. If oil is "recovered" but at a net energy loss, the operation is a failure. Also, the environmental impacts of developing shale oil, especially related to the available water supply (the headwaters of the already over used Colorado river), and the disposal of wastes, do not seem manageable, at least at the present time, and perhaps not all.

The clear implication of both of these government statements is that oil shale is a huge readily available source. Because of the enormous amount of "oil" which has been claimed that could be recovered, this gives a large sense of energy security which does not exist. For this reason it is a particularly dangerous myth.

Myth: Canada's oilsands with 1.7 trillion barrels of oil will be a major world oil supply

It appears to be true that in the Athabasca oilsands and nearby related heavy oil and bitumen deposits of northern Alberta there is more oil than in all of the Persian Gulf deposits put together.

Reality:

The impressive figure of 1.7 trillion barrels of oil is deceiving. It is likely that only a relatively small amount of that total can be economically recovered. The oil is true crude oil, but it cannot be recovered by conventional well drilling. Almost all of it is now recovered by strip mining. The overburden is removed and the oilsand is dug up and hauled to a processing plant. There the oil is removed by a water floatation process. The waste sand has to be disposed of.

Much of the oilsand is too deep to be reached by strip mining. Other methods are being tried to recover this deeper oil, but the economics are marginal. With the strip mining and refining process now in use, it takes the energy equivalent of two barrels of oil to produce one barrel. To expand the strip mining operation to the extent which could, for example, produce the 19 million barrels of oil used each day in the United States would involve the world's biggest mining operation, on a scale which is simply not possible in the foreseeable future, if ever. Canada will probably gradually increase the oil production from these deposits, but until the conventional oil of the world is largely depleted these Canadian deposits are likely to represent only a very small fraction of world production. The production will always be insignificant relative to potential demand. Oilsands are now and will be important to Canada as a long-term source of energy and income. But they will not be a source of oil as are the world's oil wells today.

Other Myths

Myth: Energy from any source is readily used

Energy can be defined as the "capacity for doing work." (Webster's Collegiate Dictionary, Seventh Edition). Alternative energy sources are sometimes thought of as easily interchangeable. Energy is energy: there are no great problems in switching from one energy source to another. *This is a **myth**.*

Reality:

An important fact, commonly ignored in discussing alternative energy sources, is that energy sources come in very different forms. Adapting these various forms to various end uses presents many problems. Electricity and gasoline can each do work, but these energy sources present very different problems when it comes to using them in particular applications. This is generally ignored by people who suggest on bumper stickers, for example, that "Solar Is The answer," or "Go Solar." Sounds simple. It isn't.

The conversion of the intermittently available very low-grade solar energy into an energy form which could be used to power the automobile as we use the automobile today is a complex process, and has not yet been satisfactorily solved. In many cases it is not possible to conveniently or easily substitute one energy source for another. Each has its own characteristics which may be useful in some circumstances and a decided problem in another situation. Coal can be used to produce electricity quite easily in a conventional coal-fired electric power plant. But using coal directly to power an airplane, or using the electricity produced by coal to power an airplane does not now, at least, seem possible, and may never be. Atomic powered airplanes also seem unlikely.

Energy from a variety of sources is not universally interchangeable in its applications. The transition from one energy source to another will in many cases be difficult, and may cause major adjustments in lifestyles.

Myth: We can conserve our way out of the energy supply problem

The movement to conserve our way out of the energy crises and supply problems has been vigorously promoted from time to time when energy shortages have occurred. In between such times, energy conservation seems to fade a bit as a general concern. But the widespread concept remains that conservation can solve the energy problem.

Reality:

Energy and mineral conservation and recycling are useful goals, but conservation is only a temporary solution to the overall problem of continued growth of energy demand from an ever-increasing population. To accommodate more and more people, each person might use less and less resources, but at some point there is a minimum amount of the resource which has to be used. Reducing the amount beyond that point is not feasible. If one uses a vehicle for business, by a careful planning of the necessary travel route, one can reduce the need for fuel, but one cannot continue indefinitely to reduce the amount of fuel needed. Eventually there is simply not enough fuel to do the job. At some point the real problem must be addressed — the demand for the resource — and this demand comes from numbers of people, and lifestyle. There is no way to ultimately conserve out of the energy supply problem against an ever-increasing population. Demand can be reduced, but, if at the same time an increase in population absorbs those savings, there is no gain. Demands cannot be reduced to zero. Conservation and recycling can only buy time in which to stabilize population to a size which can exist on a renewable resource economy, which also has to be devised.

Myth: The political campaign promise — "we will achieve energy independence"

During the 1970s and early 1980s, because of the recent oil crises, a popular political campaign promise was that a presidential candidate and his party would achieve "energy independence" for the United States. Presumably this would be accomplished in four years or no more than eight as there is a two-term limit on the U.S. presidency. Citizens look for cures to their problems, and the candidate who can most convincingly promise them may be the winner.

Without making specific reference as to which politicians (some of them were elected) made such promises, it may be noted that, win or lose, soon after the campaigns have been over, the goal of energy independence seems to have been lost in the shuffle of everyday politics as usual.

Reality:

It may be hoped that U.S. energy independence can eventually be achieved, but it will never be based on oil produced in the United States. Unless oil consumption is greatly reduced, the United States henceforth will be increasingly dependent on foreign supplies.

As part of the "energy independence" program came the headline statement from one presidential candidate, "We Will Find New Fuels". That promise was made in 1976. The candidate lost, and we have not made much progress on new fuels, now importing twice as much oil as then.

A promise made by a sitting U.S. president in the 1970s was an edict stating, "I am inaugurating a program to marshal both government and private research with the goal of producing an unconventionally powered, virtually pollution-free automobile within 5 years." As the electric car was known then (and indeed electric cars existed before gasoline-powered cars) presumably the "unconventionally powered" car would have to be something else.

That promise is now more than two decades old and the promised new era automobile has not arrived. These statements are made primarily to gain public favor — and votes. But in the process the public is led down unrealistic paths. Politicians making such statements owe it to the people who give them public trust, to more carefully examine the facts, and not simply express cheerful hopes. Political posturing and optimism will not solve the energy supply problem. However, political decisions can encourage development of alternative energy supplies, and subsidize research toward that end. This should be done.

Energy independence for the United States is at present becoming less and less a near term possibility. The economy continues to be based very largely on petroleum, and oil imports continue to increase each year. Any political candidate who states that energy independence can be achieved for the United States in any presidential term of office (or even in two or three decades) is simply either not being honest or is totally ignorant of energy supply, and the prospects for viable alternatives.

A national move toward energy independence, which has to be expressed by the citizens through their elected representatives in the Congress, has not materialized. Energy independence for the U.S. will remain a myth if the present energy course is continued. It need not be a myth but the will to make the effort, and the reorganization of society which it would take to make energy independence a reality are nowhere in sight. Also, even if there were a consensus now, it would take many years to do the things necessary to achieve energy independence, and the capital expenditures necessary to do this would be huge. Any promise of energy independence for the U.S., at least within the next several decades, remains clearly a myth, hopeful vote-luring political statements notwithstanding.

Myth: "At current rate of consumption . . ."

This is commonly used as a comforting statement to assure the public that there is no looming shortage of a given resource. "At the current rate of consumption" a given resource will last for at least X number of years. Usually, this is quite a long time. There is no problem.

Reality:

This very misleading myth is that the "current rate of consumption" does not represent the future. The rate of consumption of almost all resources, particularly energy, is increasing every year. The increase in resource consumption is caused by three factors: population growth, a demand for an increase in per capita consumption of a resource to increase living standards, and a larger number of uses found for a given resource. Oil is the classic example which illustrates increased demand from all three causes. Present demand for oil is increasing at the rate of about two percent annually, which means demand will double in 35 years. "Current rate of consumption" has no realistic relationship to the future.

Demand does not grow arithmetically, but increases exponentially. That is, it goes up as a percentage each year over the previous year. Therefore, the statement that a depletable resource will last for X number of years "at current rate of consumption" has little relation to the reality of the actual life of the resource. A resource may have a life of 100 years at the "current rate of consumption." But, at the seemingly low rate of a five percent annual increase in demand, the resource will only last about 36 years. Because almost all resources are finite, and the population has no theoretical limit to growth, ultimately the population by its exponential growth of demand will overwhelm the available resource.

That we are living in a time of exponential growth is ably presented by Lapp in his classic book *The Logarithmic Century*.(18) That the general public does not appreciate the importance of the effect of exponential growth has been pointed out by Bartlett who has written a convincing discussion of the myth of "at current rate of consumption," and the large numbers which quickly result from a seemingly insignificant annual rate of increase in use of a resource.(3) In other writings and in numerous lectures, Bartlett has pointed out, by several striking examples, that this is one of the most dangerously misleading myths to which the public is continually exposed. He states, "The greatest shortcoming of the human race is our inability to understand the exponential function."

A recent example of such a misleading statement regarding oil supplies is that made by a ranking oil industry analyst on a popular Friday night Public Broadcasting System program.(30) The statement, regarding world oil reserves, was that current supplies are "...enough to last us for 40 years at current consumption rates." This statement is grossly misleading for two reasons: First, "current consumption rates" are transitory, and demand for oil will continue to increase, as population increases. "Current consumption rates" have little relevance to the future. Second, if the statement was to be taken literally it would mean that for 40 years we would have the same amount of oil available as we have today, but in the 41st year there would be none. This also has no relation to reality.

The production of a finite resource is never a flat line. In broad form, smoothing out irregularities caused by political, economic, and technological events, the production is a bell-shaped curve. (Figure 8) It is estimated now that world oil production will continue to increase until about the year 2010 (see Ivanhoe, Chapter 28 and Figure 9), and then begin a permanent decline. There is little, if any, possibility that the amount of oil available worldwide 40 years hence will be the same as today. It will be less, and the critical point is *when world oil production begins to decline*, not when the last drop of oil is ever pumped from the ground.

One might peripherally observe that the statement made that the world has 40 years' oil supply at current rate of consumption was made in the context of being reassuring. However, 40 years hence is within the life expectancy of many, if not most people living in today's highly oil-dependent industrialized societies. The figure of 40 years is both illogical and irrelevant, and it misleads the average citizen into thinking there is no problem for at least 40 years. The reality is that a permanent world oil crisis will occur when world oil production begins to decline early in the 21st century. Most of the present world's citizens will see that time.

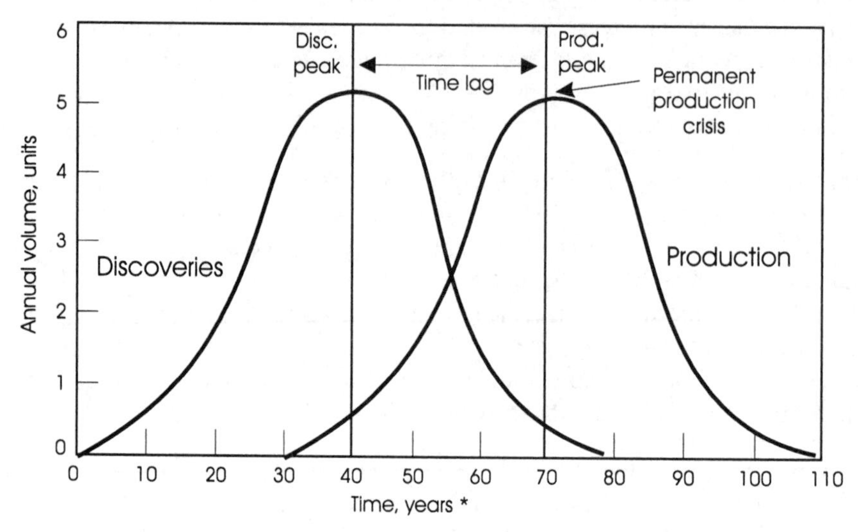

Figure 8. Curves of discoveries and production during a complete production cycle of a finite resource.

(After Hubbert, 1956)

Myth: Mining the moon

It may seem to younger persons who were not part of the time of great enthusiasm for space exploration that to suggest mining the moon is ridiculous. But older persons recall the heady days of early lunar exploration when this idea was proposed. Mining the moon was one of the seriously suggested reasons widely discussed and advocated for lunar exploration. The minerals would be brought back to Earth for processing, or mineral processing stations could be set up on the moon and the refined product brought to Earth.

Reality:

Small samples of moon rock have been brought back to Earth. Unfortunately, from the samples taken, the moon appears to be made up largely of a rock very similar to basalt here on Earth, of which there is a vast supply and which has no commercially useful mineral composition. The surface moon rocks do apparently have a slightly higher iron content than the average composition of the Earth, but going to the moon to mine iron does not seem to have attractive economics in either the near or foreseeable future. The energy cost of transportation would be astronomical.

Myth: Export the population problem to outer space

This also may seem like an idea too absurd to discuss. However, it is amazing what can be suggested even in high government circles. In those early space exploration times, some thought that the answer to the population problem was to export it from Earth. Hardin has identified the source of this myth stating: "In 1958, four years after the founding of NASA — the National Aeronautics and Space Administration — its congressional guardian, the Science and Astronautics Committee, supported the idea of space migration as an ultimate solution to the problem of a 'bursting population.'" Hardin adds, ". . . when an agency is fighting for space that counts — space at the public trough — its administrators are in no hurry to correct statements that increase the size of their budget."(14)

Regardless of their logic or otherwise, ideas of populating space persist. In 1996, an article in a national magazine proposed that most industrial plants on Earth be replaced by those built on the moon and that the population pressures on Earth be solved by colonizing Mars. Some quotes from that article follow:

> "The only way to keep the economy expanding infinitely is to expand our resource base infinitely. The universe is a big place. Human ingenuity is such that we will find innumerable ways to economically prosper in space"
>
> "We will have escaped the trap of a closed, cyclical economy; the riches of the solar system will lie before us."
>
> "The moon, with no ecosystem to damage, can become the seat of heavy industry. The earth, relieved of its population pressure and industrial burden as people migrate, can be allowed to regreen."(16)

Reality:

Just to keep even with population growth, much less reduce the people pressure on this already overcrowded planet, approximately 250,000 people a day would have to be rocketed off to "somewhere" in outer space! The only merit might be that it would generate a lot of employment in a very large aerospace industry to produce the spaceships needed daily. The amount of energy needed to propel these vehicles was never calculated or how it was to be continually obtained.

Mining the moon and sending people off into space to solve the population problem were myths at one time advocated by people who wanted to promote their special interests in the space program. That these suggestions would come from U.S. Government agencies is almost incredible.

Similar suggestions made more recently stem from a recognition that we face increasing environmental problems and demands on limited resources.(16) With this there can be little disagreement, but continuing to escort people to space to solve the problem is not reasonable, to put it mildly. The support systems necessary to keep people alive in space already seen in our current very small space program are very expensive in terms of resources. To provide such for the 250,000 people a day launched into space just to keep the Earth's population stable is almost beyond comprehension, and this would have to be done indefinitely. Humans are adjusted to the environment on Earth, and space is a vast and very hostile environment unfit for human habitation. Space does not offer a viable alternative to the environment on Earth. The dream of colonizing space will remain just that. Any credibility given it only serves to momentarily divert attention from the reality of the closed resource system which is the Earth and with which we must deal.

Cohen has stated what we may hope will be the final word on the concept of exporting excess population to outer space:

> "Let me dispense once and for all with extraterrestrial emigration. To achieve a reduction in the global population growth from say 1.6 percent to 1.5 percent

would currently require departure of 0.001 x 5.7 billion = 5.7 million astronauts in the first year and increasing numbers in each later year. To export this number of people would bankrupt the remaining Earthlings and would still leave a population that doubled every 46 years. Demographically speaking space is not the place."(10)

A final fundamental fact related to moon mining and space travel in general is the cost. At present the cost of moving the space shuttle, satellites, and other payloads into orbit is about $10,000 a pound. In 1996, Lockheed Martin Corporation was awarded a billion dollar contract by the U.S. government to develop what is called the X-33 next generation of space shuttle. One of these is expected to be operational before 2010, and could bring the cost down to $1,000 a pound or perhaps slightly less for payload transport to space. However, this too, seems excessive for an extensive use, and reinforces a view which has been expressed regarding vehicles designed to access space that it is the "most effective device known to man for destroying dollar bills."

Let us hear no more about the absurdity of space colonization. These examples of myths emphasize the continual need to use reality in examining statements made, even by government officials, with regard to our energy and mineral resources, and population problem. These are basic to our very existence, and it is most important that plans for the future, by both government and the private sector, be firmly based on realities.

Myth: The omnipotence of science and technology — it can do anything

There continues to be a belief in some circles that technology and science can indeed solve all problems of human material existence indefinitely, as noted by the following and what might be regarded as the ultimate myth.

In 1995, a large volume appeared wherein a number of scientists and others expressed some moderately positive and reasonable views of the present human condition and the future. However, the introduction contained the following statements:

> "Technology exists now to produce in virtually inexhaustible quantities just about all the products made by nature —foodstuffs, oil, even pearls, and diamonds — and make them cheaper in most cases than the cost of gathering them in the wild natural state."

> "We have in our hands now — actually, in our libraries —the technology to feed, clothe, and supply energy to an ever-growing population for the next seven billion years...Indeed, the last necessary additions to this body of technology — nuclear fission and space travel — occurred decades ago. Even if no new knowledge were ever invented after those advances, we would be able to go on increasing our population forever, while improving our standard of living and control over our environment."(25)

Reality:

If it were not for the fact that this volume was published under the auspices of a presumably creditable national research institute, these statements would not merit comment. A few brief observations are made here.

The terms "virtually inexhaustible" cannot reasonably be applied to anything on this Earth except perhaps the ocean water, and rocks. Also, to support the concept that "we would be able to go on increasing our population forever" or at least for a minimum of "seven billion years" one might assume that some sort of calculations were made to back up the statement. No calculations were presented. The author of the "7 billion years" published statement is later reported to have said it was a misprint and should have been "7 million years" of population growth. University of Colorado physicist Albert Bartlett made the calculations, however, stating in regard to the reduction in number from 7 billion to 7 million, "it is too early to breathe easily." Using the 7 million figure and based on a 1% annual population growth rate (current annual rate is 1.7%), he determined that the population after 7 million years would be $2.3 \times 10^{30409.7137}$ and added that "it is hard to imagine the meaning [of such a large number]...The number is something like 30 kilo-orders of magnitude larger than the number of atoms estimated to be in the known universe!"(5)

The editor of the book who wrote the fanciful introduction is not a scientist nor technologist. It is an observable fact that people other than scientists and technologists are frequently more confident of what these disciplines can do for the future than are the scientists and technologists themselves — the people who are aware of the basic facts of the availability of resources and what might be done with them, or to replace them.

Faith that science and technology can solve all resource supply problems is evidenced by the widely expressed public view that "you scientists will think of something." It ignores the fact that something cannot be made from nothing, and in order to have a resource one must have some material thing with which to work. This fact, however, is met with the thought that substitutions can be made. This is true, within the reality that eventually substitutions also become exhausted. Also, there are definite limits as to what substitutions can be made. There is, for example, no substitute for water. The age of alchemy is not here nor is it ever likely to arrive. Alchemy is the medieval chemical "science" which strived to turn base metals into gold. In general it is thought of as the ability to transform some common material into something more valuable. If this were possible one could make some wonderful substitutions — oil from granite. This is an absurdity. Yet there are shades of this concept expressed. In discussing copper as a resource, Simon states that there is no problem, "because copper can be made from other metals..."(24) This statement has no basis of fact, and it is highly unlikely that such will ever be possible. No scientific research suggests that this could be done on any commercial scale. Minute amounts of copper might be produced from other materials in a so-called "atom-smasher" at a huge cost of energy. The nature of matter is such that transmutation of elements is not a practicality.

However, Simon goes on, "Even the total weight of the earth is not a theoretical limit to the amount of copper that might be available to earthlings in the future. Only the total weight of the universe...would be a theoretical limit."(24)

In discussing energy, Simon states, "With respect to energy, it is particularly obvious that the Earth does not bound the quantity available to us. Our sun (and perhaps other suns) is the basic source of energy in the long run..."(23) Should an energy policy be based on the idea that we can draw on "other suns?" This astounding statement that we might be able to draw on "other suns" is by a professor in a reputable state university, and was published in the venerable magazine *Science*.

Simon also expresses faith in the ability of science and technology to supply the world with natural resources in unlimited amounts and in his book he has titled a chapter, *Can the Supply of Natural Resources Really be Infinite? Yes!* He states,

> "...we shall be compelled to reject the simple depletion theory. The revised theory will suggest that natural resource are not finite in any meaningful economic sense, mind-boggling though this assertion may be. That is, there is no solid reason to believe that there will ever be a greater scarcity of these extractive resources in the long-run future than there is now. Rather, we can confidently expect copper and other minerals to get progressively less scarce."(24)

Bartlett has written a well-reasoned review of Simon's concept that there is no meaningful limit to resource availability.(4)

Science and technology do have limits imposed by the immutable laws of physics, chemistry, and mathematics. At the present time it seems clear that if current trends continue in growth of population, the demands of the human race will soon overwhelm the ability of science and technology to solve the problems of availability of resources, which are the basis for human existence.

Alan Overton of the American Mining Congress states: "the American people have forgotten one important fact: It takes stuff to make things." Pesticides, paint, medicines, and fertilizer cannot be made from solar energy.

In 1992, the U.S. National Academy of Sciences and the Royal Society of London together issued a statement warning that "if current predictions of population growth prove accurate and patterns of human activity on the planet remain unchanged, science and technology may not be able to prevent either irreversible degradation of the environment or continued poverty for much of the world." If present trends continue, ultimately scientists and technologists will not be able "to think of something."

Ryerson, commenting on the concept of a "technology fix" with respect to population growth, states:

> "Some of the more outlandish claims of the 'technology fix' advocates — for example, that we could ship our excess people to other planets — have almost been forgotten (imagine sending aloft 90 million people per year). Yet, while extraterrestrial migration is no longer taken seriously by most people, many of the unsubstantiated claims of new technologies that will 'save the day' are still seen by many as a reason not to worry about population growth."(22)

It is important to understand that a "technology fix" is not the answer to unrestrained population growth. And future plans should not be based on unrealistic expectations.

Myth: Because past predictions of resource and population problems have proved incorrect, all future such predictions will not come true, therefore there is no need to be concerned.

This view stems in part from past predictions of disasters which did not materialize as scheduled. Notable were those by Malthus in 1798. The argument presented by those who apparently see no need now to relate population to resources is that if Malthus' predictions of two centuries ago proved so wrong, why should similar predictions be taken seriously today.

Reality: Malthus — then and now

Malthus' predictions were wrong because he did not foresee the coming industrial and scientific revolution. The Industrial Revolution provided much improved housing with adequate space heating, greatly improved sanitary facilities, and machines and the energy to run them. It provided the basis for supporting a much expanded population. Huge resources not known to Malthus were discovered and developed.

But with this much improved scene today, why should there be concern for the future?

The problem is that science and technology will not be able to continue to discover and develop the amount of new resources necessary to support a population growing at an exponential rate. And resources which might be thought of as something which could be depended on indefinitely such as soil and groundwater are being degraded. Population demands on resources are beginning to outpace the ability of science and technology to provide them. This is due to the fact that resources are not limitless. The availability of material resources to sustain the quality of life cannot keep pace with a continued exponential growth of population. Advanced exploration and production technologies have allowed geologists and engineers in less than three hundred years to discover and develop the huge store of mineral and energy resources which accumulated slowly over billions of years. In a fraction of a second in terms of the length of human existence, Earth resources basic to civilization have been brought into production in volumes never before seen.

Soils, oil, high grade metal and coal deposits and now those of lower grade, groundwater, and other resources including dam sites, are being used up at an unparalleled rate. Since 1900, world population has increased nearly four times, but the world economy has expanded more than 20 times. Fossil fuel use has increased by a factor of 30 and industrial production has grown by a factor of 50, and four-fifths of these increases have occurred since 1950. Civilization exists now in a new reality which is far different from that of Malthus's time. Population grows but mineral and energy resources do not increase. By discovery and advanced recovery technology, the immediate supply can be made to increase, but in total, minerals and energy sources with the exception of sunlight, are depletable.

Since the beginning of the Industrial Revolution, the speed of human assault upon Earth's resources has greatly increased. More petroleum, coal, and metals have been used since 1950 than in all previous human history. In the United States the high grade, easily won, low cost deposits of iron ore (hematite), copper, and petroleum have been depleted. In some other regions of the world, high grade deposits still exist but are rapidly being developed and used. There are few major dam sites in the United States on which to build large reservoirs for additional hydroelectric power, and irrigation projects. Elsewhere more do exist but are now being developed, as, for example, the huge Yangtze River project. Dam sites are non-renewable and when the reservoirs completely silt up as has already happened at some localities, that resource is gone. All over the world, groundwater tables are dropping, in many areas precipitously, as in China, India, Australia, the Middle East and in parts of western United States. In Malthus' time none of these things had occurred.

For hundreds of thousands of years the human population had made only a minor impact on mineral and energy resources. With low living standards, and little or no medical services, the population grew very slowly, and sometimes was even briefly reversed by famines and plagues. But these hazards have been largely eliminated and population has soared. It took from the beginning of human existence to approximately the year 1850 to reach the first billion in world population mark. It will take less than 10 years to increase the present five and three quarter billion by another billion.

What is different from the time of Malthus? The population in his time was small and the potential resources were large and undeveloped. Subsequently, the Industrial Revolution was rather rapidly able to produce enormous resource and material wealth in contrast to the past. It was the hare of energy and mineral development leaping ahead of the tortoise of population. In part, the population growth was tortoise in speed because of the lack of modern medicine, including vaccines and the knowledge of what caused plagues which would decimate populations. And, to a large extent, that hare of mineral and energy has kept ahead of population. This has been achieved by expanding the search, discovery and development of vital raw materials to a worldwide endeavor. That was not possible during Malthus' life.

But now with the present worldwide transportation network made possible chiefly by oil not available to Malthus, mineral and energy supplies can be searched for and produced on one area and transported great distances to another region. When one area experiences declining production, discoveries are made in other regions. Britain's metal deposits and coal resources were small but they supplied the basis for the start of the Industrial Revolution. But eventually the supply base moved to the rich undeveloped North American continent, and then oil was discovered. But now these North American metal and oil deposits have been largely developed and some are in decline. The oil development has gone more and more to the Middle East. Metal exploitation has moved to South America, New Guinea, Australia, and Africa. Worldwide, petroleum and metals are still in abundance. This tends to give a false expectation of a continual cornucopia of Earth resources, and an unjustified complacency especially in political circles toward the future.

However, we are running out of more world to explore and exploit. Only the ice-covered Antarctic continent remains untouched. In Malthus' time, the entire world's mineral and energy resources were virtually undeveloped, and the means to exploit them did not exist.

In Malthus' time, there was a small population and huge undeveloped world energy and mineral resources. The situation is now reversing. The difference is the present peaking or declining energy and mineral production in many parts of the world, and an already huge and continually expanding population. We live on a finite globe which now has been rather thoroughly explored. There are no more continents on which to continue to move as one region becomes depleted. The globe has been encircled. Malthus was simply ahead of his time.

Promotion of Myths

The media — newspapers, magazines, television, radio — report the news. But in the competitive haste to do so, sometimes they become accessory to spreading misinformation. The statements by uninformed people, politicians pursuing votes, unscrupulous promoters, or citizen groups trying to further a particular point of view may ignore realities. Too often these statements are picked up by the media and reported as fact.

Two such are cited. In a three-hour television special on energy (August 31, 1977), a CBS reporter stated in regard to how oil from shale might replace oil: "Most experts estimate that oil shale deposits like those near Rifle, Colorado, could provide more than 100-year oil supply." In another media report during the U.S. oil crisis of 1973, two young men in timber country announced that they planned to build plants using wood wastes which would be converted to gasoline and would "put the oil companies out of business." Subsequently in audiences at lectures I was giving, these two statements about alternative fuel supplies were brought up as genuine possibilities. Bartlett cites the CBS oil shale television program and convincingly points out that because the exponential factor of growth in use had been ignored, the resource could not possibly supply U.S. oil needs for 100 years.(3) Furthermore, at the time the CBS statement was made there was no evidence, just as there still is no evidence, that shale oil could replace conventional oil to any significant degree or that it could be produced at a net energy profit.

The statement was totally unrealistic, but with the 1973 oil crisis still fresh in mind, the program served to lull the public into a false sense of oil security. People like programs and statements which make them feel comfortable.

In the second situation, oil from wood wastes, very simple calculations would have shown that the volumes of wood waste available would not be even remotely sufficient to supply the raw material to provide any significant amount of gasoline in terms of U.S. consumption. The reporter on that story could have asked for some statistical data to back up the claims which he was about to print, and it would have made him a much better reporter for it.

It is perhaps too much to ask the media to thoroughly examine facts behind such statements. But there should be at least some minimal effort to do so because there is an unfortunate tendency for people not to critically read what is in the papers, or thoughtfully examine what television and radio brings them. Most do not have the background to make critical examinations. In the case of broad sweeping statements on things so vital as energy supplies, the media could at least quite quickly get a second opinion and present that also, which would give a useful balance to the reporting.

Degrees of Myths

It may be noted from the foregoing myths that there are degrees of such. Some may be regarded as marginal, and with some unforeseen technology (also to some degree a myth), the myth might become plausible. The myth of "562 billion barrels of recoverable oil" in oil shale might be regarded in this category, although at present it is definitely a myth. The myth that "We now have in our hands — actually in our libraries — the technology to feed, clothe, and supply energy to an ever-growing population for the next seven billion years"(25), and the myth that we could put our industrial facilities on the moon and that "human settlements on Mars could help to alleviate population and environmental problems"(16) plainly belong in the category of the absurd.

It is distressing to see that in many instances the general public cannot differentiate between what might be in the faint realm of possibility, from the absurd and utterly impossible. Sommers has commented on this stating in regard to our educational system that,

> "...many students now graduate from college knowing little or nothing about math or science, thus creating a void into which 'flow negative and bizarre views.'...A consensus emerged at a conference of over 200 scientists, physicians, and humanists: Scientists must speak up against the popular manifestations of irrationalism."(26) Sommers adds, "Harvard Prof. Holton has noted that parascience and pseudo-science 'became a time bomb waiting to explode' when

incorporated into political movements...A scandalously inadequate system of science education and diminished public regard for clear thinking and objective truth are just early casualties." (26)

If society is to survive, reason and clear recognition of reality must prevail, and plans made on that basis. Part of education should be directed toward that important end. The political leadership especially must be able to correctly differentiate between the possible and the absurd. This is particularly important when it comes to decisions relative to the foundations of civilization — the energy and mineral resources upon which everything else depends.

Conclusions

It has been said that "optimists have more fun in life, but pessimists may be right." Hardin has aptly noted, "If the reception of *The Limits to Growth* and *The Global 2000 Report* taught us nothing else it should have taught us that the Greeks were right. In the public relations game only optimism sells." Hardin quotes Teiresias in Euripides' *The Phoenician Woman,* "A man's a fool to use the prophet's trade. For if he happens to bring bitter news he's hated by the man for whom he works."(15) Hardin might have further noted that in political elections which are the quintessence of a public relations game, the same applies.

Regardless of the popularity of optimism over realism, the wisest route for humanity would be that plans and decisions be based on today's scientific and technological realities and reasonably visible resources, rather than on hopes for things which may never arrive. Optimism is vital in looking toward the future. One must be optimistic as a basis for making an effort. But optimism should be tempered with facts. The media and government leaders should try to learn the facts, and then have the courage to state them. Campaigns for public office should not lead the citizenry into false hopes. As civilization proceeds, it will be much more convenient and less disruptive to be pleasantly surprised along the way than unpleasantly surprised. Myths must be replaced by reality on which intelligent decisions are made.

"Facts do not cease to exist because they are ignored."

Aldous Huxley

BIBLIOGRAPHY

1 AMERICAN PETROLEUM INSTITUTE, 1995, Basic Petroleum Data Book: Washington, D. C., v. 15, n. 1, (no pagination, large volume).

2 ANONYMOUS, 1975, Fact and Fancy: The Wall Street Journal, December 12.

3 BARTLETT, A. A., 1978, Forgotten Fundamentals of the Energy Crisis: The American Journal of Physics, v. 46, p. 876-888.

4 BARTLETT, A. A., 1985, *Review of* The Ultimate Resource, by J. L. SIMON: American Journal of Physics, v. 53, n.3, p.282-285.

5 BARTLETT, A. A., 1996, The Exponential Function, XI, The New Flat Earth Society: The Physics Teacher, v. 34, September, p. 342-343.

6 BECHTEL CORPORATION, 1980, Biomass Liquefication at Albany, Oregon: Report to U.S. Department of Energy under government contract no. EG-77-C-03-1338, 18 p.

7 BECK, R. J., and BELL, LAURA, 1995, Rally in Fourth Quarter 1994 Fails to Bolster OGJ Group Profits: Oil and Gas Journal, June 12, p. 27-32.

8 BROWN, L. R., 1993, Postmodern Malthus: Are There Too Many of Us to Survive?: The Washington Post, July 18.

9 BYLINSKY, GENE, 1979, Biomass: The Self-replacing Energy Source: Fortune, September 24, p. 78-81.

10 COHEN, J. E., 1995, How Many People Can the Earth Support?: The Sciences, November/December, p. 18-23.

11 CROWE, C. T., 1981, Our Energy Fix — No Quick Fix: Quest, Spring issue, Washington State Univ., Pullman, Washington, p. 14-17.

12 DUNCAN, D. C., 1981, Oil Shale: A Potential Source of Energy: [pamphlet], U.S. Geological Survey, Washington, D. C., 15 p.

13 FOWLER, J. M., 1984, Energy and the Environment: McGraw-Hill Book Company, New York, 655 p.

14 HARDIN, GARRETT, 1959, Interstellar Migration and the Population Problem: Journal of Heredity, v. 50, p. 68-70.

15 HARDIN, GARRETT, 1993, Living Within Limits. Ecology, Economics, and Population Taboos: Oxford Univ. Press, New York, 339 p.

16 HOWERTON, B. A., 1996, Why Bother About Space? The Futurist, January/February, p. 23-26.

17 HUBBERT, M. K., 1956, Nuclear energy and fossil fuels: Drilling and Production Practices, American Petroleum Institute, p. 7-25.

18. LAPP, R. E., 1973, The Logarithmic Century: Prentice-Hall, Inc., Englewood Cliffs, New Jersey, 263 p.

19 NULTY, PETER, 1979, Shale Oil is Braced for Big Role: Fortune, September 24, p. 43-48.

20 OIL & GAS JOURNAL, 1995, OGJ Gasoline Prices: Oil & Gas Journal, July 31, p. 101.

21 ROYKO, MIKE, 1996, Gas-crisis Hysteria May Just Be a Case of Sniffing Fumes: Chicago Tribune, May 2.

22 RYERSON, W. N., 1995, Sixteen Myths About Population Growth: Focus, v. 5, n. 1, Carrying Capacity Network, Washington, D. C., p. 22-37.

23 SIMON, J. L., 1980, Resources, Population, Environment: An Oversupply of False Bad News: Science, v. 208, June 27, p. 1431-1437.

24 SIMON, J. L., 1981, The Ultimate Resource: Princeton Univ. Press, Princeton, New Jersey, 415 p.

25 SIMON, J. L., (ed.), 1995, The State of Humanity: Blackwell, Cambridge, Massachusetts, 676 p.

26 SOMMERS, C. H., 1995, The Flight From Science and Reason: The Wall Street Journal, July 10.

27 SULLIVAN, ALLANNA, 1995, Alaska Refuge Oil-reserve Estimates are Slashed: The Wall Street Journal, August 7.

28 TICKELL, SIR CRISPIN, 1994, The Future and Its Consequences: The British Association Lectures 1993, The Geological Society, London, p. 20-24.

29 UTAH [STATE OF] NATURAL RESOURCES DIVISION OF ENERGY, (no date given, approx. 1990), Oil shale: [pamphlet] Utah Department of Natural Resources, Salt Lake City, 14 p.

30. WALL STREET WEEK WITH LOUIS RUKEYSER, 1996, Public Broadcasting System, program of December 13.

CHAPTER 28

Earth Resources, the Future, and the "Sustainable" Society

Everything we have comes from the Earth. All tools, machines, and the most of the energy to run the machines that produce the necessities as well as the luxuries of modern society originally come from the rocks and soil beneath our feet, or from the oceans and beneath the ocean floor. Food comes from the Earth, and accordingly, we come from the Earth.

Living Off Our Capital

The human race inherited a storehouse of materials produced in the Earth by myriad geological events over great spans of time. It is an inheritance which nature took billions of years to accumulate. For hundreds of thousands of years the human race slowly evolved culturally and drew very little from this treasure storehouse. Generations came and went and almost no mineral or energy mineral resources were used. The Earth was left much as each generation found it. But the Industrial Revolution, combined with advances in education, technology, and medicine did two things. The rate of growth in population greatly accelerated, and the use of mineral and energy mineral resources rose exponentially. Both of these trends continue unabated.

More and more people are using more and more minerals including energy minerals at an increasing rate. Far more energy and mineral resources have been used in the world since 1900, than during all previous time. In the case of oil, the first 200 billion barrels of oil in the world were consumed between 1859 and 1968, but it only took the following 10 years to consume the second 200 billion barrels.

Among the smaller countries, it is very evident that some are living off their natural resource capital. The one-resource nations are the most visible in this regard, and the most visible of all at present is the single island nation of Nauru, that has essentially exhausted its phosphate deposits which made up most of the island. Now there is hardly space in which to live (Chapter 7, *The One-Resource Nations*).

But all the world is living off its inherited resource capital. In the case of the United States, it lived off its oil capital without having to borrow any until 1970. Now it is in a continual borrowing situation. All other industrial nations are in a similar situation. The "sustainable economy" is at present nowhere in sight.

Energy the key resource

The phenomenal rise in the production and consumption of energy mineral and mineral resources is unmatched in all history. Where it will lead, no one can really know. It is clear, however, that among the various resources we are considering beyond soil and water, energy is the most important. Without energy, other mineral resources cannot easily be obtained, nor effectively used once they have been obtained. It has been truly said that energy is the key which unlocks the storehouse of all other mineral wealth.(4)

To find a piece of native copper and by means of a stone beat it into a crude knife, as was done by many early humans, is useful only to a small degree. To mine and process millions of tons of copper ore into thousands of miles of electric wire, and many kinds of electrical equipment, and then energize these wires and equipment with power produced from copper-wound electric generators makes copper an infinitely more useful material than when it is simply hammered into a knife. Energy sources are far more important to world economies than just their simple figure of percentage of gross national product. Without energy the rest of the economy would hardly function at all.

What do we know of energy resources for the future? The petroleum interval will be brief. The United States has become a net importer of petroleum and will continue to be for the indefinite future. Russian oil production has apparently peaked. More and more oil production will be concentrated in the Middle East, which is a good reason for the aggressive pursuit by major industrial nations of alternative energy supplies. This will not be a simple task. In general, the economics of alternative energy sources are not nearly so attractive as those of drilling for oil have historically been. However, drilling for oil and mining for coal are showing a poorer ratio of energy recovered to energy used in the recovery than in the past. It is taking more energy to obtain the same amount of energy.

Currently in the United States, about 25 percent of energy produced is used to produce other energy. Energy has to be used to drill for oil, mine coal, mine uranium, cut wood, make solar energy conversion devices, and so on. As the easily recovered surface and near surface energy mineral resources (petroleum, coal, uranium) are largely exhausted, the cost in energy to produce more energy is estimated to rise to about 33 percent in the next ten years. It will probably continue to rise after that. This is a trend in energy costs which can be fatal, for if it ultimately takes as much in energy as the energy produced by the effort, there is no energy profit to use for other energy consuming sectors of the economy. At that point the game is up.

Varied energy sources: petroleum dominates

The world now has a varied energy mix. Initially, the energy mix was hardly a mix at all. It was chiefly wood, and other biomass such as grass, twigs, leaves, and dung. But gradually, coal, oil, gas, hydroelectric power, wind, solar, geothermal, and nuclear power have been drawn upon, although not in the same ratios in various countries. For example, in the industrialized nations, petroleum (oil and gas) is now the major energy source. In some countries wood continues to be the most important fuel. France gets 70 percent of its electricity from nuclear plants, whereas New Zealand uses no atomic power but instead relies chiefly on hydro sources (dams) and geothermal energy to generate electricity.

The current United States energy mix is: oil 44 percent, coal 22 percent, natural gas 21 percent, nuclear power 5 percent, hydroelectric power 4 percent, and "other" (wind, solar, geothermal, wood) 4 percent. In 1984, nuclear power surpassed hydroelectric power in percentage in the United States. Hydroelectric power is likely to continue to shrink as a percentage of the energy supply as U.S. energy demand grows, because there are almost no major hydroelectric sites still available to be developed. There are also other limitations of hydroelectric power becoming evident. The impact of dams of the largest single hydroelectric system in the United States, Bonneville Power System on the Columbia River, on the once great salmon runs has been so severe as to suggest that power production may actually have to be reduced by spilling water rather than running it through the turbines to facilitate the salmon migration.

Fifty years from now, and probably sooner, the energy mix of the United States will be substantially different from what it is today. This will be true of all countries. The reason is the foreseeable time of the peak of world oil production, as oil is at present the world's largest single source of energy, and it supplies 97 percent of all energy for transportation uses. But oil is finite, and in fifty years supplies will be greatly reduced from what they are today.

Time of world oil production peak projected — a signal event

In 1956, M. King Hubbert predicted that the United States would peak in oil production approximately in 1970. It did just that. This was a crucial event for the United States, as its most important energy source gradually then began to pass into foreign hands.

However, of even greater importance is the recent study by Ivanhoe (1996) who states "Starting with Hubbert-type curves for discoveries and production history, estimates of the decline side show world [oil] supply may peak before 2010."(44) The study is based on the methodology used by Hubbert which proved to be uncannily accurate. It might be noted that when Hubbert made his prediction it, very unfortunately, was widely ignored or discounted. The same should not happen to Ivanhoe's projection.

The importance of Ivanhoe's updating of the Hubbert curves to assert that the peak of world oil production will occur probably within the first decade of the 21st Century cannot be overemphasized. It is, in fact, more important than Hubbert's determination for the United States in that Ivanhoe's prediction involves the entire world. For this reason his study is cited here, rather than in Chapter 13, *The Petroleum Interval,* as we are here considering the "sustainable society." Oil is what currently sustains the major portion of the energy base of world society.

If it is asked at all, the common question posed by the public with regard to oil supplies is "When does oil production end?" But that is the wrong question. As Riva (see Chapter 13) has pointed out, some oil will be produced far into the 21st Century and perhaps beyond. And Ivanhoe notes "Obviously oil fields do not die all at once."(44) Some oil will be produced 100 years from now but not in significant amounts.

The critical question is the one which Ivanhoe presumes to answer — "When will the peak of world oil production occur?" The scenario which might be seen is that by the time of peak oil production, demand and available oil supplies will be equal. And world economies will even more than now be dependent on oil as a major energy source as additional nations will have come further into the industrial age, particularly India, Southeast Asia, and China. Oil will be produced as fast as it possibly can be, and then the irreversible decline begins. Who will want to cut back? The contest for the remaining oil reserves will be on, which could have serious military aspects. Those nations which can continue to have access to whatever oil exists, either by military or economic strength or a combination of both, will be able to maintain their oil-oriented economies longer than other weaker nations, but eventually the energy mix of all nations will have to change, as oil supplies decline. Figure 9 illustrates the projected curves of world oil discovery and production.

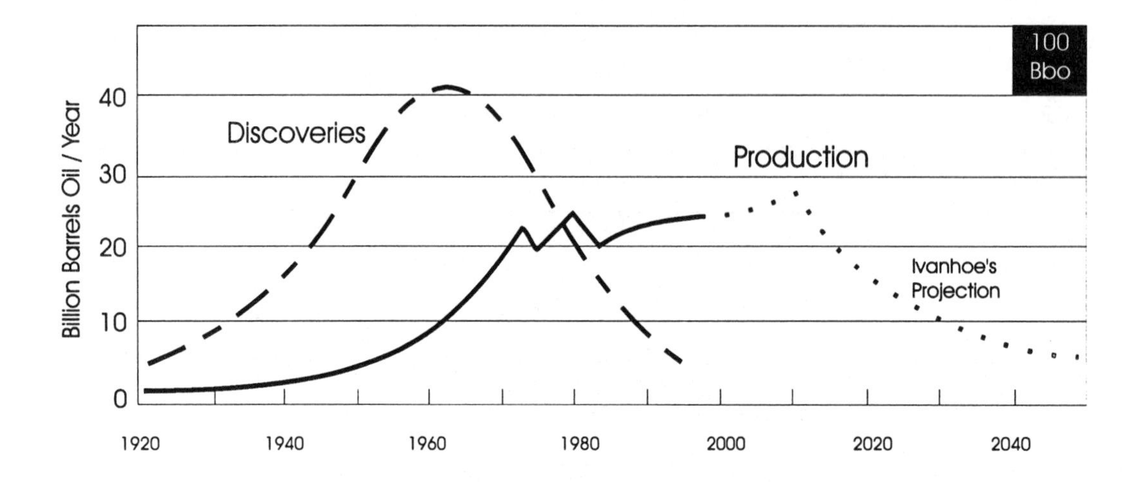

Figure 9. World oil supply: discoveries and production.
(Source: L. F. Ivanhoe, 1996, by permission)

At the present time world oil production meets the demand and could exceed demand by a small amount. OPEC now tries to maintain oil production quotas to support the price. But production quotas will finally be abandoned at the time world oil production peaks, and oil subsequently becomes permanently in short supply — probably within the next 15 years.

There is a fatal and fundamental difference between the situation when the United States peaked out in oil production in 1970, and when the peak of *world* oil production will be reached —approximately the year 2010. When the U.S. peaked out it made up the difference between domestic supply and demand with imports, chiefly from the Middle East, Venezuela, and Nigeria. When the world oil supplies peak out, there is nowhere else to go. The oil crisis will be worldwide, and permanent. The impact on the now greatly oil-dependent industrial societies, as well as on the equally oil-dependent agricultural activities, and on the lifestyles of all citizens will be immense.

When oil supplies are unable to meet demand, the need for alternative energy sources will become critical. However, present trends in both industrialization and technology are not encouraging toward an easy transition to a non-oil economy. Populations continue to increase, and more and more countries are becoming industrialized, the combination of which now pushes oil consumption ever higher, with no sign of slackening. All the world wants to get "wheels". At the same time, technological advances moving toward alternative energy sources, particularly those for transportation needs, are proceeding very slowly — too slowly at present to fill the gap which will begin to occur when world oil production peaks only a few years from now. From that time there will be a permanent global oil crisis. If the events of the 1973 and 1979 oil crises in the United States are any indication of things to come, the peak of world oil production, and then the start of an irreversible decline will cause worldwide economic and social disruptions. It is important to realize this will occur within the lifetimes of most people now living.

Ivanhoe has summarized his study:

> "It is concluded that the critical date, per USGS discovery records, when global oil demand will exceed world production, will fall sometime between 2000 and 2010, and may occur very suddenly due to unpredictable political events. This is within the lifetime of most people now alive; and this foreseeable energy crisis will affect everyone on earth. Governments will have first call on oil supplies during global shortages." (44)

It is unsettling to observe that the problems of both domestic and worldwide oil supplies receive little if any attention in ruling circles. In the 1996 U.S. presidential campaign no mention was made of the fact that the U.S. now imports more than half its oil, nor was there any discussion of developing a U.S. energy policy which might be able to offset this trend. Political agendas are notably short-sighted.

Nevertheless, the world energy mix, now dominated by oil, within two decades will begin a major change. The energy mix will continue to change until all nations have reached a stage where total energy supplies are from renewable sources.

This changing energy mix will present a number of challenges. Can one energy source effectively replace another both in terms of volume, and in convenience for end use? For example, can electric power really replace gasoline for cars on a large scale? Where will these substitute energy sources come from? If they are not produced domestically, how can a nation pay for the imported energy it needs? Can foreign sources be depended upon? And, for how long? What kinds of greater or lesser changes in lifestyles will occur as a result of changing to other energy sources? And what sort of restructuring of industries and societies must take place to adjust to different energy sources?

Total effect of loss of self-sufficiency

One of the more important aspects of energy and mineral supply is the total effect upon a nation as it loses its ability to produce its own energy and mineral resources. The American public has only recently begun to recognize that, as more and more resources have to be imported (oil being the best example), industries and jobs, and the jobs peripheral to those industries are all lost. This not only increases the balance of payments problem, but an erosion of the value of the currency results if exports cannot ultimately balance this drain on a nation's monetary position. A nation cannot print money indefinitely to pay for imported material things. Sooner or later the suppliers will resist weak money. And, with a weak currency goes a general decline of economic influence and world prestige. It is a slow process, and the citizenry may fail to notice what is happening for a time, but eventually the standard of living must suffer and will be felt by all. In this way also mineral resources affect the destiny of the nation. This began to be apparent to U.S. citizens in the 1990s, as the international balance of payments reached record deficits, and the value of the dollar declined.

The United States imports more and more oil and has the world's largest deficit in international balance of payments. It is clearly living beyond its means. At one time, not long ago, it did not. It was self-sufficient in oil, and had a positive balance of payments with the rest of the world. This changed dramatically in just two decades. But the U.S. is not willing to reduce its standard of living to these realities. The United States has been living on borrowed money, and perhaps borrowed time. At some time, this imbalance must be faced. The challenge will be to do it gradually, without undue social and economic turmoil.

Energy source and lifestyle

The past hundred years have seen great changes in lifestyle. The widespread use of oil, especially to power transport vehicles, has been responsible for a substantial part of this change. The daily massive migration of vehicles into and out of city centers such as San Francisco, Seattle, New York, Houston, Los Angeles, and Chicago is a phenomenon (in some places approaching a nightmare) which could never have been imagined the day Col. Drake struck oil. But we are already seeing some changes. The one and one-half ton automobile in the United States is gradually being replaced by a more modest car, many imported from Japan where energy economy lessons came early. Mass transit is making a reappearance. Jet airplanes, surely one of the wonders of the modern age, have grown greatly in size, which, in effect, is a facet of the move toward mass transport.

Buildings will be remodeled or entirely replaced with more energy-efficient features. Our buildings — offices, homes, factories — not the transportation system, are the largest consumers of energy in the United States, about 40 percent of the total. Savings from rebuilding the United States and much of the rest of the world to more energy efficient standards will be considerable. It will have to be done. In myriad ways energy costs and the major kinds of energy in future energy mixes will affect the basic living habits and economies of all nations. Except for water and soil, energy minerals are the most critical world mineral resource. Increasingly, technology, money, and political attention must be directed toward this vital concern.

One thing about energy is certain: Almost all alternative energy sources, as well as the principal energy supplies currently used — coal, oil, and natural gas —will cost more and in some cases much more, in the future than they do today. Energy will take an increasing percentage of governmental and personal budgets. Success of fusion technology may have the potential to change this outlook, but that is unlikely for decades to come, if it can be accomplished at all.

Energy gets increased attention

We have been emphasizing the importance of energy for those who use it. For the producers of energy, it is destined by geology of the rocks which generated the oil, and the rocks which now preserve the Earth's oil and gas resources that the Persian Gulf countries will increasingly be the focus of world attention. War may occur again, but inevitably various forms of political maneuvering and economic warfare will persist in that region for as long as substantial supplies of oil and gas remain. This probably means for the next hundred years or perhaps more, but rising in severity within the next 50 years as oil supplies elsewhere run out. Oil will pour out of the Middle East and money will pour in. The challenge will be to keep problems from escalating to military action. We have already seen this happen in the Gulf War of 1991. For the next several decades at least, the Persian Gulf nations will be increasingly prominent in world affairs. The massive transfer of wealth from the industrialized nations to the oil-producers will continue, and will have profound effect on both parties.

The fall of oil

Beyond those times, another chapter of history will be written. That chapter will tell of the changes in the nations of the world, as oil supplies and oil revenues decline until they become virtually non-existent. The human race at times believes itself to be beyond the laws of nature, but nature will ultimately prevail. The story of petroleum will be the same story of resource discovery, abundance, use, and decline which has been repeated for other resources in other areas in the past. An abundance of a life-sustaining resource is discovered. The consumers flourish and multiply and multiply again. Eventually, however, the region is depleted of its resource and the consumers must reduce their numbers.

The scenario is very much like that of abundant use of fertilizer on a garden. The plants grow and multiply and flourish as long as the fertilizer is applied, but when it is no longer available, the garden goes back to its normal basic ability to support the plants. The challenge for Persian Gulf and other now oil-rich nations (such as Brunei) is to establish economies, technologies, and investments which will survive the end of oil. The eventual depletion of oil is also a challenge for the consumers to put in place new technologies and social organizations which can adapt to alternative energy sources which must be used. Alternative energy sources and their probable higher costs will cause great changes in societies, and in their industrial bases.

There is no parallel in history for such a rapid development and use of a resource as in the case of oil, and the profound changes it has brought worldwide. But it is well to keep in mind that changes work both ways, with the coming of oil — and the going of oil.

What is true about oil also applies to other minerals, especially for nations that are one-resource countries in terms of what they can export to obtain foreign exchange. What happens when this source of money to buy the things not produced locally no longer exists? Smaller nations face this problem acutely, for in most cases manufactured goods, especially specialized products such as medical equipment and sophisticated electronics, must be imported. It will be necessary for the wealthier nations to deal with this on a humanitarian basis, rather than strictly economic factors.

Shift of supply sources

In the case of both energy resources and hard minerals, as these resources are depleted in one area production will shift to other regions. There will not only be a change in energy mixes and mineral mixes (as substitutes are introduced) in consuming nations, but there will also be changes in sources of supplies. As the economies in nations with raw materials mature, these nations will more and more tend to upgrade their raw materials at home. This will impact refining industries in the importing countries. Current examples are the rapid growth of the petrochemical industries in the nations of the Persian Gulf region, and the development of aluminum plants in Venezuela close to the source of the ore and energy to process it. For many years, even though the United States did not have much bauxite itself, this ore of aluminum was shipped to the United States for processing, and the U.S. was an exporter of the refined metal. But increasingly the refining of aluminum ore is done by the countries which have the big deposits such as Australia. The United States is now a net importer of that metal, and some aluminum plants in the Pacific Northwest of the United States lie idle. They will probably remain so, in part because of the energy shortage in the Columbia River basin, and the problems of dams and the environment, especially the marked decline of salmon populations. With this shift in aluminum ore processing to other countries, along went a number of well-paying jobs which the Columbia River aluminum plants had provided. Also, having to import refined aluminum increased the U.S.' negative balance of international payments.

Another example of job movement to the resource site is the development of gold refining capabilities in Zimbabwe in 1988. Previously, that country had to ship out its gold ore concentrate for refining. This was done by Western Development Corporation at Perth Mint in Australia.

Nations which produce mineral resources are destined by their geological endowment to be able to gradually take away industries and jobs from the consuming nations. In effect, the consuming nations which for a time were processing the raw materials from the producing nations, simply were borrowing the resource, which was destined to be reclaimed by those nations which had the resource. This will continue to be an ongoing series of adjustments for both parties, with the nations which have the minerals tending to gain economically at the expense of the mineral importing nations.

More nations will use more resources

The largest use of energy minerals and minerals has been markedly concentrated in the past and to the present time, in a relatively few nations representing a relatively small fraction of the world's population. At the present time world crude oil production is now about 62 million barrels a day. The United States, with five percent of the world's population, uses about 27 percent of the world's oil, and a similar percentage of the world's total energy supplies.

Per capita consumption of energy in the various countries of the world shows an exceedingly wide range. The United Nations has issued statistics based on energy equivalent in kilograms of coal per capita. For some selected countries these figures are: United States 10,127, West Germany 5,377, Japan 4,032, Mexico 1,689, China 810, Brazil 798, India 307, Indonesia 274, and Bangladesh 69. In simple comparison, each U.S. citizen uses more than 12 times the energy of a resident of China, 33 times that of a resident of India, or 147 times that of a resident of Bangladesh. Part of the difference between U.S. energy consumption and that of other nations can be attributed to the fact that the U.S. is heavily industrialized and makes products, not only for itself, but for other nations, so those nations which import the products benefit from U.S. energy consumption. Also, the U.S. is a large country, populations are widely spread out, and transport to these regions and populations take more energy than it does in more compact nations such as Japan and those in Europe. However, even adjusting for these considerations, the U.S. does enjoy individual energy consumption beyond any other nation. Can this continue?

In the industrial world, the chief metals used are aluminum, chrome, cobalt, copper, iron, lead, manganese, magnesium, molybdenum, nickel, platinum, tin, tungsten, vanadium, and zinc. Just since 1950, the use of these materials has increased between 100 and 500 percent. With somewhat less than 30 percent of the world population, the industrialized countries used more than 80 percent of these metals. It is unlikely that these percentages will be maintained. The trend in the raw material producing countries is to keep more and more of their resources at home for their own growing industries, and to upgrade the exported materials before shipment. This creates increased competition for the industrialized countries for new minerals, and for their processing. Oil products are a particularly good example. The Persian Gulf nations illustrate this trend. They now have a great affection and demand for cars so more and more of their oil production is used internally, and increasing amounts are also processed locally into refined products, which previously were shipped abroad in crude form. Combining this trend with the more rapidly growing populations in these countries as compared with the industrialized nations, the conclusion is that there will be an increased diffusion of industrial development, and of use of energy and energy mineral resources in the future.

For decades, raw material producers were content to be just that. But much changed after World War II. First, the British Colonial Empire came apart. Then U.S. and British oil and mining interests were expropriated by various countries.

This is not to unduly criticize these events. They were an inevitable outgrowth of the end of colonialism, and the rising sense of national identity of these countries. It was also a recognition of what their mineral resources could be worth to them. Some of the actions were quite arbitrary, and in a few cases were outright theft. But it was inevitable. The new governments as owners of these operations may or may not be as efficient or even as good employers as were the previous owners, but the pride of the country is on the line, and now "our minerals are now ours."

Population — the Basic Problem

There is a continuing thread which runs through much of this book concerning mineral and energy mineral resources. That crucial thread is population, its rapid growth, and the problem this presents. The "population problem" is becoming increasingly visible in many contexts. It is expressed in the pollution of rivers, lakes, oceans, and the atmosphere. It is visible in the greatly increased rates of extinction of organisms where human habitation has displaced wildlife habitat. Even the best of intentions such as the setting aside of game preserves, in Africa, can be thwarted by population growth. When people are hungry or economically hard pressed, these sanctuaries will be invaded and wildlife destroyed.

To provide even a small amount of the simple energy source, wood, forests are being cut down in South America, Asia, and Africa. As a result, more and more destructive floods occur in many areas. Deforestation of the land is causing severe erosion, and in many areas the damage is irreparable in the foreseeable future.

In terms of energy and other mineral resources, the problem is to provide the world's people with a dependable supply of basic commodities of a scale enabling them to live in some degree of comfort. Ultimately, this must be accomplished in an essentially renewable recyclable steady-state economy. Accomplishing this requires that the global population be stabilized. Sustainable resource use cannot be achieved when the population continues to be a moving target in an upward direction. It has been truly said, "whatever your cause is, it is lost without population control." Without population control there will sooner or later be an inevitable collision between the human race and what the Earth can provide. This impact will be felt in many ways and many places and it will come. In some areas this seems to have occurred already, particularly in Africa, and in other places such as Haiti.

Although some countries are making strong efforts and are doing reasonably well to control population growth, the world in total does not seem likely to achieve zero population growth in the near future. Recently, China, which has been vigorously trying to control the growth of its population, has had to admit partial defeat in achieving this goal.

The dire fact is that world population now grows at the rate of nearly100 million people a year and this number is rising. There seems no early end in sight to this trend. Although it is sometimes reported that world population is beginning to level off, this is only true in some industrial countries. It is not the reality for the world as a whole. The World Bank predicts that the current world population of five and one-half billion will rise to 10 billion by 2050.

Beyond just the sheer number of people now on Earth, of even more concern is the rate at which population growth has been occurring as shown by the following statistics:

Years Needed for Human Population to Reach . . .	
1st billion	2,000,000
2nd billion	105
3rd billion	30
4th billion	15
5th billion	12

Indications are that the 6th billion will take only nine years.

Cohen states "the immense momentum of human population growth resembles the very long stopping time of a fully loaded truck."(23) Population will continue growing for many years to come, and reducing its size as seems necessary for human survival in reasonable living conditions, will take even longer. There is a race between the size of population and resources. So far, indications are that population is winning the race, but the world's standard of living will be the loser for it. Cohen further observes, "Within the next 150 years or so, and possibly much sooner than that a drastic but not necessarily abrupt decline in global population growth is inevitable."(23) Resource availability makes this prediction destined to come true.

Competition for Resources

The direct effect of an increasing population is that the demand for energy minerals and other minerals will continue to grow. Soon a world of relative plenty will become a land with more modest and more expensive mineral and energy resources, and increased competition for them.(77) In some aspects that is true now. The U.S. imports more oil than it can produce. In the Middle East the water supply situation is already at a critical point in many areas because of increased demands by growing population, and conflicts already are arising among countries which have to share limited common water resources.

The spread of affluence

The future will not only see a considerable increase in world population, but also the parts of the world which do not now enjoy much affluence will continue to try to obtain more of the global mineral and energy mineral resources as the basis for an improved physical standard of living. Foremost among these regions is Asia — particularly China. As has been noted before, with 1.2 billion people, if China used oil in the per capita quantities in which it is used in the United States, the Chinese alone would use considerably more oil than is produced in the entire world today. Clearly, China will never enjoy such affluence, but it is moving in that direction. Automobile production is to be greatly increased. China's energy needs including its demand for oil is predicted to rise sharply.(1) Some production increase can come from within China, but not nearly enough to meet their requirements. This is one reason China is laying claim to disputed potential oil areas in the South China Sea. Also, with increasing manufacturing capacity to produce a great variety of consumer goods, China is earning substantial amounts of foreign exchange by which to bid for oil in the world market.

The United States now runs a substantial current exchange deficit with China. Thus China has U.S. dollars (in which oil is priced) by which to bid for oil.

Barring some major political upheaval in the Middle East, there will probably be no immediate great shortage of minerals or petroleum, although the price of oil will probably rise substantially within the next decade, and beyond. There will be a gradual industrialization of some of the lesser developed nations. The major oil producing nations will expand their operations to upgrade their petroleum to more valuable end products, offering increased competition to similar industries in the western nations. Minerals and energy supplies will become more widely in demand, and the lesser developed countries may increasingly be able to bid against the industrial nations for these resources.

Noting the economic growth of the Far East, Reifenberg writes: "The traditional markets of Europe and North America just can't compete with the likes of India, where oil demand is growing at 6.5%, or South Korea, where gasoline consumption has shot up 20% every year for the past five."(70)

"With total Asian demand likely to grow by seven million barrels a day in fewer than 10 years, the region after the turn of the century will dethrone North America as the world's biggest oil consumer, says Fereidun Fesharaki, director of the resources program at the East-West Center in Honolulu."(70)

Teitelbaum, viewing coming world oil demand, writes: "Led by China (population 1.2 billion), the Third World's appetite for petroleum is increasing at a furious rate, and as consumerism reaches full frenzy around the globe, that increase is likely to go exponential."(78)

Tunali has provided some interesting statistics and comments:

> "At the mid-point of the 20th century, when there were 2.6 billion people on Earth, there were 50 million cars. Now, as we near the end of the century, the human population has more than doubled, but the car population has increased *ten*-fold — to 500 million. Today, everyone seems to want a car. And within another 25 years there may be 1 billion cars on the world's roads. It is not in already car-dominated Los Angeles, Paris, or Rome where this is happening, but in the booming cities of the developing world. In China, for example, there are just 1.8 million passenger cars today — one for every 670 people. In ten years there may be twice as many, and by 2010, that number is projected to rise to 20 million. Such a scenario seems likely to be repeated in most parts of the developing world."(80)

It may be noted by comparison that there is now one car for each two people in the United States. Currently countries that make up 16 percent of world population, which is North America, Europe, and Oceania, own 81 percent of all cars. The United States alone accounts for 35 percent of world car ownership.(80)

But with the increasing affluence of the developing countries, derived in part from exports to the industrialized countries, oil demand will soar in developing countries, and competition for the world's oil will increase dramatically. At the same time, the ability of the world to increase oil production will be leveling off. In this regard, Russia is an interesting example. Once the world's largest oil producer, Russia now barely produces enough for its own internal use. But the Russians are developing a huge demand for cars forecast to soar by at least 50 percent within the next five years. "The market is boundless" says Vladimir Kadannikov, former president of Russia's biggest car maker AvtoVAZ Inc. American, French, German, and Italian car manufacturers are negotiating to build plants in Russia. If Russia gets "wheels" as anticipated, it will soon become an oil importer, and the relatively nearby Persian Gulf will be of increasing interest to the Russians.

Adapt and survive

The only constant in all of this is change. The human race is an extension of the long arm of nature. The great lesson learned from the fossil record is that those organisms which adjust to change, survive, and those which do not, become extinct. Adaptation for survival will be required in terms of energy and mineral production and consumption. As populations increase against resources, which in some areas are now declining in both quantity and quality, the problems of adaptation to changing circumstances will increase much more rapidly than in the past. The coming demands of the present rising population as compared with the past are now so very large in terms of resource consumption that we will in a few decades be faced with mega-events previously unknown. We have gone through an era of unprecedented abundance and quality of energy mineral and mineral resources inherited from the literally billions of years of geological events that it took to form these resources. We are consuming them in a geological instant. We have used and are using the best resources in terms of both quality and ease of recovery. Resource by resource the trend is toward a less affluent situation.

In brief, the industrialized nations of today are having to play on a different economic and political field than in the past. This field has been created by a rising tide of nationalism, and an increased industrialization of the lesser developed nations, who can now more effectively compete for vital resources. And the industrialized nations have already depleted, to a considerable degree, many of the Earth resources they originally had. It is a profound change and one which is reshaping the world economies and the futures of all nations.

Global Inter-Dependence

One of the pronounced effects of greater demand for mineral resources and the fact that the various components of industrialized nations' mineral supplies must come from widely diverse sources is to emphasize the increasing inter-dependence of nations. By the same token, nations must now be concerned with global problems beyond what they have been in the past. What happens in one nation several thousand miles away may be of vital interest to a country. The concern shown by nations such as Japan, Italy, France, and the United States with regard to events in the Persian Gulf region in the 1980s — the war between Iraq and Iran, and ultimately the Gulf War in 1990-1991, and the free flow of oil — is an example. Thus mineral inter-dependence may make us "one world" beyond what any political statements can accomplish.

The "Sustainable" Society

Central to any discussion of the future of humankind is the concept of "sustainability." What are the limits to growth? How long can society be sustained at a given standard of living? This seems to have become only a relatively recent concern, particularly in the United States and Canada, the last two countries with large and varied natural resources to be settled.

A hundred years ago most people saw no visible limits to the resources of North America. Pearce and Turner note that:

> "Between 1870 and 1970, mainstream economists (with some notable exceptions) appeared to believe that economic growth was sustainable indefinitely. After 1970 a majority of economists continued to argue that economic growth remained both feasible (a growing economy need not run out of natural resources) and desirable (economic growth need not reduce the overall quality of life). What was required was an efficiently functioning price system. Such a system was capable of accommodating higher levels of economic activity while still preserving an acceptable level of ambient environmental quality. The 'depletion effect' of resource exhaustion would be countered by technical change (including recycling) and substitutions which would augment the quality of labour and capital, and allow for, among other things, the continued extraction of lower quality non-renewable resources."(63)

However, a careful study of history shows that in the past and certainly at the present time, the rise in the standard of living, economic growth, and the growth in population are closely related. All are dependent on Earth resources. The energy and mineral resources which have been used to accomplish this to date are, for the most part, non-renewable.

"Sustainable" defined

The term "sustained" (sustainable) is widely used in various economic and environmental writings. The simple definition of "sustainable" is "...capable of being kept going on an indefinite basis — not until the end of the week, or the end of the decade, or even the end of the next century, but *indefinitely*."(68)

The U.S. President's Council on Sustainable Development appointed by President Clinton in 1993 (final report issued February 1996) has accepted the definition of sustainable development as "to meet the needs of the present without compromising the ability of future generations to meet their own needs."

When one gets down to specifics on how "sustainable" might be accomplished, the problem becomes complicated. Among the questions which arise are:

- Sustainable for how long: Fifty years, one hundred years, indefinitely?
- Sustainable in what lifestyle and at what standard of living — that at present in the United States and other industrialized nations or at a subsistence level?
- How many people can be sustained at whatever standard of living is chosen?

Interestingly enough, none of these questions was addressed in the final report by the President's Council on Sustainable Development.

The matter becomes further complicated by trying to determine exactly what technologies could reasonably be expected to be available. What energy and mineral resources would be used and in what quantities to achieve this "sustainable" society. It is impossible to achieve complete recycling of materials; some small amount is irretrievably dissipated. Furthermore, to recycle most things takes energy, and even if solar energy is involved, it takes both energy and raw material to build solar energy devices, which themselves deteriorate with age and weathering, and must be replaced.

It is an interesting exercise to ask someone, or a group, to precisely define "sustainable" and then to describe in some detail how this might be accomplished. Individuals and organizations freely use the term "sustainable" but when the details are requested, there are no reasonably precise answers. The generality of a "sustainable economy" is bandied about frequently in environmental discussions of the future. It is a nice concept and just thinking about it seems to make people feel more comfortable. But when the matter is pursued by requesting some specifics of how this might be done and for what sized population, the discussion turns a bit vague.

The thought of a "sustainable" society is nice to contemplate. Many of the increasingly numerous books on the environment deal with the question of "sustainable," but only as a generalized concept.(6,55) There does not seem to be a blueprint for how many people could be "sustained" by rather precisely what kinds of energy sources, and with what soil fertility preservation programs, and using what kinds of minerals and in what quantities. In regard to minerals, some are quite rare, have no substitute, and are essential to our present economy, for example cobalt and yttrium. How this problem of rare, but vital materials would be handled in a "sustainable" society seems not to have been considered. Some would inevitably be lost, even if they could be recycled.

Population, precise numbers, and technologies?

Predictions of the ultimate number of people the Earth could reasonably support may range widely, nevertheless it is useful to make an attempt at some assessment. This would be by way of putting present resource use and population size and trends in some perspective. Using simply the known technologies and applying them to the apparent available mineral and energy resources which are the basics of civilization, some figures could be derived. Some studies have been made, and from these it is already rather clear that presently known resources and technologies are not available to bring the rest of the world up to the living standards now enjoyed by the industrialized nations. Facing this fact points out the need to more vigorously pursue two objectives: devise sustainable renewable energy sources, and stabilize and probably reduce population size. Neither of these objectives is being pursued with sufficient vigor or success to prevent major crises in the not too distant future.

The growing population factor

There have been generalized attempts to present systems which might be "sustainable"(6), but the factor of the continuing rising population tends to defeat coming to any clear conclusions. If the argument is made that population size can at least be stabilized at "sometime" in the future, the "sometime" seems so indefinite as to preclude using that as a fixed point for calculations. Some definite population size needs to be assumed to make reasonable judgement as to what resources and known technology might be employed to sustain that population.

At present, the concept of "sustainability" in whatever context it is used, seems to be a vague and hopeful generality. An example of this might be that given by Meadows, et al.: "There are many ways to define sustainability. The simplest definition is: A sustainable society is one that can persist over generations, one that is far-seeing enough, flexible enough,and wise enough, not to undermine either its physical or its social systems of support."(55)

In personal correspondence, Lester Brown of the Worldwatch Institute, which organization has been in the forefront of worldwide ecological studies particularly related to population growth, writes: "There is no empirical calculation of sustainable that I know of. Conceptually, it is that development which satisfies the needs of the current generation without jeopardizing the prospects for future generations."

This seems to be a correct statement as to where the definition of "sustainable" stands today. There is no empirical calculation of "sustainable."

Perhaps future research will be able to provide specific answers to avert the harsh solution which nature has historically provided in the past to the problem of organisms which exceed the carrying capacity of their environment. However, there is no certainty that this will happen. Some target figure of population size along with numerical details related to resource use and the known proven technologies to be employed is an essential goal to be put forth. Realistically, even if such a plan is presented, the will and consensus to put it into effect must be found to make it successful. Accomplishing this worldwide would be most difficult, particularly as attitudes about population size (family size) differ widely among ethnic and religious groups.

Pimentel and Pimentel (65,67) and their associated research group at Cornell University have made perhaps the most comprehensive and quantitative study of "sustainability" in relating it to a resource base. The 1994 study is an outstanding summary of this matter.(65) Two vital parts of our resource base are soil and water, and here in personal correspondence David Pimentel has offered some firm figures in regard to sustainability. "For erosion under agricultural conditions, we should be losing less than 1 ton/hectare/year for sustained farming. For groundwater resources we should be pumping no more than 0.1% of the total aquifer for use. In very few locations are we managing our soil and water resources based on these criteria."(66) Giving some fairly exact figures, as is done in this case, to the concept of sustainability for other resources would be most welcome. Pimentel et al., have further suggested "the human population would have to be much smaller than the present 5.5 billion" as the sustainable population for the Earth, at a satisfactory standard of living, and about 200 million (present population 267 million) for the United States.(65) Other figures for the United States go as low as 16 million. In any case, the significant fact is that the number is considerably less than the present, and growing population.

There are suggestions as to what might be involved in establishing a sustainable society. One brief outline states:

> "(a) Utilise renewable resources at rates less than or equal to the natural rate at which they can regenerate.
>
> (b) Optimise the efficiency with which non-renewable resources are used, subject to substitutability between resources and technological progress." (63)

The terms "sustainable development" and "sustainable growth" appear in a number of publications. Are "sustainable development" and "sustainable growth" the same? They seem to be more or less interchangeable. But how either of these themes is related to population, or population growth is rarely stated. Does "sustainable development" mean taking care of the world's increasing population at the present standard of living (which now differs widely from place to place)? Does it mean improving the standard of living of the presently existing total population: Does it mean improving the standard of living of a substantially reduced number of people which could be supported more or less indefinitely with existing or reasonably foreseeable technologies? Because of the great dependency which civilization now has on non-renewable energy and mineral resources inherited from the past, it seems clear that "sustainable development" cannot imply taking care of the world's increasing population indefinitely. Sustainability must be defined with some specifics attached, particularly with regard to the concept of "growth."

Growth

This is the key word. Growth means things get larger — more things, bigger cities, a larger population, more subdivisions, more highways, and a bigger and bigger gross domestic product which is the sum total of all things produced by a country.

It is the symbol of success that communities grow, that companies grow. If a manufacturer produces more cars over the previous year, the year is said to be a success for the company. Towns grow and chambers of commerce encourage new industries to come to provide more jobs for a growing population. Growth is also measured to some extent by the increase in the physical standard of living. This can be measured in various ways including the size and quality of housing — how many square feet each individual has to live in, and what things go with the living spaces such as electrical appliances and heating and cooling facilities. It also includes quality and variety of foods, good clothing, and adequate transportation for the needs of the individual, whether it be by bus, train, car, airplane, or simply a bicycle.

"Growth" is a continuing current economic objective. The cover of *Business Week* for July 8, 1996, in big bold letters proclaimed that "America needs faster growth." No politician would ever be elected on a "no growth" or "slow growth" platform. But "growth" in one fashion or another requires more and more minerals and energy minerals. To make these things "grow" in total quantity to match the growth in population, or to simply provide more "things" for the existing population it takes more Earth resources. "Growth" in these terms has been phenomenal the past 100 years. Physicist Ralph Lapp has termed it "The Logarithmic Century."(50)

The final report of the President's Council for Sustainable Development, titled *Sustainable America,* lists 16 things in which the council believes, and has as objectives. Number 14 states: "A growing economy and healthy environment are essential to national and global security." The report does consider population stating: "...The Council...wants to make clear that it seeks to move toward voluntary population stabilization at the national level." But even if population were stabilized, the statement is still made that the Council wishes to see a "growing economy." (Their item 14). But there are no limits to growth stated. An economy cannot grow indefinitely. Also, a growing economy uses more and more resources and inevitably resource exploitation has an environmental impact. How these matters can be accommodated is not stated. There are statements about how to maintain a sustainable agriculture in its present form, and "restoration of fisheries," but it is not stated how these objectives can be maintained against a "growing economy." Growth means more of something, and neither agriculture nor fisheries can be sustained against the demand to produce more and more. Also, the report makes no mention of how to meet the problem of the rapid use of non-renewable resources — petroleum, coal, or metals. "Sustainable development," "sustainable growth," or "economic growth" continue to be pleasant platitudes which so far have not been defined nor addressed in specific realistic terms. And firmly grappling with the problem of both internal population growth and immigration seems not to be for the political stomach. It is a very touchy subject to be treated only very lightly, as indicated by riots which have already occurred in the United States over immigration matters. The President's Commission looks toward "voluntary population stabilization." This is nowhere in sight. And population is critical to all resource problems and the future.

But more recently, books and articles have inaugurated a serious debate over just how much the economy can grow, even in the resource rich countries. The Arab oil embargo against the United States in 1973 made many people realize, as they sat in long lines of cars at gasoline stations, that there were limits to resources and growth. In that respect the oil embargo served a good purpose. Reality arrived, if only briefly.

In regard to using the term "growth," it has been pointed out that it is inappropriate to use the term "growth" when it may be based on use of non-renewable resources. Mineral and energy mineral resources are "natural capital" and when this is spent to achieve "growth" is there really growth or are we simply transferring wealth from one form to another, and perhaps in the process losing some through entropy? (83)

Balancing population against available mineral and energy resources must be the ultimate objective. Renewable energy sources must be developed in quantity. This, together with technology for efficiently recycling metals with only a modest amount of new material annually drawn upon from the Earth might lead to a "sustainable" balance between some optimum size of population and the available renewable resources. The key is population. As long as world population continues to increase as it has been doing, even at the seemingly insignificant rate of about two percent a year, planners for a sustainable future are faced with having to shoot at a moving target, and no realistic plans can be made. But what does "only a two percent growth rate" mean?

"Only" a two percent growth rate"

To illustrate how fast the human population target moves, and the inability of material resources to ultimately keep up with the demand from such a growth rate, the late geochemist, Harrison Brown, is credited with the calculation that if the world's population continued to increase at the rate of two percent annually, in two thousand years the Earth would be a solid mass of people expanding out into the universe at the speed of light.(15) In just six hundred years (not really long in terms of human history), the Earth would pass the standing room only situation of five square feet per person, including covering both the continents and the oceans. This is what "only a two percent growth rate" means. In the short range of 35 years which many people now living will see, a two percent growth rate means the world population will double.

Population growth to date

History has already shown an alarming record of exponential growth in population. Population equates to a demand for Earth resources. The term "sustainable growth" (or sometimes the term "sustainable economic growth") seems to imply maintaining a continually growing population at least at the present standard of living, presumably indefinitely, as there is no stated limit to "growth." In such a case, the term "sustainable growth" as Bartlett has noted, is clearly an oxymoron.(9) Porritt reinforces this view with some additional comments about groups and individuals who use this term and seem to believe in it:

"Indeed the concept of 'sustainable growth' is a contradiction of terms: exponential growth (in either human numbers or volumes of production and consumption) *cannot* be sustained indefinitely off a finite resource base. A growth rate of 3 percent implies a doubling of production and consumption every twenty-five years. Nobody actually disagrees with that, not even the most manic growthists. But professional Micawbers that they all are, they just go on hoping that something will turn up before their bluff is finally called."(68)

A recent study by Bouvier and Grant makes some of the strongest statements yet regarding the situation in the United States: "...the evidence is clear and convincing: the United States cannot maintain even a semblance of its current quality of life without some reduction in population size. The natural resources available to the nation are not sufficient for ever-increasing numbers of people. At current levels of consumption, they are not even sufficient for the existing population."(12) Jackson has similarly stated, "I consider it very improbable that the earth could continually support, on a nondestructive basis, a population as large as the present one."(46)

Lester Brown and associates at the Worldwatch Institute, make the flat statement that, "as a result of our population size, consumption patterns, and technology choices, we have surpassed the planet's carrying capacity."(19)

In some parts of the world the population has clearly gone much beyond what is sustainable in terms of the carrying capacity of the land, or what might even be imported by means of foreign exchange which could be earned. Haiti is an example. With nearly seven million people living on about 10,000 square miles with very few resources, Haiti's economy is unable to generate any significant amount of foreign exchange. Yet the population is growing at a rate which will double in about 25 years — an absolute catastrophe. Currently, Haiti receives significant support from the United States. This cannot be the ultimate solution. But there are few voices raised in either the United States or Haiti regarding the population problem which is the crux of the matter. Haiti's present trend is on collision course with disaster, and if one visits Haiti it is apparent that the time of impact is very near. It may have already arrived. Continuing to ship food into countries which do not have the capacity to support themselves is not an ultimate solution. In the longer term, as population increases, it simply makes matters worse. The U.S. Undersecretary of State for Global Affairs, Timothy Wirth, has said that Haiti's population is "already unsustainable by every measure."(52) In such a circumstance, what is the world community supposed to do? Continuing to send resources into such a situation only allows the country to avoid coming to grips with its own problems, and simply exports the problems to other countries which are, at the moment at least, willing and able to help. Unfortunately, such situations are increasing around the world. Consideration of this problem in its several aspects is made by Lucas and Ogletree.(51)

How many people can the Earth support?

Cohen has compiled an impressive comprehensive review and study of world populations under the title, *How Many People Can the Earth Support?* (23) He considers the natural resource limitations together with cultural attitudes which may affect population numbers, and suggests that the variables involved are so complex that there is probably no way to make any very precise predictions at to sustainable numbers. The number would vary with changes in social attitudes and with technologies then currently available. He does make the historical observation that, "...since 1600, the human population increased from about a billion to nearly six billion. The *increase* in the last decade of the twentieth century exceeds the *total* population in 1600...Within the lifetime of some people now alive, world population has tripled; within the lifetime of everyone over 40 years old, it has doubled — yet never before the last half of the twentieth century had world population doubled within the life span of any human." He offers the tentative conclusion that, "...the possibility must be considered seriously that the number of people on the Earth has reached, or will reach within half a century, the number the Earth can support in modes of life that we and our children and their children will choose to want."(23)

Perhaps then, instead of "How many people can the Earth support" the question could be better stated "How many people *should* the Earth support." This is a matter related to how people want to live. Recently Yosemite National Park in California had to be closed temporarily twice because of the press of people. For those who could not get in, there were already too many people. This is a simple example to suggest that even in the USA Cohen may be right that the number of people has reached the maximum that people wish to have for their mode of life. There are others who believe that the number which can be sustained in a reasonable standard of living has already been exceeded. Yet still others are opposed to any suggestions as to family size limitation, or significant controls in immigration.

Sustainable numbers probably less than at present

As presented elsewhere in this volume, the current mining of groundwater, loss of arable land, depletion of high grade energy sources, and siltation of reservoirs strongly imply that the number of people who might be sustained on this Earth at a comfortable existence has indeed been exceeded. Comprehensive studies by Brown, Ehrlich, Hardin, Pimentel, and others have reached this conclusion. The study by Pimentel et al., 1994, is an excellent and concise summary of the relationship between natural resources and human population. It is recommended reading particularly for its quantitative approach to the problem. Basic figures are presented which are frequently lacking in other estimates of Earth human carrying capacity.(65)

And, as a final thought, is there any quality of life which would be improved by increasing the size of the population which we now have? Would perhaps some qualities of life be enhanced with a reduced population?

Living off an inheritance

The basic reason even the present world population size cannot be sustained indefinitely is that we most recent inhabitants of the Earth are living to a large extent off a huge but decreasing inheritance of resources from the geologic past which cannot be renewed. Hardin makes the clear statement that, "...once we have come to the end of our fossil fuels the sustainable population will probably be far less than the present 5 billion human beings. Our species is living on borrowed time. Unfortunately academic economists are almost completely unprepared to deal with a steady-state world, although it may be 'right around the corner'"(41) Examination of the eventual available alternatives to the resource base on which our present population exists gives every reason to support Hardin's conclusion.

In the case of the most valuable and important mineral complex of all, soil, it is very evident that we are losing it faster than it is being formed. We are concerned with balancing the national budget and there is great furor in the Congress about that. But a vastly more important legacy we could give our descendants would to balance the world soil budget. Little is heard about that in the legislative halls. Political agendas tend to be very short term.

Geological inheritance versus current income

The Industrial Revolution with all its technology, inventions, and machines enabled humans to draw heavily on non-renewable resources through mining and drilling, and use the resources thus obtained to build and feed the present huge industrial complex and growing population. The ability to replace human energy with other energy, largely from fossil fuels, together with the products from the industrial machine greatly raised the standard of living for some of the world.

Our industrial economy now draws very heavily from the past, but a "sustainable" economy would require everyone to live on current energy and mineral income. This is certain to cause a marked change in lifestyles from the present. Solar energy and fusion (if it can be accomplished commercially) appear to be the only two major sources of energy which can be sustained for the indefinite future. An economy based on one or a combination of them would look much different from that of today. And neither of these energy sources can replace valuable topsoil, increase groundwater supplies, restore groundwater aquifers which have collapsed, or cause more water to flow in rivers where demand already exceeds supply such as the Nile, the Jordan, and the Colorado. All this strongly suggests that not only must population be stabilized, but it should be reduced as the writings of such researchers as Bouvier and Grant, Brown, Hardin, Jackson, and Pimentel logically argue. But there is no such movement promoted or evident in ruling political circles around the world. Government officials and politicians are plainly afraid to face the problem. One might cite China as a possible exception, but for whatever the population policy has been, the fact is that China reached the 1.2 billion population mark in 1995, a figure which had not been expected until 2000. The impression sometimes held that their population is stabilizing is not correct. It continues to grow at the current rate of more than 37,000 people a day.

Many people in the so-called "Third World" now are living on the current income from the land rather than from a geological inheritance. They live at a much lower standard of living than does the industrialized world which draws on oil, gas, coal, nuclear power, and synthetic fertilizers. Seeing their situation should stimulate the presently affluent segment of the world to more serious efforts to prepare for the coming "current resource income age" which will inevitably arrive. However, the world, for the most part, now lives for the present much as the grasshopper and the coming winter.

The logarithmic century

Physicist Albert Bartlett has stated that, "the greatest shortcoming of the human race is our inability to understand the exponential function."(7) This cannot be emphasized enough. Small percentages of growth look deceptively harmless, but yield enormous numbers in a relatively short time. Lapp, Meadows, et al., Brown, Pimentel, Hubbert, and others have written extensively on this very important and inescapable mathematical fact of the very large numbers which just a small percentage of annual growth will rather quickly produce. The "logarithmic century" about which Lapp writes is visible in many ways. For example, "measured in constant dollars, the world's people have consumed as many goods and services since 1950 as all previous generations put together."(28)

The curves of consumption have risen not arithmetically but logarithmically, and population is also rising logarithmically. The world population doubled between 1950 and 1992, and it did it in the shortest doubling time in history. During this same time period, the world smelted more iron and steel than in all previous history, and it consumed two-thirds of all the coal and oil and natural gas that had ever been taken from the Earth.(64) In the case of the United States, President Kennedy noted in his inaugural address of 1962 that in the previous 30 years the United States alone had consumed more energy and energy mineral resources than the entire world had previous to that time.(31) This exponential consumption of resources continues.

At what would appear to be the insignificant annual growth rate of "just" 1.1 percent, the population of the United States will double from its size in 1996 by 2050. The U.S. is greatly dependent on oil, but produces less than half of what it consumes, and domestic production will continue to decline. What will the vital oil situation be in 2050, less than one lifetime from the present? At the present time, the U.S. consumes more than 19 million barrels of oil a day. If U.S. population doubles by 2050, to maintain its "oil standard of living" the U.S. will have to obtain more than 38 million barrels of oil a day. It is very unlikely that this can be done.

Two centuries ago when the British ruled India, that country had a population of 60 million. Today it is 850 million and still growing at the rate of two percent a year. This means that the population will be 1.7 billion in just 35 years, if the rate continues. Poverty is already widespread in India. Soil is being lost far faster than it is being formed, and groundwater supplies are being overdrawn. Thus, two basic resources are declining while at the same time the population keeps growing logarithmically. To further show what this logarithmic growth would mean: if continued at two percent, the population of India would be 3.4 billion by 2060, which is more than half the total population of the world today. By 2101, barely more than 100 years from now, the population of India would be 6.8 billion, or more than the present entire world's population. It is not possible in terms of resources to support such a population. How will the major changes in population growth trend which will certainly occur, be accomplished — by reason or disaster?

At the present time, the world's population is growing at the astounding amount of nearly one-quarter of a million a day. In 1995, world population increased by 90 million, a greater number than in any previous year in history. At that rate every 2.9 years a population the size of that of the United States is being added to the world. It is a situation which is not sustainable. As Lapp has pointed out, the Earth cannot support two logarithmic centuries back to back.(50) Unfortunately, the human race appears to be proceeding as though it can. It cannot. The only question is whether adjustment to the carrying capacity of the Earth will be accomplished by intelligent actions or by the harsh methods nature has used in the past to solve this problem.

Brown, in an assessment of the future, clearly recognized the problem stating: "...we are faced by the fact that there are indeed limits to growth. We might argue among ourselves as to exactly where those limits lie. But we must recognize that neither population nor affluence can continue to grow forever. Unless we willfully stop their growth ourselves, nature will stop this growth for us. And she has many ways of doing it."(15) The human race has the benefit of intellectual reason. Will it use it in time? The time is now.

Clearly, the challenge of population and resource balance is awesome. Humanity would do well to put its energies and resources into the common goal of a reasonably affluent life for populations all over the world which now exist, and for those which are to come, rather than into strife and division as is the present situation. However, there now appears no adequate means by which to accomplish this. In fact, against all logic, there seems not to be even a consensus to do so. There are many people and organizations, both religious and ethnic, who see no need for population control, and indeed strive against it. All of which suggests that it may be that rather than rationally solving the problem, it will be solved by nature as it has in the past, by scourges and starvation. In ancient times, when carrying capacity of the land was exceeded, and people were forced to face reality, populations could not draw on resources of distant lands as is done today and they perished. Having greatly depleted its own oil resources, the United States today draws much of this vital energy source from "other" areas. In parts of the world where famine is endemic, food supplies are drawn from "other" areas. But the resources of "other" areas are also limited and cannot be drawn upon indefinitely. Certain regions of the world are blessed with particular resources, and by world trade, these geographic anomalies can be adjusted for.

But it is not possible to continue to borrow from each other indefinitely. Our planet is a finite resource system including even sunlight, for only a fixed amount falls on the Earth. World populations must eventually live with the total renewable carrying capacity of the Earth. Hardin, in a recent publication, has a comprehensive review of many of the fallacious attitudes which are extant today with regard to growth, the availability of resources, and population. His book is an excellent statement of reality.(41)

A realistic, simple, but accurate analysis of the problems of energy supply, and of the numerous fallacious statements made about it are also in the article by Bartlett.(7) He also suggests a system which he calls "sustainable availability."(8) This is defined as "having the rate of use of a finite nonrenewable resource that guarantees that the resource will last forever."(8) In practical terms this would mean that only a percentage of a resource would be used at any time rather than using up 100 percent of it. To follow this out, eventually the amount available would be exceedingly small and getting smaller each time it is used, but the amount would never go to zero. However, the quantity of the resource available would eventually be so small that it would not be useful in practical terms. But, pursuing this program would gradually demonstrate what problems will come from the loss of a resource, and perhaps stimulate a serious and persistent course of action to do something about it.

The world in many aspects of resources is finite. We are indeed living off our capital at an increasingly rapid rate, with no adequate replacements in sight for many basics which now support civilization. Alternative replacements need to be developed to determine which are really adequate to the task ahead.

How Much Is Enough?

At various places in this book the terms "physical" standard or "material" standard of living are used. This includes the essential human need for food, shelter, clothing, and a command both directly and indirectly, of some amount of energy beyond our own muscles. But above a certain basic level of existence, there are myriad things which may be included in the physical standard of living. Inventors in the industrialized nations together with manufacturing and marketing forces have combined to make available a phenomenal spectrum of goods. All these are using up resources at an exponential rate.

At some point, perhaps even now, it is important to consider "how much is enough." People are better off now than those a century ago because great progress in medicine has eliminated plagues, provided vaccines against such scourges as polio and typhoid fever, and given advanced sanitary and heating arrangements to many of us. But at what point do simply more goods really make people happier?

It has been estimated that people in the United States, "...are on the average now four-and-a-half times richer than their great-grandparents were at the turn of the century, but they are not four-and-a-half times happier. Psychological evidence shows that the relationship between consumption and personal happiness is weak. Worse, two primary sources of human fulfillment — social relations and leisure — appear to have withered or stagnated in the rush to riches."(28) Some people are now turning back to simpler lives and say they are happier for it. To some, the emotional cost and stress of high pressure jobs, and commuting in the traffic snarls are not worth the increased luxury goods which such activities may provide. The opposite of over consumption which is abject poverty is not desirable either. Somewhere between lies balance.

A Perspective

It is not within the scope of this book to become overly philosophical, but one may observe that the important relationships of family and friends, and an occasional return to the natural world of forests and plains, and lakes and mountains from which we came as a human race, are not necessarily enhanced by a mountain of material goods. Some people believe they enjoy a better life without many of these things. What people do, both legally and illegally, to obtain material things beyond their basic needs and reasonable affluence, suggests a set of values which in the long run, and sometimes the short run, may not lead to happiness. J. Paul Getty, the oil multimillionaire, wrote a fascinating book titled *How to be Rich.* Some bought it because they misinterpreted the title as "How to *GET* Rich." But in it Getty told what he thought to be "rich" really meant, and it did not include being an oil multimillionaire. He stated that after five attempts he would give up all his riches for "one successful marriage."(33)

As the demand for resources increases and their costs rise, perhaps an adjustment in values as to what is important and worthwhile can compensate for a reduced amount of material things. Can a sustainable lifestyle be developed which in balance combines a physical standard of living made up of having basic needs and some amenities, together with a way of life which also provides a satisfying emotional and spiritual standard?(5)

Where each nation places its values relative to its demand for, and use of energy and mineral resources, in particular those which have to be imported, will affect its future.

How much is enough?

Some in the affluent industrialized world are giving this some thought.

Entering the Future

Entering the future is not a well defined point. Each moment we enter into the future to a minute degree. There is no abrupt event. It is an on-going process, which in the longer-term will probably involve great changes from the scene today. One summary view of the way the future may unfold is offered by Richards, in the editor's note to the *Cousteau Almanac,* who writes:

> "Perhaps the most important error in our thinking about the future, however, is the notion that we are going to confront merely a *crisis* or a series of *crises*. The word "crisis," as the British philosopher Eric Ashby has noted, suggests a temporary situation. He proposes to substitute the word "climax," in its ecological sense, suggesting a community of plants and animals that has reached a state of equilibrium through successful adaptation to the environment. We have come to that period when growth is banging into limits —of space, of resources — and this

finiteness of the planet cannot be overcome. The situation will persist. The old challenges of frontiers and exploitations are largely past, and the new challenges are mostly ones of adaptation. The work of the coming age will be that of reinvention and reexploration; the systems must be retooled for the long haul, for the generations to come. Whatever develops, the coming age will be one of transition, and history suggests that periods of great change are never easy."(71)

In whatever way the future may unfold, it will have to be, as Richards states, within the limits of resources. As it always has been and always will be, the materials of the Earth, soil, water, metals, and energy supplies, will be the base for civilization and control its destiny.

BIBLIOGRAPHY

1 ANONYMOUS, 1995, China's Energy Demand, Investment Needs to Soar: Oil and Gas Journal, August 10, p. 30-32.

2 ABERNATHY, VIRGINIA, 1991, Festschrift in Honor of Dr. Garrett Hardin: Population and Environment. A Journal of Interdisciplinary Studies, Human Sciences Press, Inc., New York, v. 12, n. 3, p. 189-349.

3 ABERNETHY, V. D., 1993, The Demographic Transition Revisited: Lessons for Foreign Aid and U.S. Immigration Policy: Ecological Economics, v. 8, p. 235-252.

4 AYRES, EUGENE, and SCARLOTT, C. A., 1952, Energy Sources — The Wealth of the World: McGraw-Hill Book Company, Inc., New York, 344 p.

5 BABBITT, DAVE, and BABBITT, KATHY, 1993, Downscaling. Simplify and Enrich your Lifestyle: Moody Press, Chicago, 220 p.

6 BARTELMUS, PETER, 1994, Environment, Growth and Development. The Concepts and Strategies of Sustainability: Routledge, New York, 163 p.

7 BARTLETT, A. A., 1978, Forgotten Fundamentals of the Energy Crisis: American Journal of Physics, v. 46, n. 9., p. 876-888. (Note: This interesting article has been reprinted in full as a separate publication in its monograph series, by Population-Environment Balance, 2000 P Street N. W., Suite 210, Washington, D. C., 20036-5915. Copies may also be obtained for $2 each, the cost of copying and mailing, directly from Prof. Albert Bartlett, Department of Physics, Campus Box 390, University of Colorado, Boulder, Colorado 80309-0390. Also, a 65 minute video titled, "Arithmetic, Population, and Energy," based on Bartlett's article and narrated by Dr. Bartlett can be purchased from the Academic Media Services, Stadium Building, Room 310, Campus Box 379, University of Colorado, Boulder, Colorado 80309-0379.)

8 BARTLETT, A. A., 1986, Sustained Availability, a Management Program for Non-renewable Resources: American Journal of Physics, v. 54, p. 398-402.

9 BARTLETT, A. A., 1994, Reflections on Sustainability, Population Growth, and the Environment: Population and Environment, v. 16, n. 1, p. 5-35.

10 BENDER, F. K., 1986, Mineral resources availability and global change: Episodes, v. 9, n. 3, p. 150-154.

11 BORGSTROM, GEORG, 1969, Too Many. An Ecological Overview of Earth's Limitations: Collier Books, New York, 400 p.

12 BOUVIER, L. F., and GRANT, LINDSEY, 1994, How Many Americans? Population, Immigration and the Environment: Sierra Club Books, San Francisco, 174 p.

13 BOYER, W. H., 1984, America's Future. Transition to the 21st Century: Praeger Publications, Westport, Connecticut, 168 p.

14 BROOKS, D. A., and ANDERSON, P. W., 1974, Mineral Resources, Economic Growth, and World Population: Science, July, p. 13-18.

15 BROWN, HARRISON, 1978, The Human Future Revisited. The World Predicament and Possible Solutions: W. W. Norton & Company, Inc., New York, 287 p.

16 BROWN, L. R., 1978, The Twenty-Ninth Day. Accommodating Human Needs and Numbers to the Earth's Resources: W. W. Norton & Company, Inc., New York, 363 p.

17 BROWN, L. R., et al., 1991, Saving the Planet. How to Shape an Environmentally Sustainable Global Economy: W. W Norton & Company, New York, 224 p.

18 BROWN, L. R., 1993, Postmodern Malthus: Are There Too Many of Us to Survive?: The Washington Post, July 18.

19 BROWN, L. R., and KANE, HAL, 1994, Full House. Reassessing the Earth's Population Carrying Capacity: W. W. Norton & Company, New York, 261 p.

20 CAMERON, E. N., 1986, At the Crossroads. The Mineral Problems of the United States: John Wiley & Sons, New York, 320 p.

21 CARTLEDGE, BRYAN, (ed.), 1993, Energy and the Environment: Oxford Univ. Press, New York, 170 p.

22 CETRON, MARVIN, and O'TOOLE, THOMAS, 1983, Encounters with the Future: A Forecast of Life into the 21st Century: McGraw-Hill, New York, 308 p.

23 COHEN, J. E., 1995, How Many People Can the Earth Support?: W. W. Norton & Company, New York, 532 p.

24 DALY, H. E., (ed.), 1973, Toward a Steady State Economy: W. H. Freeman and Company, San Francisco, 332 p.

25 DALY, H. E., 1990, Sustainable Growth — An Impossibility Theorem: Development, v. 3, n. 4, 45-47.

26 DALY, H. E., and TOWNSEND, K. N., (eds.), 1993, Valuing the Earth. Economics, Ecology, Ethics: MIT Press, Cambridge, Massachusetts, 387 p.

27 DURNING, ALAN, 1991, Limiting Consumption. Toward a Sustainable Culture: The Futurist, July/August, p. 11-15.

28 DURNING, A. T., 1992, How Much is Enough? W. W. Norton & Company, New York, 200 p.

29 EHRLICH, P. R., and EHRLICH, A. H., 1990, The Population Explosion: Simon and Schuster, New York, 320 p.

30 ELLIOT, J. A., 1994, An Introduction to Sustainable Development: Routledge, New York, 128 p.

31 FLAWN, P. T., 1970, Environmental Geology: Conservation, Land-use Planning, and Resource Management: Harper & Row, Publishers, New York, 313 p.

32 GALL, NORMAN, 1986, We are Living Off Our Capital: (Interview with JOSEPH P. RIVA, Jr.), Forbes, September 22, p. 62, 64-66.

33 GETTY, J. P. 1965, How to be Rich: Playboy Press, Chicago, 264 p.

34 GEVER, JOHN, et al., 1986, Beyond Oil. The Threat to Food and Fuel in the Coming Decades: Ballinger Publishing Company, Cambridge, Massachusetts, 304 p.

35 GIAMPIETRO, MARIO, et al., 1992, Limits to Population Size: The Scenarios of Energy Interaction Between Human Society and the Ecosystem: Population and Environment, v. 14, p. 109-131.

36 HAMMOND, A. L., (ed.), 1992, World Resources 1992-1993. Toward Sustainable Development: The World Resources Institute, New York, 385 p.

37 HARDIN, GARRETT, 1968, The Tragedy of the Commons: Science, v. 162, p. 1243-1248.

38 HARDIN, GARRETT, 1973, Exploring New Ethics for Survival. The Voyage of the Spaceship Beagle: Penguin Books Inc., Baltimore, 273 p.

39 HARDIN, GARRETT, and BADEN, JOHN, (eds.), 1977, Managing the Commons: W. H. Freeman and Company, San Francisco, 294 p.

40 HARDIN, GARRETT, 1981, Dr. Pangloss Meets Cassandra. A Review of *The Ultimate Resource, by J. L. Simon;* The New Republic, October 18, p. 31-34.

41 HARDIN, GARRETT, 1993, Living Within Limits. Ecology, Economics, And Population Taboos: Oxford Univ. Press, New York, 339 p.

42 HARDIN, GARRETT, 1995, The Immigration Dilemma. Avoiding the Tragedy of the Commons: Federation for American Immigration Reform, Washington, D. C., 140 p.

43 HAWKEN, PAUL, 1993, The Ecology of Commerce. A Declaration of Sustainability: HarperCollins Publishers, New York, 250 p.

44. IVANHOE, L. F., 1996, Updated Hubbert Curves Analyze World Oil Supply: World Oil, November, p. 91-94

45. IVANHOE, L. F., 1997, Get Ready for Another Oil Shock!: The Futurist, January-February, p. 20-23.

46 JACKSON, WES, 1971, Man and the Environment: Wm. C. Brown Company Publishers, Dubuque, Iowa, 322 p.

47 JOHNSON, WARREN, 1978, Muddling Towards Frugality: Sierra Club, Shambhala Publications, Inc., Boulder, Colorado, 252 p.

48 KATES, R. W., 1994, Sustaining Life on the Earth: Scientific American, v. 271, n. 4, p. 114-122.

49 KIDDER, R. M., 1989, Reinventing the Future. Global Goals for the 21st Century: MIT Press, Cambridge, Massachusetts, 194 p.

50 LAPP, R. E., 1973, The Logarithmic Century: Prentice-Hall, Inc., Englewood Cliffs, New Jersey, 263 p.

51 LUCAS, G. R., Jr., and OGLETREE, T. W., (eds.), Lifeboat Ethics. The Moral Dilemmas of World Hunger: Harper & Row, Publishers, New York, 162 p.

52 MANN, JUDY, 1994, Biting the Environment that Feeds Us: The Washington Post, July 29.

53 McMANUS, H. J., and WILCOX, LYLE, (eds.), 1978, Alternatives for Growth. The Engineering and Economics of Natural Resources Development: Published for the National Bureau of Economic Research, Inc., by Ballinger Publishing Company, Cambridge, Massachusetts, 256 p.

54 MEADOWS, D. L., et al., 1972, The Limits to Growth: Universe Book Publishers, New York, 207 p.

55 MEADOWS, D. H., et al., 1992, Beyond the Limits. Confronting Global Collapse. Envisioning a Sustainable Future: Chelsea Green Publishing Company, Post Mills, Vermont, 300 p.

56 MILBRATH, L. W., 1989, Envisioning a Sustainable Society. Learning Our Way Out: State Univ. of New York Press, Albany, New York, 403 p.

57 MILES, R. E., Jr., 1976, Awakening From the American Dream. The Social and Political Limits to Growth: Universe Books, New York, 246 p.

58 MILLER, G. T., Jr., 1982, Living in the Environment: Wadsworth Publishing Company, Belmont, California, 500 p.

59 MOFFITT, DONALD, (ed.), The Wall Street Journal Views America tomorrow: Amacon, New York, 184 p.

60 OPHULS, WILLIAM, 1977, Ecology and the Politics of Scarcity: W. H. Freeman and Company, San Francisco, 303 p.

61 PARK, C. F., Jr., 1968, Affluence in Jeopardy. Minerals and the Political Economy: Freeman, Cooper & Company, San Francisco, 367 p.

62 PARK, C. F., Jr., 1975, Earthbound: Minerals, Energy, and Man's Future: Freeman, Cooper & Company, San Francisco, 279 p.

63 PEARCE, D. W., and TURNER, R. K., 1990, Economics of Natural Resources and the Environment: The Johns Hopkins Univ. Press, Baltimore, 378 p.

64 PIEL, GERHARD, 1992, Only One World. Our Own to Make and to Keep: W. H. Freeman and Company, New York, 367 p.

65 PIMENTEL, DAVID, et al., 1994, Natural Resources and An Optimum Human Population: Population and Environment, v. 15, n. 5, p. 347-369.

66 PIMENTEL, DAVID, 1995, letter dated May 10.

67 PIMENTEL, DAVID, and PIMENTEL, MARCIA, 1996, Food Energy and Society (Revised Ed.): University Press of Colorado, Niwot, Colorado, 363 p.

68 PORRITT, JOHNATHAN, 1993, Sustainable Development: Panacea, Platitude, or Downright Deception? in CARTLEDGE, BRYAN, (ed.), Energy and the Environment, Oxford Univ. Press, New York, 170 p.

69 REED, C. B., 1975, Fuels, Minerals, and Human Survival: Ann Arbor Science Publishers Inc., Ann Arbor, Michigan, 200 p.

70 REIFENBERG, ANNE, 1995, Saudis Hope to Jump-start Oil Industry. OPEC Turns to Refining, Marketing Ventures in Asia: The Wall Street Journal, December 6.

71 RICHARDS, MOSE, (ed.), 1980, The Cousteau Almanac: Doubleday & Company, Inc., Garden City, New York, 838 p.

72 SCHMOOKLER, A. B., 1991, The Insatiable Society. Materialistic Values and Human Needs: The Futurist, July/August, p. 17-19.

73 SCHURR, S. H. (project director), 1979, Energy in America's Future. The Choices Before Us: Resources for the Future, Washington, D. C., 555 p.

74 SIMON, J. L., 1981, The Ultimate Resource: Princeton Univ. Press, Princeton, New Jersey, 415 p.

75 SKINNER, B. J., (ed.),. 1980, Earth's Energy and Mineral Resources: William Kaufmann, Inc., Los Altos, California, 196 p.

76 SKINNER, B. J., 1989, Resources for the 21st Century: Can Supplies Meet Demand? Episodes, v. 12, n. 4, p. 267-275.

77 TANZER, MICHAEL, 1980, The Race for Resources. Continuing Struggles Over Minerals and Fuels: Monthly Review Press, New York, 285 p.

78 TEITELBAUM, R. S., 1995, Your Last Big Play in Oil: Fortune, October, p. 88-104.

79 TOBIAS, MICHAEL, 1994, World War III. Population and the Biosphere at the End of the Millennium: Bear & Company Publishing, Santa Fe, New Mexico, 608 p.

80 TUNALI, ODIL, 1996, A Billion Cars: The Road Ahead: Worldwatch, January/February, p. 24-33.

81 VIEDERMAN, STEPHEN, 1994, Sustainable Development: What is It and How Do We Get There?: Current History, v. 93, p. 180-185.

82 WALDRON, INGRID, and RICKLEFS, R. E., 1973, Environment and Population. Problems and Solutions: Holt, Rinehart and Winston, Inc., New York, 232 p.

83 ZACHARY, G. P., 1996, A 'Green Economist' Warns Growth May be Overrated: The Wall Street Journal, June 25.

CHAPTER 29

The Ultimate Resource — Can It Secure Our Future?

The importance of mineral and energy mineral resources to nations and individuals cannot be overstated. But one might note that countries such as Japan and Switzerland enjoy a high standard of living, yet have almost no mineral or energy mineral resources. They have a highly educated society, well versed in modern technology, which produces many useful things. The ultimate resource of these mineral-poor nations and all nations is the educated creative human mind.

However, one cannot make something out of nothing. Switzerland and Japan must use steel to make their turbines. Japan could not be a world-class automobile manufacturer if it were not for imported iron ore, and virtually all the other materials which go into an automobile, because Japan has almost none of these resources. The history of civilization has been marked by the discovery, development, and use in a great variety of ways of Earth resources to advance the human race. But to do this, resources in some form must be used.

What has moved civilization ahead has been the innovative and educated human mind combined with Earth materials. Oil can be used for illustration.

Native Americans in east Texas and on Alaska's North Slope walked over the two greatest oil fields in North America. Nomads of the Arabian Peninsula rode camels over the largest oil deposits in the world. But in each case, what could they have done about them even if they had known they existed? Lacking knowledge which could unlock and utilize these resources left these people poor. But the educated human mind changed all that. Combining the educated human mind with material resources makes the difference.

In more detail, the same situation applies. If crude oil is obtained from underground reservoirs, or at natural oil seeps as occur many places in the world (almost all major oil fields exhibit oil seeps), it still has relatively few applications in its unprocessed form. Its use has been erratic and site specific. A principal Middle East use of crude oil for several thousand years was the treatment of camel mange. But when technology is applied to the raw material and the crude oil is put through a refinery, and then some of these refined products are further processed, the end result is literally thousands of items which make for better living. The list of products is huge, including paint, a great variety of plastics, medicines, pesticides, special coatings of all kinds, inks, and dyes.

Similar situations exist among the metals. Iron is a good example. Occurring very rarely in native form, iron was first found in meteorites. Swords fashioned from this hard material were very highly prized in battle. They were termed "swords of heaven." Because of the very high melting point of iron, the metallurgy of iron was discovered and developed at a rather late date. But, finally, what had been huge deposits of unusable iron ore in many parts of the world were valuable resources to be exploited. The resource was there all the time but had to be combined with the educated human mind to make the iron available and useful. After the discovery that iron could be obtained from previously worthless rocks, it was further found that the addition of vanadium, chromium, tungsten, molybdenum, and other minor metals would give iron a variety of valuable properties, producing alloys for many important specialized purposes.

On the Mesabi Iron Range in Minnesota, the rich hematite ore became exhausted, but very large quantities of lower grade ore called taconite remained. This low grade ore is now crushed, the iron concentrated into pellets, and then shipped to steel mills. The uniform iron content of the pellets compensates in part for the initial lower grade ore by allowing blast furnace operations to be more efficient than when using raw but somewhat variable quality higher grade ores. Despite competition from foreign high grade ores, technology at least partially compensates for the depletion of the high grade ores of Minnesota. This allows the area to continue to be a competitive iron ore source, although iron mining is substantially reduced from what it once was. It is the educated human mind which allows the survival of this domestic industry.

"You Scientists Will Think of Something"

The educated human mind has done great things for our society in providing material wealth. Population growth has followed the expansion of knowledge. Knowledge has shown how to produce more food with the use of energy, and has provided the energy to do so. Medical knowledge has allowed more people to survive and live longer, with the result that the world's population is at an all time high and expanding exponentially.

More people are alive today than at any previous time in history, and we also have more scientists and engineers alive now than at any time in the past, and probably during all past time. Technical journals have increased to the point where libraries across the world urgently need more shelves. Knowledge is expanding, but oil production is declining in the U.S., which in 1950 produced more than half the world's oil. Soil is being eroded, groundwater tables are falling, and some aquifers have collapsed and cannot be restored.

In the course of giving lectures on mineral and energy issues, and pointing out the reality that, for example, the supply of oil is limited, someone in the audience almost always remarks, "You scientists will think of something." This is quite a flattering thought, and one is tempted for the moment to perhaps believe it as many people obviously do. This faith by the public in "science" is derived from the many scientific successes in the past. These range from jet airplanes, the discovery of the Salk vaccine against polio, landing men on the moon, and the ability to harness the power of the atom. Scientists and technologists have achieved great things.

The notion that "you scientists will think of something" is a placebo which allows people to ignore unpleasant facts. Bartlett has observed aspects of this and writes, "There will always be popular and persuasive technological optimists who believe that population increases are good, and who believe that the human mind has unlimited capacity to find technological solutions to all problems of crowding, environmental destruction, and resource shortages.

These technological optimists are usually not biological or physical scientists. Politicians and business people tend to be eager disciples of the technological optimists."(3)

What Bartlett is saying, is that "we scientists might *NOT* be able to think of something" to save humanity from its own follies. This view that science may not always be able to rescue them will disturb many people. Fortunately for scientists who want to share their concerns regarding resource supplies and population growth today, we have gotten beyond the time in society when the bearer of bad news to the King was beheaded. However, telling a Chamber of Commerce meeting that there are limits to growth, usually does not get the speaker invited again.

For a politician to campaign on truths which are unpleasant is suicide. No President would be elected stating the fact that when the world runs out of oil there is no apparent substitute visible in the ways and volumes in which we now use that vital material.

Unfortunately, both the popular press, and politicians perpetuate the idea that somehow science will think of something. This is because if the reality is less cheery, it neither sells newspapers nor gets people elected. Government has a vested interest in seeing the future as bright as possible, and in being "solution" oriented. The press also editorially has an optimistic bias, or readers will turn away from the products which are advertised. Two examples of promoting the idea that science will think of something: *Time* magazine in an article on energy sources states, "The supply [of coal] is adequate to carry the U.S. well past the transition from the end of the oil and gas era to new, and possibly not yet discovered sources of energy in the 2000s."(1) The implication here is clearly that there is a suitable replacement for oil, and there are some yet undiscovered energy sources. That "scientists will think of something" is essentially promised to the reader, who can then cheerfully turn to the sports section knowing that there will be no energy problems. But the reality is that the energy spectrum from sunlight and wood all the way to fusion which runs the Sun is known. There are no even faint indications of other significant energy sources. But the *Time* article hopefully cites "undiscovered energy sources," implying such exist, but with no basis of fact. It just makes cheery reading, and sells magazines.

In another example of depending on scientists to come to the rescue, in 1976, W. L. Rogers, Special Assistant to the Secretary of the Interior stated, "Coal could help fight a rear-guard action to provide time for scientific breakthroughs which will move the world from the fossil fuel era of wood, gas, oil, and coal to the perpetual energy era of infinitely renewable energy resources."(19) Here again the public is clearly promised by a prominent member of the U.S. government that scientists will think of something. In fact they will bring the world into "the perpetual energy era of infinitely renewable energy resources." That promise should get any administration re-elected. Aside from the error of fact in the quotation identifying wood as a fossil fuel, there is a large error in logic in both statements. They imply that coal can replace oil as an energy source. It is not possible to convert enough coal to oil to replace the more than 19 million barrels of oil used daily just in the United States, much less replace the 71 million barrels of liquid fuel (crude oil plus other liquid fuels) used each day in the world.

If coal were to replace both oil and gas, mining the amount of coal which would have to be used to provide that equivalent energy would be the biggest mining operation the world has ever seen, and is an even larger impossibility (if it is possible to have a larger impossibility!) It would involve tearing up many square miles of land every day. Also, from the energy recovered would have to be subtracted the energy costs involved in mining, and the energy costs of reclaiming the land. These two costs in total are much larger than the energy costs of producing oil from most oil wells where no overburden on the resource has to be removed and no land regrading has to be done after the resource is extracted. Unfortunately, rosy forecast solutions to future energy problems tend to be taken by the general public as accomplished fact when seen in print, or uttered by government officials. Occasionally, however, a cautionary note comes from high sources. Richard E. Bennett, of the U.S. Department of State, has also expressed concerns about the ability of technology to save the day, stating, "While it is true that technology has generally been able to come up with solutions to human dilemmas, there is no guarantee that ingenuity will always rise to the task. Policymakers must contend with a nagging thought: what if it does not, or what if it is too late?"

Pimentel and Giampietro have brought the same cautionary message — that "Technology cannot substitute for shortages of essential natural resources such as food, forests, land, water, energy, and biodiversity. Technology can help us in making a better use of some of these resources. However, we must be realistic as to what technology can and cannot do to help humans feed themselves and to provide other essential resources."(18)

There is a limit to how much the ingenuity of the educated human mind can compensate for the scarcity or exhaustion of raw materials. So far, however, the human intelligence has been doing quite well with resources even though some are tending to be of lower and lower grade. But there are limits to that trend also. Non-renewable resources become depleted and alternatives must be found. But alternatives also have limits, and frequently are more costly than are the materials which they presume to replace.

Geological and geophysical explorations during the past century have given us a good inventory of the Earth's mineral resources, which we are consuming at an exponential rate. Since 1900, the human race has used more minerals and energy mineral resources than were used during all previous history. This consumption trend cannot be sustained. The time will come when oil is no longer available in the quantities we are now using. This will also be true of other conventional resources such as copper, lead, zinc, tungsten, cobalt, antimony, and many other vital materials. Admittedly, iron and aluminum are in huge supply (aluminum, for example, makes up eight percent of the Earth's crust), but obtaining these metals from their ores takes large amounts of energy. The other metals cited are also still in reasonably ample supply, but as the higher grade and more accessible deposits are used, the cost in energy in recovering more of these metals will greatly increase. In some cases also, the ore bodies do not grade out, but end abruptly. In this circumstance when the well-defined edge of the ore body is reached there is no lower grade ore on which technology can be applied.

Mind over Matter?

The question is whether it will be possible for the educated human mind to find adequate substitutes for these critical materials or be able to recycle them with such a low net loss of material as to make the supplies stretch far out into the future? In the energy field, can fusion ever produce energy commercially? Or, can solar energy be made to do it all in the reasonably near or even distant future?

If not, can we continue to use today's conventional energy and mineral resources long enough to give us time to adjust our population size and restructure our present industrialized societies to satisfactorily enter a different energy and mineral use paradigm? We must get on with the program, but it looks as though we are behind schedule. The stork is getting ahead of science. Population is beginning to outrace the ability of "scientists to think of something."

In terms of the present spectrum of metals and major energy sources (coal, oil, gas) each generation now uses the highest quality and most accessible deposits. The next generation inherits what is left, after the cream has been skimmed. Can technology continue to supply more and more people from resources which are less accessible and of diminishing quality? In terms of mined groundwater and soil which is lost by erosion, these resources are gone. For oil, gas, coal, and metals, at some point technology cannot continue to supply more and more. A different economy with a balance between population and renewable resources will have to emerge.

Science and technology have freed large segments of the population from a subsistence existence, and given time and opportunity to solve the clearly apparent problems of resources and population. This window of opportunity must be used before it closes.

Wealth we can continue to accumulate

As our mineral wealth and conventional energy resources, chiefly petroleum and coal, are being depleted, there is one resource we can continue to accumulate rather than dissipate. It is knowledge. Like the Greeks who were prevailed upon to forego their annual tribute of silver from the mines at Laurium in order to build the fleet which defeated the Persians at the Battle of Salamis, so the human race must apportion a part of the current wealth toward the development of "shiploads" of information which can carry humanity into a sustainable future economy with a reasonable standard of living.

Needed: Better Use of Time, Talent, and Money

As time passes and natural resources diminish, human resources must be increasingly directed toward providing for the future. Government financing of basic research through direct grants or tax incentives must be increased. It is encouraging that some universities have now inaugurated "Future Studies." There is merit in teaching history. We can learn much from the past, but we will live in the future. It is an interesting commentary on society that history has been taught for many centuries, but the study of the future is just recently being considered. We are aware of the past, and we live mostly for the present. We have tended to let the future take care of itself. It is going to need more attention.

Allocation of human resources

Currently there is in the United States a distorted allocation of educational resources, and a misplaced emphasis. Japan trains 1,000 engineers for every 100 lawyers. In the United States, the ratio is just reversed. Every year in the United States, three times as many lawyers are graduated as exist in all of Japan, and there are now in Washington, D. C. alone, three times as many lawyers as there are in all of Japan. Japan is producing; the United States is suing. The United States is the most litigious nation in the world. The insurance group, Lloyds of London, reports that only 12 percent of its insurance business is in the United States, but 90 percent of its insurance claims are there. It is doubtful, however, that all the lawyers in the United States will ever discover an adequate substitute for oil, or how to reach the goal of commercially economic fusion power.

Another large and relatively non-productive drain on the time, talents, and money of U.S. citizens is the tax system. This not only discourages savings but involves a huge amount of legal and court time. At present, about 300,000 of some of our brighter minds are involved in advising Americans on tax problems. More than 80,000 tax lawyers and accountants interpret the complex tax laws for citizens. American taxpayers annually spend more than 300 million hours filling out tax forms, yet 40 percent found they still needed a tax specialists to complete the task. There is nothing comparable to this expenditure of time and money on tax matters anywhere else in the world. The annual cost of compliance with the federal tax system, as it is now, has been estimated by the Tax foundation, a Washington, D. C. research organization, to be $200 billion.(11) This is a tremendous drain on the economy, and it does not include the psychological cost of taxpayer frustration.

Government should emphasize and support more basic research. The $200 billion spent on the now almost totally unintelligible and hugely complex tax system would pay for a lot of that sort of research, ultimately having much more lasting value than filling out myriad tax forms. The often frivolously spent money on a great variety of pet projects of elected representatives for their districts also would go a long way toward promoting more worthwhile causes if properly used.

Human resources must be put to more productive use, building basic information for the future, and spending much less time with lawsuits and tax forms. The United States has had the "luxury" of doing this up to now because it had all the mineral and energy mineral resources it needed. However, in the future, attention will have to increasingly focus on providing the basic resources by which we all live, as population increases against our diminishing resource base. Hopefully the human mind will eventually devise a system which will give us an almost completely renewable, recyclable economy and sustainable-sized population. But it is highly unlikely that this will be developed in a courtroom or in a tax accountant's office.

Although some populations are now clearly outstripping their available resources, in some cases drastically so, thus far the industrialized world has won the race between resources and the demand for them. The natural bounty inherited from the geological events of the past that produced the minerals and energy minerals we use today, combined with educated people has allowed much of the world, and industrialized nations in particular, to progressively raise their living standards.

Up till now, knowledge of how to obtain and use our mineral and energy mineral resources has increased beyond basic needs. This has resulted in a higher and higher standard of living in developed countries with energy and mineral supplies to use beyond those required just for subsistence. But the easy technologies have been discovered and have been used on the most easily recovered and richest mineral resources. But can knowledge in the future increase as fast as the need for knowledge to solve the increasingly difficult resource/population problems?

In the case of energy, we have in less than 200 years, gone from wood burning to atomic fission. But the step from fission to fusion is a very large, probably as great, or greater in terms of technology than the total technological distance from wood burning to fission. There is even some doubt, despite the "cold fusion" furor of the late 1980s, that fusion will ever be commercially accomplished in any foreseeable time frame. Do we then, if we cannot achieve fusion, see the end of major advances and discoveries in energy sources?

If we must depend on known energy sources without once again being able to discover and draw on major new ones such as we did in 1859 when the Petroleum Interval was brought forth, can the human mind solve this problem? It is a great deal more challenging a situation than we have faced in the past. However, there is no other tool that will be more useful than the human mind combined with Earth resources. The carrying capacity of the Earth at a reasonable standard of living can be maintained if the intelligent human mind is applied properly.

The statesman, Winston Churchill, perhaps noting the demise of the once geographically far-flung British Empire, observed "The frontiers of the Twentieth Century are the frontiers of the mind." It will be so from now and into the future, but it is still necessary to have some basic resources with which to work. Minerals and energy will be important indefinitely. The human mind cannot conjure something out of thin air, and even in recycling there is an inevitable degradation and dissipation of a certain amount of substance and energy. Energy minerals and mineral resources must be efficiently and economically managed to provide the basics for continued human existence at an acceptable level of living. But there is more beyond simply intelligence.

Constructive use of knowledge and resources

There is another facet to the use of the human mind applied to Earth resources. Resources can be used for both contructive and destructive purposes. The use of uranium is an example. Two nuclear physicists, Wiesner and York, during the so-called "Cold War," observed the arms race where one side produced an atomic bomb, the other side produced a more destructive nuclear weapon, and so on. Wiesner and York concluded, "It is our considered professional judgement that this dilemma has no technical solution."(27)

Technology by itself does not hold the answer to achieving a bright human future. The intelligent human mind must also have good judgment as to how resources may best be employed. Warfare uses valuable and irreplaceable mineral resources in gigantic quantities. This is entirely avoidable if the human intellect is properly employed, especially at the national leadership level where foreign policies are determined. Scientists and technologists can produce resources from the Earth but the decision as to how to use them has for the most part not been their province. Lapp has emphasized this, "It is a waste of time to hunt witches and seek to pillory scientists and technologists as responsible for national ills; they have never been in the driver's seat. If technology is to be controlled, if it is to be used wisely, and if its ill effects are to be avoided, the nation needs not castigation but leadership that combines the technical skills of specialists with the responsible management of those charged with our political administration." (14)

The Premier of Sweden, Oxensterna, in the late 1600s, as he sent forth his son into the world said, "Son, you will be amazed to find with what little wisdom the world is governed."

Brobst makes this observation, "Geologic reasons...suggest that exponential growth at the anticipated rates in the world use of mineral and fossil fuel materials cannot continue indefinitely, although national governments of this world appear committed to domestic and international politics that look forward to growth on a business-as-usual basis in the years ahead. Much of the reasoning for these policies of an expanding economy is founded on the premise that the past is the key to the future. The fact that we have done extremely well in attaining the goals of an affluent society by these policies in the past is, however, no guarantee that they will be just as completely successful in the future."(5)

Ability to Visualize the Future

We now know a great deal more about the basic resources which control our lives — the energy and mineral supplies which influence our destinies. We now can visualize much better than people 400 years ago, where present trends in growth of population and mineral resource use may be taking us. Human use of resources must be viewed as part of the Earth's macro-ecology. We are an extension of the long arm of nature, subject to established natural laws which apply to all living things in their relationship between resources and survival. We ignore these relationships at our increasing peril. The carrying capacity of the environment on a sustained basis cannot be exceeded or the ultimate result will be disaster to any over abundant and consuming species, human or otherwise. We must adjust our numbers and resource use to the geo-determined limits of our geological environment. The scientists and technologists who have this basic information should be brought more fully into government circles, and future national decisions based on realities rather than on politics. We have the information by which to do good planning for the future.

Pimentel and Giampietro have put it very well: "The continued disregard for the problem of overpopulation is generated by the current imperialistic attitude of human species toward the rest of nature that sees humans as distinct from the natural realm. However, for their own sake, humans have to learn as soon as possible how to include in their cultural identity, values and moral obligations which are based on respect for the rest of the biosphere to which they belong. In fact, this is the only way to guarantee a fair opportunity to future generations for a quality of life."(18)

A recognition of the destinies of nations and individuals which the existence and the distribution of energy and mineral resources has imposed on the human race is the principal objective of this book. A second objective is to realistically look at these resources in terms of their shorter and longer term availability and their ultimate control over the human carrying capacity of this planet. Unlimited growth cannot continue in a finite system which the Earth is.

We plunge headlong through life, mostly living for the moment. But with an increased understanding of how much civilization has advanced, and that it now lives largely on the basis of a rich but non-renewable mineral and energy mineral heritage, some planning for the future is in order. The future will have to increasingly be based on "current income," in contrast to drawing on a depleting resource inheritance. Again, the matter of population is foremost in this perspective. People share a "commons" which is spaceship Earth, with its closed systems of water, soil, and other minerals.(8) Only sunlight is beyond this system, and even it is limited. Only so much is received each day.

The future will be chaos if populations continue to grow as they have in the past several decades. Some places already appear to have reached population induced chaos. Stabilization, or, an orderly reduction of population although needed will be difficult to accomplish. Meadows, et al. have stated, "Once the population and economy have overshot the physical limits of the Earth, there are only two ways back: Involuntary collapse caused by escalating shortages and crises, or controlled reduction of throughput by deliberate social choice."(16) A simple statement seen on a bumper sticker which makes the point of individual responsibility reads: "Control your local stork."

Two thoughts on population: one by Toynbee: "Maximum welfare, not maximum population, is our human objective," and one by Dr. George Wald, "We need to make a world in which fewer children are born, and in which we take better care of them."

The Twentieth Century has seen an unparalleled exponential growth of both population and the use of resources. It cannot continue. Lapp remarks upon these times: "Ours may well remain unique — the only logarithmic century ever to spin itself out upon a defenseless planet."(14) It need not be that way if resource use and population can be brought to a sustainable balance by use of the educated human mind. It is up to humanity to save itself from itself. H. G. Wells wrote: "Human history becomes more and more a race between education and catastrophe."(25) Hardin updated that concept: "Wells died in 1946, about a generation before the ecological perspective began to enter into the educational process. 'Education' is now understood in a broader sense than Wells ever dreamed of. A large fraction of six billion people are woefully in need of education for survival. Do we have time enough to change the balance? We don't know. We can only do our best."(10)

Developing an international consensus for the rational route is the challenge. Can it be done in this diverse racial, ethnic, political, and religious world in which we live? The future of all nations and all people depend on this.

To give this process at least a reasonable chance for success, people must be educated to the facts of the resources basic to our existence, and where we are now in regard to these resources. At the same time and perhaps more important, individuals must develop the responsibility to effectively use this knowledge. Present knowledge and available technology, including that which is related to population control, can be applied to secure a future without undue hardship. As in many aspects of today's social problems, individual attitudes are important. If society does not provide a responsible setting and attitude in which technology can be effective, then the technology is useless. Even technology cannot provide a sustainable future against an ever-increasing population. It may not even be able to do so with the present population. Either we begin to relate population size to the sustainable carrying capacity of the Earth, or we will, as Johnson has suggested, "muddle toward frugality."(12) The "muddling" means decisions forced on us by crises such as famines or fuel shortages. It would be much less pleasant than a voluntary, orderly move toward a sustainable future.

A "sustainable future" cannot really be visualized in the sense of not using up some materials from the Earth. Some Earth materials, minerals even if they are only in the form of rocks or soil, have to be used for civilization to survive. Humans cannot live in dwellings built just of sunlight and air, nor exist only on sunlight and air. Your very bones are made of minerals. But presumably some economy might be developed which would use so little of Earth's materials that the term "sustainable" could be accepted.

However, the present trends in population and economic growth are not sustainable. Charles F. Park, Jr., former Dean of Mineral Sciences at Stanford University, has made the following observation:

> "Our study of minerals and their place in the political economy has brought us inevitably to the conclusion that a constantly expanding economy, keeping step with a growing population, is impossible because mineral resources and cheap energy are not available on this earth in unlimited quantities. The affluence of modern civilization is indeed in jeopardy."(17)

The Ultimate Resource

The will to do

The key to the human future is to use knowledge, and use it in the right way. That is not a matter for scientists and technologists but for the political leadership and the public at large which must use the information to make the critical decisions. The place where decisions are finally implemented must be with the thoughtful and responsible individual, and there must be the individual will to make proper use of knowledge. Knowing is one thing, doing is another. Rossman's essay *The Will To Do What We Ought To Do* is a most perceptive statement of the problem.(20) Sir Crispin Tickell, in considering the future and the need for decisions states: "I was recently asked if I was an optimist or a pessimist. The best answer was given by someone else. He said that he had optimism of the intellect but pessimism of the will. In short we have most of the means for coping with the problems we face, but are distinctly short on our readiness to use them. It is never easy to bring the long term into the short term. Our leaders, whether in politics or business, rarely have a time horizon of more than five years...It is still within our power to make the future congenial to ourselves and future generations. The choice is ours."(22) Bartlett has put it concisely stating that the human mind is the ultimate resource but *"only if it is used."* [italics mine].(4)

Natural resources, the educated human mind, and, very importantly, the will to use the educated mind effectively, combine to make the ultimate resource on which we must depend. Growth of technical knowledge to very rapidly exploit the accumulated energy mineral and mineral resources of billions of years has brought us to the exponential predicament in which we find ourselves. But even as growth of knowledge to exploit the Earth has created problems, so also growth of knowledge, and its peaceful, intelligent use offers a path to humankind's successful survival, limited by the geodestined requirements of resource renewability.

Kennedy has stated about the future, "What is clear is that as the Cold War fades away, we face not a 'new world order' but a troubled and fractured planet, whose problems deserve the serious attention of politicians and publics alike...The pace and complexity of the forces for change are enormous and daunting; yet it may still be possible for intelligent men and women to lead their societies through the complex task of preparing for the century ahead. If these challenges are not met, however, humankind will have only itself to blame for the troubles and disasters that could be lying ahead."(13)

Realism, Pessimism, OPTIMISM

It may be a perception that this volume is unduly pessimistic about the future. It is the intention only to be realistic on the premise that proceeding on any other basis will not build a foundation for a stable future. It may be that a realistic assessment of present resource use and future availability for the needs of a currently very rapidly growing population leads to pessimism. Many present trends including heavy dependence on non-renewable resources, especially petroleum, the irreplaceable loss of topsoil, and the mining of groundwater in many areas are not encouraging. But the only suitable view is to be optimistic, and then take the course of action which can justify such expectations.

Viederman has stated, "Pessimism being a self-fulfilling prophesy, optimism is the only course. We must believe we have the political and moral will to change directions, and act to make that belief a reality. We cannot be complacent; with each passing year the windows of opportunity for change are narrowing. We must accept the moral imperative to preserve the planet for all its inhabitants —human and nonhuman — now and in the future."(23)

With the knowledge we now have and the unique ability above all other organisms to visualize what predicaments the present trends of resource use and population growth will lead to, we should have the will and capability to adopt a course which will lead to a sustainable future at a reasonable standard of living for all. We can combine the educated human mind and the will to use it with available Earth resources to successfully adapt to the resource environment which the geology of the Earth has established for us, and in which we are inevitably destined to live.

But to meet the challenges now upon us, we must live together in peace and not waste time, effort, intellectual ability, and valuable Earth resources in conflicts. Humanity has too many problems in common to be divided. As a beginning, let us individually be thoughtful and kind to one another.

BIBLIOGRAPHY

1 ANONYMOUS, 1971, Superbrain's Superproblem: Time, p. 58-64, 67.

2 BARTLETT, A. A., 1985, Review of the Ultimate Resource, J. L. SIMON, 1981, Princeton Univ. Press: American Jour. Physics, v. 53, n. 3, p. 282-285.

3 BARTLETT, A. A., 1994, Reflections on Sustainability, Population Growth, and the Environment: Population and Environment, v. 16, n. 1, p. 5-35.

4 BARTLETT, A. A., 1996, The Exponential Function, XI: Off the Scale: The Physics Teacher, September, v. 34, p. 342-343.

5 BROBST, D. A., 1979, Fundamental Concepts for the Analysis of Resource Availability: in SMITH, V. K., (ed.), Scarcity and Growth Reconsidered: published for Resources for the Future, by The Johns Hopkins Univ. Press, Baltimore, p. 106-142.

6 BROWN, HARRISON, 1978, The Human Future Revisited. The World Predicament and Possible Solutions: W. W. Norton & Company, Inc., New York, 287 p.

7 DALY, H. E., and TOWNSEND, K. N., (eds.), 1993, Valuing the Earth: MIT Press, Cambridge, Massachusetts, 387 p.

8 HARDIN, GARRETT, and BADEN, JOHN, 1977, Managing the Commons: W. H. Freeman and Company, San Francisco, 294 p.

9 HARDIN, GARRETT, 1993, Living Within Limits. Ecology, Economics, and Population Taboos: Oxford Univ. Press, New York, 339 p.

10 HARDIN, GARRETT, 1995, Letter dated May 28.

11 HERMAN, TOM, 1995, Briefs: The Wall Street Journal, July 12.

12 JOHNSON, WARREN, 1978, Muddling Toward Frugality: Sierra Club, Shambhala Publications, Inc., Boulder, Colorado, 252 p.

13 KENNEDY, PAUL, 1993, Preparing for the Twenty-First Century: Random House, New York, 428 p.

14 LAPP, R. E., 1973, The Logarithmic Century: Prentice-Hall, Inc., Englewood Cliffs, New Jersey, 263 p.

15 LUCAS, G. R., Jr., and OGLETREE, T. W., (eds.), 1976, Lifeboat Ethics. The Moral Dilemmas of World Hunger: Harper & Row, Publishers, New York, 162 p.

16 MEADOWS, D. H., et al., 1992, Beyond the Limits. Confronting Global Collapse. Envisioning a Sustainable Future: Chelsea Green Publishers, Post Mills, Vermont, 300 p.

17 PARK, C. F., Jr., 1968, Affluence in Jeopardy. Minerals and the Political Economy: Freeman, Cooper and Company, San Francisco, 368 p.

18 PIMENTEL, DAVID, and GIAMPIETRO, MARIO, 1994, Food, Land, Population and the U.S. Economy: Carrying Capacity Network, Washington, D. C., 81 p.

19 ROGERS, W. L., 1976, Rogers, quoted in the Denver Post, November 19.

20 ROSSMAN, PARKER, 1994, The Will To Do What We Ought To Do: The Futurist, July-August, p. 63.

21 SIMON, J. L., 1981, The Ultimate Resource: Princeton Univ. Press, Princeton, New Jersey, 415 p.

22 TICKELL, SIR CRISPIN, 1994, The Future and its Consequences: The British Association Lectures 1993, The Geological Society, London, p. 20-24.

23 VIEDERMAN, STEPHEN, 1994, Sustainable Development: What It Is and How Do We Get There?: Current History, v. 93, n. 581 (March), p. 180-185.

24 VKLLEE, C. A., Jr., (ed.), 1985, Fallout From the Population Explosion: Paragon House Publishers, New York, 263 p.

25 WELLS, H. G., 1920, The Outline of History: Garden City Publishing Company, Inc., New York, 1171 p.

26 WHITMORE, F. C., Jr., 1981, Resources for the 21st Century: Summary & Conclusions of the International Centennial Symposium of the U.S. Geological Survey: U.S. Geological Survey Circular 857, Washington, D.C., 41 p.

27 WIESNER, J. B., and YORK, H. F., 1964, National Security and the Nuclear-test Ban: Scientific American, v. 211, n. 4, p. 27-36.

Epilogue

From the several billion years of Earth's history and geological processes, the human race inherited a great storehouse of mineral and energy resource wealth. Very slowly at first, mineral and energy resources were used in the Stone Age and then on through the Copper, and Bronze ages. The ages are appropriately named because these mineral resources had a significant influence on peoples in each time.

Then came the Iron Age, followed by the Industrial Age with a great variety of new technologies by which Earth resources could be exploited quickly and in great volumes. The availability of these mineral and energy resources in abundance allowed civilization to achieve, in about two hundred years, affluence never before imagined. Accompanying this rise in wealth has been a huge increase in population, made possible by medical advances, and the availability of abundant metals and cheap energy to put millions of new acres of land into cultivation to feed the multitudes.

To increase agricultural output, mineral fertilizers are now used which enable us to harvest crops from what have been termed "ghost acres" of land to produce more than the land could naturally do. In the same way, civilization is living on "ghost centuries" of distant geological ages. We consume coal and oil and gas derived from organisms which absorbed the energy of the Sun over past millennia to produce these energy sources. We are now using this long-accumulated and stored geologic heritage in only a few hundred years. These times will be recorded as only a brief instant in human history.

Similarly, we use metal deposits which formed during past millions of years in the Earth's interior. It subsequently took more millions of years of uplift and erosion to bring these valuable deposits to or near the Earth's surface within the reach of development by humans.

We have been using this great geological inheritance at an exponential rate. Yet, against a diminishing fossil fuel, metals, and soil base, the population continues to expand. Soon, when we have spent Earth's resource savings account which took so many millennia to accumulate, we will face the enormous problem of living on current resource income.

The mineral and the energy deposits we use today, metals, oil, gas, and coal, do not grow. They only decline. Population grows. Bartlett's calculation cannot be ignored — that if the present rate of growth of world population ("only" 1.9%/yr) continues, the population would reach a density of one person per square meter of dry land surface of the Earth (excluding Antarctica) in 550 years, and the mass of people would equal the mass of the Earth in just

1620 years.(2) Obviously this will not happen. The question is why will it not happen. Will it be the result of intelligent matching of an optimum population size with a sustainable natural resource base, or by a chaotic or even catastrophic series of events? Can the human race use its intellect combined with a will to make an orderly transition to a sustainable society?

The statement, "scientists will think of something" is a popular public placebo by which to ignore the facts. Pimentel and Giampietro have warned "Technology cannot substitute for shortages of essential natural resources such as food, forests, land, water, energy, and biodiversity...we must be realistic as to what technology can and cannot do to help humans feed themselves and to provide other essential resources."(3)

Our geological inheritance of mineral and energy resources brought civilization to its present affluence. Past geological events control the destinies of nations and individuals in myriad ways which are now playing out upon the world scene. Understanding the inevitability of geodestinies — that Earth resources control our futures — is the basis for using these resources wisely toward the objective of an orderly transition to a sustainable society. The present use of non-renewable resources simply allows us to buy time during which to accomplish this. Time appears to be running out in this regard, especially against the background of a continually and rapidly growing population.

History records many conflicts over possession of non-renewable resources —both metals and energy. The Romans fought for the possession of the tin mines in Wales by which they could make bronze weapons. Germany, in the Franco-Prussian War of 1871 took the Alsace-Lorraine area from France to obtain coal and iron deposits. Japan entered World War II to secure oil and minerals in Southeast Asia, and the Gulf War between Iraq and the western world was over oil. But conflict over resources does not produce a long term stable situation. What non-renewable resources exist must be allocated in a peaceful, economical, and intelligent manner while at the same time a renewable, sustainable economy is built.

Geological processes established the framework in which nations and individuals now must exist.(4,5) Knowing the geodestined facts of both the possibilities and limitations of our mineral and energy resource environment, and living intelligently and with foresight is the challenge for attaining a truly sustainable future.

BIBLIOGRAPHY

1 LAPP, R. E., 1973, The Logarithmic Century: Prentice-Hall, Inc., Englewood Cliffs, New Jersey, 263 p.

2 BARTLETT, A. A., 1978, Forgotten Fundamentals of the Energy Crisis: American Jour. of Physics, v. 46, n. 9, p. 876-888.

3 PIMENTEL, DAVID and GIAMPIETRO, MARIO, 1994, Implications of the Limited Potential of Technology to Increase the Carrying Capacity of our Planet: Human Ecology Review, Summer/Autumn, v. 1, p. 248-251.

4 POSS, J. R., 1975, Stones of Destiny: Michigan Technological Univ., Houghton, Michigan, 253 p.

5 RAYMOND, ROBERT, 1986, Out of the Fiery Furnace: The Impact of Metals on the History of Mankind: The Pennsylvania State University Press, University Park, Pennsylvania, 274 p.

Index

C

D

E

F

G

Q

R

S

Y

Z

ISBN 0-89420-299-5